Improved Oil Recovery by Surfactant and Polymer Flooding

Academic Press Rapid Manuscript Reproduction

Improved Oil Recovery by Surfactant and Polymer Flooding

Editors

D.O. SHAH
University of Florida
Gainesville, Florida

R.S. SCHECHTER
University of Texas at Austin
Austin, Texas

ACADEMIC PRESS, INC.   New York   San Francisco   London   1977

A Subsidiary of Harcourt Brace Jovanovich, Publishers

ACADEMIC PRESS, INC.
111 Fifth Avenue, New York, New York 10003

United Kingdom Edition published by
ACADEMIC PRESS, INC. (LONDON) LTD.
24/28 Oval Road, London NW1

Library of Congress Cataloging in Publication Data

AIChE Symposium on Improved Oil Recovery by Surfactant
        and Polymer Flooding, Kansas City, Kan., 1976.
    Improved oil recovery by surfactant and polymer
flooding.

    1. Oil field flooding—Congresses.   2. Surface
active agents—Congresses.   3. Polymers and polymerization—Congresses.   I. Shah, Dinesh Ochhavlal, 1938–
II. Schechter, Robert Samuel.   III. American Institute
of Chemical Engineers.   IV. Title.
TN871.A13 1976   622'.33'82   77-22627
ISBN 0-12-641750-4

PRINTED IN THE UNITED STATES OF AMERICA

Contents

Preface

It was a great personal pleasure for us to edit and compile the papers presented at the AIChE Symposium on Improved Oil Recovery by Surfactant and Polymer Flooding, held in Kansas City in April 1976. The papers presented at the symposium were all invited papers, so that a broad spectrum of topics related to enhanced oil recovery by surfactant and polymer flooding could be discussed. Therefore, the symposium included papers ranging from an introduction to petroleum reservoirs, interfacial tension, and molecular forces to molecular aspects of ultralow interfacial tension, the structure, formation, and phase inversion of microemulsions, thermodynamics of micellization and related phenomena, adsorption phenomena at solid/liquid interfaces and reservoir rocks, to flow through porous media studies on polymer solutions, microemulsions, and soluble oils. Hence, the symposium covered molecular, microscopic, and macroscopic aspects of oil displacement in porous media by surfactant and polymer solutions and related phenomena. The authors of these chapters were requested to write their papers in the style of a critical review of the current state of the art in a given area. We are very happy that our suggestion has been closely followed. The literature cited in this book also forms a comprehensive list of references in relation to improved oil recovery by surfactant and polymer flooding. In this rapidly expanding and important area of enhanced oil recovery, we hope that this book will be useful to experts as well as to novices entering this field of research.

The editors wish to convey their sincere thanks to Academic Press, the contributing authors, Mrs. Jeanne Ojeda for her excellent typing of the entire book, and Miss Laura Taylor for her skilled editorial assistance. The permission to reproduce diagrams from previously published papers by authors and other publishers are gratefully acknowledged. This book will have served its intended purpose if the readers find it a useful reference book in the coming years for enhanced oil recovery research.

Contributors

Dr. V.K. Bansal, Department of Chemical Engineering, University of Florida, Gainesville, Florida 32611

Mr. R.J. Brugman, Department of Chemical Engineering, University of Florida, Gainesville, Florida 32611

Mr. K.S. Chan, Department of Chemical Engineering, University of Florida, Gainesville, Florida 32611

Dr. J.G. Dominguez, Department of Chemical and Petroleum Engineering, University of Kansas, Lawrence, Kansas 66044

Dr. W.B. Gogarty, Marathon Oil Company, Denver Research Center, Littleton, Colorado 80120

Professor K.E. Gubbins, Chemical Engineering Department, Cornell University, Ithaca, New York 14850

Professor J.M. Haile, Department of Chemical Engineering, Clemson University, Clemson, South Carolina 29631

Dr. H.S. Hanna, School of Mines, Columbia University, New York, New York 10027

Dr. R.N. Healy, Exxon Production Research Company, P.O. Box 2189, Houston, Texas 77001

Dr. L. W. Holm, Union Oil Company, Union Research Center, P.O. Box 76, Brea, California 92621

Dr. W.C. Hsieh, Phillips Petroleum Company, Exploration and Production Research, Bartlesville, Oklahoma 74004

Dr. K.J. Lissant, Petrolite Corporation, Tretolite Division, 369 Marshall Avenue, St. Louis, Missouri 63119

Dr. E.W. Malmberg, Sun Oil Company, 503 North Central Expressway, P.O. Box 936, Richardson, Texas 75080

Professor A.B. Metzner, Department of Chemical Engineering, University of Delaware, Newark, Delaware 19711

Dr. V. Mohan, Department of Chemical Engineering, Illinois Institute of Technology, 3200 South State Street, Chicago, Illinois 60616

Professor J.C. Morgan, Department of Chemistry, University of Texas at Austin, Austin, Texas 78712

Professor J.P. O'Connell, Department of Chemical Engineering, University of Florida, Gainesville, Florida 32611

Dr. R.L. Reed, Exxon Production Research Company, P.O. Box 2189, Houston, Texas 77001

Dr. B.B. Sandiford, Union Oil Company, Union Research Center, P.O. Box 76, Brea California 92621

Professor R.S. Schechter, Department of Petroleum Engineering, University of Texas at Austin, Austin, Texas 78712

Professor D.O. Shah, Department of Chemical Engineering, University of Florida, Gainesville, Florida 32611

Dr. L. Smith, Sun Oil Company, 503 North Central Expressway, P.O. Box 936, Richardson, Texas 75080

Professor P. Somasundaran, School of Mines, Columbia University, New York, New York 10027

Dr. G.L. Stegemeier, Shell Development Company, Bellaire Research Center, Houston, Texas 77001

Dr. S. Trushenski, Amoco Production Company, P.O. Box 591, 4502 East 41st Street, Tulsa, Oklahoma 74102

Professor W. Wade, Department of Chemistry, University of Texas at Austin, Austin, Texas 78712

Professor D.T. Wasan, Department of Chemical Engineering, Illinois Institute of Technology, 3200 South State Street, Chicago, Illinois 60616

Professor G.P. Willhite, Department of Chemical and Petroleum Engineering, University of Kansas, Lawrence, Kansas 66044

Dr. L.A. Wilson, Jr., Gulf Research and Development Company, P.O. Drawer 2038, Pittsburgh, Pennsylvania 15230

PHYSICO-CHEMICAL ENVIRONMENT OF PETROLEUM
RESERVOIRS IN RELATION TO OIL RECOVERY SYSTEMS

L. A. Wilson, Jr.
Gulf Research and Development Company

I. ABSTRACT

The physico-chemical environment of petroleum reservoirs
is reviewed with emphasis on how this environment can influence the behavior of oil recovery processes. The chemical
flooding processes are reviewed and the significance of the
character of the oil on the processes is emphasized. Examples
are given of interstitial water analyses and reservoir rock
mineral analyses and the influence of constituents of these
on the recovery processes is discussed, along with the influence of the geology, lithology and reservoir temperature.
Methods for circumventing some of the problems are described.

II. SCOPE

The Enhanced Recovery oil potential in the United States
has been variously estimated at up to about 60 billion barrels.
Geffen, Hasiba and Wilson, and Sharp (1), (2), (3) have published these estimates and projected that, of this potential,
about 60 percent may derive from the chemical flooding processes. For the purposes of this paper, chemical flooding is
meant to include the "water-based" processes such as micellar
and surfactant floods, alkaline waterfloods and polymer floods.
These processes are all sensitive to certain environmental
elements or factors that are commonly present in oil reservoirs. The purpose of this paper is to review the interactions of the process systems with the reservoir environment
and the possible effect on the successful application of the
recovery method.

III. CONCLUSIONS AND SIGNIFICANCE

This review examines five elements of the reservoir environment that can influence the chemical flooding processes.
These are the in-place oil and water, the mineralogy, the
geology/lithology and the temperature. It has been observed
that the oil can influence the process selection and the
composition of the chemical slugs. The water can influence

1

the injection sequence of the fluids, the longevity or attenuation of the caustic and surfactant slugs, the efficiency
of the slug in mobilizing oil, the mobility of the surfactant
and polymer solutions and, consequently, the mobility control.

The mineralogy influences the divalent ion availability
and the adsorption of surfactant, caustic and polymers. Both
the ions and the adsorption are factors in the efficiency of
the alkaline water and surfactant processes by virtue of their
influence on slug attenuation and mobility control. The
geology/lithology influences the injectability of the fluids,
the behavior of the polymer solutions in that resistance factor and shear degradation can be factors of the lithology, and
the mobility control by virtue of the effect of inaccessible
pore volume on polymer flow. Reservoir temperature can influence phase stability of the surfactant slug, the chemical
stability of the polymer solution, and the reaction rate and
wettability reversal of the alkaline waterflood.

These, then, are among the problems to which solutions
are sought in the current research activity in chemical
flooding.

IV. REVIEW OF PROCESSES

Figure 1 is a schematic representation of the processes
we want to discuss.

A. Surfactant Flooding

We will use the "generic" name of surfactant flooding to
encompass those processes which commonly are referred to in
the literature as surfactant floods, detergent floods, sulfonate floods, microemulsion floods, emulsion floods, micellar
floods and soluble oil floods (Figure 1-a). The basic purpose

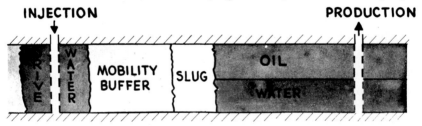

Fig. 1-a. Surfactant Flood

of the process is to inject into the reservoir a slug of surface active material that is capable of mobilizing residual
oil that can be displaced and produced. The surfactant slug,
representing only a fraction of the total pore volume, is
driven through the reservoir by a subsequent slug of thickened
water (polymer solution), which is in turn displaced by water
or brine. The mobilities of each of these slugs are adjusted

to minimize by-passing and channeling and to improve the volumetric coverage of the process. Ideally, the displacement
by surfactant flooding approaches a miscible displacement.

It is important that the integrity of the slugs be maintained for as long as possible. They are sensitive to certain
elements of the reservoir, and these factors can contribute
to early attenuation.

B. Polymer Flooding
As mentioned, in a surfactant flood the polymer solution
is injected to aid in obtaining a good volumetric sweep of
the reservoir by the process. A polymer solution may also be
used in conjunction with a water flood to achieve the same
purpose. This is illustrated in Figure 1-b. It is as if we

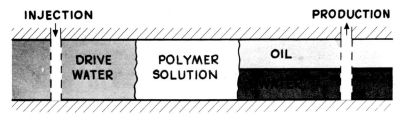

Fig. 1-b. Polymer Flood

are looking at only the back portion of the surfactant flood
with no mobilization of residual oil downstream of the polymer
solution. The intent is to reduce the mobility of the water,
thus forcing the water to flow through more flow channels in
the rock than would be the case with water injection alone.

C. Alkaline Waterflooding
In an alkaline waterflood, a slug of water containing
caustic is injected into the reservoir and followed by water
or brine (Figure 1-c). The slug might contain up to 5 percent

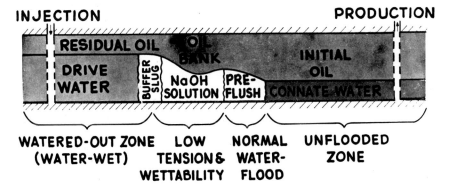

Fig. 1-c. NaOH Waterflood

sodium hydroxide and approximate about 15 percent of the pore volume. The caustic effects an increase in oil recovery by one or more of the following mechanisms: (1) a favorable change in the wettability of the rock, (2) a low tension displacement, and (3) improved sweep.

V. ENVIRONMENTAL PARAMETERS OF RESERVOIRS

We will address ourselves to five elements of the reservoir environment which can be strong influences on the processes. These are the oil, the interstitial water, the geology/lithology, the mineralogy, and the temperature.

A. Oil

The character of the oil can influence the processes in several ways. The viscosity can influence process selection. In general, as illustrated by Poettman (4), the chemical floods are applicable to oils whose viscosities are such that they would be, or have been, candidates for waterflood. The upper limit of viscosity is extended somewhat over that of a waterflood but, as a rule, the applicable oils are those that exhibit API gravities greater than 20° (spec. grav. < 0.934 60°/60°) and reservoir viscosities less than 100 centipoise (cp). In this range of viscosities, polymer flooding is the only one of the chemical floods which is applicable for viscosities greater than about 50 cp.

The chemical character of the crude is important for the surfactant flood and alkaline waterflood. The surfactant process requires low interfacial tension between the slug and contacting fluids. The chemical nature of the oil is important, therefore, to the extent that it influences the interfacial tension and, consequently, the surfactant selection and slug composition. Foster (5) points out, for example, that with a particular petroleum sulfonate the minimum tension between a salt solution containing the sulfonate and an intermediate-paraffinic crude would occur at higher sodium chloride concentrations and lower surfactant concentrations than for a typical naphthenic crude. Recently, Cash et al. (6) have reported additional data relating oil type and interfacial tension. This relationship is illustrated in Figure 2 which is an interfacial tension contour map for a petroleum sulfonate–sodium chloride–water system against an intermediate paraffinic crude. If the oil were naphthenic, the region of minimum tension would move down and to the right as indicated by the arrow.

In an alkaline waterflood the oil must have acidic components that can react with the caustic to form surface active agents. Ehrlich and Wygal (7) have shown that oil recovery by an alkaline waterflood can be increased significantly over

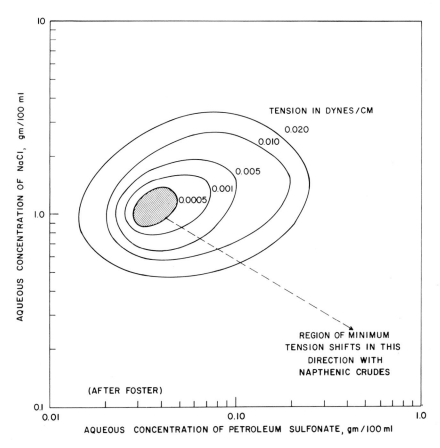

FIGURE 2

INTERFACIAL TENSION CONTOUR MAP FOR
PETROLEUM SULFONATE–NaCl–WATER SYSTEM
AGAINST INTERMEDIATE–PARAFFINIC CRUDE

a waterflood for crude oils with acid numbers greater than
0.1–0.2 mg KOH/g of oil, but little additional recovery
results with oils exhibiting lower acid numbers. Jennings
(8) correlated the surface activity of crude oils against
caustic solution with acid number, API gravity and viscosity.

Dissolved acid gases such as H_2S and CO_2 that are present in the oil can result in reaction with caustic during an alkaline waterflood and consume part of the caustic that otherwise would be available for beneficial reactions.

B.   Interstitial Water

Certain chemical constituents in the interstitial water can strongly influence the behavior of chemical floods and the composition of this water can vary considerably. Shown on Table I are analyses of samples for formation water from various locations and depths. The intent of these data is to show the substantial variation in content of reservoir brines. There is no implication of a unique relationship between water analyses and geographical location. Of particular interest is the variation in concentrations of the mono-, di- and tri-valent ions. The boron is of special significance with some polymers and the acidic components can react with caustic as was just pointed out for the acids in the oil.

In alkaline waterflooding, divalent ions such as Ca^{++} and Mg^{++} could react with the caustic to precipitate the insoluble hydroxides with a resulting loss in alkalinity. The calcium ion can have a more profound effect, however. At the same time the presence of NaCl appears to enhance the lowering of interfacial tension that occurs with caustic. Jennings et al. (9) have demonstrated these effects as shown in Figure 3. Note the marked increase in tension in the presence of Ca^{++} and the reduction in tension when NaCl is in the water. H. Jennings et al. (10) and Cooke et al. (11) give good discussions of the influence of water composition on alkaline waterflooding.

The surfactant flood and polymer flood are both sensitive to the chemistry of the interstitial water. Both the mono- and divalent metal ions can affect the surfactant solution. The concentration of these ions can influence the interfacial tension, viscosity and phase stability of the solutions. Hill et al. (12) have shown that there appears to be an optimum sodium chloride concentration for a particular crude oil-surfactant solution system. If the NaCl concentration should be changed by mixing with interstitial water, tension could increase and oil recovery might suffer. Holm and Josendal (13) suggest that concentrations of NaCl up to 2 percent may be satisfactory. Hill et al. (12) and Dauben and Froning (14) tend to confirm this. Healy and Reed (15) and Healy, Reed and Stenmark (16) discussed the role of optimal salinity in providing for prolonged "locally-miscible" displacement as evidenced by minimum multiphase region in the ternary diagram.

TABLE I

ANALYSES OF PRODUCED WATER

(After Case and Crawford) (33),(34)

Source (Depth) Component (ppm)	Okla. 3875'	Okla. 4150'	Okla. 5728'	Okla. 7084'	Kansas 3310'	N. Mex. 3841'
Na^+	4869	66933	7060	20100	7500	3085
Ca^{++}	278	17120	2177	1269	876	1308
Mg	102	2248	373	175	370	676
SO_4	171	380	115	664	825	674
Cl	7800	139720	23879	37306	13150	7446
HCO_3	636	75	52	222	746	1891
CO_3	nil	nil	0	0	nil	nil
Fe^{++}	trace	30	trace	--	trace	trace
B	--	--	5	187	--	--
H_2S	yes	No	--	--	Yes	Yes
Organic Acids as acetic	--	--	1562	48	--	--

7

TABLE I (Continued)

Component (ppm)	Michigan 2840'	Colorado 3130'	Montana 3000'	Wyoming 5250'	Wyoming 7350'	Louisiana 15225'	Miss. 8100'
Na^+	75709	873	2835	10012	649	29765	76200
Ca^{++}	33088	8	35	486	340	1600	37700
Mg	7670	0	trace	103	78	117	2410
SO_4	86	279	28	10946	1523	360	64
Cl	197570	2092	1566	8316	222	49612	190100
HCO_3	trace	610	4900	355	795	451	--
CO_3	nil	trace	--	trace	0	0	--
Fe^{++}	14	--	--	--	--	--	298
B	--	--	--	--	--	66	--
H_2S	No	--	--	--	--	--	--
Organic Acids as acetic	--	--	--	--	--	--	

Mn 105
Pb 101
Zn 367
Sr 2290

INFLUENCE OF NaCl AND Ca⁺⁺ ON INTERFACIAL TENSION

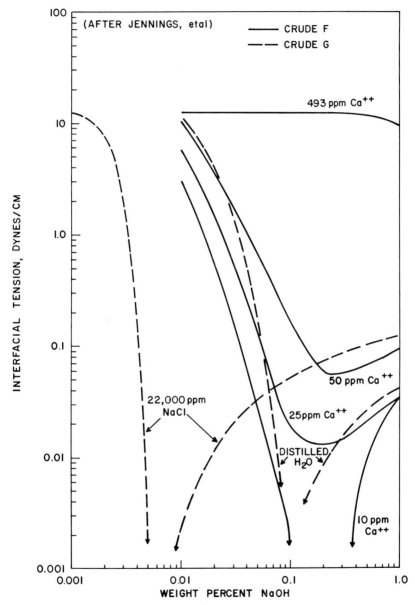

FIGURE 3

The NaCl influence is illustrated in Figure 4. For a
fixed sulfonate composition the solution might be optimum,
of increasing or decreasing activity, or two-phase depending
upon the NaCl concentration.

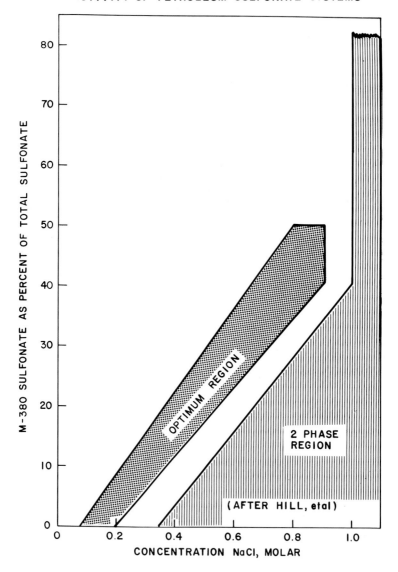

**INFLUENCE OF NaCl ON INTERFACIAL
ACTIVITY OF PETROLEUM SULFONATE SYSTEMS**

FIGURE 4

Calcium and magnesium ions can precipitate the sulfonates and Bernard (17) attributes a portion of the reduction in oil-displacing effectiveness to this mechanism. They can, however, cause other less predictable and, perhaps, more disturbing effects. Hill et al. (12) have shown that there is also a sharp tendency of interfacial activity on the Ca^{++} concentration, just as with NaCl. In Figure 5 there is defined a region of optimum interfacial activity which is clearly a function of the Ca^{++} content of the surfactant solution.

**INFLUENCE OF Ca⁺⁺ ON INTERFACIAL
ACTIVITY OF PETROLEUM SULFONATES**

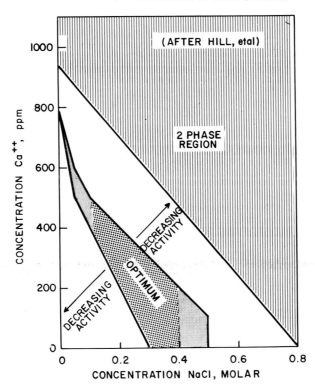

FIGURE 5

Relatively small changes in the Ca^{++} concentration can result in significant changes in the viscosity of the surfactant solution. Figure 6 illustrates this with data reported by Trushenski et al. (18). This increase in viscosity could result in a breakdown of the mobility control of the system.

The polymers that are usually considered for oil recovery are either polyacrylamides or biopolymers. Their function is to reduce the mobility of the water and improve volumetric sweep.

The mobility reduction with polymers is obtained by virtue of an increased viscosity or a decreased permeability, or both. As pointed out by MacWilliams et al. (19) the nature

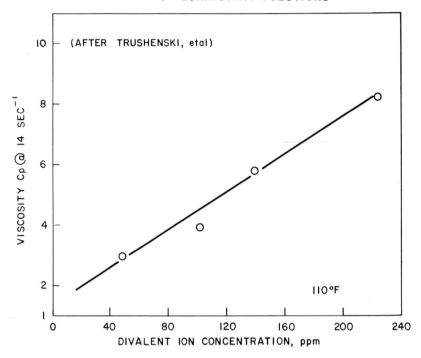

INFLUENCE OF Ca⁺⁺ ON VISCOSITY
OF SURFACTANT SOLUTIONS

(AFTER TRUSHENSKI, etal)

110°F

VISCOSITY Cp @ 14 SEC⁻¹

DIVALENT ION CONCENTRATION, ppm

FIGURE 6

of the polymer and the electrolyte content determine largely
which mechanism dominates. An index of the mobility reduction of the polymer solution is the resistance factor which
is defined as the ratio of the mobility of the water to the
mobility of the water-polymer solution in the same rock.
Both types of polymers are sensitive to the mono- and divalent ions but to different degrees. Any deterioration of
the polymer solution may result in poor performance.

It is well known that sodium chloride causes a loss of
viscosity with both polymer types. MacWilliams et al. (19)
suggest that the polyacrylamides lose much more viscosity
than do the biopolymers. Viscosity is not as important for
mobility reduction for the polyacrylamides, however, and much
of the resistance factor can be maintained even with the loss
in viscosity. This is illustrated on Figure 7. Note that in
Figure 7-a the indicated change in viscosity is for 3 percent
NaCl. The viscosity drops sharply with NaCl concentration
and about 85 percent is lost at one-half percent. The

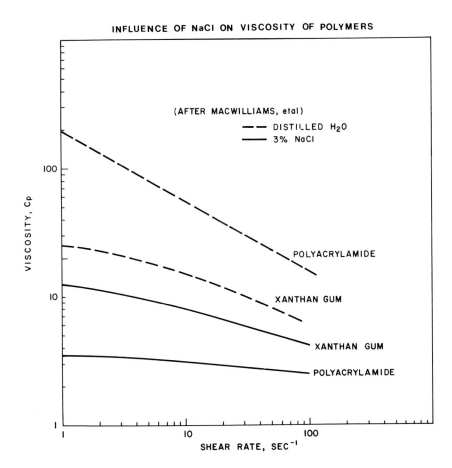

INFLUENCE OF NaCl ON VISCOSITY OF POLYMERS

FIGURE 7-a

resistance factor for the polymer in 3 percent NaCl is shown
in Figure 7-b and approaches a minimum value of ten, even
with the marked change in viscosity shown in Figure 7-a.
The divalent ions affect the polyacrylamides similarly, but,
as Lipton (20) notes, the biopolymers are usually much less
sensitive to the divalent ions in the pH range normally
found in reservoir water. He does note, however, that certain of the trivalent ions can cause gel formation with

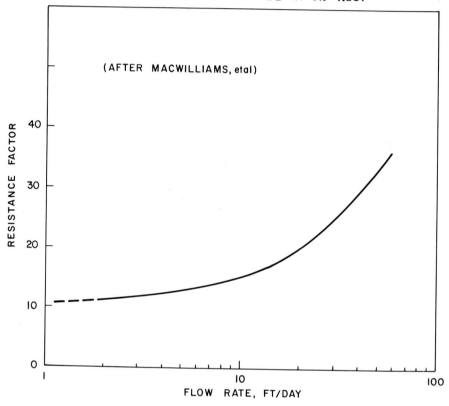

RESISTANCE FACTOR OF 0.05% POLYACRYLAMIDE IN 3% NaCl

(AFTER MACWILLIAMS, etal)

FLOW RATE, FT/DAY

FIGURE 7-b

biopolymers in the normal range of reservoir pH. This gela-
tion can occur in the presence of both ferric iron and certain
coordination complexes of boron.

The above reactions of injected fluids with interstitial
fluids presupposes that displacement of interstitial fluids
is not piston-like and that some mixing occurs. Even in a
completely uniform rock with uniform displacement there would
be mixing resulting from diffusion and dispersion. Beyond
this, the extent of the mixing is a function of the heterogeneity of the rock and the mobilities of the fluids, but it
is highly probable that mixing will be of concern in most
practical cases. Holm and Josendal (13) have illustrated
the influence of heterogeneity as shown schematically in
Figure 8. More rapid movement of the displacing surfactant

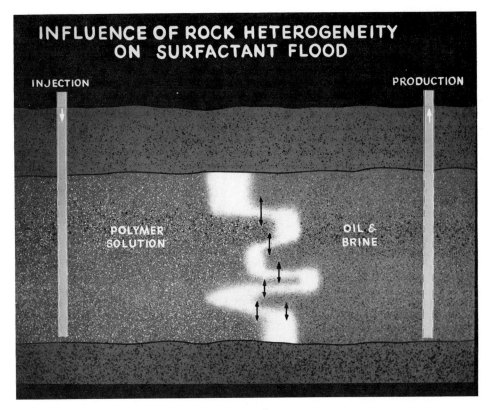

INFLUENCE OF ROCK HETEROGENEITY
ON SURFACTANT FLOOD

INJECTION   PRODUCTION

POLYMER
SOLUTION   OIL &
                            BRINE

Figure 8

through some zones result in regions in which the surfactant
slug in one zone can mix (1) with the downstream brine, or
water in one adjacent zone, and (2) with the polymer solution
in another. Both could result in slug attenuation and/or loss
of mobility control.

C. Mineralogy

    The chemical flood processes can also be strongly in-
fluenced by the mineralogy of the reservoir rock. Table II
shows the mineral makeup of several samples of rock from
various reservoirs. The high surface area of clays make
them particularly significant components, as Cooke et al.
(11) suggest. In alkaline waterflooding the clays that are
initially in equilibrium with the interstitial water will,
by ion exchange, tend to equilibrate with the injected alkaline water. In so doing, hydrogen, calcium and/or magnesium
ions are transferred to the flood water resulting in a loss
of alkalinity and reduction in the effectiveness of the

TABLE II

MINERAL ANALYSES OF RESERVOIR CORES

Component, %	Source						
	Berea Outcrop	Louisiana 8700'	W. Texas 3100'	W. Texas 2900'	Tex. Coast 3000'	W. Tex. 900'	Okla. 1600'
Quartz	75-80	80-85	60-65	-	45-50	80-85	75-80
Feldspar	1-5	1-5	1-5	1-5	1-5	5-10	5-10
Dolomite	-	-	10-15	80-95	-	-	-
Anhydrite or Gypsum	-	-	-	5-10	-	-	-
Calcite	-	-	-	-	20-25	-	-
Clay	10-15	5-10	10-15	-	15-20	10-15	10-15
Mont.	-	-	35-40	-	10-15	25-30	-
Kaol.	75-80	80-85	20-25	-	70-75	55-60	40-45
Ill.	15-20	10-15	30-35	-	5-10	1-5	10-15
Chlor.	-	-	(20-25)	-	-	1-5	35-40

process, as previously discussed. Ehrlich et al. (21) have pointed out that anhydrite ($CaSO_4$) or gypsum ($CaSO_4 \cdot 2H_2O$) are particularly detrimental to the alkaline waterflood process by virtue of its reactivity with the caustic to form the insoluble $Ca(OH)_2$ and resultant loss of alkalinity. This is illustrated in Figure 9. The core used for this test con-

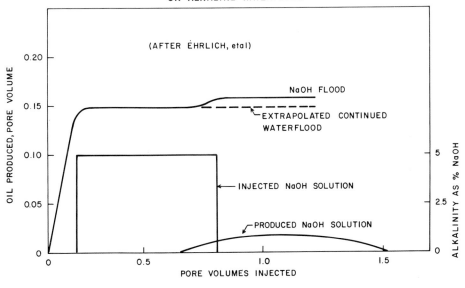

INFLUENCE OF REACTIVE MINERALS ON ALKALINE WATERFLOOD

FIGURE 9

tained a substantial amount of anhydrite. An 0.8 pore volume slug of 5 percent NaOH produced only about 0.01 pore volume of additional oil after a waterflood. The consumption of the caustic by anhydrite is evidenced by the loss of alkalinity and the poor recovery.

Clays in the rock can affect the surfactant flood in ways that are mechanistically similar to those cited above for the alkaline waterflood. Bernard (17) has suggested that by ion exchange divalent ions are transferred to the surfactant solution from the clays resulting in precipitation of the surfactant and loss of surfactant in the displacing fluid. Holm and Josendal (13) have demonstrated that the surfactant solutions will "extract" Ca^{++} and Mg^{++}. Figure 10 shows their data for flow of surfactant solutions through various cores. These cores contained fresh water prior to injection of the surfactant; however, similar extraction was observed from cores initially containing 9.4 percent brine and 7000

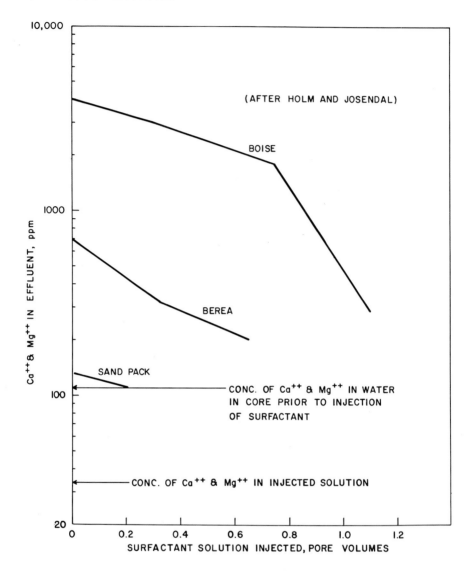

FIGURE 10
EXTRACTION OF DIVALENT IONS
FROM CLAYS BY SURFACTANT SOLUTION

ppm Ca^{++} and Mg^{++}. The high Ca^{TT} and Mg^{++} concentration in the effluent relative to that injected in the surfactant solution indicates a transfer of the divalent ions from the rock.

As discussed by Bernard (17), Trushenski et al. (18), and Gale and Sandvik (22), adsorption can also result in

loss of surfactant and reduced effectiveness of the slug.
Table III shows data from Gale and Sandvik (22) that illustrates the dependency of adsorption on both clay type and
equivalent weight of the surfactant. Note that for a given

TABLE III

SULFONATE ADSORPTION ON CLAYS

(After Gale and Sandvik) (22)

Equivalent Weight	mg/gm Ca Montmorillonite	mg/gm Kaolinite
233	-2.0*	3.7
310	-1.0	4.6
333	1.7	6.0
342	8.2	9.2
400	13.3	10.8

25 ml of 0.5 percent surfactant solution in 2 percent
Na_2SO_4; 7 gm Clay.

*More water adsorbed than surfactant. Equilibrium solution
enriched in surfactant.

clay type the adsorption is an increasing function of the
equivalent weight.

The polymer solutions are also influenced by the mineralogy. Dissolved Ca^{++} and Mg^{++} reduce the viscosity of polyacrylamide solutions, as mentioned previously, and these
polymers can be adsorbed on the clays resulting in increased
mobility of the polymer solution and a loss of mobility
control. Bilhartz and Charlson (23) describe a pilot field
injection test which demonstrated the dissolution of Ca^{++} and
subsequent loss in viscosity. Table IV shows these data for
tests in which polymer solutions were injected and subsequently produced back. The increase in Ca^{++} was attributed
to dissolution of gypsum.

The overall influence of the divalent ions is typically
summarized in Figure 11. These are the results of displacement tests with surfactant solutions and polymer drives.
Clearly the divalent ions can be detrimental to the process.

TABLE IV

Ca^{++} DISSOLUTION BY INJECTED POLYMER

(After Bilhartz and Charlson) (23)

Test	Average Injection Rate, BPD	Range of Ca^{++} Content, mg/l Injected	Produced
1	1740	2-3	13-107
2	998	3-13	62-159
3	998	–	30-129
4	454	6	10-131

Various techniques have been suggested for minimizing, or circumventing, the problems associated with the chemistry of the interstitial water and the mineralogy of the rock. One technique is to inject a preflush slug of water that would displace the interstitial water and, if necessary, condition the clays by ion exchange using saline water. This would ideally minimize contact of the injected fluid with the initial interstitial water and also reduce Ca^{++} and Mg^{++} dissolution in the surfactant and polymer solutions by ion exchange. This may not always prove successful, however. Pursley et al. (24) reported on a field test in which a fresh water preflush did not adequately displace the resident brine resulting in contact of the surfactant with intolerably high salinities. French et al. (25) argue that in some applications a preflush appears to be satisfactory and Knight et al. (26) argue that preflushes may not necessarily be justified.

Another solution to the problem is to add to the injected fluid a compound that will remove or sequester the divalent ions. Sodium carbonate and sodium tripolyphosphate have been suggested for this purpose and their use tends to "protect" the surfactants and polymers from the multivalent ions. Curve B in Figure 11 are data from a test in which Na_2CO_3 was added to the polymer solution. An increase in recovery was obtained.

Cheap sacrifical chemicals have been suggested to reduce adsorption of surfactant and polymer on the clays. These would be adsorbed instead of the more expensive chemicals. As a result less surfactant and polymer would be required and the economics could be improved.

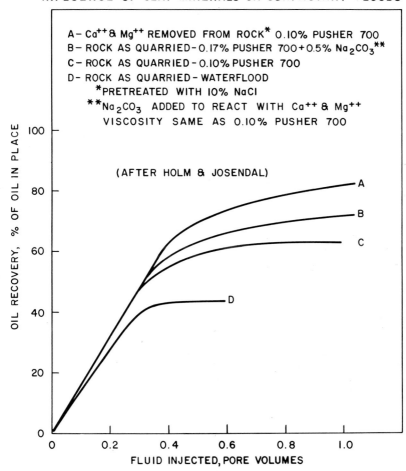

INFLUENCE OF CLAY MINERALS ON SURFACTANT FLOODS

A – Ca^{++} & Mg^{++} REMOVED FROM ROCK* 0.10% PUSHER 700
B – ROCK AS QUARRIED – 0.17% PUSHER 700 + 0.5% Na_2CO_3**
C – ROCK AS QUARRIED – 0.10% PUSHER 700
D – ROCK AS QUARRIED – WATERFLOOD
    *PRETREATED WITH 10% NaCl
    **Na_2CO_3 ADDED TO REACT WITH Ca^{++} & Mg^{++}
    VISCOSITY SAME AS 0.10% PUSHER 700

(AFTER HOLM & JOSENDAL)

FIGURE II

D. Geology/Lithology

The nature of the reservoir rock can also be an environmental constraint to the successful application of chemical
flooding. Clearly, if the reservoir is fractured, the injected chemicals will by-pass the bulk of the rock with
little or no response to the injection. If the reservoir
permeability is low it may not be possible to inject the
surfactant and polymer solutions at a sufficient rate. In
general, the resistance factor of polymers increases inversely
to the permeability, as discussed by R. Jennings et al. (9),
and at sufficiently low permeability the resistance can be
prohibitive for practical purposes.

The polyacrylamides can be adversely affected by shear degradation with a loss in viscosity and resistance factor. Maerker (27), and Hill et al. (28) have shown this with laboratory studies, and Bilhartz and Charlson (23) provide data from a field pilot test. The biopolymers exhibit much less tendency to shear degrade (28). The breaking of the molecules by shear can occur in the surface equipment, well-bore, perforations, and in the reservoir in the vicinity of the wellbore owing to the high velocities through small pore channels. The shearing increases with decreasing permeability such that the permeability can be a constraint. Figure 12 shows data from Maerker (27) that relates the loss of screen factor and viscosity to flow rate through several cores of different permeability. (Screen factor correlates

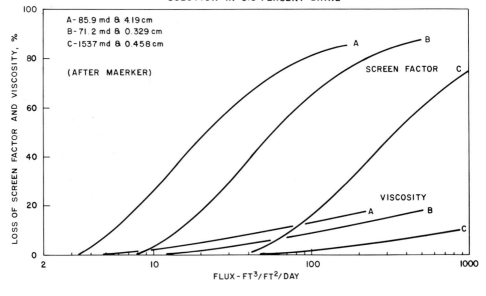

SHEAR DEGRADATION OF 600 PPM POLYACRYLAMIDE SOLUTION IN 3.3 PERCENT BRINE

FIGURE 12

with resistance factor.) As shown, the degradation increases with decreasing permeability.

Another element of the lithology that can influence the behavior of the polymers is the inaccessible pore volume. Inaccessible pore volume is defined as that fraction of the pore volume of the rock that is excluded for polymer flow only. Thus, in surfactant flooding, the polymer tends to by-pass some flow channels and overrun the surfactant. As a result proper mobility relations between the polymer and surfactant can be lost. Trushenski et al. (18) confirmed

earlier work described by Dawson and Lantz (29) and suggested
the need to increase surfactant slug size to compensate for
penetration of the slug by polymer as a result of the inaccessible pore volume.

E. Temperature

The temperature of the reservoir is another environmental factor that can influence chemical flooding. Mungan (30)
found the alkaline waterflood process to be temperature-
dependent. He found for a particular crude oil that the process was ineffective at 70°F and effective at 160°F. Cooke
et al. (11) have pointed out that in an alkaline waterflood
the higher temperature results in a higher rate of consumption of the base by the reactive reservoir solids and possible
increased costs for chemicals.

Increased temperature can detrimentally affect the surfactant floods. Dauben and Froning (14) have shown that,
depending upon salt concentration, phase separation can occur at 150°F with a solution that is stable at room temperature.

An additional problem posed by increased temperature is
the chemical stability of the polymer solutions for the long
time periods involved in reservoir flow. Knight (31) concluded that dissolved oxygen promotes degradation of the
polyacrylamides and that this could be adequately controlled
up to 140°F with the use of oxygen scavengers such as sodium
hydrosulfite, or formaldehyde. More recently, Martin (32)
reported on laboratory tests up to 210°F with similar conclusions and added glyoxal as a scavenger. The issue raised
by Hill et al. (28) appears to remain valid, however. Is
there any direct evidence that any polyacrylamide has retained a significant part of its mobility control activity
when it has been in a reservoir environment, at any temperature, for several years?

Hill et al. (28) reported on stability studies with the
biopolymers. Their work suggests that there is even less
assurance of long-term chemical stability at elevated temperatures in reservoir environments than with the polyacrylamides. There is, however, some evidence that appropriate
additives (sodium sulfite) may be effective.

VI. LITERATURE CITED

1.   Geffen, Ted M., "Improved Oil Recovery Could Help Ease
       Energy Shortage," World Oil 177, No. 5, 84-88 (1973).
2.   Hasiba, H. H., and Wilson, L. A., Jr., "The Potential
       Contribution of Enhanced Recovery Technology to
       Domestic (U.S.A.) Crude Oil Reserves," Erdoel-
       Erdgas-Zeitschrift 91, 77-80 (1975).

3. Sharp, James M., "The Potential of Enhanced Oil Recovery Processes," Paper SPE 5557, presented at the 50th Annual Fall Meeting of the Society of Petroleum Engineers of AIME, Dallas, Texas, Sept. 28–Oct. 1, 1975.

4. Poettman, F. H., "Microemulsion Flooding," Secondary and Tertiary Oil Recovery Processes, Interstate Oil Compact Commission, Oklahoma City, Oklahoma, 1974.

5. Foster, W. R., "A Low-Tension Waterflooding Process," J. Pet. Technol. XXV, 205–210 (1973).

6. Cash, R. L., Cayias, J. L., Fournier, G. R., Jacobson, J. K., Schares, T., Schecter, R. S., and Wade, W. H., "Modeling Crude Oils for Low Interfacial Tension," Paper SPE 5813, presented at Society of Petroleum Engineers of AIME Symposium on Improved Oil Recovery, Tulsa, Oklahoma, March 22–24, 1976.

7. Ehrlich, R., and Wygal, R. J., Jr., "Interrelation of Crude Oil and Rock Properties with the Recovery of Oil by Caustic Waterflooding," Paper SPE 5830, presented at Society of Petroleum Engineers of AIME Symposium on Improved Oil Recovery, Tulsa, Oklahoma, March 22–24, 1976.

8. Jennings, Harley Y., Jr., "A Study of Caustic Solution-Crude Oil Interfacial Tensions," Soc. Pet. Eng. J. 15, No. 3, 197–202 (1975).

9. Jennings, R. R., Rogers, J. H., and West, T. J., "Factors Influencing Mobility Control by Polymer Solutions," J. Pet. Technol. XXIII, 391–400 (1971).

10. Jennings, H. Y., Jr., Johnson, C. E., Jr., and McAuliffe, C. D., "A Caustic Waterflooding Process for Heavy Oils," J. Pet. Technol. XXVI, 1344–1352 (1974).

11. Cooke, C. E., Jr., Williams, R. E., and Kolodzic, P. A., "Oil Recovery by Alkaline Waterflooding," J. Pet. Technol. XXVI, 1365–1374 (1974).

12. Hill, H. J., Reisberg, J., and Stegemeier, G. L., "Aqueous Surfactant Systems for Oil Recovery," J. Pet. Technol. XXV, 186–194 (1973).

13. Holm, L. W., and Josendal, V. A., "Reservoir Brines Influence Soluble-Oil Flooding Process," The Oil and Gas J. 70, No. 46, 158–168 (1972).

14. Dauben, D. L., and Froning, H. R., "Development and Evaluation of Micellar Solutions to Improve Water Injectivity," XXIII, 614–620 (1971).

15. Healy, R. N., and Reed, R. L., "Physico-Chemical Aspects of Microemulsion Flooding," Soc. Pet. Eng. J. 14, No. 5, 491–501 (1974).

16. Healy, R. N., Reed, R. L., and Stenmark, D. G., "Multiphase Microemulsion Systems," Paper 5565, presented at 50th Annual Fall Meeting of the Society of Petroleum Engineers of AIME, Dallas, Texas, Sept. 28–Oct. 1, 1975.

17. Bernard, George W., "Effect of Clays, Limestone and
     Gypsum on Soluble Oil Flooding," J. Pet. Technol.
     XXVII, 179-180 (1975).
18. Trushenski, S. P., Dauben, D. L., and Parrish, D. R.,
     "Micellar Flooding-Fluid Propagation, Interaction
     and Mobility," Soc. Pet. Eng. J. 14, No. 6, 633-645
     (1974).
19. MacWilliams, D. C., Rogers, J. H., and West, T. J.,
     "Water-Soluble Polymers in Petroleum Recovery,"
     Polymer Science and Technology, Vol. 2 (Water-
     Soluble Polymers), edited by N. M. Bikales, Plenum
     Press, New York, 1973.
20. Lipton, Daniel, "Improved Injectability of Biopolymer
     Solutions," Paper SPE 5099, presented at the 49th
     Annual Fall Meeting of the Society of Petroleum
     Engineers of AIME, Houston, Texas, Oct. 6-9, 1974.
21. Ehrlich, R., Hasiba, H. H., and Raimondi, P., "Alkaline
     Waterflood for Wettability Alteration - Evaluating
     a Potential Field Application," J. Pet. Technol.
     XXVI, 1335-1343 (1974).
22. Gale, W. W., and Sandvik, E. I., "Tertiary Surfactant
     Flooding: Petroleum Sulfonate Composition-Efficacy
     Studies," Soc. Pet. Eng. J. 13, No. 4, 191-199 (1973).
23. Bilhartz, H. L., Jr., and Charlson, G. S., "Field
     Polymer Stability Studies," Paper SPE 5551, presented
     at 50th Annual Fall Meeting of the Society of Petro-
     leum Engineers of AIME, Dallas, Texas, Sept. 28-Oct. 1,
     1975.
24. Pursley, S. A., Healy, R. N., and Sandvik, E. I., "A
     Field Test of Surfactant Flooding, Loudon, Illinois,"
     J. Pet. Technol. XXV, 793-802 (1973).
25. French, M. S., Keys, G. W., Stegemeier, G. L., Weiber,
     R. C., Abrams, A., and Hill, J. H., "Field Test of an
     Aqueous Surfactant System for Oil Recovery, Benton
     Field, Illinois," J. Pet. Technol. XXV, 195-204 (1973).
26. Knight, B. L., Jones, S. C., and Parsons, R. W., "Dis-
     cussion to Trushenski et al.," Soc. Pet. Eng. J. 14,
     No. 6, 643-645 (1974).
27. Maerker, John M., "Shear Degradation of Polyacrylamide
     Solutions," Soc. Pet. Eng. J. 15, No. 4, 311-322
     (1975).
28. Hill, H. J., Brew, J. R., Claridge, E. L., Hite, J. R.,
     and Pope, G. A., "The Behavior of Polymers in Porous
     Media," Paper SPE 4748, presented at Society of
     Petroleum Engineers of AIME Symposium on Improved
     Oil Recovery, Tulsa, Oklahoma, April 22-24, 1974.
29. Dawson, R., and Lantz, R. B., "Inaccessible Pore Volume
     in Polymer Flooding," Soc. Pet. Eng. J. 12, No. 5,
     448-452 (1972).

30. Mungan, N., "Certain Wettability Effects in Laboratory
     Waterfloods," J. Pet. Technol. XVIII, 247-252 (1966).
31. Knight, Bruce L., "Reservoir Stability of Polymer Solu-
     tions," J. Pet. Technol. XXV, 618-626 (1973).
32. Martin, F. D., "Laboratory Investigation in the Use of
     Polymers in Low Permeability Reservoirs," Paper SPE
     5100, presented at the 49th Annual Fall Meeting of
     the Society of Petroleum Engineers of AIME, Houston,
     Texas, Oct. 6-9, 1974.
33. Case, L. C., "Water Problems in Oil Production," The
     Petroleum Publishing Company, 211 So. Cheyenne,
     Tulsa, Oklahoma.
34. Crawford, J. G., "Waters of Producing Fields in the
     Rocky Mountain Region," Trans. AIME, 179 264-287
     (1949).

OIL RECOVERY WITH SURFACTANTS: HISTORY AND A CURRENT APPRAISAL

W. Barney Gogarty
Marathon Oil Company

I. ABSTRACT

Enhanced recovery with surfactants is being pursued as a means of increasing the U.S. energy supply. Laboratory and field work is under way on both low- and high-concentration processes to optimize the method of injecting surfactant. Critical factors in economic projections of surfactant recovery processes are oil recovery, oil price, and tax load.

II. SCOPE

Enhanced recovery methods are being actively pursued by industry and through government projects as a means of increasing the U.S. energy supply. An additional 59 billion bbl of oil from known reservoirs are estimated by Geffen (1) to be recoverable with existing techniques, including surfactant methods, thermal techniques, and CO_2 flooding. Of the potential reserves, about 60 percent are estimated to be amenable to chemical flooding with surfactants.

This paper considers chemical flooding that uses surfactants for oil recovery. Background information is presented showing the development of both low- and high-concentration surfactant processes. Laboratory results are discussed briefly to illustrate the different aspects of chemical flooding and to indicate trends of research. Field testing over the past 13 years is considered from the standpoint of using project results to predict full-scale development economics. The main thrust of the paper deals with steps necessary to commercialize a surfactant flooding process. The important question of chemical supply is discussed. Economic examples are presented based on Marathon Oil Co.'s experience in Illinois. The effects of crude oil price and loss of statutory depletion on profitability are presented, along with capital requirements for a 6,000-acre, 10-year development program. Results show that proper economic incentives, including such things as favorable tax measures and high crude prices, will be needed before surfactant flooding processes can be used to develop large volumes of tertiary reserves.

III. CONCLUSIONS AND SIGNIFICANCE

1. Essentially, two different methods have developed for using surfactants to enhance oil recovery. One uses a large pore volume of a low-concentration surfactant solution. The other uses a small pore volume of a high-concentration surfactant solution.

2. Laboratory results reported in the literature indicate that, with low-concentration surfactant injection, oil production is sustained at a lower level for a longer period of time than with high-concentration surfactant injection.

3. The use of different salts and mixtures of petroleum sulfonate with broad equivalent weights have been reported as methods for reducing adsorption in low-tension floods with surfactant solutions.

4. Oil recovery in the laboratory is improved through proper formulation of micellar solutions; a part of this improvement is attributable to minimizing surfactant adsorption on rock surfaces.

5. Mobility control is necessary for enhanced oil recovery using both high- and low-concentration surfactant flooding processes.

6. Based on published data, field results indicate that higher values are obtained with high-concentration-surfactant, low-pore-volume systems than with low-concentration, high-pore-volume systems.

7. Considerably increased production capacity for both petroleum sulfonates and polymers will have to be developed if significant quantities of tertiary oil are to be obtained by surfactant methods.

8. The key factors in any economic projections for enhanced oil recovery processes using surfactant are investment requirements including chemical and development costs, oil recovered, the time required to obtain that oil, oil price, and tax load. The uncertainties introduced by long-term projections of these factors significantly cloud any economic projection.

9. Recoveries obtained in tertiary projects will significantly affect the economic projections. Until recoveries from commercial-scale projects become available, a high risk will be involved in using recovery values obtained from the smaller-sized projects.

IV. CONCEPT DEVELOPMENT

Surfactant use for oil recovery is not a recent development in petroleum technology. Water-soluble surfactants were described by De Groot (2,3) as an aid to improve oil recovery in patents filed in the late 1920's and early 1930's. These patents taught the use of such water-soluble surfactants as a polycyclic sulfonic body and wood sulfite liquor in concentrations of 25 to 1000 ppm. Other water-soluble compounds have been suggested by Holbrook (4) for surfactant flooding; these compounds include organic perfluoro compounds, fatty acid soaps, polyglycol ether, salts of fatty or sulfonic acids, and polyoxyalkylene compounds. Laboratory results were presented showing that these solutions reduced interfacial tension and enhanced oil recovery. Publications since then have stressed coupling different salts with surfactants to reduce the interfacial tension to a minimum value and to prevent the adsorption of surfactants within the reservoir. These techniques have given rise to the low-tension surfactant flooding processes. In low-tension floods, much of the reservoir pore volume is filled with surfactant solution of a relatively low concentration. For example, a 30-percent PV slug containing less than 2-percent surfactant might be used.

In 1959, Holm and Bernard (5) filed for a patent in which they proposed injecting 0.1- to 3-percent surfactant dissolved in low-viscosity hydrocarbon solvent. This procedure reduced surfactant adsorption in water-wet formations. In 1961, Csaszar (6) filed for a patent specifying the use of a mixture of an anhydrous soluble oil and a nonaqueous solvent containing up to about 12-percent surfactant. These patents have given rise to the soluble-oil flooding process.

Blair and Lehmann (7) filed for a patent in 1942 on a well stimulation process. This patent described the injection of transparent emulsions into production wells to remove objectionable waxy solids. This appears to be the first mention of using transparent solutions (sometimes called microemulsions, swollen micelles, fine emulsions, micellar dispersions, etc.) in the petroleum recovery literature. In 1962, Gogarty and Olson (8) filed for a patent describing the use of microemulsions in a new miscible-type recovery process known as Maraflood.[TM] The microemulsions contain surfactant, hydrocarbon, and water. Cosurfactants or electrolyte may be added as defined by Gogarty and Tosch (9). The Gogarty and Olson patent teaches the injection of a small fraction of the pore volume of micellar solution containing a surfactant concentration greater than about 5 percent. Patents have been issued to Jones (10,11) claiming the use of high-water-content, oil-external microemulsions and water-external micellar dispersions in oil recovery. Jones' patents, along with those of Reisberg and Cooke (12-14), teach the use of a relatively high surfactant concentration in various aqueous systems and micellar dispersions for the recovery of oil.

[TM]Trademark of Marathon Oil Company.

Essentially two different concepts have developed for
using surfactants to enhance oil recovery. In the first concept, a solution containing a low concentration of a surfactant is injected. The surfactant is dissolved in either water
or oil and is in equilibrium with aggregates of the surfactant
known as micelles. Large pore volumes (about 15 to 60 percent
or more) of the solution are injected into the reservoir to
reduce interfacial tension between oil and water and, thereby,
increase oil recovery. Oil may be banked with the surfactant
solution process, but residual oil at a given position in the
reservoir will only approach being reduced to zero after
passage of large volumes of surfactant solution. In the
second process, a relatively small pore volume (about 3 to 20
percent) of a higher-concentration surfactant solution is injected into the reservoir. With the higher surfactant concentration, the micelles become a surfactant-stabilized dispersion of either water in hydrocarbon or hydrocarbon in
water. The high surfactant concentration allows the amount
of dispersed phase in the microemulsion to be high as compared with the low value in the dispersed phase of the
micelles in the low-concentration surfactant solutions. The
injected slug is formulated with three or more components.
The basic components--hydrocarbon, surfactant, and water--are
sufficient to form the micellar solutions. A cosurfactant
fourth component (usually alcohol) can be added. Electrolytes,
normally inorganic salts, form a fifth component that may be
used in preparing the micellar solutions or microemulsions.
The high-concentration surfactant solutions displace both oil
and water and rapidly displace all the oil contacted in the
reservoir. As the high-concentration slug moves through the
reservoir, it is diluted by formation fluids and the process
reverts to a low-concentration flood.

Work is under way in the laboratory and the field to
select the optimum method of injecting surfactant to enhance
oil recovery. Each group of workers (low concentration and
high concentration) feels that its process is the best.
Clearly, much more work is needed before the question can be
resolved satisfactorily.

V.   PERFORMANCE IN THE LABORATORY

Laboratory results of studies made with both low- and
high-concentration surfactant systems have been published.
Different aspects of the two types of processes, including
displacement, adsorption, mobility control, and scaling, have
been discussed. Some of these results are summarized in this
section to illustrate how oil recovery is enhanced with surfactants.

A. Displacement with Surfactant Systems

Displacement results from three different laboratories as reported by Gale and Sandvik (15), Hill et al. (16), and Davis and Jones (17) are presented in Table I. All floods took place in Berea cores and used petroleum sulfonate as the surfactant. Since these results came from different laboratories, some of the variables, such as flooding rate and amount of polymer, were not the same. Floods 1 and 2 are indicative of low-concentration, large-pore-volume surfactant flooding. Flood 3 is representative of a high-concentration, small-pore-volume flood. Polymer was used for mobility control in Floods 2 and 3. The use of polymer in Flood 2 may account for its better recovery over that of Flood 1.

The recoveries given in Table I indicate that both types of surfactant processes are capable of recovering significant quantities of tertiary oil. Process efficiency can be related to the ratio of oil recovered divided by the amount of surfactant injected. On a relative basis, the amount of surfactant is simply the product of the surfactant concentration and pore volume injected. Based on this definition of surfactant injection, the efficiency ratio for Floods 1, 2, and 3 is 0.70, 2.62 and 4.09, respectively. Flood rate, core dimensions, and the degree of mobility control will affect these ratios to some degree; but, based on published data, it is doubtful whether the low-concentration surfactant efficiency ratios would increase to that of the high-concentration surfactant process. The results in Table I indicate that surfactant utilization is best when injected at high concentrations in a small-pore-volume slug.

Figure 1 compares the tertiary production history of Flood 1 with Flood 3 of Table I. Note that oil production starts earlier in Flood 1 (about 0.1 PV) than in Flood 3 (about 0.24 PV) and that the total recovery of 70 percent for Flood 1 is reached after producing about 1.6 PV as compared with about 1.1 PV for Flood 3, at which point 85 percent is recovered. These results indicate that with low-concentration surfactant injection, oil production starts sooner and is sustained at a lower level for a longer period of time than with high-concentration surfactant injection. The early breakthrough and late recovery in Flood 1 may be partly caused by inadequate mobility control. If these same results were obtained in the field, economic comparisons based on the time value of money would favor high-concentration surfactant injection.

Results in Figure 2 compare the oil fraction in produced fluids from Floods 2 and 3. The initial increase in oil cut for the low-concentration surfactant flood occurs at essentially the same point as the high-concentration flood. Both

TABLE I

LABORATORY DISPLACEMENTS WITH SURFACTANTS

Flood	Surfactant Concentration (percent)	Pore Volume Injected (percent)	Core Dimension Diameter (in.) x Length (ft.)	Rate/Pressure	Tertiary Oil Recovery (percent)	Efficiency Ratio (Oil Recovery Sur-factant injected)
1--Gale & Sandvik (15)	2	50	2 x 3	3 psi	70	0.70
2--Hill et al. (16)	1.3	25	2 x 16	1.4 ft/D	85	2.62
3--Davis & Jones (17)	10.4	2	2 x 4	4 ft/D	85	4.09

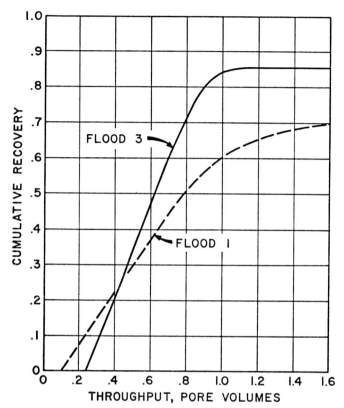

*Fig. 1. Tertiary production history for low- and high-
concentration surfactant floods*

these floods are typical of tertiary recovery operations where
oil is banked during early slug injection and only water is
produced from the reservoir, which is at essentially residual
oil. Production of the oil bank at about 0.22 PV for the low-
concentration surfactant Flood 2 as compared with 0.11 PV for
Flood 1 probably reflects the use of polymer for mobility
control in Flood 2. Note in Figure 2, as in Figure 1, that
oil production continued for a longer period of time but at
a lower oil cut with the low-concentration surfactant flood.

Work appears to be under way in industry and at univer-
sities to define the displacement mechanism of both low- and
high-concentration surfactant flooding. The paramount question relates to the optimum way of injecting a given quantity
of surfactant into a reservoir to recover the maximum amount
of oil. From an economic standpoint, total oil recovery needs
to be obtained with a minimum total pore volume of fluid injection. A better understanding of both low-tension and

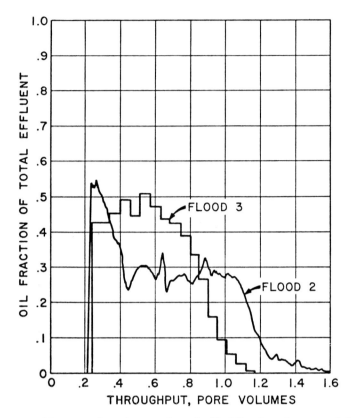

Fig. 2. *Oil fraction in produced fluids for low- and high-concentration surfactant floods*

high-concentration displacement mechanisms, including fluid-rock interactions, will help to answer these questions.

B. Adsorption

The question of adsorption is always present with oil recovery methods using surfactants. Various methods have been suggested for reducing surfactant loss in low-tension flooding: these methods include the use of different salts and mixtures of petroleum sulfonate with broad equivalent-weight distributions. Figure 3 shows the effect of using petroleum sulfonate mixtures with different equivalent-weight distribution on tertiary recovery from studies by Gale and Sandvik (15). The broad equivalent-weight distribution is reported to increase recovery because the central portions of the distribution act as sacrificial adsorbates. This adsorption minimizes the loss of the high equivalent-weight fraction that is most efficient in lowering the interfacial

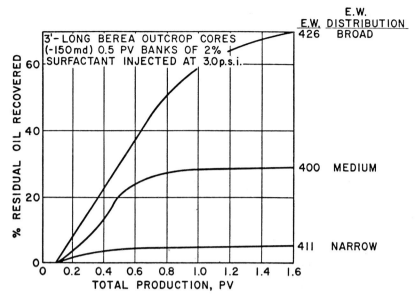

Fig. 3. Effect of surfactant equivalent-weight distribution on oil recovery

tension. Certainly, additional research is needed to find other ways to minimize adsorption in the low-tension flooding processes. The goal here should be to determine the best way of effectively propagating surfactants through the formation.

Apparently, the adsorption mechanism associated with the high-concentration solutions is quite complicated. The results shown in Figure 4 indicate that surfactant loss for one high-concentration system reaches a minimum value near the critical micelle concentration of 30,000 ppm as reported by Trushenski et al. (18). Preflush with sodium chloride brine appears to reduce the surfactant loss. Undoubtedly, the high-concentration micellar slug containing up to five components interacting with the in-place oil, water, and reservoir rock affects the surfactant-loss mechanism. Judicious selection of micellar solution components and composition appears to be one method to control the degree of adsorption and, thereby, allow more efficient transport of surfactant through the rock.

As reported by Gogarty and Davis (19), Figure 5 shows how oil recovery can be improved through proper formulation of the micellar solution. The flooding results shown in this figure were obtained in Berea cores saturated with essentially the same oil and water. The slugs were formulated with different types of hydrocarbon, petroleum sulfonate, cosurfactants, and electrolytes. With a 3-percent PV slug, the recovery was increased ninefold. This increase in recovery certainly represents an improvement in utilization of surfactant, and some

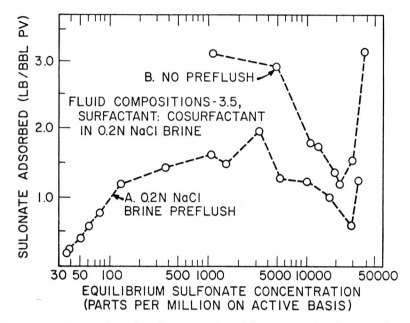

Fig. 4. *Adsorption isotherms of a Mahogany petroleum sul-*
fonate on Berea sandstone at 110°F

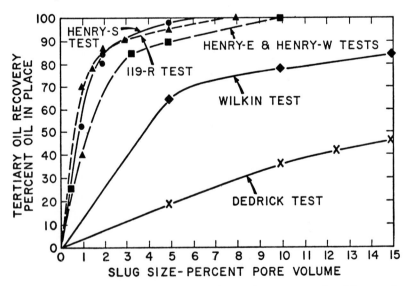

Fig. 5. *Improved oil recovery through proper micellar solu-*
tion formulation

part of this improvement is undoubtedly the result of reduced
surfactant adsorption. As with low-concentration surfactant
flooding, more work is needed to better understand the mechanism whereby high-concentration surfactant solutions are
transmitted through a reservoir.

C. Mobility Control
     Mobility control is important in both low- and high-
concentration surfactant flooding. Conditions have been
described by Gogarty et al. (20) for obtaining mobility control with miscible-type waterfloods using micellar solutions.
High-molecular-weight, water-soluble polymers have been used
for mobility control in both types of surfactant processes.
Gogarty and Davis (19) report that oil-in-water emulsions also
have been used for mobility control with high-concentration
surfactant flooding.

     With low-concentration surfactant injection, mobility
control of the surfactant slug is accomplished by dissolving
polymer in the surfactant solution. Results reported by
Gogarty and Tosch (9) indicate that, generally, the mobility
of the high-concentration surfactant slug is fixed by adjusting the composition of such micellar components as the
cosurfactant and electrolyte. Continued stability in both
low- and high-concentration displacements requires the use
of a mobility buffer. Most often, solutions of polyacrylamides or polysaccharides have been used as mobility buffers.
The polyacrylamide achieves part of its mobility control by
permeability reduction of the reservoir, in contrast to an
increased viscosity. On the other hand, polysaccharide solutions obtain mobility control mostly through increased viscosity, with little permeability reduction.

     The effect of mobility control on oil recovery in high-
concentration surfactant flooding is illustrated in Figure 6
from a paper by Gogarty and Davis (19). Both the linear cores
and the two-dimensional slabs were waterflooded to residual
oil before injecting a 5-percent PV micellar slug. These
laboratory results show that (1) little additional tertiary
oil is recovered with mobility buffers greater than 50 percent PV and (2) below 50 percent PV, favorable mobility
control is more critical in two-dimensional than in linear
flooding. Larger pore volumes of mobility buffer probably
are needed in actual reservoirs where geometries are more
complex.

     The absence of mobility control in low-concentration
surfactant flooding can be inferred from the data in Table I.
As mentioned above, polymer was used in Flood 2 and was not
used in Flood 1. Apparently, the use of mobility control in
Flood 2 resulted in a higher recovery even though the surfactant concentration was lower.

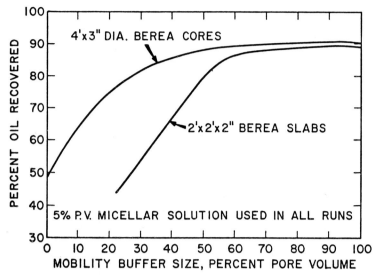

Fig. 6. *The effect of mobility-buffer size on recovery*

Many problems remain to be solved in the area of mobility control. Presently, polymers seem to have the inside track, but long-range research on other mobility-control systems is needed. Cost effectiveness should be the over-all goal in mobility control research. Systems need to be developed for use in both high- and low-permeability reservoirs. Research should be aimed at systems that are relatively unaffected by either temperature or salinity. The interaction between mobility-buffer fluids and surfactant solutions needs further study.

D. Scaling

Scaling of laboratory results to the field is a difficult problem with any recovery process. The displacement mechanism, associated adsorption, and mobility control with surfactant systems make scaling particularly difficult. Much research will be needed to determine acceptable methods for translating laboratory data to accurate field designs. Parsons (21) has discussed the problem of determining proper laboratory rates to best simulate injection in field patterns. His work indicates that frontal velocities less than 1 ft/D should be used in the laboratory. Healy et al. (22) have made similar observations on the rates to be used in laboratory flooding experiments. These workers indicate that microemulsion flooding is not rate-sensitive as long aş the micellar-solution slug is intact. After the slug dissipates and the process reverts to a low-concentration surfactant flood,

recovery depends on rate. Studies by Taber et al. (23,24) indicate that low-tension flooding is governed by $\Delta p/L\gamma$, or its
equivalent, $v\mu/\gamma$, as reported by Foster (25). The single-well
field testing procedure described by Jones and McAtee (26) is
an initial step toward engineering full-scale field projects
from smaller field tests, but additional work will be needed
to develop these procedures. Over-all work on designing
large field surfactant floods from laboratory and small-scale
field test results is in its infancy. Indeed, no commercial-
scale field surfactant floods have been attempted.

VI. FIELD TESTING
     In an article by Bleakley (27), a survey showed that the
oil industry was operating or had terminated 177 separate
projects to enhance oil recovery. This report listed details
for field trials of both high- and low-concentration surfactant slugs. Responses from 16 companies to the Gulf Univer-
sities Research Consortium (GURC) questionnaire of December
1974, as reported by Sharp (28), showed 10 active and nine
planned tests using surfactant methods. In addition to tests
within the petroleum industry, the U.S. Energy Research and
Development Administration (ERDA) has a contract with Cities
Service Oil Co. to conduct a field test in Kansas on improved
oil recovery by micellar flooding. The total cost of the
project is given as $8.1 million. Phillips Petroleum Co. and
ERDA will conduct a $9.7 million micellar polymer project in
Oklahoma as reported in an Oil and Gas Journal article (29).
These industry and government activities indicate an increased
interest in field testing of surfactant processes.
     The cost of the ERDA projects points to the expense asso-
ciated with field development of any oil recovery process using
surfactants. Marathon also has had this kind of experience in
developing its process for enhanced oil recovery. After 13
years' laboratory and field work, with a sizable expenditure
of over $40 million, the technical feasibility of the process
has been demonstrated in limited shallow reservoirs. Our
original goal was to use small-scale, low-cost pilot results
to prove the technical feasibility and obtain data to predict
large-scale project economics. Practically, we have found the
need to use larger-scale projects that contain repeated patterns to determine the extent commercial application is possible. The larger projects allow a more realistic field
determination of drilling and development cost, along with
associated fluid costs. Investments associated with these
larger-scale projects are high.
     Since 1962, Marathon, alone and in conjunction with other
companies, has conducted some 17 field trials of its Maraflood
oil recovery process. These tests have ranged in size from

less than 1 acre to about 40 acres. Marathon's process is
representative of a high-concentration-surfactant, low-pore-
volume process. Results of many of the field tests have been
reported in detail by Gogarty and Davis (19), Earlougher et al.
(30), Danielson et al. (31) and Gogarty and Sukarlo (34). The
test conditions and recovery results from a report by Earlougher
et al. (30) are summarized for the projects in terms of barrels
per acre-foot in Table II. These projects have produced
400,000 bbl of tertiary oil. Cumulative oil production to
July 1975 for the 119-R Project was about 248,700 bbl from the
total pattern. The oil production curve for the 119-R Project
is given in Figure 7. The lower curve shows the production
from the seven confined producers of the test. One of the objectives of the test was to determine the recovery factor from

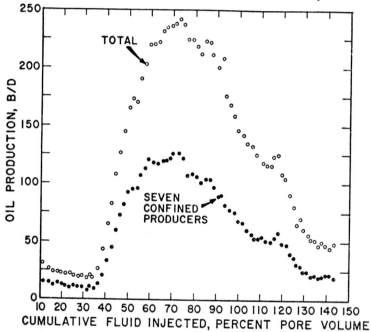

Fig. 7. Oil production for the 119-R Project

the confined portion. The estimated ultimate recovery from
this confined area is about 124,000 bbl, or 282 bbl/acre-ft.
The recovery value from the confined area in July 1975 stood
at 275 bbl/acre-ft. These recoveries compare with a field-
wide waterflood recovery of about 180 bbl/acre-ft and a primary recovery of 300 bbl/acre-ft.

The other large project in Table II is the 47-acre Bingham
expansion. Its oil producing curve is shown in Figure 8 from
a paper by Danielson et al. (31). The current predicted

TABLE II

SUMMARY OF MARAFLOOD OIL-RECOVERY PROCESS FIELD TESTS

Test Name	Test Dates		Pattern		Pore Volume Slug Injected (percent)	Pore Volume Mobility Buffer Injected (percent)	Recovery (bbl acre-ft)
	Start	End	Type*	Size Acres			
Illinois							
Dedrick	Nov. 1962	Dec. 1964	R-5	2.5	3.5	6.6	384
Wilkin	Jan. 1964	Jan. 1965	R-5	2.5	3.5	6.6	86
Henry-W	Nov. 1965	April 1967	I-5	0.75	9.0	200.0	383
Henry-E 0.75 acres	Nov. 1965	June 1966	I-5	0.75	40.0	44.0	169
Henry-E 1.5 acres⁺	June 1966	Jan. 1967	I-5	1.5	20.0	83.0	205
Henry-E 3.0 acres	June 1967	June 1968	I-5	3.0	10.0	55.0	159
119-R	Sept. 1968	—	LD	40.0	7.0	100.0	—
Henry-S	Oct. 1969	March 1970	MT	0.2	4.8	175.0	350
118-K	Sept. 1969	Dec. 1971	R-5	2.4	3.5	87.0	200
Aux Vases	May 1970	July 1972	R-5	4.3	2.5	24.0	—
MT No. 1	March 1973	June 1973	MT	0.2	7.0	93.0	284
MT No. 2	Oct. 1973	—	MT	0.2	7.0	93.0	—
Pennsylvania							
Bingham 533	Dec. 1968	May 1971	I-5	0.75	10.0	200.0	230
Bingham Expansion	Jan. 1971	—	16I-5	47.0	5.0	95.0	—
Goodwill Hill	May 1971	—	9I-5	10.0	5.0	95.0	—

*Pattern-Type Codes:
R-5 — Regular five-spot, producer surrounded by four injectors
I-5 — Inverted five-spot, injector surrounded by four producers
MT — Mini-test, single injector, no producers
LD — Direct line-drive pattern
16I-5 — 16 inverted five-spots.

Recovery or displacement, bbl acre-ft
The Henry East pattern started as a 0.75 inverted five-spot. It was expanded to 1.5 acres then to 3.0 acres by shutting in old producers and opening new producers. Production data are cumulative.

41

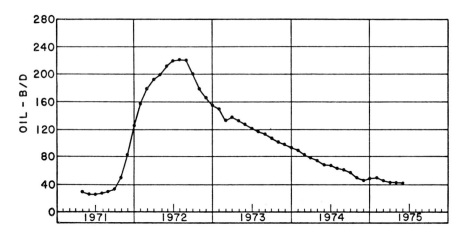

Fig. 8. Oil production for the Bingham expansion project

recovery within the confined area of the project is 5,000 bbl
of oil per acre, or about 312 bbl/acre-ft. This recovery
factor is based on the affected rather than the gross reservoir.
     The test results for Illinois and Pennsylvania have pro-
vided proof of the technical feasibility of using high-concentration surfactants in micellar solutions to displace and
recover oil in previously waterflooded reservoirs. Caution
should be used in extrapolating recovery factors from small-
scale to field-wide projects. Changes in reservoir characteristics, spacing, fluid properties, etc., will affect performance on a field-wide scale. The recovery factor needs
to be determined more precisely from large-scale field projects before the economic potential of this, or any, process
can be obtained. Commercial-scale projects are needed to
fulfill these conditions. Marathon has started commercial-
scale operation by developing a 120-acre tract in south-
eastern Illinois. Expansion of this first tract to several
thousand acres is planned over a period of years.
     In addition to the results shown in Table II, some
details have been published, as reported by Foster (25),
Knight and Baer (32), Pursely et al. (33), and Hill et al.
(16), on a limited number of field tests using petroleum
sulfonate for surfactant flooding. Tabulated results are
presented in Table III for six tests. The two Marathon tests
have been included from Table II for comparison. A strict
comparison may not be entirely valid since these two tests
represent the extreme in pattern size of projects using the
Maraflood process in Illinois; but the results do seem to
indicate that percentage of recovery decreases as slug and
mobility buffers decrease and as pattern type and size

TABLE III

SUMMARY OF SURFACTANT FIELD TESTS

Company	Field/Project	Reference	Type of Process	Pattern Size (acres)	Slug Size (percent PV)	Mobility Buffer Size (percent PV)	Oil Saturation Before Flood (bbl/acre-ft)	Recovery (bbl/acre-ft)
Marathon	119-R Project	Earlougher et al (30)	High-concentration surfactant —Maraflood oil recovery process	40.0	7.0	100.0	599	282[1]
Union	Higgs Unit	Knight and Baer (32)	High-concentration surfactant Uniflood process	8.23	4.0[2]	67.8[2]	409	257
Marathon	Henry-W	Gogarty and Surkalo (34)	High-concentration surfactant —Maraflood oil recovery process	0.75	9.0	200.0	621	391
Exxon	Loudon field	Pursley et al (33)	Low-concentration surfactant	0.625	40.0	30.0	488	75
Shell	Benton field	Hill et al (16)	Low-concentration surfactant	1.0	111.1[3]	330.0[3]	461[4]	150[1]
Mobil	Loma Novia field	Office of Oil and Gas Energy Resource Div., FEA (35)	Low-concentration surfactant	5.0	12.0	—	—	[5]

[1] Estimated ultimate recovery based on confined pattern area. Estimated ultimate recovery for total pattern — 230 bbl acre-ft

[2] Based on total volume of fluid injected. 50 to 65 percent of injected fluids effective in pilot area

[3] Assumes all fluid injected went into Zone B of 1-acre pilot area

[4] Assumes oil produced from Zone B only

[5] No oil was produced but oil-cut increases were observed at observation wells. Post analysis indicated a need for mobility control

increases (repeated 2.5-acre, direct-line drive vs isolated, 0.75-acre, inverted five-spot). Increased amounts of confined area in commercial operations should improve percentage of recovery by reducing the relative amount of off-pattern losses from the unconfined areas. Additional field tests will be needed to determine which of these three factors affect recovery most. Results with the Uniflood process in the Higgs Unit represent the only other published results of a high-concentration surfactant process.

Of the three low-concentration surfactant tests, only two recovered oil. Mobil reported that oil cuts increased at observation wells in their low-tension process, and they attributed the lack of production to inadequate mobility control. There also may be some question as to whether sufficient volume of the surfactant slug was used in Mobil's test.

A comparison of the recovery results in Table II indicates that higher values are obtained with higher-concentration-surfactant, low-pore-volume systems than with the low-concentration, high-pore-volume systems. In terms of efficient surfactant use, the amount of oil recovered needs to be related to the amount of surfactant injected within the pattern from where the oil is displaced. In isolated, nonrepeatable pattern tests, this is a difficult task. The results in Table III need to be considered from this standpoint. Only in the case of the 119-R Project were repeated confined patterns used. Because of this, the recovery may be representative of full field development in reservoirs of equal quality, equal spacing, and with the same types of patterns. Undoubtedly, many large-scale field tests are needed for both the low- and high-concentration surfactant processes before recovery factors can be compared with any degree of certainty.

VII. COMMERCIAL APPLICATION

Commercial application of any surfactant flooding process depends on economic projections indicating an adequate return on investment. Factors such as oil recovery, well and equipment costs, and the cost of chemicals are important in making these projections. Projected chemical costs are a function of the quantity used. Equipment and drilling costs depend on the project size, formation depth, and spacing. Field operating problems and the corresponding expense increase as the developed acreage per project increases. Economics are presented here to illustrate the development costs on potential returns for the Maraflood process in southern Illinois.

Chemical supply and cost must be considered as a part of economic projections. Large-scale applications of micellar-solution flooding or any like process will require large volumes of surfactant and polymer and smaller volumes of co-surfactants. Table IV shows chemical requirements for

TABLE IV

CHEMICAL REQUIREMENTS SURFACTANT FLOODING

Basis: 100-million bbl/year oil production

	Marathon Oil Co. (million lb/year)	Exxon Co. U.S.A. (million lb year)
Surfactant	1,700	680 to 1,360
Cosurfactant	123	550 to 1,100
Total	1,823	1,230 to 2,460
Polymer	102	98

surfactant flooding as reported at the Federal Energy Administration's Symposium on Enhanced Oil and Gas Recovery by the
Office of Oil and Gas Energy Resource Division (35). The
total petroleum sulfonate manufactured in the U.S. during
1972 was about 388 million lb, including barium, calcium, and
sodium salts. This figure, when compared with the values in
Table IV, shows that additional sulfonate manufacturing capacity will have to be developed if significant quantities of
tertiary oil are to be obtained from surfactant methods.
Marathon has taken steps in this direction by constructing a
5,000-B/D crude-oil, sulfonate-slug manufacturing facility at
the Company's Robinson, Ill., refinery as reported by
Earlougher et al. (30). This facility will be completed and
placed on stream during 1975. The plan is to use production
from this facility to initiate injection into the first project during 1975 and then develop Marathon's properties thereafter to use the slug facility output.

     Similar chemical supply problems occur in the availability
of polymers for mobility control. The total polymer produced
for water treatment and for the paper industry now stands at
about 50 million lb/year. This is about one-half the value
given in Table IV. Until chemicals are available abundantly,
costs will probably remain high and long-term chemical cost
projections will be difficult to make.

     Economic projections for emerging tertiary oil recovery
processes of this type are most difficult to make since the
industry has no experience with large-scale field projects.
Critical factors in an economic analysis include the amount
or fraction of *oil recovered* and the *time* required to recover
that oil. Estimates of these parameters now must be made by
extrapolating information from laboratory models and limited
small-scale field projects. Fundamental scaling laws have
not been developed. Not only is scaling to large projects a
problem, but field testing has been confined to a very few
reservoirs, and extrapolation to other reservoirs is even
more speculative. *Oil price* and *tax* load are as critical as
recovery in making economic predictions about the viability

of any tertiary recovery process. The uncertainties introduced by long-term projections of oil price and tax load
significantly cloud any economic projection. Inflationary
pressures on the price of goods and services also are a key
factor in the ultimate economics of a tertiary recovery
project.

Because of the political climate, price seems to be the
most important key to enhanced oil recovery. From a report
in the Oil and Gas Journal (36), industry and government
experts seem to agree that enhanced recovery has a potential
of producing from 25 to 60 billion bbl of oil. Industry
spokesmen believe that only minor volumes will be produced
at a price of $7.50/bbl now considered in the U.S. Senate.
L. E. Elkins of Amoco Production Co says that a price of
$11/bbl would produce more oil. Calculations given in this
section of the paper indicate that the price must be higher
than $11/bbl if large quantities of tertiary oil are to be
recovered by surfactant flooding methods. Higher recoveries
offset, to some degree, the need for higher crude oil prices.
Recovery results from projects presently under way will help
to answer the recovery question.

To obtain an idea of the total cost of tertiary crude
and the investment requirements for a large-scale development, a 6,000-acre, 10-year development of the Robinson
sandstone has been postulated. Starting in 1975, a 600-acre
block would be drilled and developed each year for 10 consec-
utive years. Slug injection would take place in the year
after development. Polymer solution and drive water would
be injected continuously until completion in each 600-acre
block. Reservoir depth is about 1,000 ft. Drilling and
equipment costs are based on actual field operations. Calculations were made for both 2.5- and 5-acre spacing with a
range of recoveries. Injection rates for polymer solution
and drive water were taken as 5.1 and 4.8 B/D per foot of
sand for 2.5- and 5-acre cases, respectively. These rates
are based on waterflood experience in the area. Lower and
higher injectivity rates change the project life and increase or decrease correspondingly economic parameters tied
to the time value of money. The shapes of the assumed injection and oil production schedules for a 600-acre block
developed on 2.5-acre spacing and having a recovery of 242
bbl/acre-ft are shown in Figure 9. Schedules for the other
cases were similar. Basic economic and reservoir input parameters are given in Table V for the cases using 2.5-acre
spacing. Parameters such as the number of wells, project
life, field costs, and well operating expenses were scaled
for the 5-acre-spacing cases. In all calculations, an average thickness of 21 ft was used for all 6,000 acres. As with
any recovery process, thicker sections improve the economics.

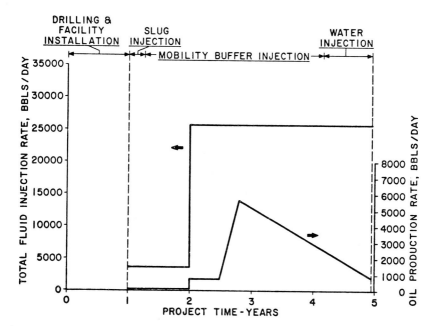

Fig. 9. Projected injection and production rates, 600-acre block, for Maraflood process in Illinois

The economic parameters in this study are based on 1975 costs, and no provision is made in the analysis for escalation of either oil price or investment and operating costs with time caused by inflation. Because of continuing cost increases, most of the costs listed in Table V are lower than actually would be experienced when the project gets under way. If the real price of crude escalates faster than goods and services, then the actual economics would be better than presented here. Should crude oil escalate at a slower rate than goods and services, the opposite will be true.

Figure 10 summarizes results of calculations for three different crude prices and the two spacings with the dis- counted cash flow rate of return values being presented as a function of recovery of tertiary oil. All calculations are on an after tax basis. No application of the proposed wind- fall profit tax is included in this analysis. The imposition of such a tax without other compensating relief would further burden the economics of the project.

Figure 10 shows that the economics of micellar solution flooding strongly depend on oil recovery as determined by the amount of oil in place before initiation of a tertiary recov- ery operation and the recovery efficiency. No large-scale projects have been run on which to base good estimates of

TABLE V

ILLINOIS 6,000-ACRE MARAFLOOD PROCESS DEVELOPMENT,
BASIC ECONOMIC INPUT PARAMETERS FOR EACH 600-ACRE BLOCK

Conditions

1975 costs and prices
No cost or price escalation with time

Field Data

Area = 600 acres, developed on 3-acre spacing
Thickness = 21.0 ft
Porosity = 19.5 percent
Oil saturation = 40.0 percent (at the start of tertiary recovery operations)
Oil cut = 3 percent (at the start of tertiary recovery operations)
Injection wells = 240
Production wells = 272

Fluid Injection Data

Micellar solution slug = 7 percent PV (slug contains 24-percent hydrocarbon-like components)
Mobility buffer = 105 percent PV
Average polymer concentration in mobility buffer = 594 ppm
Drive water injection = 38 percent PV
Project life = 4.96 years (including 1 year drilling and site preparation)

Field Costs

Drilling and completion costs = $8,966,100
Fluid handling and injection equipment = $2,940,000
Salvage credit = $1,816,300
Lease cleanup = $1,407,400

Unit Costs

Crude oil price = $11.90/bbl
Micellar solution slug = $7.69/bbl
Slug injection expense = 18.4¢/bbl
Polymer = $1.02/lb
Polymer solution injection expense = 4.1¢/bbl
Water injection expense = 2.6¢/bbl
Water disposal expense = 2.6¢/bbl
Well operating expense = $251/well/month

Royalty and Tax Data

Royalty = 12.5 percent
Income tax rate = 50 percent
Investment tax credit = 7.0 percent (one-half of equipment installed will be used, thus the effective rate is 3.5 percent)

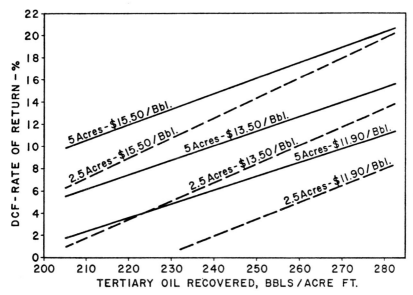

*Fig. 10. Projected economics, 6,000-acre Maraflood process
development in Illinois; after-tax projections
showing effect of recovery efficiency, spacing,
and crude price on rate of return*

recovery efficiency. Analysis of the 119-R Project (Table
III) indicates a recovery factor from 230 bbl/acre-ft for
the total project to 282 bbl/acre-ft for the confined elements of the pattern.

The actual recoveries for Robinson sandstone reservoirs
will be determined from commercial-scale projects. The range
of recoveries can be used with Figure 10 to obtain an indication of process economics. The price of stripper oil in
Robinson reservoir for the first half of 1975 was $11.90/bbl.
As of July 1, 1975, the price increased to $12.33/bbl.
Depending on recovery, additional price increases will be
necessary before this type of development would be more than
marginally economic at 2.5-acre spacing.

Results in Figure 10 might indicate that 5-acre spacing
rather than 2.5-acre spacing would be a more economic way to
develop the 6,000 acres. Actually, this will be true only if
equal recoveries are obtained using both spacings. Frontal
velocities will be lower with 5-acre spacing; and since
recoveries decrease at the lower flooding rates, the 2.5-acre
spacing still may be better. Also, slug and fluid system
integrity are favored by the shorter travel distance of 2.5-
acre spacing. The recovery trade-off is illustrated in Figure
10 by using the 2.5- and 5-acre curves for the $15.50/bbl

crude price at a 14-percent rate of return. To obtain this
same rate of return, the 2.5- and 5-acre development must
recover 248 and 234 bbl/acre-ft, respectively.

Figure 11 indicates the influence that statutory depletion would have on the economics of enhanced oil recovery for
the 2.5-acre case. The curves on this figure show the rate
of return for the proposed development project as a function
of oil recovery. The solid lines are for the existing zero
statutory depletion case and the dashed lines assume a value
of 22-percent statutory depletion. Note that to obtain a
12-percent rate of return with a crude price of $13.50/bbl, a
recovery of 272 bbl/acre-ft is required without depletion as
compared with a recovery of 237 bbl/acre-ft with depletion.
These values show that higher recoveries must be obtained
without depletion or other tax incentives if large quantities
of tertiary oil are to be obtained by surfactant flooding.
Alternatively, Figure 11 shows that if recovery is constant,
oil price must be increased without depletion for the same
return on investment to be obtained.

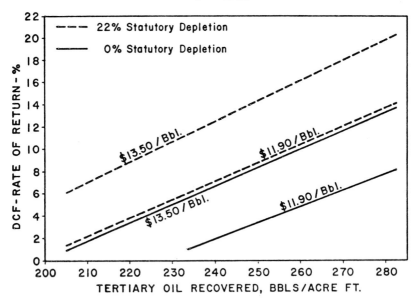

Fig. 11. Projected economics, 6,000-acre Maraflood process
          development in Illinois; after-tax projections
          showing effect of recovery efficiency, crude price,
          and statutory depletion on rate of return

Results presented in Table VI show the costs associated
with the projected 6,000-acre development. The values depend
on oil recovery; the 242 bbl/acre value may or may not be
typical of field recovery but it was used to give an

TABLE VI

ILLINOIS 6,000-ACRE MARAFLOOD PROCESS
DEVELOPMENT ECONOMIC SUMMARY

Crude oil price — $11.90/bbl

	2.5-Acre Spacing	5-Acre Spacing
Oil recovery, bbl/acre-ft	242	242
Gross production, million bbl	30.50	30.50
Net production, million bbl	27.09	27.09
Revenue, million dollars	322.34	322.34
Investment, million dollars	260.21	203.26
Expense	49.73	58.48
Total cost per net barrel (before federal income tax)	11.44	9.66
Project life	13.96	17.27

indication of expected values. With this recovery, the maximum daily production for both spacings is about 8,360 B/D. As shown, an investment between $203 and $260 million is required for the 6,000-acre development. Investment includes the drilling and equipment costs and the cost of injected chemicals. For the assumed recovery, the investment develops about 27 million net barrels of reserve. The total cost per net barrel, as indicated in Table VI, will decrease with increased recovery, but the values given are an indication of the high cost of tertiary oil.

The calculations presented are for projects in the Robinson sandstone reservoir where there is considerable testing experience. Extrapolation or generalization to other reservoirs or conditions must be done with care and with an understanding of the uncertainties involved. Drilling and completion costs for wells directly depend on reservoir depth and drilling difficulties encountered in a given area. In most areas, some existing wells would be used in the tertiary operations. The other wells probably would be replaced, and those used would require remedial work. New wells may be desirable to change the flooding pattern or pattern spacing. The cost of surface facilities to mix and handle injected fluids is much less sensitive to specific reservoir selection but is influenced by well spacing or pattern density. Injected fluid costs, particularly costs of the polymer solution used as a mobility buffer, will vary depending on reservoir

temperature and reservoir rock and fluid characteristics.
Maraflood process fluids are designed to provide mobility
control within a reservoir to prevent unstable displacement
or fingering. The amount of mobility control necessary is
determined by the characteristics of each reservoir. If
more mobility control is required, more polymer must be used.
These polymers are also less effective at elevated temperatures, thus requiring more polymer to yield a given level of
control at higher temperatures.

Micellar solution flooding requires a high initial investment, and rate-of-return calculations are highly dependent on project life. Project life is determined by well
spacing and by the rate that fluids can be injected. Well
spacing is a function of reservoir thickness and depth.
Spacing on 2.5 or 5 acres can be considered in shallow
reservoirs of a given thickness, but such spacings will become uneconomic at greater depths. Injection rate depends
on reservoir characteristics and depth. The nature of the
reservoir rock, crude oil properties, and the character of
the formation water all strongly influence the recovery
efficiency of a micellar solution; and these factors must
be considered in extrapolating performance to other reservoirs.

VIII.   LITERATURE CITED
1.   Geffen, T. M., "Here's What's Needed to Get Tertiary
         Recovery Going," World Oil 53-57 (March 1975).
2.   De Groot, M., "Flooding Process for Recovering Oil from
         Subterranean Oil-Bearing Strata," U.S. Patent No.
         1,823,439 (1929).
3.   De Groot, M., "Flooding Process for Recovering Fixed Oil
         from Subterranean Oil-Bearing Strata," U.S. Patent No.
         1,823,440 (1930).
4.   Holbrook, O. C., "Surfactant-Water Secondary Recovery
         Process," U.S. Patent No. 3,006,411 (1958).
5.   Holm, L. W. and Bernard, G. G., "Secondary Recovery
         Waterflood Process," U.S. Patent No. 3,082,822 (1959).
6.   Csaszar, A. K., "Solvent-Waterflood Oil Recovery Process,"
         U.S. Patent No. 3,163,214 (1961).
7.   Blair, C. M., Jr. and Lehmann, S., Jr., "Process for
         Increasing Productivity of Subterranean Oil-Bearing
         Strata," U.S. Patent No. 2,356,205 (1942).
8.   Gogarty, W. B. and Olson, R. W., "Use of Microemulsions
         in Miscible-Type Oil Recovery Procedure," U.S. Patent
         No. 3,254,714 (1962).
9.   Gogarty, W. B. and Tosch, W. C., "Miscible-Type Water-
         flooding: Oil Recovery with Micellar Solutions,"
         J. Pet. Tech. 1407-1414 (Dec. 1968); Trans., AIME, 243.

10. Jones, S. C., "High Water Content Oil-External Micellar Dispersions," U.S. Patent No. 3,497,006 (1967).

11. Jones, S. C., "Use of Water-External Micellar Dispersions in Oil Recovery," U.S. Patent No. 3,506,070 (1967).

12. Reisberg, J., "Secondary Recovery Method," U.S. Patent No. 3,330,344 (1964).

13. Reisberg, J., "Surfactants for Oil Recovery by Waterfloods," U.S. Patent No. 3,348,611 (1965).

14. Cooke, C. E., Jr., "Microemulsion Oil Recovery Process," U.S. Patent No. 3,373,809 (1965).

15. Gale, W. W. and Sandvik, E. I., "Tertiary Surfactant Flooding: Petroleum Sulfonate Composition-Efficacy Studies," Soc. Pet. Eng. J. 191-199 (Aug. 1973).

16. Hill, H. J., Reisberg, J. and Stegemeier, G. L., "Aqueous Surfactant Systems for Oil Recovery," J. Pet. Tech. 186-194 (Feb. 1973); Trans., AIME, 255.

17. Davis, J. A., Jr. and Jones, S. C., "Displacement Mechanisms of Micellar Solutions," J. Pet. Tech. 1415-1428 (Dec. 1968); Trans., AIME, 243.

18. Trushenski, S. P., Dauben, D. L. and Parrish, D. R., "Micellar Flooding--Fluid Propagation, Interaction, and Mobility," Soc. Pet. Eng. J. 633-642 (Dec. 1974); Trans., AIME, 257.

19. Gogarty, W. B. and Davis, J. A., Jr., "Field Experience with the Maraflood Process," SPE Paper No. 3806 presented at the SPE-AIME Improved Oil Recovery Symposium, Tulsa, Oklahoma, April 16-19, 1972.

20. Gogarty, W. B., Meabon, H. P. and Milton, H. W., Jr., "Mobility Control Design for Miscible-Type Waterfloods Using Micellar Solutions," J. Pet. Tech. 141-147 (Feb. 1970).

21. Parsons, R. W., "Velocities in Developed Five-Spot Patterns," J. Pet. Tech. 550 (May 1974).

22. Healy, R. N., Reed, R. L. and Carpenter, C. W., Jr., "A Laboratory Study of Microemulsion Flooding," Soc. Pet. Eng. J. 87-100 (Feb. 1975); Trans., AIME, 259.

23. Taber, J. J., "Dynamic and Static Forces Required to Remove a Discontinuous Oil Phase from Porous Media Containing Both Oil and Water," Soc. Pet. Eng. J. 3-12 (March 1969).

24. Taber, J. J., Kirby, J. C. and Schroeder, F. U., "Studies on the Displacement of Residual Oil: Viscosity and Permeability Effects," paper 47b presented at the AIChE Symposium on Transport Phenomena in Porous Media, 71st National Meeting, Dallas, Feb. 20-23, 1972.

54   W. B. GOGARTY

25. Foster, W. R., "A Low-Tension Waterflooding Process," J. Pet. Tech. 205-210 (Feb. 1973); Trans., AIME, 255.

26. Jones, S. C. and McAtee, R. W., "A Novel Single-Well Field Test of a Micellar Solution Slug," J. Pet. Tech. 1371-1376 (Nov. 1972).

27. Bleakley, W. B., "Journal Survey Shows Recovery Projects Up," Oil and Gas J. 69 (March 25, 1974).

28. Sharp, J. M., "GURC Report No. 140-S, Final Report on a Survey of Field Tests of Enhanced Recovery Methods for Crude Oil," presented to the National Science Foundation and the Federal Energy Admin., Dec. 27, 1974.

29. Oil and Gas J., "ERDA, Phillips Plan Oklahoma Tertiary Pilot," 43 (May 26, 1975).

30. Earlougher, R. C., Jr., O'Neal, J. E. and Surkalo, H., "Micellar Solution Flooding: Field Test Results and Process Improvements," SPE Paper No. 5337 presented at the SPE-AIME Rocky Mountain Regional Meeting, Denver, Colo., April 7-9, 1975.

31. Danielson, H. H., Paynter, W. T. and Milton, H. W., Jr., "Tertiary Recovery by the Maraflood Process in the Bradford Field," SPE Paper No. 4753 presented at the SPE-AIME Improved Oil Recovery Symposium, Tulsa, Okla., April 22-24, 1974.

32. Knight, R. K. and Baer, P. J., "A Field Test of Soluble-Oil Flooding at Higgs Unit," J. Pet. Tech. 9-15 (Jan. 1973).

33. Pursley, S. A., Healy, R. N. and Sandvik, E. I., "A Field Test of Surfactant Flooding, Loudon, Illinois," J. Pet. Tech. 793-802 (July 1973).

34. Gogarty, W. B. and Surkalo, H., "A Field Test of Micellar Solution Flooding," J. Pet. Tech. 1161-1169 (Sept. 1972).

35. Office of Oil and Gas Energy Resource Div., Federal Energy Admin., Enhanced Oil and Gas Recovery Symposium, Washington, Dec. 4, 1974.

36. Oil and Gas J. "Price Seen Key to Enhanced Recovery," 160-161 (May 5, 1975).

MECHANISMS OF ENTRAPMENT AND MOBILIZATION
OF OIL IN POROUS MEDIA

G. L. Stegemeier
Shell Development Company

I. ABSTRACT

Trapping and release of oil in natural petroleum reservoirs occur under a wide variety of interrelated initial and
applied conditions. Factors which determine microscopic displacement mechanisms are: 1) geometry of the pore network;
2) fluid-fluid properties, such as interfacial tension,
density difference, bulk viscosity ratio, and phase behavior;
3) fluid-rock properties, including wettability, ion exchange,
and adsorption; and 4) applied pressure gradient and gravity.

Several mechanisms of oil displacement can be significant
in dynamic, multicomponent frontal processes such as surfactant and polymer flooding. Opposing viscous and capillary
forces appear to be the controlling mechanism for simple two-
phase systems. Other mechanisms include interphase mass
transfer, interface aging effects, wettability changes, and
emulsification.

II. PREFACE

This review was prepared by the author at the request of
the American Institute of Chemical Engineers. As such, it
does not necessarily reflect results of current research
programs at Shell Development; rather, it is a compendium of
published information, and of the author's own ideas on the
subject of residual oil.

III. SCOPE

Trapping and release of fluids from porous media have
been the subject of extensive study by the petroleum production industry in the past and are currently of even
greater interest as a result of the critical need to improve
recovery efficiency from petroleum reservoirs. Literature
related to the subject is large, and yet incomplete, since
the mechanisms are complicated by interrelated properties of
complex rock pore structure, fluid properties and applied
conditions. Furthermore, the variability of oil reservoir
rocks and fluids is so great that most generalized conclusions have limited applicability. For this reason the

present review will necessarily simplify and limit conditions in an attempt to demonstrate the broad principles which control these processes. For a specific system, trapping behavior is controlled by 1) the pore geometry of the rock matrix, 2) fluid-rock properties, in particular, wettability, and 3) fluid-fluid interactions including viscosity ratio, density difference, interfacial tension, and partition coefficients.

Because most theoretical studies of trapping address single aspects of the phenomena, there appears to be little quantitative theory to explain the overall effect, that is, the recovery efficiency of a trapped fluid by a displacing fluid. The readily usable quantitative methods which exist rely on equilibrium conditions since effects of dynamic phemomena (1) arising from surface viscosity, surface tension, diffusion, and emulsification have not yet been satisfactorily incorporated into predictive techniques.

Experimental studies have detected dynamic effects when a fluid is trapped from a continuous interconnected phase. For release of isolated residual phases, however, most experimental observations can be accounted for by equilibrium considerations. For this reason and to reduce the scope, this paper with only a few exceptions, is limited to trapping and release as determined by equilibrium conditions.

For further simplification, initial considerations are that the pore structure is an aggregation of variable pore body and pore entry sizes such as found in typical reservoir sands or sandstones. A nonwetting oil phase and a wetting water phase will be considered the normal configuration of a two-liquid phase system. Some variations on these basic assumptions, which are known to exist in natural reservoirs, will be discussed in relation to possible extensions of the basic theory.

IV.   CONCLUSIONS AND SIGNIFICANCE

Mechanisms which control the entrapment of oil, water and gas in natural oil reservoirs are defined by the interrelation of rock and fluid properties with applied forces.

A.   The character of the rock structure itself, including pore shape and size, largely determines the amount and distribution of trapping. Some properties of these natural pore systems are:

1)   Trapping of fluids occurs in unique and reproducible patterns which are controlled by capillary forces.

2)   Nearly complete networks of interconnected, equal sized pores exist throughout the pore size distribution.

    3) Individual pores have good accessibility with adjacent pores, thereby allowing alternate paths of flow around isolated immobile phases.

    4) Fluids can be trapped at pore constrictions for all degrees of wetting, including neutral wettability.

    5) Nonwetting phases are trapped in discontinuous masses whose lengths are largely determined by interfacial tension and potential gradient.

    6) Wetting phase trapping is continuous and is determined by the capillary pressure-saturation relationship. Even small contact angles isolate a preferential wetting phase in pendular rings.

  B. The magnitude of trapped nonwetting phase saturation is not well defined theoretically. Undoubtedly, it is closely related to the ratio of pore body to pore neck radii. Direct measurement by injection and withdrawal porosimetry adequately quantifies nonwetting trapping.

  C. Trapped phases can be released either by increasing the ratio of viscous to capillary forces or by changing relative phase volumes. The latter may affect recovery either by swelling and mobilization of an immobile phase or by solution into the mobile phase. Factors which define release of trapped phases by viscous/capillary mechanisms are: rock matrix properties, including porosity and Leverett Number, fluid-rock properties, including relative permeability, pore shape factor, and wettability, and a filament geometric constant independent of physical properties of either fluid or rock.

  D. A review of experimental studies of removal of trapped phases reveals:

    1) Isolated nonwetting phase recovery from natural sandstones corresponds reasonably well with viscous/capillary theory, indicating desaturation at $N_{VC} \cong 10^{-4}$ for most rocks.

    2) At low viscous/capillary ratios, continuous phases under dynamic conditions trap at lower saturations than the equilibrium saturations of isolated phases. At high viscous/capillary ratios desaturation of continuous and isolated phases are comparable.

    3) Wetting phases require significantly greater forces for removal of the final saturations than nonwetting phases.

  E. In oil recovery processes more than one mechanism is often operative. Mechanisms which predominate in commonly used processes are:

1) In surfactant-polymer flooding, reduction in interfacial tensions and increased water viscosity are important factors. At the extremely low interfacial tensions needed, however, large phase volume changes, which often simultaneously occur, and can be significant.

2) In solvent flooding, as in the previous case, a swelling mechanism is preferable to solution mechanism since it generates a clean oil bank and is more efficient.

3) In emulsification processes, water-in-oil emulsions act as an oil swelling mechanism and oil-in-water emulsions behave as a solution type.

4) In gas-oil-water systems in which oil spreading occurs, the oil swelling mechanism is operative.

V.   NATURE OF TRAPPING

Two types of hysteresis in fluid saturations occur in porous media. The first arises from hysteresis in contact angle of advancing and receding interfaces and can be observed even in straight capillary tubes. The second type results from variations in shape of the containing walls of the pore system.

This latter effect of the rock framework geometry has been the subject of a number of studies (2-12). Because a knowledge of the properties of the rock pore structure is basic to an understanding of trapping it will be reviewed in some detail.

A.   Rock Matrix Pore Structure

1.   *Properties of Single Pores*

Single pore models can be useful in describing behavior in natural systems which are too complex for direct study. Conclusions drawn from the single pore models, however, can be misleading since the simplest models often behave quite differently than complete pore networks. For example, two-phase flow in a straight or variable diameter capillaries exhibit additive resistance from multiple interfaces, commonly known as the "Jamin effect." This effect is much larger in a "series" pores system than in a pore network of packed particles in which many alternate "parallel" paths exist. The extent of these degrees of freedom is of great importance to trapping and will be discussed more fully in descriptions of pore networks.

The straight capillary tube model, in spite of its obvious shortcomings in describing two-phase flow, is a useful concept for describing single fluid movement within a given

pore. The validity of this model for single-phase flow can
be demonstrated experimentally by comparing functional relationship of flow rate to average pore radius for bead packs
with that for straight capillary tubes. For uniform size bead
packs ranging in diameter from at least 0.003 cm to 0.3 cm,
permeability is proportional to the square of bead diameter
(and consequently average pore radius).

$$k = C \, \phi \, d_e^2 \tag{1}$$

$$C = \text{constant} \cong 0.002$$

This dependence of flow velocity on the square of the average
pore radius in both Darcy's and Poiseuille's equations will
be used in subsequent trapping theory calculations.

The capillary doublet (13-16) is a somewhat more complex
pore which allows one alternate flow path. Although the pore
doublet is still too simple to explain all the phenomena in
irregular porous media, it does demonstrate the unequal competition of viscous and capillary forces. In this model,
capillaries of two sizes are connected at the pore inlet and
pore outlet (Figure 1). An analytical expression for the
force balance between capillary and viscous forces is obtained by combining the Poiseuille and Darcy equations together with the Laplace equation of capillary pressure (17).
Expressions for flow velocities and pressures in various
flow regimes are summarized in Table I for the unit viscosity ratio case. The first column in Table I is a dimensionless factor, $C_p(B)$, which, together with the capillary pressure in the large pore, P_{c_2}, defines the net pressure drop,
ΔP, across the entire doublet. Notice that this pressure is
negative for most conditions. The second column is a dimensionless factor, $C_u(B)$, which is a measure of the net flow
into and out of the doublet. Its value is positive when
wetting phase is displacing nonwetting phase and negative for
the inverse displacement. At typical oil reservoir flow
velocities, $C_u(B)$ is very close to zero relative to the
other values in column 2. At these flow rates the pore
doublet is "starved" for fluid since the wetting phase is
imbibed at extremely high rates compared to average fluid
front velocity. Therefore, positive flow will occur only
in the narrow capillary; that is, water will imbibe into
the narrow pore and oil will be trapped in the larger one.
Although net flow is positive, the usual pressure gradient
across a doublet containing an oil/water interface is
strongly negative as a result of the enormously greater

TABLE I

CONDITIONS FOR DIRECTION OF FLOW IN DOUBLET (17)--(FIGURE 1)

Net Pressure Difference Factor $C_p(B)$	Net Flow Velocity Factor $C_u(B)$	Doublet Flow Direction	Wetting Displacing Nonwetting	Nonwetting Displacing Wetting
$0 \quad \genfrac{}{}{0pt}{}{+}{-}$	$\dfrac{(B^3+1)}{(B^2+1)} \quad +$	Both Pores $+$	Wide Pore(r_2) Breakthrough First	
$-\dfrac{1}{(1+B)}$	$\dfrac{B}{(B+1)}$		Narrow Pore(r_1) Breakthrough First	
-1	$\dfrac{B^3(1-B)}{(B^2+1)}$	Narrow Pore Positive	Narrow Pore Breakthrough Only	
$-\dfrac{(B^3+1)}{(B^4+1)}$	$0 \quad \genfrac{}{}{0pt}{}{+}{-}$	Wide Pore Negative		Wide Pore Breakthrough Only (In Negative Direction)
$-\dfrac{1}{B}$	$-\dfrac{(1-B)}{(B^2+1)}$	Both Pores Negative		Wide Pore Breakthrough First (In Negative Direction)

$\Delta p = P_0 - P_3$

$\Delta p = P_{c_2} \cdot C_p(B)$

$P_{c_2} = P_{o_2} - P_{w_2}$

$P_{c_2} = \dfrac{2\sigma}{r_2}$

$u = q_o/A$

$u = \dfrac{\phi\sigma}{4\mu} \dfrac{r_2}{\ell} \cdot C_u(B)$

$B = r_1/r_2$

60

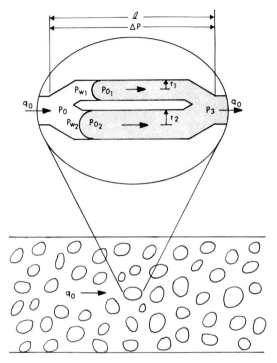

FIG. 1 DOUBLET PORE MODEL

76/069/1

capillary than viscous forces at this flood front. Inspection of the equations in Table I shows that this generalization is valid over practically all radii ratios of the capillary pairs and all pore radius/length ratios that could reasonably be expected in beds of packed particles. During imbibition into the small capillary, the equations predict countercurrent flow in the large pore. With no space to receive this counterflow except back into the small pore, the oil in the large capillary remains essentially motionless until wetting phase reaches the pore outlet. Then, depending upon pore outlet geometry, the oil may be trapped. Such behavior would explain the relative insensitivity of trapping to viscosity ratio of the fluid pair. The quantity of non-wetting phase trapped in a pore doublet, being the ratio of the large capillary volume to the total volume, is always greater than 50 percent and typically is considerably larger. For example, if the large pore radius were twice that of the small one, 80 percent of the nonwetting phase would be trapped.

A simple geometric shape such as a torus, is a generally applicable model of pore outlet geometry. A nonwetting phase drop emerging from such a constriction becomes unstable and separates when the capillary pressure at the neck, $P_{c_n} = \sigma(1/r_n - 1/r_t)$, exceeds the capillary pressure of the interface front, $P_{c_f} = 2\sigma/r_f$ (63) (see Figure 2). Note that in this model the curvatures at the front of the drop are both positive, whereas those at the pore neck assume the positive value of the neck radius and the negative value of the torus cross-section. A pore composed of stacked tori will not trap nonwetting phase unless the radius of the torus cross-section (r_t) is many times the pore opening radius (r_n). This relative incapacity for a single pore to trap exists because an entering nonwetting phase drop will be constrained by the pore wall of the downstream torus before the curvature can expand enough to cause snap off at the pore neck.

For a pore model composed of cubic packed spheres, the cross-sectional geometry is similar to the previous one; however, the curvature at the front of the drop may assume both positive and negative values. Capillary pressure as a function of nonwetting phase penetration into the pore is shown in Figure 3. After the nonwetting phase exceeds the pore

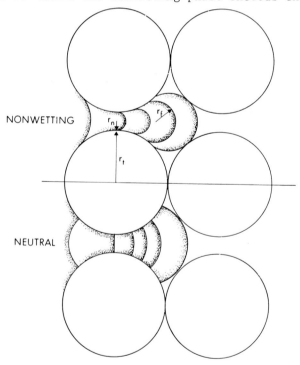

FIG. 2 TOROIDAL PORE MODEL

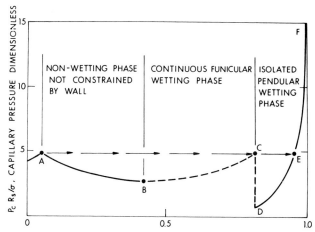

FIG. 3 FRACTIONAL VOLUME OF NON WETTING PHASE
IN PORE BETWEEN CUBIC PACKED SPHERES, R_s =1

76/069/3

entry pressure (A) and enters the pore, it expands freely at
diminishing capillary pressure until it is constrained by the
pore wall (B). As the nonwetting phase further occupies the
pore, the capillary pressure increases as the interface is
forced into the narrower parts of the pore. Both curvatures
remain positive until the critical pore saturation is reached
at which the capillary pressure discontinuously drops (C → D)
as the wetting phase snaps into pendular rings having one
negative interface curvature. Thereafter, the capillary pressure continuously increases (D → F) as the wetting phase re-
treats further and further toward the sphere contact points.
The passage of an interface from pore entry (A) to pendular
ring takes place in one rapid movement (A → E) because all
intermediate saturations are below the pore entry pressure.
These violent movements are known as "Haines jumps" or
"rheons" (6). Nonwetting saturation at equilibrium after
entry into a pore of cubic packed spheres is about 95 percent.
In the reverse direction, expansion of the pendular rings by
imbibition of the wetting phase, can result in capillary pressures less than the pore neck pressure at greater than 80 percent nonwetting phase. However, this situation is operative
over a very narrow saturation regime and for regularly packed
beds single pores are usually very ineffective traps. Exceptions to this are vuggy or oomoldic type porosity in which
pores are interconnected by extremely narrow necks.
2. *Properties of Pore Networks*
     Trapping is not limited to single pores or doublets. In
fact, single pores often have at least one alternate flow path

with an insufficient constriction for trapping. Consequently, most traps are composed of a number of pores, such as shown in Figure 4.

The pore networks of natural reservoir rocks, such as sandstones or limestones contain a nearly continuous distribution of pore sizes. The nature of the pore network largely controls the amount of trapping and the ease of release. Important parameters are the pore size distribution, pore body/ pore neck ratio, and the degree of interconnection between pores. Measurements of these quantities are usually made with fluid intrusion and extrusion porosimetry (9,10,11) and photomicrography (18,19,20). Measurement of pore dimensions by statistical scans of microphotographs of rock cross-sections is a direct technique for obtaining an average pore body size. The specific pore neck dimensions which control a specific three-dimensional pore body, however, are not obtained. Furthermore, the complex geometry of multipore traps is not readily translated into quantitative trapped volumes. For these reasons, an empirical approach using direct measurement of trapping volumes by fluid intrusion and withdrawal porosimetry is commonly used to quantify trapping (10). Nonwetting phase injection capillary pressure is the basic tool for determining pore entry radii of porous media as a function of fluid saturation (see Figure 5). Procedures such as mercury injection into a vacuum or air injection into a liquid-filled pore space depend upon interconnection of pore networks within narrow intervals of pore entry radii. For example, if large pores at the interior of a rock sample were completely surrounded by small pores, the fluid intrusion at just above the

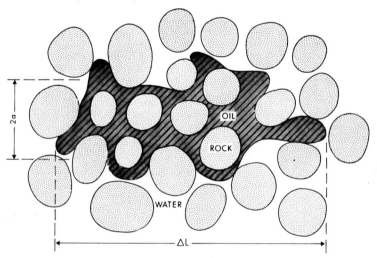

FIG. 4 MULTIPORE RESIDUAL OIL FILAMENT

76/069/14

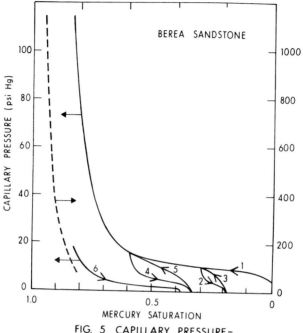

FIG. 5 CAPILLARY PRESSURE-
RESIDUAL INITIAL SEQUENCE

76/069/4

large pore entry pressure would occupy only large pores near
surface of the rock sample. Actually it appears that pore
networks in natural sandstones are largely interconnected
since 1) capillary pressure exhibits little dependence of
sample area/volume (provided the sample is large compared to
multipore trap length) and 2) continuous flow paths for a nonwetting phase exist at low saturations. This latter property
is demonstrated by finite mercury permeability across a rock
after only small amounts of injection (3). Thus, these data
suggest a pore structure composed of a number of parallel pore
networks of varying entry pressures with good accessibility
within each network. Such a pore network is illustrated
schematically in Figures 6a, 6c, and 6e.

If all rocks were geometrically similar the capillary
pressure-saturation relationship could be expressed by a
single dimensionless function known as the Leverett Number (5),

$$N_{Le}(S) = \frac{P_c(S)}{\sigma T(\theta)} \sqrt{\frac{k}{\phi}} \quad . \tag{2}$$

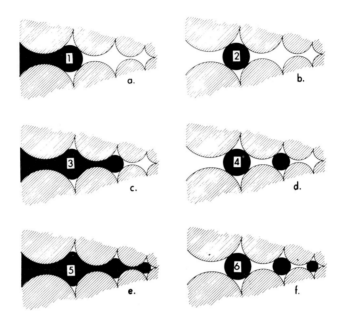

FIG. 6 PORE NETWORK CROSSECTION
NORMAL TO DIRECTION OF FLOW

By defining

$$T(\theta) = \frac{P_c(S)\, r_n}{2\sigma} \, , \tag{3}$$

the Leverett Number, which is a function of saturation, can be
expressed in terms of the <u>pore entry radii</u> and the average
permeability. $T(\theta)$ will be discussed more fully under Fluid-
Rock Interactions. The concept of the Leverett Number being
a function of pore entry radii rather than particle radius is
essential to the subsequent development of theory for release
of trapped phase. The pore entry size implicitly appears on
the RHS of Equation (2) in the capillary pressure, but not in
the permeability since permeability is a measure of some average of the pore dimensions. Because the dependence of pore
neck radius to permeability is not known, Leverett simply used
an average permeability of the whole rock. Thus, N_{Le} is a
dimensionless capillary pressure rather than a constant.
Average values of N_{Le} in natural sandstones are often quite
similar to those obtained by Leverett on unconsolidated sands.
Typically for natural sandstones, N_{Le} equals 0.25 at initial
entry of nonwetting phase, and N_{Le} equals 0.45 at 50 percent
nonwetting phase saturation.

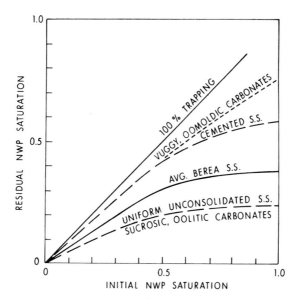

FIG. 7 TYPICAL NON-WETTING PHASE
TRAPPING CHARACTERISTICS OF RESERVOIR ROCKS

and perhaps not even reproducible; however, this is not the
case. Photographic studies (22) of fluid in glass bead pore
networks display a remarkable similarity in pore-to-pore
configuration of trapped fluids during successive floods. If
capillary forces control trapping, each trap will be determined
by the rock structure, the interface at the smallest pore opening breaking first and the largest breaking last. Although
the length of a specific pore group trap depends upon its
pore geometry, it has been proposed (17) that the average
length, ΔL, of the pore group trap which is effective, is
determined by fluid properties and applied gradients. This
relationship can be derived by equating the force resulting
from the pressure drop across a nonwetting phase mass to the
interfacial tension resisting separation.

For a cylindrical oil mass of average radius, a, such as
shown in Figure 4, the average length is:

$$\Delta L = \frac{2\sigma}{a\nabla\Phi} \quad . \tag{4}$$

Although experimental data at low viscous/capillary ratios
comparable to oil reservoir processes are not presently available, Equation (4) is qualitatively supported by photographs
(22) of glass bead packs at relatively low and high rates,
which illustrate the reduced size of isolated trapped nonwetting phase as flow rate is increased.

B. Fluid-Rock Interactions

The complementary configurations of the wetting and the nonwetting phases on a microscopic scale have a profound effect on macroscopic behavior of the two fluids. The well-known asymmetry of relative permeabilities as functions of saturations attests to this (see Figure 8). In many rocks, however, the residual saturations (S_{wr} and S_{or}) at which wetting and nonwetting phases cease to flow are almost equal even though the "trapping" mechanism is entirely different.

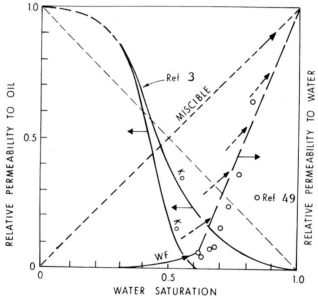

FIG. 8 WATER PERMEABILITY BELOW S_{OR}

76/069/6

1. *Wetting Phase Trapping*

In the case of wetting phase trapping, residual fluid is held in pendular rings, interconnected with only thin water layers such as described by Reisberg and Doscher (23). If there is pressure continuity between rings, the wetting phase is immobile and in a sense is "trapped" for applied potential gradients which result in less than the capillary pressure of the trapped phase. For higher applied potential gradients and capillary pressures, the rings will drain to lower saturations.

In practice, the residual wetting phase (water) after a nonwetting (oil) flood is larger than predicted by isolated ring theory using single pore models such as shown in Figure 3. In linear core floods, nonwetting oil will displace wetting

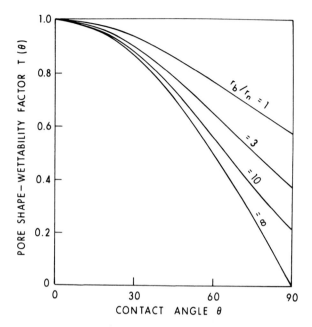

FIG. 9 PORE SHAPE-WETTABILITY FACTOR

76/069/7

$$T(\theta) = \frac{\cos(\theta - \eta)}{1 + (r_t/r_n)(1 - \cos\eta)} \qquad (5)$$

The maximum interface curvature exists at

$$\eta_m = \theta - \arcsin\left|\frac{\sin\theta}{1 + (r_n/r_t)}\right| \qquad (6)$$

$T(\theta)$ is a more general expression for the effect of contact angle than the commonly used $\cos\theta$, for a straight capillary. Since the toroidal pore neck contains all the pore wall angles possible at a constriction, $T(\theta)$ has wider applicability than the simplicity of the model would suggest. For a fixed pore geometry, $T(\theta)$ is independent of the absolute value of pore size, and in the range of shapes common to clean sandstones $(r_t/r_n = 3$ to $4)$, it is only slightly dependent on the ratio of r_t/r_n. This ratio for most rocks appears to fall in the range of 3 to 10. Unlike the cos θ which vanishes at $\theta = 90°$, implying no capillary resistance, $T(\theta)$ remains greater than 0.2 for these common pore shapes. Therefore, the overall effect is a moderation in the influence of contact angle on trapping and release.

For all intermediate wetting, the pendular rings are isolated at their individual trapping pressure. Either fluid then can be trapped in a manner similar to that described for the complete nonwetting phase. Because residual saturations at the extremes of wetting are often similar and because of the compensating effect of $T(\theta)$, the amount of trapping should not be very sensitive to the degree of intermediate wetting.

4. *Wettability of Natural Reservoirs*

A wide variety of wetting conditions from water-wet to preferentially oil-wet have been reported for oil reservoirs (25-36). Complete oil wetting was not observed in any of these studies. Treiber et al. (36) measured contact angles of crude oil from fifty oil reservoirs with synthetic brines on flat quartz and calcite surfaces. These samples mainly represent Midcontinent reservoirs. Most systems had finite angles with only a few completely water-wet. On the average the quartz surfaces, representing sandstone reservoirs, were slightly preferentially water-wet, and the calcite surfaces, representing carbonate reservoirs, were slightly preferentially oil-wet. The authors also observed the extreme sensitivity of the wettability to trace amounts of contaminants. Only 10 ppm of copper or nickel salts in brine reversed wetting on a quartz surface from water-wet ($\alpha = 0°$), to preferentially oil-wet ($\alpha = 170°+$). In most of the other studies imbibition and flooding techniques or capillary pressures were used with natural reservoir rocks and fluids to describe wetting on a semiquantitative scale. Results are not easily summarized because of conflicting data on effects of core handling and storage, and the variety of reservoirs studies. Overall, a pattern emerges of some water-wet, and many preferentially water-wet or preferentially oil-wet cores. More recently Salathiel (35) has presented a mixed wettability model in which parts of the mineral surface in contact with oil in the large pores are oil-wet, or preferentially oil-wet and the fine pores are water-wet. If the continuity of the oil-wet regions remains complete, the oil will exhibit typical wetting phase trapping characteristics, such as rate sensitivity and significant oil flow after waterflood breakthrough.

VI. MECHANISMS OF RELEASE OF TRAPPED PHASES

A multitude of processes having several distinct mechanisms have been developed to remove residual oil. These can be grouped into two predominant types. The first are processes in which viscous/capillary force ratios are changed,

usually by reducing capillary forces. The second are processes in which fluid phase volumes are altered by substantial
interphase mass transfer.

A.   Processes Which Alter Viscous/Capillary Force Ratios
     This first category has been extensively studied both
experimentally and theoretically for surfactant flooding
applications. Mobilization of residual oil can be accomplished by either applying sufficient viscous drag or by
reducing capillary forces.
1. *Theory*
     A number of authors have used the concept of competition
of viscous and capillary forces in describing displacement of
residual phases. These studies have generated a variety of
dimensionless numbers including those of Leverett (5),
$(P_c/r_n \nabla P)$; Brownell and Katz (37), $(k \nabla \Phi / \sigma)$; Moore and Slobod
(13), $(u \mu / \sigma \cos \theta)$; Foster (38), $(u \mu / \phi \sigma)$; Lefebvre du Prey (39),
$(\phi \sigma / u \mu)$; and Abrams (49), $[u \mu_w (\mu_w / \mu_o)^{0.4} / \phi \Delta S \sigma \cos \theta)$. Results of these and other studies (40-44,17) all demonstrated
that higher viscous/capillary force ratios are needed to
recover residual phases than are commonly encountered in
ordinary oil production processes. An examination of the
microscopic picture reveals the reason for this difficulty.
The minute distance over which a pressure gradient can be
applied to an isolated drop, when multiplied by even large
gradients, yields only a small total force on the drop.
Opposing this are large capillary forces which increase as
pore size decreases. For a nonwetting immobile (oil) phase,
such as that shown in Figure 10 for a single trap, a simple
force balance (17) demonstrates that the oil mass will be
displaced if the applied pressure exceeds the net restraining
capillary pressure. That is,

$$\Delta P_A = \nabla \Phi \cdot \Delta L > \frac{2 \sigma}{r_n} \cdot \psi = \Delta P_c \qquad (7)$$

The applied potential gradient is defined,

$$\nabla \Phi = dp/dl + \Delta \rho g (1 + G) \quad ,$$

where the first term is the applied pressure gradient and the
second contains gravity and other acceleration terms.
     The LHS of the inequality is comprised of the pressure
gradient and the length of the alternate flow path around the
trap. This can be defined as a dimensionless length in terms
of the pore entry radius:

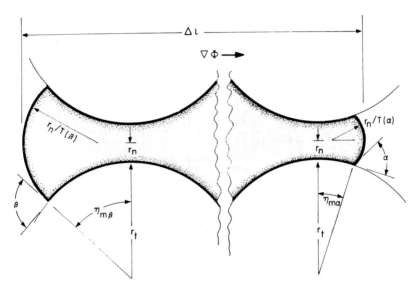

FIG. 10 SCHEMATIC MULTIPORE MODEL

76/069/8

$$m = \Delta L/r_n \quad . \tag{8}$$

The capillary equation of Laplace which appears in the RHS of inequality (7) is not exact when curvature is small; however, both experimental (45) and theoretical (62) work have shown the effect to be small for pore radii greater than 0.1 micron. Permeabilities in the range of 0.01 to 0.1 md are expected for this pore size.

The correction term, ψ, for the interface front with an advancing angle, α, and the back with a receding angle, β, is similar to that given by Melrose and Brandner (7):

$$\psi = T(\alpha) - \frac{T(\beta)}{r_b/r_n} \quad . \tag{9}$$

This functional relationship of pore body/pore neck radii ratio, r_b/r_n, is comparable to the first factor in Dullien's (46) "Difficulty Index." For example, the Difficulty Index is defined:

$$D = \left| \frac{1}{\overline{r}_n} - \frac{1}{\overline{r}_b} \right| \int_{\infty}^{0} \int_{r_b}^{0} \hat{\alpha}(r_b, r_n) \cdot dr_n \cdot dr_b \quad .$$

For the special case of $\alpha = \beta = 0$, $\psi = r_n[(1/r_n) - (1/r_b)]$.
     The fluid-rock term, ψ, only partially describes the
structure of the rock. Pore inlet size distribution can be
expressed by combining Equations (2) and (3):

$$r_n(S) = \frac{2\sqrt{k/\phi}}{N_{Le}(S)} \quad . \tag{10}$$

Substituting Equations (8), (9), and (10) into Equation (7)
gives

$$\left| \frac{1}{N_{Le}^2(S)} \right| \cdot \left| \frac{1}{\psi(S)} \right| \cdot \left| 2m(S) \right| \cdot \left| \frac{k}{\phi} \frac{\nabla \Phi}{\sigma} \right| \geq 1 \quad . \tag{11}$$

     The dimensionless alternate path length, m, being some
function of pore dimensions, is not readily obtainable; therefore, we define a dimensionless length of the entire multipore
trapped oil mass as

$$f = \Delta L/a \quad . \tag{12}$$

This ratio is a fundamental geometric constant applicable to
all fluids and independent of pore dimensions. It defines the
maximum length at which a stable filament can exist without
breaking into segments. Calculated values of "f", determined
by comparing a cylindrical filament with an equivalent volume
sphere for various assumed conditions, are shown in Table II.
     For static conditions Plateau (47) has shown a filament
will become unstable at a length 2π times its radius; assuming
dynamic conditions, Rayleigh (48) determined the value to be
approximately 9.0. Experimental values for oscillating jets
of water in air range from 2π to 9, but approach the higher
value for vibration-free systems. Upper and lower limits for
cylindrical filaments in space, assuming constant capillary
pressure and constant area from cylinder to sphere, are also
given. Applicability of these data to oil masses in porous
media have not been determined; however, observations by Taber
(42) and others suggest an approximate value of 10.
     When Equation (12) is combined with Equation (4) an ex-
pression for the alternate flow path in terms of interfacial
tension and applied pressure gradient is

TABLE II

GEOMETRIC VALUES OF CYLINDRICAL FLUID FILAMENTS
AND EQUAL VOLUME SPHERES

Equilibrium Condition	$f = \Delta L/a$	Sphere/Cylinder Ratios		
		Radii a_s/a_c	Curved Area A_s/A_c	Capillary Pressure P_{c_s}/P_{c_c}
Constant Curved Area	4.50	1.5	1.0	1.33
Maximum Static	6.28	1.676	0.894	1.19
Maximum Dynamic	9.015	1.891	0.793	1.06
Constant P_c	10.667	2.00	0.749	1.00

74

$$\Delta L = \sqrt{\frac{2\sigma f}{\nabla \Phi}} \quad , \tag{13}$$

and using Equations (8), (13), and (10), the dimensionless alternate flow path is

$$m(S) = [N_{Le}(S)] \cdot \left[\frac{f}{2}\right]^{1/2} \cdot \left[\frac{\phi \sigma}{k \nabla \Phi}\right]^{1/2} \quad . \tag{14}$$

The dependence of the average length of an isolated oil mass on applied gradient or interfacial tension is a characteristic distinct from saturation dependence of viscous to capillary force ratio. At a given value of $\nabla \Phi / \sigma$, the configuration of residual oil may be visualized as adjacent isolated and trapped oil masses closely fitting together in a three-dimensional "jigsaw puzzle" pattern. Trapping at higher ratios of $\nabla \Phi / \sigma$ will result from separation at both original and at weaker filament locations. The new pattern will have smaller, but still closely fitting pieces, so that in the low viscous/capillary ratio region, residual saturation may be practically unchanged.

By combining Equations (11) and (14), the dimensionless ratio of viscous to capillary forces originally proposed by Brownell and Katz (37) is equated to three properties of the rock-fluid system:

$$N_{BK}(S) = \frac{k \nabla \Phi}{\sigma} \geq [\phi N_{Le}^2(S)] \cdot [\psi^2(S)] \cdot [1/2f] \quad . \tag{15}$$

The first term on the RHS of the inequality defines the geometry of the rock pore network which is independent of type of fluids present; the second term defines pore body/pore neck radii and its interaction with contact angles of advancing and receding interfaces; and the third term is a constant fluid geometric property independent of rock structure.

N_{BK} provides an approximate desaturation scale for most rocks since the RHS of Equation (15) often is in the range of 10^{-3} to 10^{-4}; however, N_{BK} is not entirely satisfactory for comparing various rock types since one rock property, permeability, remains on the LHS. Another dimensionless number, N_{VC}, is obtained by substituting Darcy's Law into Equation (15) and placing relative permeability in the rock-fluid property term:

$$N_{VC}(S) = \frac{u\mu}{\sigma} \geq [\phi N_{Le}^2(S)] \cdot [k_{rw}(S) \ \psi^2(S)] \cdot [1/2f] \quad . \tag{16}$$

This number, which differs from N_{BK} by a factor of k_{rw}, segergates all rock properties to the RHS, and thereby provides a good measure of the comparative ease of recovery from different rocks. Various authors have used either u, u/ϕ, or u/$\phi\Delta$S for the velocity term. Because porosity is a rock property, it has been placed on the RHS with the other rock properties, and Darcy velocity, u, rather than a "frontal" velocity is used.

Although $N_{Le}(S)$ in Equation (16) was derived as a function of initial saturations, it can be expressed in terms of a normalized residual oil, $S_R = S_{orc}/S_{or}$, because residual saturation, (S_{orc}), itself is a single valued function of initial saturation, (S). S_{or} is defined to be either the maximum trap saturation as determined from residual-initial measurements or the residual saturation at an infinitely small N_{VC}. Regardless of the method, S_{or} is a rock property only, rather than a rock-fluid property. The relative permeability to water, k_{rw}, can also be expressed in terms of normalized saturation. These functions when substituted into Equation (16) relate normalized residual oil and capillary number:

$$N_{VC}(S_R) = \frac{u\mu}{\sigma} \geq [\phi \cdot \hat{N}_{Le}^2(S_R)] \cdot [\hat{k}_{rw}(S_R) \cdot \psi^2(S_R)] \cdot [1/2f] \ . \quad (17)$$

Experimental values of N_{Le} and S_R as functions of initial saturation can be obtained by standard procedures described previously in Section VA-2.

The relationship of relative permeability, k_{rw}, to oil saturation below S_{or} is not well defined, since the values are beyond the usual range of two phase flow for normal water-flooding. A pore model calculation of k_{rw}, assuming a bundle of capillaries and using fluid injection porosimetry data, predicts a slightly S-shaped function which is fairly well approximated by the straight line, suggested by L. W. Lake (private communication),

$$k_{rw}(S_R) = [1 - (1 - k_{rw@S_{or}})(S_R)] \ . \quad (18)$$

Experimental data by Abrams (49) on Dalton and Berea sandstones and Gilliland et al. (50) on the Second Wall Creek Sandstone exhibit a somewhat greater curvature than theoretical. Values are lower at high saturations and higher at low saturations than the straight line given in Equation (18) (see Figure 8).

The value of the pore shape-wettability function, $\psi(S)$, is the least well-defined term of N_{VC}. Like N_{Le} and k_{rw}, ψ

also changes with pore size (i.e., initial saturation); however, experimental techniques for measurement are not available. A possible method might be based upon determination of
pore body size distribution by NML measurement (51) and pore
neck size by previously described capillary pressure measurement. For the present, an average value for each rock will
be used. Probably the best method for determining an average
value of ψ is to: 1) directly measure pore body/pore neck
radii by photomicrography and injection porosimetry, 2) measure advancing and receding angles on flat surfaces, and 3)
use the toroidal geometrical factor. If photomicrographic
data for the pore body radius are not available, the ratio of
r_b/r_n can be roughly approximated by equating Poiseuille's and
Laplace's equations and by using formation resistivity factor,
F,* capillary pressure, and permeability to obtain an average
"hydraulic radius." Maximum pore body radius is then assumed
to be a constant factor, R, times this radius. The resulting
equation is

$$r_b/r_n = R \sqrt{2kF} \cdot \frac{\overline{P}_c}{\sigma\, T(\theta)} \qquad (19)$$

This method is not applicable when r_b/r_n is large; however,
for large values, ψ is insensitive to the ratio. A reasonable estimate of the ratio is obtained by assuming R = 2.5 and
using $\overline{P}_c$ at S = 0.5.

2. *Calculation of Residual Phase Saturations*

Residual nonwetting saturation as a function of rock and
fluid properties can be calculated by first using R-I data to
obtain normalized residual saturations, S_R, as a function of
S and by using the equations derived above in the sequence
(2), (19), (18), (6), (5), and (9) to obtain values of
$[\phi \cdot N_{Le}^2(S) \cdot k_r(S) \cdot \psi^2/2f]$ for comparable values of S. S_R as a
function of this group, or its equivalent in Equation (16),
$u\mu/\sigma$, is shown in Figure 11 for a typical Berea sandstone
having properties as shown in Table III. Starting at ordinary
waterflood conditions $u\mu/\sigma \cong 10^{-7}$, a factor of 100 in rate or
pressure gradient is predicted to initiate any oil movement
in a strongly water-wet rock. Thereafter desaturation would
occur over roughly another factor of 100.

A similar calculation can be made for preferentially
wetting phase residual saturation. In this case, the concept
of isolated wetting phase trap lengths is less obvious and
residual wetting phase saturation cannot be normalized by a
rock property such as S_{or}. By arbitrarily normalizing to

---

*F = ratio of resistivity of fluid saturated rock/
resistivity of fluid.

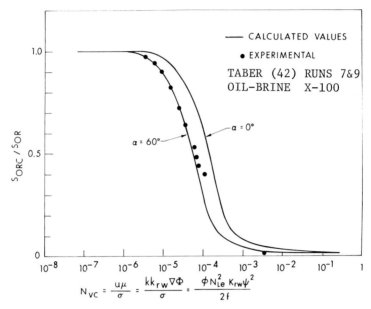

$$N_{vc} = \frac{u\mu}{\sigma} = \frac{kk_{rw}\nabla\Phi}{\sigma} = \frac{\phi N_{Le}^2 k_{rw}\psi^2}{2f}$$

FIG. 11   CALCULATED & EXPERIMENTAL RESIDUAL OIL

TABLE III

PROPERTIES OF BEREA SANDSTONE

$$\phi = 0.20$$
$$k = 0.4 \text{ D}$$
$$F = 18$$

R-I - Figure 7

P_c - Figure 5

k_{rw}
          Figure 8
k_{ro}

$(1 - S_{or})$ a desaturation relationship obtains which at least qualitatively demonstrates differences from the nonwetting case. First, at low S_{wrc}, capillary pressures, and consequently $N_{Le}(S)$, are much higher than at low S_{orc}. Relative permeabilities to oil, $k_{ro}(S)$, also are higher. The result is a considerably greater requirement of applied force to effect final removal of wetting phase. Second, desaturation is directly related to the continuous capillary pressure so that movement of "trapped" wetting phase should occur continuously as applied gradients are increased.

The calculated wetting phase desaturation function for a second drainage after a previous high nonwetting saturation predicts the trapped wetting phase to become mobile at $N_{VC} \cong 10^{-6}$, and to continuously desaturate up to values greater than 1. Of course, for completely wetting fluids at saturations high enough for good continuity, the isolated wetting phase calculation is not valid since ΔL becomes the length of the flow system.

3. *Review of Experimental Data*

A large amount of experimental data has been published which relates removal of residual oil to applied flow conditions and fluid and rock properties. Porous media used includes sands, sandstones, glass beads, and sintered Teflon particles; fluids are normally oil and water, sometimes with surfactant or alcohol additives to lower interfacial tension.

a. *Berea sandstone*. Studies have been made with Berea Sandstone for which sufficient basic rock properties are given to permit comparison with calculated values. For the following comparisons, data are limited to approximately unit viscosity ratio experiments. Specific capillary pressure, residual-initial, relative permeability, and pore shape factors were not given for each study; however, all Berea samples had about the same permeability and porosity and other rock properties are expected to be similar.

The calculated nonwetting desaturation curve shown in Figure 11 for typical Berea sandstone is compared with one experiment by Taber (42) in which complete removal of residual oil was achieved. Other experimental studies (49,13) on Berea sandstone have been performed at different initial conditions. These are compared in Figure 12.

The experiments of Taber most closely approach conditions assumed in the derivation of Equation (17), i.e., isolated, trapped nonwetting oil. In Taber's Runs 7 and 9, an oil-water-Triton X-100 system was progressively subjected to increasing pressure gradients. The shape of the experimental curve is very similar to calculated values, but desaturation occurs at a factor of 2.5 times less than predicted for a completely water-wet rock (see Figure 11). The correspondence of calculated and experimental residual oil, S_R, is as good as might be expected with the available data. Uncertainties in data include 1) capillary pressure, R-I, and relative permeability for the specific rock samples used, 2) pore body to pore neck ratio as a function of pore size, and 3) advancing and receding contact angles. Additional uncertainties are 1) the validity of the oil filament geometric constant in porous media, and 2) the adequacy of the toroidal pore neck constant. A match of Taber's Runs 7 and 9 is obtained by either assuming an advancing and receding contact angle of 60° through the

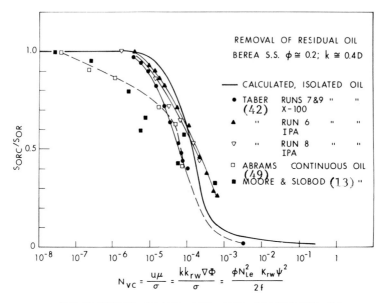

FIG. 12 EFFECT OF ISOLATED AND CONNECTED OIL

76/669/15

wetting phase, or by changing the toroidal constant by a factor of 1.5, or by changing the filament constant or interfacial tension by a factor of 2.5. Additional experimental work, with rock and fluid properties precisely defined might resolve some of these uncertainties. In Taber's Runs 6 and 8, isopropyl alcohol was added to an oil-water system, and again pressure gradient was progressively increased. The match is closer at high saturations, but final desaturation is not as readily effected as is predicted.

In contrast to these results on <u>isolated</u> trapped oil, Abrams (49) has shown that trapping under dynamic, and at least partially interconnected oil conditions, will result in initial desaturation with less change in N_{VC}. At lower trapped oil saturations the dynamic values approach the isolated oil saturations. These dynamic experiments were performed with simultaneous oil-water displacements at varying flow rates.

Moore and Slobod (13), in their original paper on viscous/capillary theory, reported saturations at breakthrough of waterflood for several fluid systems in Berea sandstone. Although these are transient data, they are similar to the dynamic experiments of Abrams and also exhibit the early response at high oil saturation.

b. *Other porous media*. Experiments with diverse porous media are so varied in experimental conditions that detailed comparisons are not valid. Some aspects of trapping can be observed by examining the broad ranges of results of several investigations (see Figure 13). Wagner and Leach (44), using a methane, n-pentane fluid system, measured reduction in trapped gas saturation in a 500 md Torpedo sandstone. By operating at high pressure near the critical point they achieved moderately low interfacial tensions. Nevertheless, N_{VC} values were low and only a small reduction in gas saturation occurred.

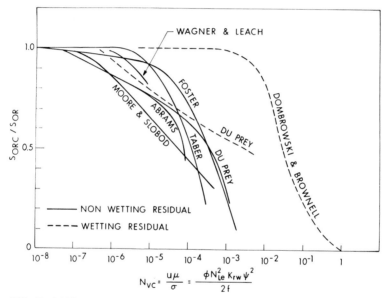

FIG. 13 AVERAGE EXPERIMENTAL RECOVERIES OF RESIDUAL PHASES

76/069/12

Lefebvre du Prey (39) combined a large number of measurements with various fluid systems on sintered Teflon particles and stainless steel and alumina. Data for Teflon "cores" are shown in Figure 13. Most of these fluid systems were preferentially oil-wet with sizable contact angles. Both nonwetting (aqueous) and wetting phase (oil) removal are given. Flooding conditions are dynamic and show characteristics similar to those discussed previously, i.e., early trapped phase movement at low N_{VC} and continued desaturation over an extremely wide range of N_{VC}.

Another wetting phase study by Dombrowski and Brownell (40), unlike Du Prey's, was carried out on liquids trapped in isolated pendular rings. Desaturation data were collected

by centrifugation in air of small glass sphere packs. Liquid
saturations were all less than 7.5 percent of the pore volume,
which is far below the lowest saturation in Du Prey's wetting
phase (33%).

The Dombrowski and Brownell data illustrate the great
retentive forces trapping the last part of a wetting phase.
Compared to a nonwetting phase, an additional factor of more
than 100 times in flow rate is required to remove the wetting
phase.

B. Processes Which Alter Phase Volumes

Two mechanisms predominate in processes (52,53) in which
interphase mass transfer is operative. The first depends upon
swelling and mobilization of the trapped phase; the second
depends upon solution of trapped phase into the flowing stream.

1. *Mobilization by Swelling of Trapped Phase*

Nonwetting oil trapped by a low viscous/capillary number
process, such as waterflooding, has the maximum oil saturation
which can remain immobile. Any increase in oil volume will
reconnect and mobilize part of the oil at a high apparent
frontal velocity. This can be observed experimentally by
injecting a small fraction of pore volume of oil into a linear
core which is at residual oil after waterflood. Resumption of
waterflood will result in early recovery of an equivalent
amount of oil at a small fractional pore volume of subsequent
water injection. Chemical or miscible processes in which an
added component is soluble in both phases can significantly
swell the oil phase if the additive preferentially partitions
into the oil phase, or if water is solubilized into the oil
phase. An alcohol flood with tertiary butyl alcohol (53) is
an example in which oil swelling predominates. The phase
behavior is shown in Figure 14A. A typical path of phase
compositions during an alcohol flood is shown from initial
conditions at "1" to breakthrough at "2" and finally to miscibility at "3". With the plait point to the aqueous side of
this path, a large amount of oleic phase swelling occurs as
alcohol concentration increases. Ahead of the alcohol front,
a stabilized oil bank forms in which saturations and fractional flow are determined by 1) material balance of oil released behind the alcohol front and banked oil at $S_o > S_{or}$,
and 2) the relative permeability-saturation relationship of
the rock. For predominantly "oil swelling" displacements, the
stabilized oil bank exhibits higher oil fractional flow than
for "oil solution" type floods. Behind the stabilized oil
bank two-phase flow continues until the residual wetting phase
is reached. Thereafter relatively small amounts of water and
oil are dissolved from the immobile wetting phase into the
flowing oleic (TBA) stream. The point at which the concentration path crosses the two-phase envelope is ordinarily far

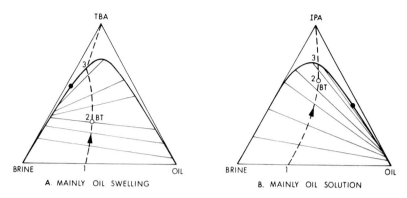

A. MAINLY OIL SWELLING   B. MAINLY OIL SOLUTION

FIG. 14 PROCESSES WHICH ALTER PHASE VOLUMES

76/069/13

enough from the plait point, where $\sigma = 0$, that interfacial tensions are high compared to those required for reduction in residual oil from viscous/capillary ratio considerations.

2. *Solution of Trapped Phases*

In the second type of interphase mass transfer mechanism, a trapped phase is recovered by extraction into the mobile phase. An alcohol flood using isopropyl alcohol is largely this type of process (see Figure 14B). In this case the plait point lies to the oleic side of the alcohol dilution path. A lesser amount of alcohol partitions into the oil than in the previous case; however, swelling is still important. Again a stabilized bank is produced because some swelling occurs, but most of the oil is recovered from solution of the oil following alcohol breakthrough. As before, the phase compositions of produced fluids are not close to the plait point, and little desaturation of oleic phase can be expected from viscous/capillary effects.

C. Mechanisms Operative on Oil Recovery Processes

Oil recovery processes can be classified by the predominant mechanism that is operative. Mutual solvent processes, such as alcohol or CO_2 floods, depend upon both phase volume effects, swelling and solution; whereas miscible processes depend upon oil swelling since only the hydrocarbon phase is soluble.

The contribution of phase behavior of micellar surfactant-oil-water systems to oil recovery is generally obscured by the accompanying large changes in capillary properties at surfactant concentrations sufficient to cause the phase volume effects. Undoubtedly phase volume mechanisms are active at low concentrations of surfactants compared to mutual solvents (54).

Healy et al. (55,56) have reported on the properties of surfactant systems and have outlined conditions which control
phase volumes. Occurrence of the third, micellar phase
greatly complicates theoretical treatment. In general, changes
in conditions which result in partition of surfactant toward
the oleic phase, e.g., increasing salinity, decreasing temperature, increasing surfactant molecular weight, increasing oil
aromaticity, and increasing divalent ion concentrations tend
to alter phase behavior from the second (oil solution) type to
the first (oil swelling) type.

Emulsification processes (57) with dilute electrolyte
form oil-in-water emulsions which can remove trapped oil in a
manner similar to oil solution by a solvent. Water-in-oil
emulsions (58), which tend to occur at higher electrolyte concentrations, have the effect of oil swelling but also often
have high viscosity and even interfacial rigidity. Processes
which utilize dynamic interphase mass transfer (59), such as
spontaneous emulsification of oil into a water phase, are
reported to be more effective than nonemulsifying systems at
comparable viscous/capillary numbers (60). Applicability of
these experiments, when scaled to field distances and time,
has not been demonstrated.

In emulsification or film formation processes, an under-
lying intent is the establishment of resistance which will
increase the pressure drop. Use of polymers in conjunction
with low interfacial tension floods can improve oil recovery
by further increasing capillary number, provided pressure
gradients can be increased at least locally in the low interfacial tension region.

Addition of a third gaseous phase also increases flow
resistance in a manner similar to the emulsification processes,
especially if surfactants are used to make foams. Gas can also
mobilize trapped oil by swelling the oil phase. Swelling results either from solution of the gas in the oil or the existence of gas as a separate phase inside the oil phase (61).
As with solvent floods, large interfacial tension, even near
the critical point, minimizes capillary desaturation.

In summary, the processes which utilize phase volume
changes are distinctly different from those which depend upon
changes in viscous/capillary ratio. The viscous/capillary
processes: 1) are dependent upon rock properties, particularly permeability and pore shape, 2) are dependent on fluid
properties, especially interfacial tension and to a lesser
extent wetting characteristics of the fluids, 3) exhibit
lower final residual saturations when trapping results from
deposition of a continuous phase compared to removal of an
isolated phase, and 4) require extremely large ($> 10^4$)
changes in viscous/capillary forces to effect complete
removal of residual phases.

In contrast, processes which depend upon phase volume changes are: 1) independent of applied pressure gradients, and 2) less dependent of rock properties (excepting adsorption). Swelling of the residual phase is the more desirable of the phase mechanisms since it tends to give rapid response and a stable displacement; whereas, solution type behavior leads to delayed recovery and unstable displacements.

VIII. ACKNOWLEDGEMENT

The author acknowledges the significant contributions by B. F. Swanson in characterization of petrophysical properties of reservoir rocks and by R. C. Nelson in phase volume concepts. I also thank P. A. Good for review of the manuscript. Finally, I express appreciation to Shell Development Company, Houston, for permission to publish and for preparation of this paper.

IX. NOTATION

A	area, $[1^2]$
a	average radius of multipore trapped oil, $[1]$
B	ratio of pore doublet small radius/large radius, Table I, [dimensionless]
C	constant in Equation (1) = 0.002, [dimensionless]
$C_p(B)$	net pressure difference factor in Table I, [dimensionless]
$C_u(B)$	net flow velocity factor in Table I, [dimensionless]
d	diameter, $[1]$
F	formation resistivity factor = ratio of electrical resistivity of fluid filled rock/resistivity of fluid, [dimensionless]
f	multipore oil filament length to radius ratio = $\Delta L/a$, [dimensionless]
G	centrifugal force, [G's]
g	gravitational constant, $[1t^{-2}]$
k_{ro}	relative permeability to nonwetting phase, [dimensionless]
k_{rw}	relative permeability to wetting phase [dimensionless]
$\hat{k}_{rw}(S_R)$	modified relative permeability function, Equation (17), [dimensionless]
m	ratio of length of multipore oil mass/radius of pore neck = $\Delta L/r_n$, [dimensionless]
J	curvature, reciprocal of sum of interface radii, $[1^{-1}]$
k	permeability, $[1^2]$
L	length, $[1]$
ΔL	length of multipore oil mass, $[1]$
ℓ	doublet length, $[1]$
N_{BK}	Brownell and Katz Number = $k\nabla\Phi/\sigma = u\mu/k_r\sigma$, [dimensionless]

N_{Ca} Foster Capillary Number = $u\mu/\phi\sigma$, [dimensionless]

N_{Le} Leverett Number = $J_d\sqrt{k/\phi}$, [dimensionless]

$\hat{N}_{Le}(S_R)$ modified Leverett Number function, [dimensionless]

N_{MS} Moore and Slobod Number = $u\mu/\sigma \cos\theta$, [dimensionless]

$N_{VC}(S)$ viscous/capillary number = $u\mu/\sigma = kk_r\nabla\Phi/\sigma$, Equation (16), [[dimensionless] function of saturation]

P_c capillary pressure [$ml^{-1}t^{-2}$]

p pressure, [$ml^{-1}t^{-2}$]

q flow rate, [l^3t^{-1}]

r radius, [l]

R ratio of maximum pore radius/average pore radius determined by resistivity (Equation 19), [dimensionless]

S saturation of nonwetting phase during injection cycle, [dimensionless]

S_R normalized residual oil saturation = S_{orc}/S_{or}, [dimensionless]

$T(\theta)$ geometrical factor for contact angle of interface passing through toroidal pore. Defined by Equation (5), [dimensionless]

u Darcy velocity = q/A, [lt^{-1}]

$\hat{\alpha}(r_n,r_b)$ Dullien's pore volume distribution function, [l^{-2}]

α advancing contact angle of oil/water interface, measured through water, [°]

β receding contact angle of oil/water interface, measured through water, [°]

$\nabla\Phi$ potential gradient = [$dp/dL + \Delta\rho g(1 + G)$], [$ml^{-2}t^{-2}$]

ϕ porosity, [dimensionless]

η pore angle of interface, [°]

$\Delta\rho$ density difference, [ml^{-3}]

σ interfacial tension, [mt^{-2}]

θ advancing contact angle of fluid used to determine rock property [°]

ψ geometrical factor for curvature of front and back of trapped oil mass (defined by Equation 9), [dimensionless]

Subscripts

b   pore body
c   cylinder
d   drainage
m   maximum
n   pore neck
or   immobile nonwetting phase at maximum trapping
orc   immobile nonwetting phase below maximum trapping
wr   immobile wetting phase at maximum trapping
wrc   immobile wetting phase below maximum trapping

rw   relative permeability to wetting phase water
s   sphere
t   torus cross-section

Conversion Factors

P dynes/cm^2 = 68944 x p psi
k darcies   = 1.013 x 10^8 x k cm^2

X. LITERATURE CITED

1.   Slattery, J. C., "Interfacial Effects in the Entrapment and Displacement of Residual Oil," Am. Inst. Chem. Eng. J. 20, 1145-1154 (1974).

2.   Dullien, F.A.L., Batra, V. K., "Determination of the Structure of Porous Media," Ind. Eng. Chem. 62, 25-53 (1970).

3.   Dullien, F.A.L., "New Network Permeability Model of Porous Media," Am. Inst. Chem. Eng. J. 21, 299-307 (1975).

4.   Gardescu, I. I., "Behavior of Gas Bubbles in Capillary Spaces," Petrol. Trans. AIME 351-370 (1930).

5.   Leverett, M. C., Lewis, W. B. and True, M. E., "Dimensional-Model Studies of Oil Field Behavior," Petrol. Trans. AIME 146, 175-193 (1942).

6.   Melrose, J. C., "Interfacial Phenomena as Related to Oil Recovery Mechanisms," Can. J. Chem. Eng. 48, 638-644 (1970).

7.   Melrose, J. C. and Brandner, C. F., "Role of Capillary Forces in Determining Microscopic Displacement Efficiency for Oil Recovery by Waterflooding," Can. J. Petrol. Technol. 13, No. 4, 54-62 (1974).

8.   Morrow, N. R., "Physics and Thermodynamics of Capillary Action," Ind. Eng. Chem. 62, 32-56 (1970).

9.   Morrow, N. R. and Harris, C. C., "Capillary Equilibrium in Porous Materials," Soc. Petrol. Eng. J. 5, 15-24 (1965).

10.   Pickell, J. J., Swanson, B. F. and Hickman, W. B., "Application of Air-Mercury and Oil-Air Capillary Pressure Data in the Study of Pore Structure and Fluid Distribution," Soc. Petrol. Eng. J. 6, 55-61 (1966).

11.   Purcell, W. R., "Capillary Pressures—Their Measurement Using Mercury and the Calculation of Permeability Therefrom," Petrol. Trans. AIME 186, 39-48 (1949).

12.   Purcell, W. R., "Interpretation of Capillary Pressure Data," Petrol. Trans. AIME 189, 369-371 (1950).

13.   Moore, T. F. and Slobod, R. L., "The Effect of Viscosity and Capillarity on the Displacement of Oil by Water," Producers Monthly 20, No. 10, 20-30 (1956).

14. Rose, W. and Witherspoon, P. A., "Trapping Oil in a Pore Doublet," _Producers Monthly_ 21, No. 2, 32-38 (1956).

15. Rose, W. and Cleary, J., "Further Indications of Pore Doublet Theory," _Producers Monthly_ 22, No. 3, 20-25 (1958).

16. Slobod, R. L., "Comments on Trapping Oil in a Pore Doublet," _Producers Monthly_ 21, No. 3, 17 (1957).

17. Stegemeier, G. L., "Relationship of Trapped Oil Saturation to Petrophysical Properties of Porous Media," SPE 4754, presented at Improved Oil Recovery Symposium, Tulsa, Oklahoma, April, 1974.

18. Batra, V. K. and Dullien, F.A.L., "Correlation Between Pore Structure of Sandstones and Tertiary Oil Recovery," _Soc. Petrol. Eng. J._ 13, 256-258 (1973).

19. Dullien, F.A.L., Dhawan, G. K., Gurak, N. and Babjak, L., "A Relationship Between Pore Structure and Residual Oil Saturation in Tertiary Surfactant Floods," _Soc. Petrol. Eng. J._ 12, 289-296 (1972).

20. Dullien, F.A.L. and Dhawan, G. K., "Characterization of Pore Structure by a Combination of Quantitative Photomicrography and Mercury Porosimetry," _J. Colloid Interface Sci._ 47, 337-349 (1974).

21. Keelan, D. K. and Pugh, V. G., "Trapped Gas Saturations in Carbonate Formations," _Soc. Petrol. Eng. J._ 15, 149-160 (1975).

22. Kimbler, O. K. and Caudle, B. H., "New Technique for Study of Fluid Flow and Phase Distribution in Porous Media," _Oil Gas J._ 55, No. 50, 85-88 (1957).

23. Reisberg, J. and Doscher, T., "Interfacial Phenomena in Crude Oil-Water Systems," _Producers Monthly_ 21, No. 1, 43-50 (1956).

24. Raimondi, P. and Torcaso, M. A., "A Study of the Distribution of the Oil Phase Obtained Upon Imbibition of Water," _Soc. Petrol. Engr. J._ 4, 49-55 (1964).

25. Amott, Earl, "Observations Relating to the Wettability of Porous Rock," _Petrol. Trans. AIME_ 216, 156-162 (1959).

26. Benner, F. C., Dodd, C. G. and Bartell, F. E., "Evaluation of Effective Displacement Pressures for Petroleum, Oil-Water, Silica Systems," Fundamental Research on Occurrence and Recovery of Petroleum, American Petroleum Institute, New York, 85-93 (1943).

27. Bobek, J. E., Mattax, C. C. and Denekas, M. O., "Reservoir Rock Wettability--Its Significance and Evaluation," _Petrol. Trans. AIME_ 213, 155-158 (1958).

28. Donaldson, E. C., Thomas, R. D. and Lorenz, P. B., "Wettability Determination and Its Effect on Recovery Efficiency," _Soc. Petrol. Eng. J._ 9, 13-20 (1969).

29. Jennings, H. Y., "Surface Properties of Natural and
     Synthetic Porous Media," Producers Monthly 21, No. 5,
     20-24 (March, 1957).

30. Kyte, J. R. and Naumann, V. O., "Effect of Reservoir
     Environment on Water-Oil Displacements," J. Petrol.
     Technol. 13, 579-582 (1961).

31. Melrose, J. C., "Wettability as Related to Capillary
     Action in Porous Media," Soc. Petrol. Eng. J. 5,
     259-271 (1965).

32. Mungan, N., "Certain Wettability Effects in Laboratory
     Waterfloods," J. Petrol. Technol. 18, 247-252 (1966).

33. Owens, W. W. and Archer, D. L., "The Effect of Rock
     Wettability on Oil-Water Relative Permeability
     Relationships," J. Petrol. Technol. 23, 873-878
     (1971).

34. Richardson, J. G., Perkins, F. M. and Osoba, J. S.,
     "Difference in Behavior of Fresh and Aged East
     Texas Woodbine Cores," Petrol. Trans. AIME 204,
     86-91 (1955).

35. Salithiel, R. A., "Oil Recovery by Surface Film Drain-
     age in Mixed Wettability Rocks," J. Petrol. Technol.
     25, 1216-1224 (1973).

36. Trieber, K. E., Archer, D. L. and Owens, W. W., "A
     Laboratory Evaluation of the Wettability of 50 Oil
     Producing Reservoirs," Soc. Petrol. Eng. J. 12,
     531-540 (1972).

37. Brownell, L. E. and Katz, D. L., "Flow of Fluids Through
     Porous Media--Part II, Simultaneous Flow of Two Homo-
     geneous Phases," Chem. Eng. Progr. 43, 601-612 (1947).

38. Foster, W. R., "A Low Tension Waterflooding Process
     Employing a Petroleum Sulfonate, Inorganic Salts,
     and a Biopolymer," J. Petrol. Technol. 25, 205-210
     (1973).

39. Lefebvre du Prey, E. G., "Factors Affecting Liquid-
     Liquid Relative Permeabilities of a Consolidated
     Porous Medium," Soc. Petrol. Eng. J. 13, 39-47
     (1973).

40. Dombrowski, H. S. and Brownell, L. E., "Residual Equi-
     librium Saturation of Porous Media," Ind. Eng. Chem.
     46, 1207-1219 (1954).

41. Mungan, N., "Interfacial Effects in Immiscible Liquid-
     Liquid Displacement on Porous Media," Soc. Petrol.
     Eng. J. 6, 247-253 (1966).

42. Taber, J. J., "Static and Dynamic Forces Required to
     Remove a Discontinuous Oil Phase from Porous Media
     Containing Both Oil and Water," Soc. Petrol. Eng. J.
     9, 3-12 (1969).

43. Taber, J. J., Kirby, J. C. and Schroeder, F. U., "Studies on the Displacement of Residual Oil: Viscosity and Permeability Effects," Am. Inst. Chem. Eng. Symposium on Transport Phenomena in Porous Media, Paper 476 presented at the 71st National Meeting, Dallas, February 1972.

44. Wagner, O. R. and Leach, R. O., "Effect of Interfacial Tension on Displacement Efficiency," Soc. Petrol. Eng. J. 6, 335-344 (1966).

45. Fedyakin, N. N., "The Motion of Liquids in Microcapillaries," Russian Journal of Physical Chemistry 36, 776-780 (1962).

46. Dullien, F.A.L., "Determination of Pore Accessibilities-An Approach," J. Petrol. Technol. 21, 14-15 (1970).

47. Plateau, J.A.F., "Statique Experimentale et Theoretique des Liquides" (1873).

48. Rayleigh, 3rd Baron, "Instability of Jets," Mathematical Society Proceedings, November, 1878.

49. Abrams, A., "The Influence of Fluid Viscosity, Interfacial Tension, and Flow Velocity on Residual Oil Saturation Left by Waterflood," Soc. Petrol. Eng. J. 15, 437-447 (1975).

50. Gilliland, H. E. and Conley, F. R., "Surfactant Waterflooding," Ninth World Petrol. Congr. Proc. 4, 259-267 (1974).

51. Loren, J. D. and Robinson, J. D., "Relations between Pore Size Fluid and Matrix Properties and NML Measurements," Soc. Petrol. Eng. J. 10, 268-278 (1970).

52. Holm, L. W. and Csaszar, A. K., "Oil Recovery by Solvents Mutually Soluble in Oil and Water," Soc. Petrol. Eng. J. 2, 129-144 (1962).

53. Taber, J. J., Kamath, I.S.K. and Reed, R. L., "Mechanism of Alcohol Displacement of Oil from Porous Media," Soc. Petrol. Eng. J. 1, 195-212 (1961).

54. Holm, L. W., "Use of Soluble Oil for Oil Recovery," J. Petrol. Technol. 23, 1475-1483 (1971).

55. Healy, R. N. and Reed, R. L., "Physicochemical Aspects of Microemulsion Flooding," Soc. Petrol. Eng. J. 14, 491-501 (1974).

56. Healy, R. N., Reed, R. L. and Stenmark, D. G., "Multiphase Microemulsion Systems," SPE 5565, presented at Fall Meeting SPE, Dallas, 1975.

57. Subkow, P., "Process for the Removal of Bitumen from Bituminous Deposits," U.S. Patent 2,288,857 (July 7, 1942).

58. Cooke, C. E., Williams, R. E. and Kolodzie, P. A., "Oil Recovery by Alkaline Waterflooding," J. Petrol. Technol. 26, 1365-1374 (1974).

59.   Sternling, C. V. and Scriven, L. E., "Interfacial
      Turbulence: Hydrodynamic Instability and the
      Marangoni Effect," Am. Inst. Chem. Eng. J. 5,
      514-523 (1959).
60.   Cash, R. L. et al., "Spontaneous Emulsification--A
      Possible Mechanism for Enhanced Oil Recovery," SPE
      5562, SPE Fall Meeting, October 1975.
61.   Holmgren, C. R. and Morse, R. A., "Effect of Free Gas
      Saturation on Oil Recovery by Waterflooding," Petrol.
      Trans. AIME 192, 135-140 (1952).
62.   Melrose, J. C., "Thermodynamics of Surface Phenomena,"
      Pure Appl. Chem. 22, 273-286 (1970).
63.   Roof, J. G., "Snap-off of Droplets in Water Wet Pores,"
      Soc. Petrol. Eng. J. 10, 85-90 (1970).

EMULSIFICATION AND DEMULSIFICATION IN OIL RECOVERY

Kenneth J. Lissant, Ph.D.
Director of Advanced Research
Tetrolite Division
Petrolite Corporation

I. ABSTRACT

In primary production, such reservoir parameters as porosity, temperature, pressure, nature and amount of connate water, type of crude, and gas-oil-water ratios must be considered in explaining the formation of oil field emulsions. Depth of well, and type of completion can also affect emulsion formation.

In secondary or tertiary recovery not only incidental emulsification need be considered but also the desirability of deliberate emulsion formation.

The selection of demulsification methods must be based on studies of the specific emulsions under consideration.

II. SCOPE

This article cannot attempt to cover an area as wide as emulsions and demulsification to any great depth. It is intended to serve as a general indication of the fact that the emulsification processes are inherent in the primary production and in secondary and tertiary recovery of petroleum. This paper does not report on new studies but is intended to serve as a general review of the field.

III. CONCLUSIONS AND SIGNIFICANCE

It can be seen that the occurrence of emulsions is a natural and all pervasive phenomena in petroleum production. In most instances, emulsions are to be avoided if possible. A considerable amount of literature has developed in techniques for resolving petroleum emulsions and this technology should be taken into consideration when emulsification is deliberately employed to achieve additional oil recovery.

IV. EMULSIFICATION AND DEMULSIFICATION IN OIL RECOVERY

A. Emulsion Types

The term emulsion has been used rather loosely over the years and has been defined in a number of ways. Most people, however, will agree that an emulsion is a polyphase system in which at least two immiscible liquid phases are present; one liquid phase being dispersed in the form of small droplets. It is usually believed that such systems are thermodynamically unstable and, theoretically speaking, this is probably true. However, emulsions can be prepared and are known to occur naturally which are stable for extended periods of time even under rather severe conditions. Recent work indicates that emulsions may be entropically stable. Emulsions are generally divided into two broad classifications called oil-in-water and water-in-oil. Most of the common emulsions in everyday life are oil-in-water emulsions with relatively low ratios of oil to water. Examples immediately come to mind such as milk and certain other foods, hand lotions, etc. We have discussed this in detail elsewhere. In oil field situations, however, the most commonly encountered emulsion is a water-in-oil emulsion and when oil-in-water emulsions occur they are referred to as "reverse emulsions."

Usually, when emulsions occur in the oil fields they are considered undesirable. Water-in-oil emulsions must be resolved before the oil can be refined. Also the presence of substantial amounts of water emulsified in the crude oil increases the costs of transporting the oil and increases corrosion and other maintenance problems. Most pipelines will not accept crude oil with significant amounts of emulsified water. We will, therefore, consider how water becomes emulsified in the crude oil, both in the formation and in the production equipment.

In primary production the source of the water for the emulsion is the connate water in the formation. This may vary widely in composition. It may be fresh water or it may contain dissolved salts especially sodium chloride. It may be saturated with carbon dioxide or hydrogen sulfide.

Assuming that an appropriate stabilization agent is present to prevent coalescence the other factor that determines emulsion stability is the settling rate of the particles. The settling rate is governed by a function essentially stated in Stokes law which says that the settling rate is proportional to the gravitic field, the density difference, the square of the radius of the droplet, and the reciprocal of the viscosity. It can be seen then that any situation which results in producing small droplets will tend to produce stable emulsions. Thus, one should avoid taking large pressure drops across orifices. Valves should be open and shut not partly open or "cracked."

To resolve emulsions then we must do two things. First, one must remove or inactivate the stabilization agent. Then one must promote settling and coalescence.

This is done by referring again to Stokes law. We can centrifuge and thus increase the gravitic field. We can heat or dilute to reduce the viscosity.

B. Emulsions in Primary Recovery

When a hole is drilled through the overlying impervious strata, and the pressure on the reservoir is relieved in the immediate vicinity of the well bore, the higher pressure in the other portion of the reservoir causes the fluids to migrate to the bore hole. In so doing, they are forced through the fine passages in the rock, sometimes at considerable velocities, and ample opportunity is provided for the oil and water to mix even if they had not been in an emulsion form previously. In some instances, as the pressure on the fluids is lowered, gas is released and the bubbles of gas may further aid in the emulsification process. After the oil-water mixture has reached the well bore, it may still encounter potential emulsion producing conditions as it passes through perforations in the drill pipe, through chokes, valves and other portions of the piping system.

When one wishes to produce emulsions in the laboratory, one adds one liquid to another, and subjects the mixture to high shear. Usually an emulsifier or dispersant is added. Emulsifiers function in two ways. They reduce the interfacial energy and thus the energy required to create more surfaces, and they prevent coalescence of the droplets once the droplets are formed. The prevention of coalescence is probably done by at least two mechanisms. First, an emulsifier can impart a charge to the droplet and thus droplets of like charge will be repelled and resist contact and coalescence. Another mechanism is for the emulsifier to form a "skin" over the droplet thus retarding coalescence if contact occurs. In many instances both mechanisms are brought into play.

Since most emulsions encountered in the oil field contain less than 30% internal phase, we will restrict our remarks to the formation and stabilization of these low internal phase ratio, usually water-in-oil emulsions. In primary recovery, oil field emulsions are stabilized by surface active ingredients occurring naturally in the crude oil or formed by heat or oxidation of the crude as it is produced and utilized. Among the types of compounds suspected of causing emulsification are porphorins, polycyclics, hemes, long chain fatty acids or esters, and unsaturates. In the case of "sour" crudes sulphur compounds may be involved.

Besides these active compounds, finely divided solids may be present. Such solids may be carbonaceous, polymers, clays, or other minerals such as elemental sulfur, iron sulfide, etc.

The API in its book on treating oil field emulsions lists
"asphalt, resinous substances, oil-soluble organic acids,
and other finely divided materials that are more soluble,
wettable, or dispersible in oil than in water." (1) They
also list iron, zinc and aluminum sulfates, calcium carbonate,
silica, and iron sulfide as causing emulsion problems. Once
the emulsions have been produced and collected in a storage
tank above ground, many conventional techniques are now available for breaking these emulsions. Most of the emulsions, as
mentioned previously, are water-in-oil emulsions containing
anywhere from one-half of a percent to 30 or 40% water, higher
internal phase ratios occur but are not as common. Many of
these emulsions are only temporarily stable and can be resolved
by simply allowing the fluids to stand until the water separates.
This gravity settling process can be speeded by heat which
serves both to reduce the viscosity of the crude and, in some
instances, to lessen the efficiency of the naturally occurring
emulsifiers. Separation can also be implemented by centrifugation, by electrical coalescence and by chemical treatment.
For more details see Franke (2), Goswame (3), Monson (4) and
Wiggins (5).

Centrifugation is not used extensively in the oil industry
although it has received some attention in the reclamation of
the oil from tar sands. In general, the capital investment
and maintenance problems encountered in centrifugation have
made it less economically desirable than other techniques.
Probably the most of the world's crude, which needs dehydration, is treated by simple "heater treaters" aided by small
amounts of demulsification chemicals. In a heater treater,
the incoming emulsion is passed through a heat exchanger;
chemical demulsifiers may be added and the mixed, heated
material passes into a quiescent zone where the water and oil
can separate and be drawn off. Mechanical designs of heater
treaters are well known and available in the literature.
Details of such equipment are given in the API Manual "Treating
Oil Field Emulsions." (1) When coupled with appropriate choice
of chemical demulsifiers they can be very efficient and simple
to operate and maintain. The trick, in this instance, is the
selection of an appropriate demulsifier chemical. The patent
literature in this particular field is very extensive and
surface active materials of almost all conceivable types have
been recommended and used in this application.

On April 14, 1914, William S. Barnickel received U.S.
Patent No. 1,093,098 on the use of iron sulfate to resolve
"roily oil." On April 24, 1917, he received U.S. Patent
1,223,659 on the use of "suitable water softening agents."
U.S. Patent No. 1,454,617 issued to E. E. Ayres on May 8,
1923, for the use of "resin soaps."

Melvin DeGroote received U.S. Patent 1,590,617 on June 29, 1926, for the use of sulfonated oils. He and his coworkers went on to receive more than a thousand patents before his retirement.

Other key early patents were No. 1,597,700 to J. C. Walker and 1,897,021 to P. Kaplan. At that time, it was not uncommon to use a barrel of chemical per 1,000 barrels of oil. Since that time, extensive work has been done and several thousand patents have issued. Treating ratios the world around are now commonly in the range of a 1 qt. to three gallons per thousand barrels and sometimes ratios of a gallon to 10,000 barrels can be obtained.

Monson (4) tells how that in the early 1900's when water was emulsified in the oil, it was referred to as "roily oil" and was simply run into pits where it was hoped that the water would settle out. Early attempts to resolve this roily oil by heat were successful in only a limited number of cases. However, it was discovered that certain simple materials such as soaps and surfactants would resolve these emulsions. One of the first materials patented for this purpose was tincture of green soap. Turkey red oil was also used. Many derivations of castor oil and tall oil have been patented and used. The use of wetting agents to selectively wet and remove solids from the interface has been mentioned in the patent literature. The problem of selecting an appropriate demulsifier is further complicated by the wide variation in the properties and compositions of both the crude oil and the connate water. At least a thousand specific formulations are currently being used around the world.

About the time that additional chemical attempts were being made to resolve roily oil, the principle of Cottrell's electric precipitators was applied to this problem and patents issued to the Petroleum Rectifying Company of California in this area.

On March 21, 1911, U.S. Patents 987,114 thru 987,117 issued to Frederick C. Cottrell. These covered the basic concepts of using an electrical field to increase sedimentation and coalescence of water-in-oil emulsions. Subsequent patents issued to Cottrell and A. C. Wright, and Wright received U.S. Patent 1,034,668 about a year later. R. E. Laird and J. H. Raney obtained a block of patents, 1,142,759 thru 1,142,761 on June 8, 1915. Since then the patents in this field fill several large volumes. Electrical dehydration of crude oil and refinery streams is now widely practiced. This technique is particularly useful in refineries where large streams of oil are to be treated at high throughput rates. Dowd (6) mentions removal of salts by such methods. The electrical dehydration is usually supplemented or aided by chemical agents.

C. Emulsions in Secondary Recovery

   Once the primary oil production of a field has slowed
below the point of economic operation, it is necessary to
consider the use of a secondary or tertiary recovery techniques. Some decades ago, water flooding was initiated as
a secondary recovery technique and has been widely used in
the United States. In this process, either new wells are
drilled or some of the production wells are converted to
water injection wells usually in what is known as a "5 spot
pattern." Water is injected into the formation through the
injection wells in such a way as to produce a stable front
which will sweep the oil ahead of it to the producing wells.
Since the porosity of various sections of the field may vary,
certain portions of the field may not actually communicate
with the producing wells. Often, the water front overruns
some of the oil and breaks through into the producing wells
before complete recovery is obtained. Surface active agents
may be added to the water to help wash the oil out of the
formation. Techniques of this type can become quite complex.
   Systems such as this have been reported to work quite
well. However, if channeling and breakthrough does occur,
concentrated portions of the surfactant front reach and mix
with the production fluid causing severe emulsion problems.
Similarly, at the end of such a flood as the polymer front
reaches the production wells, severe emulsion problems can
occur due to the presence of both surfactant and polymer.
The selection of demulsifiers which will effectively combat
these conditions is difficult because of the abnormally high
concentrations of surfactant involved in such a system. The
polymer used to stabilize the flood front is known to com-
plicate the resolution of such emulsions.
   The selection of emulsifier and polymer for these floods
is made on the basis of several considerations. Laboratory
and simulator tests indicate what materials can be considered
and then economic or supply factors take over. A surfactant
may be chosen because it is cheap or available, in house.
This may be an economy at first but if more has to be spent
on the recovery end to dehydrate the production the overall
effect may be more costly. Before any material is added to
a secondary or tertiary recovery system consideration should
be given to how it may be removed or deactivated at the production well. For further discussion, see articles by
Johnson (1), Nelson (8) and Jewett (9).
   There are certain situations in which attempts are made
to deliberately produce emulsions. In general, water or
returned brine is the only fluid which is cheap enough to
allow it to be injected into a formation for secondary or
tertiary recovery, and in formations where the surfaces are

primarily oil wetted, efficient recovery of the oil requires
that the water displace the oil from the surfaces. To promote this displacement, surfactants are added to the water
and these surfactants will tend to emulsify the oil into
the water stream. Usually, it is not economical to treat
the entire injection water volume with surfactant and,
therefore, a concentrated "surfactant front" is injected
which is then pushed through the formation with additional
water. This surfactant flood will solubilize some of the
oil and disperse the rest of the oil in it up to a point
where a continuous oil front is developed which can be
pushed to the producing well. The boundaries of these
various solution interfaces are not, however, very distinct
and both oil-in-water and water-in-oil emulsions may be
encountered in the producing wells.

In a few rare instances, low molecular weight hydrocarbons have been deliberately emulsified in water and these
emulsions forced through cores or test flood sections. The
theory here is that the low molecular weight hydrocarbon
droplets will scavenge crude oil from the formation, reduce
its viscosity and thus move it to the producing well. Here
again the surfactants used must be quite effective so that
their use is economical and means must be found to counteract their emulsifying tendencies when they appear in the
producing fluids.

D. Some Closing Thoughts

In its natural state, petroleum almost always exists in
association with connate water. Aqueous drive fluids are
almost the only fluids cheap enough, at present, to enjoy
widespread consideration for secondary or tertiary recovery.
Thus, emulsions are going to be found and the cost of resolving them must be added to the economic picture. Emulsions are often cheaper to avoid than to break. The choice
of drive surfactants should take into account emulsion
problems.

Deliberately made emulsions may be useful as drive fluids
if resolution techniques are planned into the full system.

V. REFERENCES

1.   API,Treating Oil Field Emulsions, 2nd edition (revised)
          (1962).
2.   Franke, W., "Breaking of petroleum emulsions," Erdoel
          u. Kohle, 13, 18 (1960).
3.   Goswami, T. K., "Comprehensive study on the demulsifica-
          tion process in the petroleum industry," Petrol.
          Hydrocarbons, 7, 207 (1972).
4.   Monson, L. T., "Resolution of oil-in-water emulsions,"
          Petroleum World, 43(4), 41 (1946).

5.   Wiggins, J. L., "Treating oil-field emulsions," Petrol.
      Engr., 29(5), B47 (1957).

6.   Dowd, J. D., "Considerations in the field removal of
      salt contaminants from crude oil," J. Inst. Petrol.,
      51, 181 (1965).

7.   Johnson, C. E., Jr., "Transport of concentrated sur-
      factant slugs in water flooding," Am. Chem. Soc.
      Div. Petrol. Chem., Preprints, 5(2), A13 (1960).

8.   Nelson, T. W., "Hydrocarbon recovery by fluid injection
      and thermal methods," World Oil, 164(6), 100 (1966).

9.   Jewett, R. L. and Schurz, G. F., "Polymer flooding--
      a current appraisal," J. Petrol. Technol., 22, 675
      (1970).

RECENT ADVANCES IN THE STUDY OF LOW INTERFACIAL TENSIONS

James C. Morgan, Robert S. Schechter and William H. Wade
The University of Texas at Austin

I. ABSTRACT

The study of interfacial tensions in an oil/surfactant/ water system is greatly facilitated by use of the alkane model. In this, the crude oil is replaced by that member of the homologous n-alkane series which shows interfacial tension behavior identical to that of the crude oil. A complete scale of hydrocarbon properties is used to study system variables affecting the low interfacial tension state.

II. SCOPE

Almost 70% of the total proven crude oil reserves will still be in the ground when production by standard techniques ceases to be economic (1). A successful tertiary oil recovery process, to supplement the present methods of pumping under natural forces, followed by a secondary water flood, is now an urgent research objective. In the forefront of proposed methods is chemical flooding with surfactant solutions.

Two distinct types of surfactant flooding are possible, those employing high and low surfactant concentrations. The Maraflood process (2-4) is an example of a high concentration process which has been shown to enhance oil recovery under field conditions. A small slug of micellar solution of surfactant, such as petroleum sulfonate, is used at about 12% volume concentration. Oil displacement is initially by a miscible process in this type of system, but Healy and Reed (5-6) have shown that the single phase microemulsion of oil, water and surfactant first involved breaks down into a multiphase emulsion when mixing dilutes the surfactant slug. Further displacement is then by an immiscible process.

The alternative is to use a continuous low concentration of surfactant (2% or much less), all displacement then being immiscible. A possible advantage here is that considerably less surfactant may be required overall. Possible drawbacks are that loss of surfactant from solution by adsorption onto the rock surface, or by solubilization into the oil, may more seriously affect oil recovery. Also, the method may be more sensitive to salinity and temperature changes than is miscible displacement. However, as miscible displacement inevitably degenerates by dilution to an immiscible process, understanding and control of immiscible displacement is of paramount importance.

Melrose and Brandner (7) and Taber (8) have shown that successful immiscible oil displacement depends on the existence of a very low interfacial tension, γ, between the oil and the water phases. A value of about 10^{-3} dyne/cm or less is required to mobilize the oil. Certainly, recovery of residual oil from laboratory test cores is greatly improved for systems with ultra low interfacial tension (7). The achievement and maintenance of low interfacial tensions during chemical flooding therefore seems essential.

Practical surfactant flood systems are usually very complex: present are surfactant, oil, water, electrolyte and probably a thickening agent for viscosity control, plus co-surfactants and blocking agents to enhance or protect the main surfactant. Surfactants which combine an ability to produce a low interfacial surface tension with low cost and large scale availability are not common. Petroleum sulfonates are perhaps the main candidates. These are usually sodium salts of sulfonated crude oil. The oil is fractionated by molecular weight before sulfonation, but the surfactant produced still has a range of molecular weights which may be broad. Crude oils themselves are also very complex, and vary considerably from field to field. The salinity of formation water, and type of ions present, will change with the field, as will the temperature.

This type of practical system is difficult to use in the study of variables affecting low interfacial tension. Changing the nature of one component of the system will demonstrate if, and in what manner, it affects γ, but quantitative assessments during this approach are not straightforward. In particular, achievement of a sufficiently low γ for one system does not show immediately how the system will react if a different crude oil is used.

III. THE ALKANE MODEL

The model system introduced by Cayias et al. (9), in which a pure n-alkane replaces the crude oil, solves the above problem. Development of the alkane model began when a study of the interfacial tension of a series of pure n-alkane drops against a petroleum sulfonate saline solution revealed that only one alkane gave a really low γ. An example of this specificity is shown in Figure 1. Here a solution containing 0.2% by weight of the petroleum sulfonate Witco 10-80, plus 1%

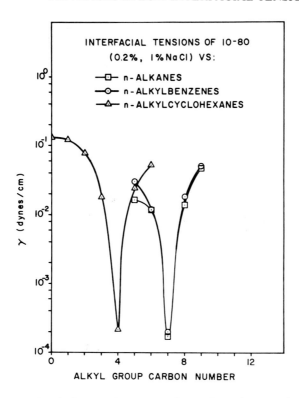

*Fig. 1.   Interfacial tensions of three homologous hydrocarbon
series with 0.2 wt.% 10-80 (1.0 wt.% NaCl)*

by weight sodium chloride, gives the lowest γ against heptane.
When a series of 1-phenyl n-alkanes is tested, the same surfactant phase gives a minimum at heptyl benzene. Similarly,
a minimum γ is found at butyl cyclohexane for this surfactant
phase against a series of 1-cyclohexyl n-alkanes (Figure 1).
Thus, heptane, heptyl benzene and butyl-cyclohexane may be
said to act as equivalent oil droplets. This equivalency is
virtually independent of the nature of the surfactant phase.
     Comparisons such as this using different surfactant
systems have been combined with a study of hydrocarbon mixtures (10) to produce the concept of the equivalent alkane
carbon number (EACN) for pure hydrocarbons and their mixtures
(11). For instance, as heptyl benzene is equivalent to heptane, the EACN of heptyl benzene is 7. The EACN of butyl
cyclohexane is similarly 7. Mixtures of hydrocarbons average
very simply. If a surfactant phase gives a low, minimum γ
against dodecane for example, it will also give a low γ against
a mixture of 0.5 mole fraction decane and 0.5 mole fraction
tetradecane. The mixture therefore has an EACN of 12. All

other mixtures of decane and tetradecane will give higher interfacial tensions against this surfactant. Any pure hydrocarbon, whether a member of a homologous series or not, may be assigned its individual EACN by testing in binary mixtures with an alkane. An average EACN may be calculated for any mixture of hydrocarbons, providing the EACN of each component is known. This is found from

$$(EACN)_{MIXTURE} = \sum_i (EACN)_i X_i \qquad (1)$$

where X_i is the mole fraction of the ith component of the mixture.

Values of EACN are not confined to integers, so that the EACN of an equal mole fraction mixture of heptane and octane is 7.5, for instance. By coincidence, all three hydrocarbon series in Figure 1 give minima almost exactly at integral carbon numbers, but in Figure 4 only two minima are at integral carbon numbers.

The EACN of a crude oil cannot be calculated directly using Equation (1), because all the hydrocarbon components of a specific crude oil have never been identified. However, applicability of Equation (1) to crude oils is inferred by its successful application to complex synthetic mixtures containing up to 29 aliphatic, alicyclic and aromatic hydrocarbons (11). Measurements of the EACN of a crude oil requires the preparation of a series of surfactant solutions giving individual minimum tensions against steadily increasing alkane carbon numbers. The crude oil is then tested against each surfactant solution, and the lowest γ found shows the EACN of the oil. By changing the surfactant series used, it may be demonstrated that the EACN of a crude oil is a property essentially characteristic of the oil, and not of the surfactant type used. EACN values varying from 6.2 to 8.6 have been reported for eight stock tank oils (11). This variation, though apparently small, is extremely important because most surfactant systems are very selective. If a surfactant gives a low tension of 10^{-3} dyne/cm or less with the 8.6 EACN oil, it will probably give a far higher tension of perhaps 10^{-1} dyne/cm with the 6.2 EACN, and vice versa.

Use of the complete alkane series instead of individual crude oils has several advantages. Firstly, it is immediately evident whether any particular surfactant system is optimally adjusted for giving a low γ with any given crude oil by comparing the alkane of minimum γ with the EACN of the oil. Also, the effect of changing one system variable on the alkane of minimum tension may be studied in detail, because a wide range of alkanes is now available. With these variable

rules complete, adjustment of a surfactant system to any
desired crude oil EACN becomes simple.

IV.   PARAMETERS AFFECTING LOW INTERFACIAL TENSIONS

Variables currently identified as important in the
achievement of the low interfacial tension state in a water/
oil/surfactant/electrolyte system are: the surfactant average molecular weight and molecular weight distribution; surfactant molecular structure; surfactant concentration; electrolyte concentration and type; oil phase average molecular
weight and structure; system temperature; and the age of the
system. This list includes variables which may be deliberately tailored to give a low-tension surfactant flood system
for a particular field, and others which may be expected to
vary during production from a field or from field to field.
Several variables are, of course, in both groups.

The effect of changing each variable on interfacial
tensions is now surveyed. All surface tensions referred to
as "low" or "minimal" have values of 2×10^{-3} dyne/cm or
lower unless otherwise stated. The standard aqueous phase
has a total surfactant concentration of 0.2% by weight and
a sodium chloride concentration of 1% by weight, the oil
phase being an alkane. No other components are present in
the surfactant phase unless mentioned. All petroleum sulfonates and xylene sulfonates have been deoiled on a silica
gel column prior to use by a method described elsewhere (12).
No preequilibration of the oil and aqueous phases was employed before measurement of the interfacial tensions, which
are made at 27°C using the spinning drop technique (13). In
early papers (10,11), this technique was not standard, and
results from these papers have been repeated before inclusion
here if comparison with later results is to be made. In fact,
differences found on repetition were usually within experimental error, which is considered to be ±0.3 of a alkane
carbon number. The molecular weights of petroleum and xylene
sulfonates quoted here are, strictly, equivalent weights,
determined by estimation of the SO_3Na content of a known
weight of surfactant.

V.   SURFACTANT AVERAGE MOLECULAR WEIGHT

A linear relationship has been found between the average
molecular weight of a mixture of two surfactants and n_{min}, the
carbon number of the n-alkane with which the mixture gives a
minimum interfacial tension (at constant total surfactant
concentration, salinity and temperature). This holds for all
but one (14) of the many binary mixtures studied to date (11,
14,15). The individual surfactants may each be of reasonably
well-defined structure, such as the two Exxon alkyl orthoxylene

sodium sulfonates (11) (Figure 2, approximate molecular structures in Figures 3(a) and 3(b)). Also, each component

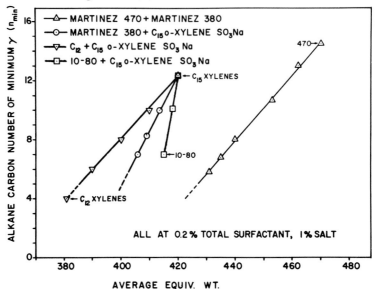

Fig. 2. *Dependence of alkane carbon number of minimum inter-facial tension (n_{min}) on the average equivalent weight of a mixture of two surfactants*

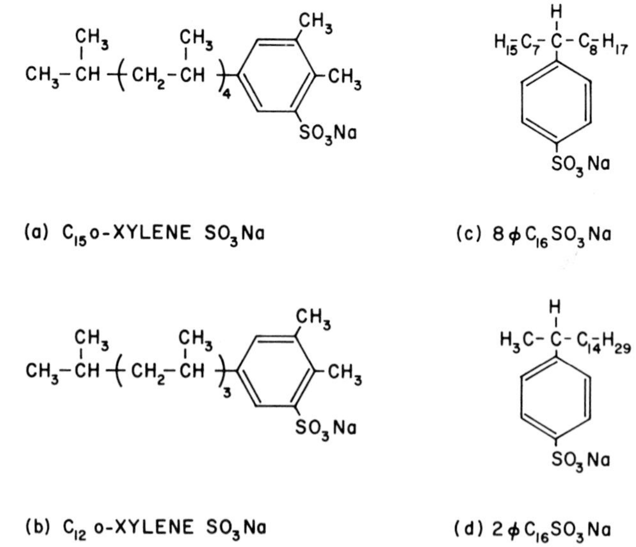

(a) C_{15} o-XYLENE SO_3Na

(b) C_{12} o-XYLENE SO_3Na

(c) $8\phi C_{16} SO_3Na$

(d) $2\phi C_{16} SO_3Na$

Fig. 3. *Representative surfactant structures (approximate for the alkyl o-xylene sulfonates)*

might contain a range of molecular weights and structures, as
with mixtures of the two Shell petroleum sulfonates Martinez
380 and Martinez 470 (11) in Figure 2. The linear molecular
weight/n_{min} relationship again holds when surfactants of completely different types are mixed, such as Martinez 380 with
the C_{15} o-xylene sulfonate (14) (Figure 2). It may not hold
for mixtures of two isomerically pure surfactants, however (14).

Of practical importance is that the composition of a surfactant mixture giving the lowest γ with any given crude oil
may now be calculated. If, for instance, the crude oil has an
EACN of 8, and a mixture of Martinez 380 and 470 is to be used
at a 0.2% by weight overall concentration in 1% by weight NaCl
solution, then Figure 2 shows that a surfactant mixture of
average molecular weight 440 is required. The surfactant
requirements can be found using,

$$\frac{\text{Wt. of Martinez 470 in g/l}}{\text{Wt. of Martinez 380 in g/l}} = \frac{470}{380} \times \frac{(440-380)}{470-440} \tag{2}$$

No other 380/470 mixture will give this low a γ against the
crude oil. Figure 2 shows that a mixture of C_{12} and C_{15} sulfonates of average molecular weight 400 will also give a low
tension with a crude oil of EACN = 8.

The linear averaging rule allows prediction of n_{min} for
any mixture of two surfactants providing the individual value
of n_{min} is known for each component. For some individual
surfactants, such as C_{12} o-xylene sulfonate and Martinez 380,
a value of n_{min} cannot be measured directly because it is not
within the liquid alkane range. Reliable values of n_{min} may
be found by extrapolation, however (14). The alkane carbon
number of minimum γ for a mixture of two surfactants is then
given by

$$(n_{min})_{MIXTURE} = X_A (n_{min})_A + X_B (n_{min})_B \tag{3}$$

Testing of Equation (3) when expanded to deal with three or
more surfactant components is not yet complete.

VI. SURFACTANT MOLECULAR WEIGHT DISTRIBUTION

To treat the surfactant molecular weight distribution as
an independent variable is not practicable at present. If the
surfactant average molecular weight is kept constant and the
range of molecular weights is changed, then the "mean structure type" is almost certain to change also, and the surfactant structure is known to be a variable to which n_{min} is
particularly sensitive. However, the presence of a distribution of surfactant molecular weights is certainly an important
factor because it introduces to the system a dependence on
overall surfactant concentration, discussed later. A

susceptibility to aging phenomena is also introduced by the presence of a range of surfactant molecular weights and structures. For example, Cash et al. (17) find that n_{min} for 10-80 shifts gradually to lighter alkanes over a period of months. The origin of this aging process is not understood, but it is only found for complex mixtures such as petroleum sulfonates.

VII.   SURFACTANT STRUCTURE

There is no single line of surfactant molecular weight against n_{min}, although lines connecting surfactant pairs of similar structure (such as two o-xylene sulfonates, or two Martinez petroleum sulfonates) have similar slopes (Figure 2). Surfactants, or surfactant mixtures, of the same average molecular weight may be far apart on the n_{min} scale. For instance, C_{15} o-xylene sulfonate and the Martinez 380-470 mixture of the same average molecular weight (420) are nine alkane units apart. Doe and Wade (16) have shown that shifts of this magnitude are due to differences in surfactant molecular structure. Shifts in n_{min} are seen when a series of isomers of n-hexadecyl benzene sodium sulfonate (molecular weight 404) are studied. There are eight possible para-isomers, of which two, the 8-phenyl hexadecane sulfonate, and the much less highly branched 2-phenyl isomer, are illustrated in Figures 3(c) and 3(d). Interfacial tensions of each isomer against the range of alkanes are shown in Figure 4. These results are for a system containing 0.07% surfactant by weight and 0.3% sodium chloride. Isopentanol (2% by volume) is also added as a cosurfactant in order to increase the solubility of the 3 and 2 isomers. The 1-phenyl isomer is not soluble. In isopentanol-free solution, the trend of Figure 4 is repeated qualitatively, but not quantitatively, for isomers remaining soluble.

Figure 4 shows that as the branching of the alkyl chain increases from the 2 isomer towards the 8 isomer, there is a steady movement of the value of n_{min} to higher alkanes (at a fixed molecular weight). This result probably explains the differences between surfactants of less well-defined structure. For example, if surfactants of molecular weight 415 are compared in Figure 2, the o-xylene sulfonate mixture is at the highest n_{min}. The o-xylene sulfonates have highly branched main alkyl groups (Figure 3(a)) because of their method of manufacture, and they also have two additional alkyl groups, albeit only methyls, on the benzene ring. Witco 10-80 has a lower n_{min} than the xylene sulfonate mix, and therefore should have a less highly branched alkyl chain structure, on the average, though this remains to be proven. Further work on structural effects must rely heavily on the study of surfactants of absolutely known structure. The structure, and even the molecular weights of the petroleum sulfonates cannot be

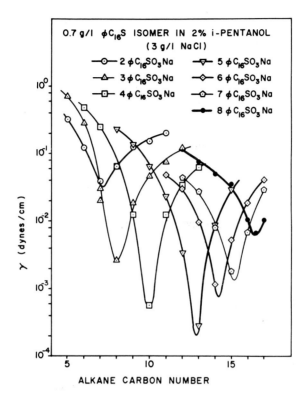

Fig. 4. Interfacial tensions of n-alkanes with isomers of phenyl hexadecane sulfonate

defined well enough--the manufacturers' quoted equivalent weights will be below the true molecular weights if disulfonates are present. This makes petroleum sulfonates largely unsuitable for study of n_{min} shifts at constant molecular weight.

The noted effect of branching is of great interest in selecting surfactants for a practical surfactant flood system. To achieve a low tension with a crude oil of given EACN, the choice lies between a high molecular weight, straight alkyl chain type of structure and a low molecular weight highly branched structure. The branched surfactant may have several advantages. It will be more soluble, so that it may be used in higher concentration if required, and its increased solubility will probably lead to decreased adsorption on the reservoir rock. It will also block a much greater surface area of rock than will an adsorbed, straight-chain molecule. Lower molecular weight surfactants are also far more tolerant of increasing salt concentration, as is discussed below.

VIII.   SURFACTANT CONCENTRATION

The concentration of surfactant in a solution containing only one surfactant molecular species may be varied over a wide range without affecting n_{min} (at constant salinity, etc.). For example, 8-phenyl n-hexadecane sodium sulfonate (Figure 3(c)) has a fixed n_{min} at any concentration down to its c.m.c. (16). For C_{15} o-xylene sulfonate at 1% salt, n_{min} stays fixed at 12.4 at all surfactant concentrations from 1% down to about 0.01% (Figure 5). The value of n_{min} at 0.01% is slightly shifted, and the minimum γ slightly raised (at 6×10^{-3} dyne/cm) when compared with results at higher concentrations. For

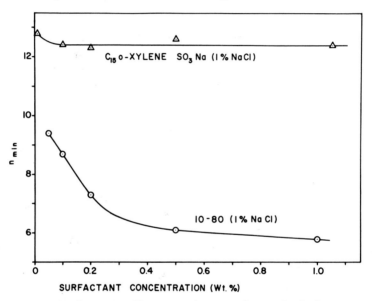

Fig. 5.   Variation of alkane carbon number of minimum tension (n_{min}) with surfactant concentration at 1.0 wt.% NaCl

C_{12} o-xylene sulfonate in 1% salt and between concentrations of 1% and 0.2%, n_{min} is at 4 (tested against butyl benzene, equivalent to butane).

As n_{min} for C_{12} o-xylene sulfonate is at 4, and n_{min} for C_{15} o-xylene sulfonate is at 12.4, a 1:1 molar mixture should give a minimum γ at 8.2, according to Equation (3). In fact, n_{min} for this mixture is at 8.0 for the concentration range 1% to 0.2%. This shows that the linear relationship between mixture average molecular weight and n_{min} holds over a wide range of concentrations. Also, if both components in the mixture are safely above some critical concentration, n_{min} for the mixture will not be dependent on the overall concentration.

In the low, overall concentration region, n_{min} for the
1:1 molar mixture of C_{12} and C_{15} xylene sulfonates becomes
concentration dependent. As the concentration decreases to
0.01%, n_{min} shifts from 8 to 12.6, which is very near the
C_{15} o-xylene sulfonate line. The C_{12} xylene sulfonate now
seems ineffective compared with the C_{15} xylene sulfonate.
This may be a reflection of the relative adsorptions at the
interface, the C_{15} o-xylene predominating because it is still
above a critical concentration required for strong adsorption,
whereas the C_{12} o-xylene sulfonate is not. This idea requires
that the "critical concentration" mentioned is higher for C_{12}
than for C_{15} xylene sulfonate, and it may be relevant here
that Klevens (18) has shown c.m.c.'s to vary inversely with
the alkyl group chain length.

The exploratory xylene sulfonate results offer an explanation of the concentration dependences found with petroleum
sulfonates. The shift of n_{min} with changing concentration of
Witco 10-80, at 1% salt, is shown in Figure 5 (data taken from
(15)). As a range of molecular weights and structures are
present, the components of 10-80 will have a range of critical
concentration values. The concentration dependence noted in
Figure 5 may be explained if, on the average, high molecular
weight species or groups of species reach their critical concentrations at lowest overall concentration. Molecules active
in the interface at the lowest concentrations then have a high
average molecular weight and, noting the trends in Figure 2,
a high value of n_{min} will be obtained. At higher overall
concentrations, steadily lower molecular weight species reach
their individual critical concentrations. The average molecular weight at the interface will slowly decrease so that
n_{min} decreases with increasing concentration. Some species,
of very low molecular weight and present only in small pro-
portion, will need a very high overall concentration before
they can reach their critical concentrations. Therefore, the
10-80 line only approaches asymptotically a concentration independent region.

IX. ELECTROLYTE CONCENTRATION AND TYPE

Several studies have now been reported on the effect of
electrolyte concentration in systems producing low interfacial
tensions (11,15,19,20), with sodium chloride being almost ex-
clusively the electrolyte used. Wilson, Murphy and Foster
(19) have found that the salinity range in which a surfactant
is interfacially most active is primarily a function of surfactant molecular weight. They suggest that if the molecular
weight is too low, the solubility may be too high for strong
adsorption at the interface, and a low γ will not be obtained
unless electrolyte is added to decrease the solubility to some

required lower level. More electrolyte will be required if
the molecular weight is low. Therefore, as the molecular
weight of the surfactant decreases, the optimum salinity for
production of a low γ against a given crude oil should increase, which is the result found by Wilson et al. (19).
Exactly the same conclusion is reached by Healy, Reed and
Stenmark (21), and by Puerto and Gale (20) in their studies
of optimum salinities for production of low γ values in
microemulsion systems.

Use of the alkane model in salinity studies (11,15) can
produce additional information, and some recent results are
given in Figure 6. Here, three alkyl o-xylene sulfonates, of
different molecular weight but of the same basic structure
(approximately as in Figures 3(a) and (b)), are compared at

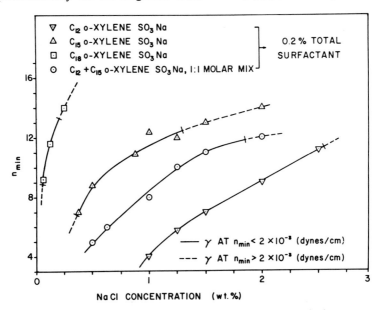

*Fig. 6.   Variation of alkane carbon number of minimum tension
(n_{min}) with concentration of NaCl at 0.2 wt.% surfactant*

0.2% concentration. In the alkane model, the equivalent to
maintaining a crude oil at a fixed minimum γ is to maintain
the value of n_{min} constant at the EACN of the oil. If n_{min}
is to be kept constant, at nonane in Figure 6, for example,
the NaCl concentration requirements to give a minimum γ with
C_{12}, C_{15} and C_{18} o-xylene sulfonates are $C_{12} > C_{15} > C_{18}$,
which is in accord with other reported results (19-21). As
crude oils have different EACN's, each will require a different
salinity at a fixed surfactant concentration and type, which
has also been noted by Wilson, Murphy, and Foster (19).

One surfactant mixture is included in Figure 6, a 1:1 mole ratio of C_{12} and C_{15} o-xylene sulfonates at 0.2% total concentration. This falls about midway between the curves of the individual components, so that the linear relationship between surfactant molecular weight and n_{min} is shown to hold, at least approximately, over a wide range of salinity.

The NaCl tolerances shown in Figure 6 are particularly interesting. The highest molecular weight surfactant, the C_{18} o-xylene sulfonate, is by far the least NaCl tolerant. Although very little NaCl is required to produce a low tension, the low γ is quickly lost as the NaCl concentration is increased further. Moreover, the slope of the curve is very steep, so that only a very small NaCl concentration fluctuation would be enough to change n_{min} drastically, and to lose the low γ with a given crude oil. The C_{15} o-xylene sulfonate, with a lower molecular weight, requires more NaCl to produce the first low tension, but low tensions are maintained over a larger NaCl concentration range, and the curve is less steep, so that the salt tolerance may be said to be increased. The C_{12} o-xylene sulfonate is tolerant of NaCl over a still wider concentration range, though the slope is not decreased further. The 1:1 molar mixture of C_{12} and C_{15} o-xylene sulfonates is approximately intermediate to the single components in behavior.

The C_9 o-xylene sulfonate is off the low end of the alkane range at low NaCl concentrations, but at high concentrations it may be moved onto the bottom part of the alkyl benzene scale. At 3% salt a minimum is seen at toluene, which is equivalent to methane as an oil phase, so that $n_{min} = 1$. At 4% salt, $n_{min} = 3$, and at 5% salt, $n_{min} = 3.4$. All these minima are only at 10^{-2} dyne/cm. An interesting feature here is that precipitation of surfactant and/or NaCl occurs in all three systems, but despite this, fairly low γ's are found, and the minimum is still shifting in its usual direction with increasing NaCl.

This may be contrasted with the behavior of the higher molecular weight o-xylene sulfonates, which lose their low tensions completely well before the precipitation stage. For instance, C_{12} o-xylene sulfonate gives a minimum tension of 1.5×10^{-3} dyne/cm at 2.5% NaCl, but 2 dynes/cm at 2.625% NaCl. It seems possible that the loss of a low γ is associated with a sudden change in the nature of the interfacial phase when the electrolyte concentration is increased beyond a certain critical value, and that the critical electrolyte concentration decreases as the surfactant molecular weight increases.

The effect of changing electrolyte type has not been studied extensively. Wilson et al. (19) find that the chloride anion can produce a lower γ than several other anions. The replacement of Na^+ by Ca^{++} by ion exchange from reservoir clays may have an effect on oil displacement, though Healy, Reed and Stenmark (21) found little effect on optimal salinity or on γ produced when a 10:1 mixture of $NaCl:CaCl_2 2H_2O$ was used instead of NaCl alone.

X.   TEMPERATURE

The effect of temperature is important because oil field temperatures range from 50°C upwards, whereas most research data is gathered at ambient laboratory temperatures. Healy, Reed and Stenmark (21) have shown that increasing the temperature increases the optimal salinity for production of a low interfacial tension with a microemulsion, at fixed surfactant molecular weight, concentration and structure. Alkane scans with a number of surfactants have shown a shift in n_{min} to lighter alkanes as the temperature is increased. Shifts measured so far include Δn_{min} = -0.08/°C for Martinez 470 over the range 27°C to 50°C, and the same for C_{15} o-xylene sulfonate over the range 27°C to 70°C. However, a smaller shift of -.04/°C has been noted for a mixture of C_{15} o-xylene sulfonate with 6-phenyl n-dodecane sodium sulfonate, so a universal temperature coefficient seems unlikely. Although it is possible that the effect of changing temperature with a crude oil/surfactant system will not be the same as with an alkane/surfactant system because the crude oil contains a wide range of molecular weights, no temperature dependence of the EACN of any crude oil has yet been observed. For example, the EACN of Horseshoe Gallup has been found to be 8.2 both at 28°C and at 50°C. This needs to be confirmed before inferences are taken from the observed shifts of alkane n_{min} with temperature. It is worth noting, however, that increasing the temperature, and moving n_{min} downwards, may be counteracted by increasing the NaCl concentration, which moves n_{min} upwards (Figure 6). This gives rise to the result of Healy et al. (21) mentioned above, that in order to maintain a fixed oil phase at a low γ while the temperature is increased, the salt concentration must be increased also. Of course, high salinity is not a necessity for a low γ at high temperatures. For instance, a surfactant with a molecular weight slightly too high to give a low γ against a particular oil at ambient temperature will give a low γ with that oil at a higher temperature. Any of the trends noted in the concluding table may be utilized in this way, and all of them may be made quantitative using the alkane model.

XI.   OIL PHASE STRUCTURE AND MOLECULAR WEIGHT

With the pure alkane series as the oil phase, the alkane carbon number of minimum γ (n_{min}) varies linearly with the

surfactant molecular weight. As n_{min} is directly proportional
to the alkane molecular weight, then a fixed ratio of surfactant molecular weight to oil phase molecular weight is required
to give a low interfacial tension. The ratio will change, of
course, if the surfactant structure or concentration, salinity
or oil phase structure type is changed.

Unlike the surfactants, which show a marked dependence of
interfacial tension behavior on the degree of branching of the
alkyl chain, the oil phase seems unaffected by this type of
isomerization. For example, iso-octane behaves exactly like
octane (11). The behavior of hydrocarbon mixtures therefore
depends only on the average molecular weight and the ratio of
aromatic:aliphatic:alicyclic hydrocarbons present. These
basic factors determine the EACN of a crude oil.

XII.   EFFECT OF COSURFACTANT (COSOLVENT)

A number of the low tension systems discussed here contain
a cosolvent, usually an alcohol (20,21). This helps to solubilize the surfactant at high concentrations. Other workers
do not use a cosolvent, because it further increases the complexity of the system (11,19). None of the alkane model systems contain a cosolvent, except in the pure isomer studies of
Doe and Wade (16), where isopentanol is used as a solubility
aid.

Generally, when alcohol is added to a system, the alkane
of minimum γ will change. Addition of a higher molecular
weight alcohol, such as isopentanol, will produce a system
with a higher n_{min} than addition of the same volume of a lower
molecular weight alcohol, such as methanol. The trends noted
in the concluding table are unaffected qualitatively by the
addition of alcohol--for example, n_{min} will shift upwards if
the salinity is increased whether alcohol is present or not.
The size of the shifts observed are changed by the addition
of alcohol, however.

Puerto and Gale (20) report that the optimal salinity for
producing a low γ is dependent on the alcohol type. If the
alcohol is highly oil soluble, such as n-pentanol, then the
optimum salinity is lower than with a highly water soluble
alcohol such as methanol, n-butanol being intermediate in
effect. This would seem to be because methanol increases the
solubility of a given molecular weight surfactant most, which
means more NaCl is required to decrease the solubility back
down to a level where strong adsorption occurs at the interface.

XIII. CONCLUSIONS

Understanding of the system variables involved has now
reached the point where choosing a surfactant phase formulation to give a low interfacial tension against any particular

crude oil is possible. If the alkane model is used, trial
and error testing is minimal. Many different surfactants
may be used with each oil, the extreme types being of high
molecular weight with unbranched alkyl chain structures, or
of lower molecular weight, with highly branched alkyl groups.
Of these, the least sensitive to slight fluctuations in system variables, and therefore the most practicable, seems to
be the highly branched, low molecular weight type.

The alkane model allows study of well-defined systems,
with considerable scope for varying system parameters. The
effect of each system variable on the carbon number of the
alkane of minimum γ (n_{min}) is reported in the text and the
results are summarized qualitatively in the following table.

TABLE I

Variable Increased	Effect on n_{min}		
	Increases	None	Decreases
Surfactant Mol. Wt.	✓		
Branching of Surf. alkyl structure	✓		
Conc. of pure surfactant		✓	
Conc. of complex surf. mixture			✓
Electrolyte concentration	✓		
System temperature			✓
Age of system with pure surfactant		✓	
Age of system with complex surf.			✓

XIV. ACKNOWLEDGEMENTS
    The authors wish to express their appreciation of the
continued interest and support of the National Science
Foundation and the Robert A. Welch Foundation.

XV. LITERATURE CITED
1.   Docher, T. M., Wise, F. A., "Enhanced Crude Oil Recovery
      Potential--An Estimate," Paper SPE 5800, presented at
      the SPE Symposium on Improved Oil Recovery, March 22-
      24, 1976.
2.   Gogarty, W. B., Tosch, W. C., J. Pet. Tech. 20, 1407
      (1968).
3.   Gogarty, W. B., Kinney, W. L., Kirk, W. B., J. Pet. Tech.
      22, 1577 (1970).

4.   Danielson, H. H., Paynter, W. T., Milton, H. W., J. Pet.
      Tech. 28, 129 (1976).
5.   Healy, R. N., Reed, R. L., Soc. Pet. Eng. J. 14, 491
      (1974).
6.   Healy, R. N., Reed, R. L., "Immiscible Microemulsion
      Flooding," Paper SPE 5817, Presented at the SPE
      Symposium on Improved Oil Recovery, March 22-24, 1976.
7.   Melrose, J. C., Brandner, C. F., J. of Canadian Petr.
      Tech. 54-62 (Oct. - Dec. 1974).
8.   Taber, J. J., Soc. Pet. Eng. J. 9 (No. 1), 3 (1969).
9.   Cayias, J. L., Schechter, R. S., Wade, W. H., "The
      Utilization of Petroleum Sulfonates for Producing
      Low Interfacial Tensions Between Hydrocarbons and
      Water," J. Coll. Int. Sci. (1976).
10.   Cash, R. L., Cayias, J. L., Fournier, G., MacAllister,
      D. J., Schares, T., Schechter, R. S. and Wade, W. H.,
      "The Application of Low Interfacial Tension Scaling
      Rules to Binary Hydrocarbon Mixtures," J. Coll. Int.
      Sci. (1976).
11.   Cash, R. L., Cayias, J. L., Fournier, G., Jacobson,
      J. K., Schares, T., Schechter, R. S. and Wade, H. H.,
      "Modeling Crude Oils for Low Interfacial Tension,"
      Paper SPE 5813, Presented at the SPE Symposium on
      Improved Oil Recovery, March 22-24, 1976.
12.   American Society of Testing Materials, Phil., Penn.,
      "ASTM Standards," D-2548-69, p. 656 (1969).
13.   Cayias, J. L., Schechter, R. S. and Wade, W. H., ACS
      Symposium Series No. 8, p. 235 (1975).
14.   Jacobson, J. K., Morgan, J. C., Schechter, R. S. and
      Wade, W. H., "Low Interfacial Tensions Involving
      Mixtures of Surfactants," Paper SPE 6002, presented
      at the SPE-AIME 51st Annual Meeting, New Orleans,
      La., Oct. 3-6, 1976.
15.   Cash, R. L., Cayias, J. L., Fournier, G., Jacobson, K. J.,
      LeGear, C. A., Schares, T., Schechter, R. S. and
      Wade, W. H., "Low Interfacial Tension Variables,"
      Proceedings of the American Oil Chemists' Society,
      Short Course at Hershey, Pa., June, 1975.
16.   Doe, P. H. and Wade, W. H., "Alkyl Benzene Sulfonates
      for Producing Low Interfacial Tensions Between
      Hydrocarbons and Water," J. Colloid Interface Sci.
      (1977).
17.   Cash, R. L., Cayias, J. L., Hayes, M., MacAllister, D. J.,
      Schares, T. and Wade, W. H., "Surfactant Aging: A
      Possible Detriment to Tertiary Oil Recovery," J. Pet.
      Tech. 985 (Sept. 1976).

18. Klevens, H. B., J. Phys. Colloid Chem. (J. Phys. Chem.)
     52, 130 (1948).

19. Wilson, P. M., Murphy, L. C. and Foster, W. R., "The
     Effects of Sulfonate Molecular Weight and Salt Con-
     centration on the Interfacial Tension of Oil-Brine-
     Surfactant Systems," Paper SPE 5812, Presented at the
     SPE Symposium on Improved Oil Recovery, March 22-24,
     1976.

20. Puerto, M. C. and Gale, W. W., "Estimation of Optimal
     Salinity and Solubilization Parameters for Alkyl
     Orthoxylene Sulfonate Mixtures," Paper SPE 5814,
     ibid.

21. Healy, R. N., Reed, R. L., and Stenmark, D. G., "Multi-
     phase Microemulsion Systems," Paper SPE 5565, Presented
     at 50th Annual Fall Meeting of the SPE, Dallas, Texas
     (1975).

MOLECULAR THEORIES OF INTERFACIAL TENSION

K. E. Gubbins and J. M. Haile
University of Florida

I. ABSTRACT

A review is given of molecular theories for the interfacial tension, and for the distribution of molecules in the
interface. A summary of the classical thermodynamics of
fluid interfaces is followed by rigorous equations relating
interfacial properties to the intermolecular potential energy.
Approximations to these rigorous results are then described.

II. SCOPE

The interfacial tension can be exactly related to the
intermolecular forces through statistical thermodynamics.
Such an equation was derived by Kirkwood and Buff in 1949 (1),
but is difficult to use because one must first know how the
molecules are distributed through the interface. The
Kirkwood-Buff equation is valid only for simple spherical
molecules. However, much more general equations have recently been derived, which apply for the more complex molecules of major interest in oil recovery processes. These
equations are again difficult to use for numerical work.

Because of these difficulties, a variety of approximate
calculation procedures have been developed, and we review
these below. These methods differ as to the amount of numerical effort involved and the degree of approximation required
(these two aspects being, in general, unrelated). The theories
reviewed are all based to some extent in statistical thermodynamics, but also involve some degree of empirical fitting
in most cases.

III. CONCLUSIONS AND SIGNIFICANCE

It is possible to write down exact relationships for the
thermodynamic properties and molecular distributions in the
interface. These expressions involve the intermolecular pair
potential and the interfacial distribution functions. Several
of the expressions have been derived only very recently, and
numerical calculations have not yet been carried out.

For liquids away from their critical points, the most
satisfactory approach is probably perturbation theory. This
gives accurate results, and the equations may be put into

119

simpler forms by introducing approximations for the distribution functions. For liquids nearer the critical point, the approach initially used by van der Waals (2), and pursued recently by others, is appropriate.

Much remains to be done to develop reliable methods for predicting interfacial tension for liquids of more complex molecules, and especially for mixtures. The computer simulation studies carried out since 1974 have been a great spur to the development of the theory; it is to be expected that such simulations will play an even more important role in the future. The next few years should see such simulations carried out for more complex polyatomic liquids, and for mixtures.

IV.   INTRODUCTION

In this paper we review the current status of the statistical thermodynamics of fluid surfaces (gas-liquid and liquid-liquid). The surfaces considered are three-dimensional (as opposed to the two-dimensional approximations commonly used for adsorption on solid surfaces), and the properties discussed are equilibrium ones. They include the surface thermodynamic properties--e.g., interfacial tension (γ), superficial internal energy (u^S), adsorption (Γ_i), etc.--and also the molecular distribution functions in the interface. For a fluid of spherical molecules the distribution functions of principal interest are the interfacial density profiles for each component, $\rho_i(z_1)$, and the pair correlation function $g_{ij}(z_1 \underline{r}_{12})$; here z is the direction perpendicular to the plane interface, $\underline{r}_{12}$ is the vector from the center of molecule 1 to molecule 2, and i and j are components of the mixture. For nonspherical molecules (including surfactants, polymers, etc.), the corresponding distribution functions are $\rho_i(z_1\omega_1)$ and $g_{ij}(z_1\underline{r}_{12}\omega_1\omega_2)$, where ω_1 and ω_2 represent angles giving the orientation of molecules 1 and 2; for linear molecules only one angle, $\omega_1 = \theta_1$, is needed, whereas nonlinear molecules require two angles, $\omega_1 = \theta_1\chi_1$. For surfactant and similar molecules it is clear that the orientation correlations will be very strong, so that $\rho_i(z_1\omega_1)$ will be sharply peaked when plotted vs. ω_1.

Surface properties may be studied by three methods--theory, computer simulation, or experiment. Only the first two methods are covered in this review, but it is of interest to briefly consider here the interaction of experiment with the other two approaches. Experimental studies to test theories for surface structure and properties fall into two classes. The first of these are measurements of the surface thermodynamic properties. Measurements of the interfacial tension alone provide only a relatively weak test of any given theory; if the measurements are extended to include such properties as the superficial internal energy, entropy, and adsorption, however, the test is much stronger. The necessary experimental measurements to obtain these properties are dictated by the classical thermodynamics of interfaces (see Section V). For a pure liquid it is only necessary to measure interfacial tension as a function of temperature [cf. Equations (21) and (22)]. The second class of experiments are those involving the scattering of light (and other types of radiation--e.g., neutrons or electrons) from the liquid surface. Such measurements provide more detailed information about the surface structure, molecular orientations, etc., and may be used to test theories for the distribution functions. The optical experiments that have been performed include: (a) measurements of the ellipticity of light reflected from the surface; the coefficient of ellipticity depends on the dielectric profile in the interface (3,4); (b) inelastic light scattering from thermal excitations in the interface (5); and (c) measurements of the reflectivity of the interface (6,7). Method (c) can, in principle, yield the density profile $\rho(z)$ provided that the measurements are carried out over a sufficient range of wavelengths.

Section V below summarizes the most important relationships of the classical thermodynamics of surfaces. The properties of interest are those associated with the surface region, or "phase", i.e., the region in which the density departs measurably from that of either of the bulk phases. In order to give a precise definition to this region and its properties the concept of the Gibbs dividing surface is introduced. Section VI gives rigorous expressions for the interfacial tension and distribution functions in terms of the intermolecular potential energy. These equations are intractable as they stand, and approximations must be introduced before numerical calculations can be made. These approximate theories are discussed in Section VII. Finally, some of the recent computer simulation studies are reviewed in Section VIII.

The discussion of the theory given here is of necessity brief. For further details, and derivation of the equations in Sections VI and VII, the interested reader should consult reviews by Buff (8), Buff and Lovett (3), Ono and Kondo (9), Croxton (10,11), and Toxvaerd (12,13).[1]

---

[1] Note added in proof: The following review has just recently appeared in print: R. C. Brown and N. H. March, "Structure and Excitations in Liquid and Solid Surfaces," Phys. Repts. (Section C of Phys. Letters), 24, 77 (1976).

V. CLASSICAL THERMODYNAMICS

In this section we summarize some of the important thermo-dynamic relationships for interfacial properties. The formulation adopted is that due to Gibbs (14); more comprehensive treatments are given in reviews (14,15,16,9). The initial discussion is for a plane interface; however, the modifications necessary for spherical interfaces are commented on at the end of this section.

Consider an equilibrium fluid mixture of m components at temperature T, volume V, containing N_1 molecules of component 1, N_2 of 2, etc. The system contains two phases, α and β (gas-liquid, liquid-liquid, or "gas-gas") separated by a plane surface of area S. The equation for the differential of the Helmholtz free energy, A, then defines the interfacial tension γ,

$$dA = - SdT - PdV + \gamma dS + \sum_{i=1}^{m} \mu_i dN_i \qquad (1)$$

where S, P and $\mu_i \equiv (\partial A / \partial N_i)_{T,V,S,N_{j \neq i}}$ are entropy, pressure and chemical potential of i, respectively. The term γdS represents the work that must be done to increase the surface by dS, keeping the volume fixed. Thus γ is obtained from the free energy as

$$\gamma = \left(\frac{\partial A}{\partial S} \right)_{T,V,N_i} \qquad (2)$$

The definition of the interfacial tension in (2) is independent of the location of any mathematical dividing surface between the two phases. However, in discussing other properties of the surface layer it is convenient to follow Gibbs in introducing the concept of such a surface. We choose a set of axes such that this surface lies in the xy plane at z = 0; the z-axis is perpendicular to the interfacial layer. The location of the dividing surface is arbitrary, but is usually chosen to lie in the transition zone. Having chosen a particular dividing surface, the total volume V may be divided into two volumes V^α and V^β. In general V^α contains the bulk phase α plus a small amount of material in the surface zone; similarly V^β contains bulk phase β plus the rest of the material in the surface. It is convenient to imagine a hypothetical system in which the surface zone is vanishingly thin--i.e., both bulk phases α and β are homogeneous right up to the dividing surface (see Figure 1). Let the number of molecules of component i in the hypothetical system be $\tilde{N}_i^\alpha$ and $\tilde{N}_i^\beta$ for the two phases, respectively, and let N_i be the total

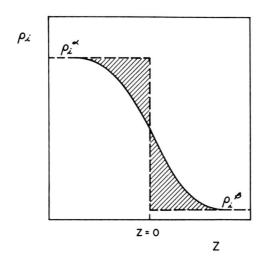

*Fig. 1.   Choice of the Gibbs dividing surface. If the
choice $\Gamma_i = 0$ is made (cf. Equation (5)) the
two shaded areas are equal*

number of i molecules in the real system. In general N_i is
not just the sum of $\tilde{N}_i^\alpha$ and $\tilde{N}_i^\beta$, but

$$N_i = \tilde{N}_i^\alpha + \tilde{N}_i^\beta + N_i^S \qquad (3)$$

where N_i^S is the number of i molecules adsorbed at the surface.
The <u>adsorption</u> of i, Γ_i is defined as this number divided by
the surface area,

$$\Gamma_i \equiv N_i^S/S = \int_{-\infty}^{0} [\rho_i(z) - \rho_i^\alpha]dz + \int_{0}^{\infty} [\rho_i(z) - \rho_i^\beta]dz \qquad (4)$$

where $\rho_i(z)$ is the number density of i as a function of z, and
ρ_i^α and ρ_i^β are the number densities of the two bulk phases.
The quantities Γ_i are nonvanishing in general and their values depend on the choice of dividing surface. However, the
dividing surface may be chosen to make one of the Γ_i vanish.
Thus, setting $\Gamma_1 = 0$ corresponds to a choice of z = 0 defined by

$$\int_{-\infty}^{0} [\rho_1(z) - \rho_1^\alpha]dz + \int_{0}^{\infty} [\rho_1(z) - \rho_1^\beta]dz = 0 \qquad (5)$$

so that the two shaded areas in Figure 1 are then equal. With this choice Γ_2, Γ_3 etc. are nonzero. It follows from (5) that for a pure fluid Γ may be set equal to zero, and such a equi-molar dividing surface is almost always used in that case.

The introduction of the Gibbs dividing surface enables clearly defined values to be assigned to thermodynamic functions for the interfacial zone. The interfacial zone may be treated as a thermodynamic phase, whose properties must obey well-defined relations. The superficial Helmholtz free energy, internal energy and entropy are defined by analogy to (3),

$$A^S = A - A^\alpha - A^\beta \tag{6}$$

$$U^S = U - U^\alpha - U^\beta \tag{7}$$

$$S^S = S - S^\alpha - S^\beta \tag{8}$$

Clearly these superficial quantities are related by

$$A^S = U^S - TS^S \tag{9}$$

We now derive the expressions relating the superficial properties A^S, U^S and S^S to interfacial tension and adsorption. The relation between A^S and γ may be found as follows. If the system is increased in volume, keeping the temperature, pressure and composition fixed, we find from (1)

$$A = - PV + \gamma S + \sum_i \mu_i N_i \tag{10}$$

From this and the corresponding expressions for A^α and A^β, together with (6),

$$A^S = \gamma S + \sum_i \mu_i N_i^S \tag{11}$$

The expression for superficial entropy is found from Equations (1), (6) and (11) as

$$S^S = - \left(\frac{\partial A^S}{\partial T}\right)_{V,S,N_i^\alpha,N_i^\beta,N_i^S} = - S \left(\frac{\partial \gamma}{\partial T}\right)_{\{x^\alpha\}} - \sum_{i=1}^{m} N_i^S \left(\frac{\partial \mu_i}{\partial T}\right)_{\{x^\alpha\}} \tag{12}$$

where $\{x^\alpha\} \equiv x_1^\alpha, x_2^\alpha \ldots x_{m-1}^\alpha$ are the mole fractions and we have taken T, x_1^α, $x_2^\alpha \ldots x_{m-1}^\alpha$ to be the independent variables that determine γ and μ_i. Since μ_i is constant throughout the equilibrium system,

$$\left(\frac{\partial \mu_i}{\partial T}\right)_{\{x^\alpha\}} = \left(\frac{\partial \mu_i}{\partial T}\right)_{P,\{x^\alpha\}} + \left(\frac{\partial \mu_i}{\partial P}\right)_{T,\{x^\alpha\}} \left(\frac{\partial P}{\partial T}\right)_{\{x^\alpha\}}$$

$$= - \bar{s}_i^\alpha + \bar{v}_i^\alpha \left(\frac{\partial P}{\partial T}\right)_{\{x^\alpha\}} \tag{13}$$

where $\bar{s}_i^\alpha$ and $\bar{v}_i^\alpha$ are the partial molecular entropy and volume, respectively, of component i in phase α. From (12) and (13)

$$S^s = - S \left(\frac{\partial \gamma}{\partial T}\right)_{\{x^\alpha\}} + \sum_{i=1}^{m} N_i^s \left[\bar{s}_i^\alpha - \bar{v}_i^\alpha \left(\frac{\partial P}{\partial T}\right)_{\{x^\alpha\}}\right] \tag{14}$$

The corresponding expression for superficial internal energy is obtained from (9), (11) and (14) as

$$U^s = S \left[\gamma - T \left(\frac{\partial \gamma}{\partial T}\right)_{\{x^\alpha\}}\right] + \sum_{i=1}^{m} N_i^s \left[\bar{h}_i^\alpha - T\bar{v}_i^\alpha \left(\frac{\partial P}{\partial T}\right)_{\{x^\alpha\}}\right] \tag{15}$$

The superficial quantities A^s, S^s and U^s are extensive and are proportional to S. It is convenient to introduce corresponding superficial quantities per unit area, which are intensive,

$$a^s \equiv A^s/S = \gamma + \sum_{i=1}^{m} \Gamma_i \mu_i \tag{16}$$

$$s^s \equiv S^s/S = - \left(\frac{\partial \gamma}{\partial T}\right)_{\{x^\alpha\}} + \sum_{i=1}^{m} \Gamma_i \left[\bar{s}_i^\alpha - \bar{v}_i^\alpha \left(\frac{\partial P}{\partial T}\right)_{\{x^\alpha\}}\right] \tag{17}$$

$$u^s \equiv U^s/S = \gamma - T \left(\frac{\partial \gamma}{\partial T}\right)_{\{x^\alpha\}} + \sum_{i=1}^{m} \Gamma_i \left[\bar{h}_i^\alpha - T\bar{v}_i^\alpha \left(\frac{\partial P}{\partial T}\right)_{\{x^\alpha\}}\right] \tag{18}$$

Equations (16) to (18) relate the superficial properties to quantities (γ, Γ_i, μ_i etc.) that may be measured experimentally, and hold for any choice of the dividing surface. A particular choice of dividing surface may be made to simplify one, but not all, of these equations. Thus if the dividing surface is chosen to satisfy

$$\sum_{i=1}^{m} \Gamma_i \mu_i = 0 \tag{19}$$

then (16) simplifies to $\gamma = a^s$. For the special case of pure fluids, (16)-(18) simplify considerably if the equimolar dividing surface is used,

$$a^s = \gamma \tag{20}$$

$$s^s = - \frac{d\gamma}{dT} \tag{21}$$

$$u^s = \gamma - T \frac{d\gamma}{dT} \tag{22}$$

For gas-liquid surfaces, γ decreases with rise in temperature for almost all substances (liquid zinc at low temperature is an exception), so that the superficial entropy is positive.

It remains to show how the surface tension and adsorption are related. Such a relation may be derived as follows. Differentiation of (11) gives

$$dA^s = \gamma dS + Sd\gamma + \sum_{i=1}^{m} \mu_i dN_i^s + \sum_{i=1}^{m} N_i^s d\mu_i \tag{23}$$

However, from (1) and the corresponding expressions for dA^α and dA^β,

$$dA^s = - s^s dT + \gamma dS + \sum_{i=1}^{m} \mu_i dN_i^s \tag{24}$$

Comparison of (23) and (24) gives the Gibbs-Duhem equation for plane surfaces,

$$d\gamma + s^s dT + \sum_{i=1}^{m} \Gamma_i d\mu_i = 0 \tag{25}$$

which relates the Γ_i to changes in the interfacial tension, temperature, and composition of one of the phases. Equation (25) is a general form of the Gibbs adsorption equation. As an example of the use of this equation, consider a binary mixture of solvent 1 with solvent 2. The dividing surface

is chosen so that $\Gamma_1 = 0$ (cf. Equation (5)), and at fixed T
(25) yields

$$\Gamma_2 = - \left(\frac{\partial \gamma}{\partial x_2^{\alpha}}\right) \left(\frac{\partial \mu_2}{\partial x_2^{\alpha}}\right)_T^{-1} \qquad (26)$$

Thus, solutes which lower the interfacial tension will be concentrated at the interface ($\Gamma_2 > 0$), and vice versa. Equation
(26) provides the connection between Γ_2 and experimentally
measurable quantities.

The above equations apply for plane interfaces, and require modification for curved surfaces. Account must now be
taken of the fact that (a) the interfacial tension may depend
on the radii of curvature, and (b) the pressures in the two
phases (and hence the chemical potentials) will no longer be
equal. The definitions of the superficial properties and also
the choice of dividing surface must thus be modified. For
details the reader should consult the review of Ono and Kondo
(9). The effect of curvature on the interfacial tension can
be neglected provided that the radius of curvature is large
compared with the thickness of the interfacial region. For
liquids well away from a critical point (gas-liquid or liquid-
liquid) the thickness of the interface is of the order 10^{-9}m,
so that the interfacial tension for drops of radius much larger
than this value should be unaffected by curvature. The fact
that the pressure differs in the two phases is readily seen,
and was apparently first shown by Laplace (17). Consider a
spherical drop of radius r (phase α) immersed in phase β, and
assume that the interfacial thickness is much less than r.
The transfer of a volume dV from phase β to α requires work
$(P^{\alpha} - P^{\beta})dV$; this must equal the work to extend the surface,
γdS. Since $dV = 4\pi r^2 dr = rdS/2$, then

$$P^{\alpha} - P^{\beta} = \frac{2\gamma}{r} \qquad (27)$$

Thus the pressure P^{α} inside the drop exceeds that in the outer
β phase by $2\gamma/r$. The fugacity (chemical potential) within the
drop will also be increased. Applying

$$\frac{\partial \ell n f_i}{\partial P} = \frac{\bar{v}_i}{kT}$$

we find

$$\ln \frac{f_i(r)}{f_i(\infty)} = \frac{2\bar{v}_i \gamma}{rkT} \tag{28}$$

where k is Boltzmann constant; $f_i(r)$ is fugacity of i in the drop of radius r, and $f_i(\infty)$ is the fugacity for a plane surface. Thus the vapor pressure of liquid within a drop increases as the radius of curvature decreases.

Buff (8) and Ono and Kondo (9) have discussed the case in which interfacial tension varies with curvature.

VI.   STATISTICAL MECHANICS OF LIQUID SURFACES

A.   Thermodynamic Functions

Several methods have been used to derive rigorous equations for the interfacial tension. The most straightforward of these is the application of Equation (2) to the statistical mechanical equation for the Helmholtz free energy. For a fluid of spherical molecules

$$\gamma = -kT \frac{\partial}{\partial S} \ln \int \cdots \int dv_1 \cdots dv_N \, e^{-U/kT} \tag{29}$$

where k is Boltzmann constant, U is the total potential energy of the system due to the intermolecular forces, dv_1 is the volume element containing molecule 1, and so on. The differentiation in (29) was carried out by Buff (18), and gives

$$\gamma = \frac{1}{2} \sum_{ij} \int_{-\infty}^{+\infty} dz_1 \int_0^{\infty} dr_{12} r_{12}^2 \int d\omega_{12} \; \rho_i(z_1)\rho_j(z_1+z_{12})$$

$$\times g_{ij}(z_1 r_{12}\omega_{12}) \frac{(x_{12}^2 - z_{12}^2)}{r_{12}} \frac{du_{ij}(r_{12})}{dr_{12}} \tag{30}$$

where $\omega_{12} \equiv \theta_{12}\phi_{12}$ is the direction of the line r_{12} joining the centers of molecules 1 and 2 and $u_{ij}(r_{12})$ is the intermolecular potential energy between a molecule 1 of component i and a molecule 2 of component j. The quantity $g_{ij}(z_1 r_{12}\omega_{12})$ is the distribution function for ij pairs in the interfacial region, and is proportional to the probability that a molecule of j is at r_{12} from i, if i is at height z_1 (a precise definition of g_{ij} is given in Section B). Equation (30) assumes that the intermolecular potential is pairwise additive; the correction for three-body potential terms has been written down (13). Equation (30) was first derived by

Kirkwood and Buff (1) by considering the general expression
for the stress in a fluid, and is known as the Kirkwood-Buff
equation. It is analogous to the pressure equation (19) for
uniform fluids. The evaluation of γ from (30) requires knowledge of the density profiles $\rho_i(z)$ for each component, and
also of the pair distribution function g_{ij}; relatively little
is known of the latter function at present, although several
approximate equations have been proposed (see Section VII).
It should be noted that the integrand in (30) vanishes except
in the interfacial region, because the fluid is then uniform
in the x and z directions.

Gray and Gubbins (20) have generalized the Kirkwood-Buff
result to liquids composed of nonspherical molecules (e.g.,
polar liquids). The result is

$$\gamma = -\frac{1}{2} \sum_{ij} \int_{-\infty}^{+\infty} dz_1 \int d\underline{r}_{12} d\omega_1 d\omega_2 \; \rho_i(z_1\omega_1)\rho_j(z_1+z_{12}\omega_2)$$

$$\times g_{ij}(z_1\underline{r}_{12}\omega_1\omega_2) \left| \overline{P_2(\cos\theta_{12}) \; r_{12} \frac{\partial u_{ij}(\underline{r}_{12}\omega_1\omega_2)}{\partial r_{12}}} \right.$$

$$\left. -\frac{3}{2} \sin\theta_{12} \cos\theta_{12} \frac{\partial u_{ij}(\underline{r}_{12}\omega_1\omega_2)}{\partial\theta_{12}} \right| \tag{31}$$

where $\underline{r}_{12} \equiv (r_{12}\theta_{12}\phi_{12})$, $\omega \equiv (\phi_i\theta_i\chi_i)$ are the Euler angles
for the orientation of molecule i, $g_{ij}(z_1\underline{r}_{12}\omega_1\omega_2)$ is the
angular pair correlation function for ij pairs in the in-
homogeneous region, and $P_2(\cos\theta_{12}) = 1/2 \; (3\cos^2\theta_{12} - 1)$ is
the second Legendre polynomial; $\rho_i(z_1\omega_1)$ is proportional to
the number density of i molecules at z_1 with orientation ω_1.

An alternative expression can be derived which relates
the interfacial tension to the direct correlation function
$c_{ij}(z_1r_{12}\omega_{12})$ for the interfacial region. This function is
defined by the generalized Ornstein-Zernike equation (21)

$$h(z_1\underline{r}_{12}) = c(z_1\underline{r}_{12}) + \int dv_3 \; \rho(z_3) \; h(z_1\underline{r}_{13}) \; c(z_2\underline{r}_{23}) \tag{32}$$

where

$$h(z_1\underline{r}_{12}) = g(z_1\underline{r}_{12}) - 1 \tag{33}$$

The direct correlation function has the advantage that it is
of shorter range than the distribution function $g(z_1\underline{r}_{12})$. The
expression for the interfacial tension is then

$$\gamma = \frac{1}{4}\, kT \int\limits_{-\infty}^{+\infty} dz_1 \int\limits_{-\infty}^{+\infty} dz_2 \int\limits_{-\infty}^{\infty} dx_{12} dy_{12} \frac{d\rho(z_1)}{dz_1} \frac{d\rho(z_2)}{dz_2}$$

$$\times\, c(z_1 \underline{r}_{12})\, (x_{12}^2 + y_{12}^2) \tag{34}$$

This equation has been derived by two different methods (22, 23). It is much more general than the Kirkwood-Buff expression, since (34) requires no assumptions about pairwise additivity or the form of the intermolecular potential; it is equally valid for monatomic or polyatomic liquids. In this sense it is analogous to the compressibility equation (19).

Equations for the other thermodynamic properties of the interfacial region may be readily derived from the usual expressions of statistical thermodynamics (19), by applying the definitions of Section V. Thus the superficial internal energy u_s (cf. Equation (7)), and the component of the pressure tension P_N normal to the interface, for a mixture of spherical molecules are given by

$$u_s = \frac{3}{2}\, kT \sum_i \Gamma_i + \frac{1}{2} \sum_{ij} \int\limits_{-\infty}^{+\infty} dz_1 \int d\underline{r}_{12}\, u_{ij}(r_{12})$$

$$\times\, |\rho_{ij}(z_1 \underline{r}_{12}) - \rho_{ij}^{\alpha\beta}(r_{12})| \tag{35}$$

$$P_N = kT \sum_i \rho_i(z_1) - \frac{1}{2} \sum_{ij} \int d\underline{r}_{12} \frac{z_{12}^2}{r_{12}} \frac{du_{ij}(r_{12})}{dr_{12}} \int\limits_0^1 d\alpha$$

$$\times\, \rho_{ij}[\underline{r}_1 - \alpha \underline{r}_{12},\, \underline{r}_1 + (1-\alpha)\underline{r}_{12}] \tag{36}$$

where $\rho_{ij}(z_1 \underline{r}_{12})$ is the interfacial pair distribution function, given by

$$\rho_{ij}(z_1 \underline{r}_{12}) = \rho_i(z_1)\, \rho_j(z_2)\, g_{ij}(z_1 \underline{r}_{12}) \tag{37}$$

and $\rho_{ij}^{\alpha\beta}(r_{12})$ is the pair distribution function for the bulk α phase when $z_1 < 0$, and the corresponding function for the bulk β phase when $z_1 > 0$. In (36) α is the Kirkwood coupling parameter. It should be noted that P_n is constant throughout the system (from hydrostatic equilibrium), and so is independent of z_1. The equation for $P_T(z_1)$, the tangential component of the pressure tensor, is obtained from (36) by replacing z_{12}^2 by x_{12}^2.

B. Molecular Distribution in the Interface

In order to calculate the surface tension from (30) the distribution functions $\rho_i(z_1)$ and $g_{ij}(z_1 r_{12})$ are needed; for nonspherical molecules these are replaced by the more general functions $\rho_i(z_1\omega_1)$ and $g_{ij}(z_1 \underline{r}_{12}\omega_1\omega_2)$. These functions are defined by the equations

$$\rho_i(z_1) = \frac{N_i \int\cdot\cdot\int dv_2 dv_3 \cdot\cdot dv_N \; e^{-U/kT}}{\int\cdot\cdot\int dv_1 \cdot\cdot dv_N \; e^{-U/kT}} \tag{38}$$

$$\rho_i(z_1\omega_1) = \frac{N_i \int\cdot\cdot\int dv_2 d\omega_2 \cdot\cdot dv_N d\omega_N \; e^{-U/kT}}{\int\cdot\cdot\int dv_1 d\omega_1 \cdot\cdot dv_N d\omega_N \; e^{-U/kT}} \tag{39}$$

$$g_{ij}(z_1 r_{12}) = \frac{N_i N_j \int\cdot\cdot\int dv_3 dv_4 \cdot\cdot dv_N \; e^{-U/kT}}{\rho_i(z_1)\rho_j(z_2) \int\cdot\cdot\int dv_1 \cdot\cdot dv_N \; e^{-U/kT}} \tag{40}$$

$$g_{ij}(z_1 \underline{r}_{12}\omega_1\omega_2) = \frac{N_i N_j \Omega^2 \int\cdot\cdot\int dv_3 d\omega_3 \cdot\cdot dv_N d\omega_N \; e^{-U/kT}}{\rho_i(z_1\omega_1)\rho_j(z_2\omega_2) \int\cdot\cdot\int dv_1 d\omega_1 \cdot\cdot dv_N d\omega_N \; e^{-U/kT}} \tag{41}$$

where $\Omega = \int d\omega_i$, and is 4π for linear and $8\pi^2$ for nonlinear molecules, respectively. By differentiating (38) and (39) with respect to z_1 it is possible to derive integro-differential equations for $\rho_i(z_1)$ and $\rho_i(z_1\omega_1)$:

$$\frac{d\rho_i(z_1)}{dz_1} = \frac{1}{kT} \sum_j \int d\underline{r}_{12} \; \rho_{ij}(z_1\underline{r}_{12}) \frac{du_{ij}(r_{12})}{dr_{12}} \frac{z_{12}}{r_{12}} \tag{42}$$

$$\frac{\partial\rho_i(z_1\omega_1)}{\partial z_1} = \frac{1}{kT} \sum_i \int d\underline{r}_{12} d\omega_2 \; \rho_{ij}(z_1\underline{r}_{12}\omega_1\omega_2) \frac{\partial u(\underline{r}_{12}\omega_1\omega_2)}{\partial r_{12}} \frac{z_{12}}{r_{12}} \tag{43}$$

These equations (known as the lowest order Born-Green-Yvon
(BGY) expressions) assume that the potential energy is pairwise additive. They can only be solved for ρ_i if the pair
distribution function for the interfacial layer is known,
or can be estimated (see Section VII). It is possible to
derive (19) analogous equations linking ρ_{ij} to the triplet
function ρ_{ijk}.

An alternative integro-differential equation has recently
been derived (24,21,25) which relates $\rho(z_1)$ to the direct correlation function,

$$\frac{d\ell n \rho(z_1)}{dz_1} = \int d\underline{r}_{12} \; c(z_1 \underline{r}_{12}) \; \frac{d\rho(z_2)}{dz_2} \tag{44}$$

(The corresponding expression for mixtures has not yet been
derived.) Equation (44) is valid independent of potential
nonadditivity or the form of the intermolecular potential,
and is a companion result to Equation (34).

VII. APPROXIMATE THEORIES FOR SURFACE PROPERTIES

The rigorous expressions given in Section VI,A for surface tension involve the density profile $\rho(z_1)$ and either
the pair distribution function $g(z_1, \underline{r}_{12})$ or the direct correlation function $c(z_1 \underline{r}_{12})$. Since these functions cannot be
calculated exactly at present, approximations are introduced.
Several of these approximate theories are reviewed in the
sections below. Particular emphasis is given to those methods
that most readily yield numerical results. Most of the results reported to date are for simple monatomic and diatomic
liquids; much less work has been done on polyatomic liquids
or mixtures.

A. van der Waals Theory

At higher temperatures the width of the interface becomes
large compared to the range of the direct correlation function
(cf. Figure 2). In such cases $c(z_1 r_{12})$ acts essentially as
a Dirac delta function, and Equation (34) simplifies to (23)

$$\gamma = k \int_{-\infty}^{+\infty} dz \left[\frac{d\rho(z)}{dz}\right]^2 \tag{45}$$

where k is a constant that depends on temperature and density.
This square gradient expression is most appropriate at high
temperature, but will be a poor approximation near the triple
point. Equation (45) was originally derived by van der Waals
(2) and later by Cahn and Hilliard (26), in both cases by

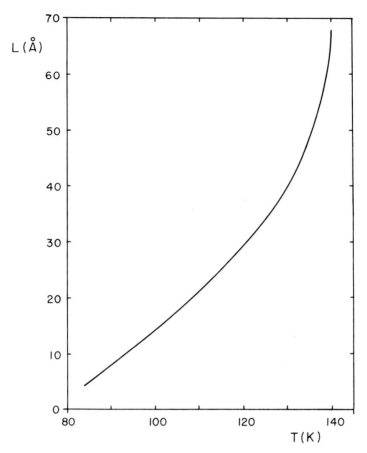

Fig. 2. *Thickness of the liquid-gas interface for argon de-*
fined by Equation (47) as calculated by Welsh and
Fitts (80) assuming a Lennard-Jones potential model
with ε/k = 119.8K, σ = 3.405Å. These estimates were
made by equating experimental values of γ with those
calculated from the Kirkwood-Buff equation, (30),
assuming a hyperbolic tangent density profile and
adjusting the thickness to ensure $\gamma_{exp} = \gamma_{calc}$. The
results using the Fisk-Widom (27) or erf (81) density
profiles were indistinguishable from those in the
figure. The results were also insensitive to the
approximation used for $g(z_1 r_{12})$. It should be noted
that the range of the pair potential for argon, and
hence of the direct correlation function $c(z_1 r_{12})$,
is of the order 10Å.

expanding the free energy density of the inhomogeneous fluid about that of a uniform fluid; for the free energy density of the uniform fluid, van der Waals used his equation of state. This expansion of the free energy density also yields an equation for the density profile (13),

$$\rho(z) = \rho_c + \frac{1}{2} (\rho_\ell - \rho_g) \tanh \left(\frac{2z}{L}\right) \tag{46}$$

where ρ_c, ρ_ℓ and ρ_g are the number densities for the critical point, liquid and gas, respectively, and

$$L = \frac{\rho_\ell - \rho_g}{\left|d\rho(z)/dz\right|_{z=0}} \tag{47}$$

is a measure of the interfacial thickness. Detailed discussion of the derivation of these equations are given in several reviews (27,12,13,28).

Recently, Bongiorno and Davis (29) have given a derivation of the van der Waals theory which is based on statistical mechanics. They obtain the free energy density in a mean field approximation, and then replace the pair correlation function $g(z_1 r_{12})$ by unity to obtain the van der Waals expression. Table I shows a comparison of experimental surface tensions with values calculated from Equation (45), using the simple van der Waals approach for $\rho(z)$ and treating k as a constant, independent of density and temperature. Agreement is surprisingly good, even at the lower temperatures, the discrepancies rarely being greater than 25%. The intermolecular potential used is the Lennard-Jones (12,6),

$$u(r) = 4\varepsilon \left|\left[\frac{\sigma}{r}\right]^{12} - \left[\frac{\sigma}{r}\right]^{6}\right|$$

Bongiorno and Davis (29) have also developed a modified van der Waals theory by introducing an improved approximation for $g(z_1 r_{12})$,

$$g(z_1 r_{12}) = g\{r_{12}; \frac{1}{2} [\rho(z_1) + \rho(z_2)]\} \tag{48}$$

where the quantity on the right hand side is the pair correlation function for a uniform fluid at the density midway between the pair of molecules. So far the theory has only been applied to argon.

TABLE I

Comparison of surface tensions predicted by original vdW
theory, Equation (45), with experiment. The coefficient
k was fitted to experiment at the temperature indicated
by *. [Reproduced with permission from V. Bongiorno and
H. T. Davis, Phys. Rev. A, 12, 2213 (1975).]

T/T_c	γ^{pred} (dyne cm^{-1})	γ^{exp} (dyne cm^{-1})
ARGON		
0.56	16.10	13.45
0.66	10.91	9.40
0.76	6.60	6.01
*0.86	2.99	2.99
0.96	0.47	0.58
0.995	0.028	0.041
BENZENE		
0.48	40.1	31.7
0.57	30.7	24.7
0.66	22.0	18.8
0.75	14.15	12.86
*0.84	7.41	7.41
0.93	2.23	2.66
PROPANE		
0.50	19.08	17.07
0.55	16.28	15.34
*0.60	13.59	13.59
0.66	11.04	11.84
0.71	8.64	10.09
0.77	6.41	8.35
CARBON DIOXIDE		
0.80	10.46	10.08
*0.83	8.06	8.06
0.90	3.90	4.34
0.95	1.49	1.90
0.98	0.36	0.57
0.997	0.03	0.07

Vargas (30) has extended the modified van der Waals theory to multicomponent systems. The surface tension for such systems is given by:

$$\gamma = \int_{-\infty}^{\infty} \sum_{ij} k_{ij} \frac{d\rho_i(z)}{dz} \frac{d\rho_j(z)}{dz} dz \qquad (49)$$

The modified van der Waals theory for multicomponent systems has been used to determine interfacial density profiles in binary liquid-liquid model systems. Vargas concludes that one method to lower surface tension in multicomponent systems is proper adjustment of the k_{ij} via introduction of appropriate additives into the system.

Fisk and Widom (27) have developed a corrected form of the classical van der Waals theory which is suitable for use near the gas-liquid critical point; their theory incorporates the correct experimental values of the critical indices (in contrast to the incorrect values given by the classical van der Waals equation of state).

B. The Fowler-Kirkwood-Buff Model

Fowler (31) suggested that the density profile could be approximated by a step function at the interface, and so obtained a simple equation for the surface tension. Kirkwood and Buff (1) introduced specific assumptions into Equation (30) and achieved the same result. These assumptions are that the interface is a gas-liquid one, that the gas phase density is negligible, and that the interfacial pair distribution function is given by

$$\rho(z_1 r_{12}) = 0 \qquad z_1 \text{ or } z_2 > 0$$
$$= \rho_\ell^2 g_\ell(r_{12}) \qquad z_1 \text{ and } z_2 < 0 \qquad (50)$$

where $g_\ell(r_{12})$ is the radial distribution function for the bulk liquid. With this assumption, Equation (30) simplifies to the Fowler-Kirkwood-Buff (FKB) result for a pure fluid,

$$\gamma = \frac{\pi}{8} \rho_\ell^2 \int_0^\infty dr_{12} r_{12}^4 \frac{du(r_{12})}{dr_{12}} g_\ell(r_{12}) \qquad (51)$$

and Equation (35) becomes

$$u_s = -\frac{\pi}{2} \rho_\ell^2 \int_0^\infty dr_{12} r_{12}^3 u(r_{12}) g_\ell(r_{12}) \qquad (52)$$

when the equimolecular dividing surface is used. A major de-
fect of the FKB model is that Equations (51) and (52) are
thermodynamically inconsistent, since they do not obey Equation (22). Despite this inconsistency, the model often gives
quite good results for γ near the triple point for simple
liquids; results for u_s, and for γ at higher temperatures, are
much poorer, however. Table II compares the FKB model with
experiment; the calculated values were obtained from Equations
(51) and (52) by a Monte Carlo procedure (32), using a Lennard-
Jones potential; the potential parameters were chosen to give
a best fit of the Monte Carlo data for bulk phase thermodynamic properties to experiment. The FKB model is seen to
give good results for the heavier inert gas liquids at the
triple point; for the polyatomics the discrepancies are
somewhat larger, probably because of the inadequacy of the
Lennard-Jones potential for these molecules. Results are
poor for u_s, and for γ at higher temperatures. More recent
Monte Carlo results (33,34,35) for a Lennard-Jones gas-liquid
surface, without the FKB approximation, suggest that the good
agreement in Table II for γ at the triple point is fortuitous;
these calculations give a γ value significantly above experiment, suggesting that the FKB approximation benefits from a
cancellation of errors due to the step-function assumption,
error in the pair potential, and neglect of the three-body
potential.
    Several authors (36,37) have attempted to improve the
FKB model by using more realistic models for ρ(z); these
have included linear, cubic and exponential profiles. Calculations for these profiles are reported by Freeman and
McDonald (32).
    Equations (51) and (52) assume spherically symmetric
potentials. The generalization of (51) to nonspherical
molecules is (20)

$$\gamma = \frac{\pi}{8} \frac{\rho_\ell^2}{\Omega^2} \int_0^\infty dr_{12} r_{12}^4 \int d\omega_1 d\omega_2 \frac{\partial u(\underline{r}_{12}\omega_1\omega_2)}{\partial r_{12}} g_\ell(\underline{r}_{12}\omega_1\omega_2) \quad (53)$$

Calculations have been made based on this equation (20,38),
using perturbation theory for g_ℓ (see Section VII,D).

C. Integral Equation Methods
    In these methods the interfacial density profile ρ(z) is
calculated by solving one of the rigorous integral equations,
e.g., (42), (44), or (36). In order to perform such a calculation, the interfacial pair correlation function $g(z_1,\underline{r}_{12})$
is approximated by some function of the corresponding quantities for the two bulk phases. For example, Toxvaerd (39) uses

TABLE II

SURFACE PROPERTIES OF SIMPLE LIQUIDS
[Reproduced with permission from K.S.C. Freeman and I. R. McDonald, Molec. Phys., 26, 529 (1973) and I. W. Plesner and O. Platz, J. Chem. Phys., 48, 5361 (1968).]

| | T(K) | γ_a (dyne cm^{-1}) | | | u^s (erg cm^{-2}) | |
		expt[a]	FKB[b]	PP[c]	expt	FKB[b]
Ar	84.0	13.4	13.7	16.5	35.0	27.6
	90.	11.9	13.2		34.4	26.0
	100.	9.4	12.2	10.7	33.3	23.3
	110.	7.1	11.1		31.7	20.9
	130.	3.0	9.0		27.1	15.3
Kr	115.8	16.4	16.7			33.8
Xe	161.3	19.2	18.7		50.1	42.2
Ne	24.6	5.7	6.8			10.6
CH$_4$	91.	17.0	15.4			32.2
	106.	14.5		17.4		
	127.	10.2		11.3		
N$_2$	63.2	12.2	10.3		27.6	23.7
	70.3	10.4		12.2		
	84.0	7.4		7.9		
CO	68.1	12.5	11.3			22.9
	74.7	11.4		13.1		
	89.5	8.0		8.5		

[a] From Buff and Lovett (1968).

[b] Calculation of Freeman and McDonald (1973) using Monte Carlo evaluation of the Fowler-Kirkwood-Buff equations, based on a Lennard-Jones potential.

[c] Calculation of Plesner and Platz (1968a) based on a Sutherland potential.

$$g(z_1 r_{12}) = \frac{\alpha \rho(z_1) + (1-\alpha)\rho(z_2) - \rho_g}{\rho_\ell - \rho_g} g_\ell(r_{12}; \rho_\ell)$$

$$+ \frac{\rho_\ell - \alpha \rho(z_1) - (1-\alpha)\rho(z_2)}{\rho_\ell - \rho_g} g_g(r_{12}; \rho_g) \qquad (54)$$

where α may take values between zero and one, whereas Salter and Davis (40) have proposed using Equation (48), i.e., the bulk liquid phase radial distribution function at a density which is the mean of the values at z_1 and z_2.

A detailed discussion of the integral equation methods is given in recent reviews by Toxvaerd (12,13), and we shall here restrict ourselves to a brief summary of these results. Toxvaerd (39) has used (54) in (42) to obtain $\rho(z)$ for a Lennard-Jones fluid, and has then calculated γ from the Kirkwood-Buff equation, (30). Quite good agreement with experimental data on argon was obtained, but the temperature dependence of γ was not reproduced well. In a more recent paper, Toxvaerd (41) has used Equation (48) in the pressure tensor equation, (36), to obtain $\rho(z)$ for both Lennard-Jones and square well fluids. Pressing and Mayer (42) have also used the pressure tensor equation to obtain $\rho(z)$.

Hill (43) has used the constancy of the chemical potential through the interface, and has solved a simplified form of the Kirkwood equation (44) for the chemical potential to obtain $\rho(z)$. Hill assumed the Sutherland potential model, and replaced the interfacial pair correlation function by the function for a bulk hard sphere fluid (spheres of diameter d) at density $\rho(z)$; this latter hard sphere function was further approximated by a step function, so that $g(z_1 r_{12}) = 1$ for $r_{12} > d$, and is zero otherwise. As a result of these approximations, only qualitative agreement with experimental surface tensions was found. Plesner and Platz (45) improved on Hill's method and applied it to the surface tension of several pure fluids, and to argon-nitrogen mixtures (46). Some of their results are shown in Table II, and in Figures 3 and 4. Figure 4 shows that nitrogen (the lighter component) is adsorbed at the interface, as required by the Gibbs adsorption equation, (26).

D. Perturbation Theory
1. *Surface Tension for Simple Liquids*
Perturbation theory relates the properties of the real fluid, in which the molecules interact with intermolecular potential energy u, to those of a reference fluid for which the potential is u_o. Thus, the potential is written

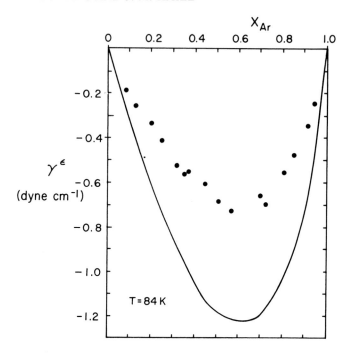

Fig. 3.   *Excess surface tension* $(\gamma^E - \gamma_{mix} - x_1\gamma_1 - x_2\gamma_2)$ *for argon-nitrogen mixtures at 84K from experiment (points) and Plesner-Platz (46) theory (line). [Reproduced with permission from I. W. Plesner, O. Platz and S. E. Christiansen, J. Chem. Phys., 48, 5364 (1968).]*

$$u = u_o + u_1 \tag{55}$$

where u_1 is the perturbation. For the case of uniform fluids such perturbation theories have been reviewed by Smith (47). These methods have been extended to nonuniform fluids composed of spherical molecules by Toxvaerd (48,13); derivations for nonuniform fluids are also given by Lee et al. (49) and by Abraham (50).

The configurational Helmholtz free energy is written as

$$A = S \int dz_1 \, \rho(z_1) \, \psi(z_1) \tag{56}$$

where $\psi(z_1)$ is the Helmholtz free energy per molecule at z_1. This free energy density is expanded about the value for the reference system, $\psi_o(z_1)$, constrained to the same density profile as the real system. To first order this expansion gives

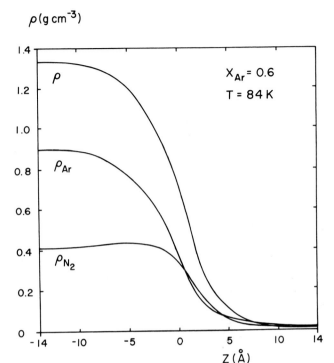

$\rho(g\,cm^{-3})$

$X_{Ar} = 0.6$

$T = 84\,K$

ρ

ρ_{Ar}

ρ_{N_2}

$Z(\mathring{A})$

Fig. 4. Interfacial density profiles for argon-nitrogen mixtures at 84K with x_{Ar} = 0.6. [Reproduced with permission from I. W. Plesner, O. Platz, and S. E. Christiansen, J. Chem. Phys., 48, 5364 (1968).]

$$\psi(z_1) = \psi_o(z_1) + \frac{1}{2} \int d\underline{r}_{12}\, u_1(r_{12})\, \rho(z_2)\, g_o[r_{12};\rho(z_1)] \quad (57)$$

Most authors have chosen the Barker-Henderson reference system (51),

$$
\begin{aligned}
u_o(r) &= u(r) & r &< \sigma \\
&= 0 & r &> \sigma \\
u_1(r) &= 0 & r &< \sigma \\
&= u(r) & r &> \sigma
\end{aligned}
\quad (58)
$$

where σ is the value of r for which $u = 0$. The functions ψ_o and g_o are then given, to good approximation, by the hard sphere functions

$$\psi_o = \psi_{HS} \qquad\qquad g_o = g_{HS} \qquad (59)$$

for the following choice of hard sphere diameter d,

$$d = \int_0^\sigma [1 - e^{-u(r)/kT}]dr \qquad (60)$$

Equations (56) to (60) provide a means of calculating the surface free energy only if $\rho(z)$ is known. This density profile
is determined as that function which minimizes A subject to
the constraint of a constant number of molecules. Then, by
locating the Gibbs dividing surface at z = 0, the surface
tension equals the superficial free energy,

$$\gamma = \lim_{h \to \infty} \left\{ \int_{-h}^h dz_1 \, \rho(z_1) \, \psi(z_1) - h[\rho_\ell \psi_\ell - \rho_g \psi_g] \right\} \qquad (61)$$

Toxvaerd (48) has extended Equation (56) to second order
and used the result to determine the interfacial density profile and surface tension of a Lennard-Jones fluid. The resulting density profile is compared with that obtained by computer
simulation in Figure 5.

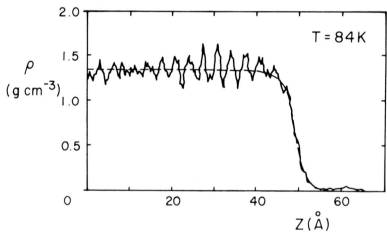

Fig. 5. *Interfacial density profile for argon near the triple
point, comparing computer simulation results (solid
line) with perturbation theory calculation (broken
line). The ripples in the computer simulation result smooth out to the perturbation theory result
as the simulation progresses. [Reproduced with
permission from J. K. Lee, J. A. Barker and G. M.
Pound, J. Chem. Phys. 60, 1976 (1974).]*

Abraham (50) has noted that the perturbation theory approach is in the spirit of the van der Waals-Cahn-Hilliard theory, with the square-gradient term replaced by integrations over the density profile. Further, since the hard sphere reference fluid does not exhibit vapor-liquid phase separation it must be the attractive part of the potential which gives rise to the vapor-liquid interface. Abraham (52) has also extended the perturbation theory to mixtures.

Lee et al. (49) found good agreement between perturbation theory and Monte Carlo results for γ for a Lennard-Jones liquid. They also attempted to improve the results compared with experimental values for argon by: a) using a more realistic pair potential due to Barker et al. (53); b) accounting for three-body interactions by including the Axilrod-Teller potential. Figure 6 indicates that inclusion of the three-body term greatly improves the perturbation theory predictions for surface tension over the two-body theory, when comparing with experiment.

2. *Surface Tension for Polyatomic Liquids*

The perturbation theory for fluids of nonspherical molecules has been developed by Gray and Gubbins (20) and by Haile et al. (38). The reference system is taken to be a nonuniform fluid of spherical molecules, whose intermolecular potential u_o is defined by

$$u_o(r_{12}) = \langle u(\underline{r}_{12}\omega_1\omega_2)\rangle_{\omega_1\omega_2} \qquad (62)$$

where

$$\langle \cdots \rangle_{\omega_1\omega_2} = \frac{1}{\Omega^2} \int \cdots \, d\omega_1 \, d\omega_2 \qquad (63)$$

i.e., an unweighted average over orientations of u. The perturbing potential u_1 in Equation (55) is thus the anisotropic part of the potential. The Helmholtz free energy A is then expanded about the reference system value A_o in powers of u_1/kT, and a series for the surface tension is obtained by applying Equation (2),

$$\gamma = \gamma_0 + \gamma_2 + \gamma_3 + \cdots \qquad (64)$$

If the anisotropic potential is of the multipolar type (i.e., it does not contain $l = 0$ spherical harmonics) the expressions for γ_2 and γ_3 are

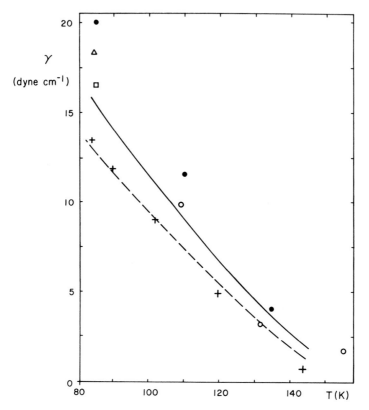

Fig. 6. Comparison of Monte Carlo and perturbation theory
results for surface tension. Crosses are experimental data for argon (82). Other points are Monte
Carlo results for the Lennard-Jones potential: □
Lee et al. (49), △ Miyazaki et al. (34), ● Chapela
et al. (33), o Liu (78). The solid line is the perturbation theory of Toxvaerd (48) using the Lennard-
Jones potential; the dashed line is the same theory
for the pair potential of Barker et al. (49) plus
the Axilrod-Teller three-body potential.

$$\gamma_2 = \frac{1}{4kT} \int\limits_{-\infty}^{\infty} dz_1 \; z_1 \int dr_{12} \; \frac{\partial \rho_o(z_1 r_{12})}{\partial z_1} \; \langle u_1^2(12) \rangle_{\omega_1 \omega_2} \qquad (65)$$

$$\gamma_3 = \frac{-1}{12(kT)^2} \int_{-\infty}^{\infty} dz_1 \, z_1 \int d\underline{r}_{12} \, \frac{\partial \rho_0(z_1\underline{r}_{12})}{\partial z_1} \, <u_1^3(12)>_{\omega_1\omega_2}$$

$$- \frac{1}{6(kT)^2} \int_{-\infty}^{\infty} dz_1 \, z_1 \int d\underline{r}_{12} \int d\underline{r}_{12} \, \frac{\partial \rho_0(z_1\underline{r}_{12}\underline{r}_{13})}{\partial z_1}$$

$$\times \, <u_1(12)u_1(13)u_1(23)>_{\omega_1\omega_2\omega_3} \tag{66}$$

where $u_1(12) = u_1(\underline{r}_{12}\omega_1\omega_2)$, etc. In these equations $\rho_0(z_1\underline{r}_{12})$ and $\rho_0(z_1\underline{r}_{12}\underline{r}_{13})$ are pair and triplet distribution functions for the inhomogeneous reference system at temperature T; in this approach these functions are not constrained to be those of the real system. Haile et al. (38) suggest a simple Padé approximant for Equation (64),

$$\gamma = \gamma_0 + \gamma_2 \left[1 - \frac{\gamma_3}{\gamma_2} \right]^{-1} \tag{67}$$

Tests against computer simulation results indicate that this equation gives good results even for strongly polar or quadrupolar liquids (54). Haile et al. (38) have given the expressions for γ_2 and γ_3 in the Fowler-Kirkwood-Buff model (Section VII,B), and have used (67) to calculate the effect of polar and quadrupolar forces on γ. Some of these results are shown in Figure 7.

3. *Density-Orientation Profile for Polyatomic Fluids*
     Haile et al. (55) have used perturbation theory to calculate the function $\rho(z_1\omega_1)$, which gives the distribution of molecular orientations (relative to the z-axis) in the interfacial region. Using the reference potential of Equation (62),

$$\rho(z_1\omega_1) = \rho_0(z_1)/\Omega + \rho_1(z_1\omega_1) + \cdots \tag{68}$$

where ρ_1 is zero for multipolar interactions, but for the commonly used models of anisotropic overlap or dispersion interactions for linear molecules it is (83)

$$\rho_1(z_1\theta_1) = c \, P_2(\cos \theta_1) \, \rho_0(z_1) \int d\underline{r}_{12} \, r_{12}^{*-n} \, P_2(\cos \theta_{12})$$

$$\times \, \rho_0(z_2) \, g_0(z_1\underline{r}_{12}) \tag{69}$$

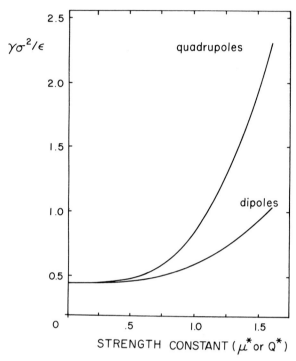

*Fig. 7. Effect of dipolar and quadrupolar intermolecular
forces on surface tension at kT/ε = 1.273, ρσ³ =
0.85, from Equation (67) using a Lennard-Jones
reference system and using the Fowler model for
γ_2 and γ_3. Here $\mu^* = \mu/(\varepsilon\sigma^3)^{1/2}$, $Q^* = Q/(\varepsilon\sigma^5)^{1/2}$.
[Reproduced with permission from J. M. Haile, C. G.
Gray and K. E. Gubbins, J. Chem. Phys. 64, 2569
(1976).]*

where θ_1 is the angle between the z-axis and the molecular
axis, subscript o indicates a reference fluid property, P_2
is the second-order Legendre polynomial, and $r^* = r/\sigma$. For
the anisotropic overlap potential $c = -8\delta\varepsilon/\Omega kT$, $n = 12$, while
for the London model of anisotropic dispersion $c = 4\kappa\varepsilon/\Omega kT$,
$n = 6$; here δ is a dimensionless overlap parameter, κ is the
dimensionless anisotropy of the polarizability, and ε and σ
are isotropic potential parameters.

   Figure 8 shows a calculation of $\rho(z_1\theta_1)$ based on this
equation for a liquid with anisotropic dispersion forces,
with $\kappa = 0.2$. In this calculation the Lennard-Jones potential was used as a reference, and Equation (54) was used for
$g_o(z_1\underline{r}_{12})$; $\rho_o(z)$ was then obtained from the solution of the
BGY equation, (42). In Figure 8 the curves are at different

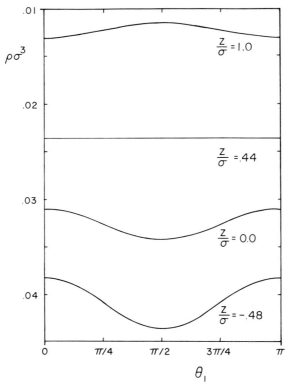

Fig. 8. The density-orientation profile for a fluid of axially symmetric molecules interacting with a dispersion model anisotropic potential (55), with $\kappa = 0.2$, $kT/\varepsilon = 0.85$. [Reproduced with permission from J. M. Haile, K. E. Gubbins and C. G. Gray, J. Chem. Phys. 64, 1852 (1976).]

heights in the interface ($z_1 = 0$ is the equimolar dividing surface). The figure indicates that on the liquid side of $z_1 = 0$ ($z_1 < 0.44\sigma$) the molecules tend to lie in the plane of the interface ($\theta_1 = 90°$). On the gas phase side ($z_1 > 0.44\sigma$) there is a slight preference for molecules to take an orientation perpendicular to the interface ($\theta_1 = 0$). For anisotropic overlap interactions of rodlike molecules ($\delta > 0$) the opposite behavior is found; the molecules tend to stand perpendicular to the interfacial plane on the liquid side of the interface; the same is true for platelike molecules, such as benzene ($\delta < 0$).

E. Semiempirical Methods

Empirical methods for estimating surface tension have been discussed by Reid and Sherwood (56). In this section we consider several methods, which, while based in statistical mechanics, involve considerably more empiricism than the approaches treated previously.

1. *Corresponding States Correlations*

Simple two-parameter corresponding states theory applies only for molecules having nearly spherical symmetry (e.g., inert gases, CH_4) (3,20). The surface tension of the inert gas liquids (Ar, Kr, Xe) is given by

$$\gamma_r \equiv \frac{\gamma}{kT_c} \left(\frac{V_c}{N_A}\right)^{2/3} = 2.4724T_r^2 - 7.5918T_r + 5.0748 \qquad (70)$$

where k is Boltzmann's constant, N_A is Avogadro's number, and $T_r = T/T_c$ is reduced temperature. One method of extending the principle to polyatomic and polar liquids is by inclusion of a third parameter. Thus Brock and Bird (57) have found the correlations

$$\frac{\gamma}{(P_c^2 T_c)^{1/3}} = \left(-0.951 + \frac{0.432}{z_c}\right)(1 - T_r)^{11/9} \qquad (71)$$

$$\frac{\gamma}{(P_c^2 T_c)^{1/3}} = (-0.281 + 0.133\alpha_K)(1 - T_r)^{11/9} \qquad (72)$$

where $z_c = P_c V_c / RT_c$ is the critical compressibility factor, and $\alpha_K = (d\ln P^s/d\ln T)_c$ is the Riedel factor. Equations (71) and (72) hold within three percent for many simple inorganic and a large number of inorganic substances; however, they do not work well for very light molecules (H_2, He), molten salts, highly polar inorganic liquids (H_2O, HCl, NH_3), associating liquids (alcohols, etc.), or fused salts.

Equations (71) and (72) may be extended to mixtures in an ad hoc way by using pseudocritical constants. The simplest procedure is to assume that P_{cm}, T_{cm}, and α_{km} are all mole fraction averages, i.e.,

$$P_{cm} = \sum_i x_i P_{ci} , \qquad etc. \qquad (73)$$

This technique is claimed to give errors of less than 7% for many nonaqueous solutions (56).

Patterson and Rastogi (58) have proposed a corresponding states principle for polymeric substances that is based on Prigogine's general corresponding states method (59). Their correlation involves the isobaric thermal expansion coefficient α and isothermal compressibility β as reducing parameters,

$$\gamma \; \beta^{2/3} \; \alpha^{1/3}/k^{1/3} = f(\alpha T) \qquad (74)$$

where f is a universal function of αT for all liquids. Figure 9 shows some of the results.

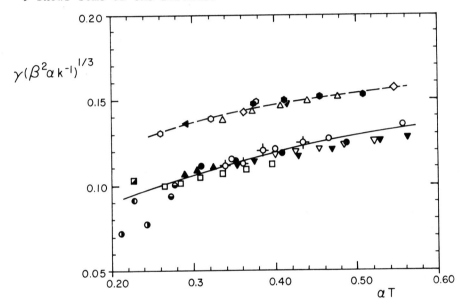

Fig. 9. *Correlations of Patterson and Rastogi (58) for surface tension of polyatomic and polymeric fluids: ● Ar; ▼ O_2; ◇ N_2; ◔ C_1; △ C_3; ● C_4; O C_5; ▽ C_6; ▼ C_8; □ C_{16}; ▲ polymethylene; ◑ polyisobutylene; ◒ polydimethylsiloxane; ○ dimer, trimer, tetramer and pentamer of dimethylsiloxane; ◓ polyoxypropylene glycol; ◪ polyoxyethylene glycol. [Reproduced with permission from D. Patterson and A. K. Rastogi, J. Phys. Chem. 74, 1067 (1974) Copyright by ACS.]*

2.   *Correlations Based on Perturbation Theory*

Useful equations for the temperature and composition dependence of surface tension may be derived from the perturbation theory expressions by introducing a van der Waals or similar approximation for the distribution functions (60). For example, if Equation (64) is terminated at the γ_2 term, if $\rho_o(z_1 \underline{r}_{12})$ is approximated by the Fowler model, and if $g_\ell(r_{12})$ is approximated by a constant, then (64) becomes

$$\gamma_r = \gamma_{or} + \frac{a\rho_r^2}{T_r} \tag{75}$$

where γ_{or} is given by Equation (70), $\rho_r = \rho/\rho_c$, and a is a constant. Figure 10 shows a test of Equation (75) for several polyatomic and polar liquids; in each case the constant a was obtained by fitting Equation (75) to one experimental point (see Table III). For temperatures up to $0.92T_c$ this simple correlation usually gives errors below 20%. It breaks down in the critical region because the perturbation term does not vanish as the critical point is approached.

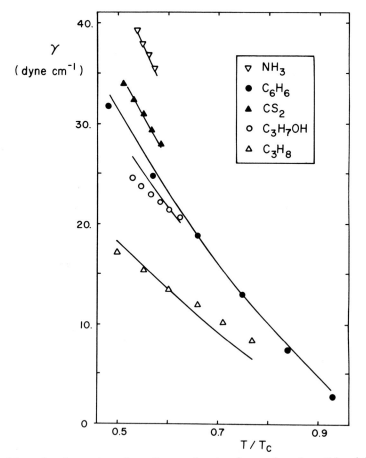

Fig. 10. *Surface tension for polyatomic and polar liquids. Points are experiment, solid lines are Equation (75).*

TABLE III

CONSTANTS IN EQUATION (75) FOR VARIOUS POLYATOMIC FLUIDS

Substance	a	T_r used to fit a	T_r Range	Maximum Error (%)
Acetone	.0146	.605	0.59 - .64	2.2
Ammonia	-.0009	.526	0.49 - .575	2.6
Benzene	.0225	.66	0.48 - .93	21.5
Carbon Dioxide	.0137	.855	0.71 - .98	14.0
Carbon Disulfide	.0059	.55	0.51 - .59	1.2
Ethyl Bromide	.0176	.582	0.56 - .60	1.4
Methanol	-.0285	.59	0.55 - .65	8.8
Methyl Chloride	.0069	.70	0.68 - .73	0.4
Propane	-.0111	.60	0.50 - .77	21.3
Propanol	.00447	.60	0.53 - .68	10.5
Water	-.0173	.50	0.44 - .58	10.8

3. *Correlation Between Surface Tension and Isothermal Compressibility*
    An estimation of pure component surface tension may be obtained from the fact that the product of surface tension and isothermal compressibility (β) is approximately constant for a wide variety of liquids near the triple point

$$\gamma\beta \simeq 0.32\text{\AA} \tag{76}$$

Thus Egelstaff and Widom (61) found that $\gamma\beta$ was in the range 0.17 to 0.47 for thirty liquids, including liquid metals, molten salts, water, simple nonmetallic liquids, hydrocarbons, polar liquids, etc. For these same liquids, γ and β separately vary by a factor of 150. Explanations of Equation (76) have been proposed which are based on the van der Waals approach (61), scaled particle theory (62), and on the Fowler-Kirkwood-

Buff model (63). In using Equation (76), if values of β are not readily available, a corresponding states correlation may be used (64).

4. *Regular Solution Theory*

Sprow and Prausnitz (65,66) have developed a method for predicting surface tension of simple nonpolar liquid mixtures based on regular solution theory (67). They treat the surface region as a separate uniform phase, distinct from the bulk liquid and vapor phases, of composition x_1^s, x_2^s $\cdots$ x_m^s. The activity coefficients of this phase are then estimated from regular solution theory. Excellent results are obtained for a variety of cryogenic liquid mixtures.

5. *Scaled Particle Theory*

By assuming the molecules to contain a rigid core of diameter a, Reiss et al. (68,69) were able to develop the following approximate expression for surface tension,

$$\gamma = \frac{kT}{4\pi a^2} \left| \frac{12y}{1-y} + \frac{18y^2}{(1-y)^2} \right| - \frac{1}{2} Pa \qquad (77)$$

where P is pressure and $y = \pi\rho a^3/6$ is reduced density. A variety of methods may be used to estimate the single adjustable parameter a. The equation gives surprisingly good results for a variety of liquids, including simple inorganics, organics, polar liquids, fused salts, liquid metals, and simple liquid mixtures (70-74,62,68,69). For these liquids the calculated γ is usually within 30% of the experimental value. The temperature dependence of the surface tension is not predicted well unless a is allowed to vary with temperature (72).

6. *Interfacial Tension*

Semiempirical methods for predicting the interfacial tension between two immiscible liquid phases have been reviewed by Gambill (75) and by Good and Elbing (76). These methods usually relate the interfacial tension (γ) to the gas-liquid surface tensions (γ^α and γ^β) for the separate (saturated) phases against a common gas. The oldest and simplest of these equations is Antonov's rule (75),

$$\gamma = |\gamma^\alpha - \gamma^\beta| \qquad (78)$$

which works well for many mixtures (75).

VIII. COMPUTER SIMULATION STUDIES

Both the Monte Carlo (32-34,49,77,78) and molecular dynamics (79) methods have been used to study the vapor-liquid interface of a fluid in which the molecules obey a Lennard-Jones potential. Such computer simulation studies can provide both the surface tension (and other thermodynamic functions) and also the distribution functions $\rho(z_1)$ and $g(z_1 r_{12})$; comparisons of the results with theory provides a test free of assumptions about the potential model, whilst comparisons of simulation and experimental results tests the potential model used in the simulation.

The surface tension calculations of Lee et al. (49) and Chapela et al. (33) are based on the Kirkwood–Buff equation (30). Surface tensions calculated in this way are subject to some uncertainty because of large fluctuations in the integrand. Miyazaki et al. (34) used a modified technique in which they separated the liquid into slabs, and thus directly calculated the work to form the surface. This procedure results in higher accuracy. Figure 6 shows the computer simulation results, together with perturbation theory and experimental data on argon. The simulation results for the Lennard-Jones fluid lie significantly above the experimental data for argon. This is believed to arise from inaccuracy in the pair potential model, and also from neglect of the three-body forces (49) (see dashed line in Figure 6).

The interfacial density profile $\rho(z_1)$ has also been evaluated by computer simulation. Early results suggested a layered structure near the interface (49), but these oscillations were later shown to be spurious (33,78). An example of the results is shown in Figure 5. The density profile obtained from computer simulation agrees well with that obtained from the perturbation theory. Of particular value is the use of computer simulation to determine the interfacial pair correlation function $g(z_1 r_{12})$ since almost nothing is known of this function either from theory or experiment. Unfortunately, little attention has yet been given to $g(z_1 z_2 r_{12})$. Opitz (79) in a short note has presented a few values for this function (see Figure 11).

IX. CONCLUSION

The theory of liquid surfaces is in a much cruder state than that of bulk phase liquids. However, there have been several notable developments in the last two or three years. The development of perturbation theories for surface tension and structure, and the extension of such methods to polyatomic liquids, should eventually provide more reliable predictive methods. Of particular importance are the computer simulation studies, which provide both a direct test of the various theories, and also a means for determining the distribution of molecules through the interface. These methods should soon be extended to polyatomic liquids and mixtures.

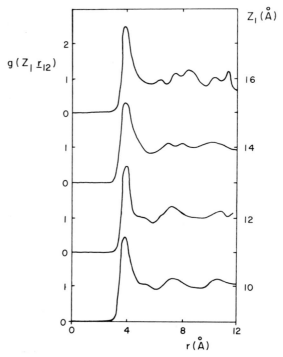

Fig. 11. Interfacial pair correlation function $g(z_1 r_{12})$ for a Lennard-Jones fluid at 124K from molecular dynamics simulation. For the correlation functions shown, r_{12} is aligned in the + z direction. The z = 0 plane is placed in the bulk liquid phase such that $z \simeq 20Å$ locates the Gibbs dividing surface. [Reproduced with permission from A.C.L. Opitz, Physics Letters 47A, 439 (1974).]

X.   ACKNOWLEDGMENTS
     It is a pleasure to thank D. D. Fitts for providing the unpublished calculations for Figure 2, and M. Wertheim and R. Lovett for sending manuscripts prior to publication. We thank the National Science Foundation for support of this work.

XI.   REFERENCES
1.   Kirkwood, J. G. and Buff, F. P., "Statistical Mechanical Theory of Surface Tension," J. Chem. Phys. 17, 338 (1949).
2.   van der Waals, J. D., "Thermodynamische Theorie der Kapillarität unter Voraussetzung Stetiger Dichteänderung," Z. Physik. Chem. 13, 657 (1894).

3.   Buff, F. P. and Lovett, R. A., "The Surface Tension of
     Simple Fluids," in "Simple Dense Fluids" (H. L. Frisch
     and Z. W. Salsburg, Eds.), p. 17. Academic Press,
     New York, N.Y., 1968.

4.   Rusanov, A. J., "Recent Investigations on Thickness of
     Surface Layers," in "Progress in Surface and Membrane
     Science" (J. F. Danielli, M. D. Rosenberg and D. A.
     Cadenhead, Eds.), Vol. 4, p. 57. Academic Press,
     New York, N.Y., 1971.

5.   Zollweg, J., Hawkins, G. and Benedek, B., "Surface Ten-
     sion and Viscosity of Xenon Near its Critical Point,"
     Phys. Rev. Letters 27, 1182 (1971).

6.   Huang, J. S. and Webb, W. W., "Diffuse Interface in a
     Critical Fluid Mixture," J. Chem. Phys. 50, 3677
     (1969).

7.   Wu, E. S. and Webb, W. W., "Critical Liquid-Vapor Inter-
     face in SF_6. I. Thickness of the Diffuse Transition
     Layer," Phys. Rev. A 8, 2065 (1973).

8.   Buff, F. P., "The Theory of Capillarity," Handbuch der
     Phys. 10, 281 (1960).

9.   Ono, S. and Kondo, S., "Molecular Theory of Surface Ten-
     sion in Liquids," Handbuch der Phys. 10, 134 (1960).

10. Croxton, C. A., "Statistical Thermodynamics of the
     Liquid Surface," Adv. in Physics 22, 385 (1973).

11. Croxton, C. A., "Liquid State Physics--A Statistical
     Mechanical Introduction," Chap. 4, Cambridge Uni-
     versity Press, Cambridge, 1974.

12. Toxvaerd, S., "Surface Structure of Simple Fluids,"
     Progr. Surface Sci. 3, 189 (1972).

13. Toxvaerd, S., "Statistical Mechanics of Surfaces," in
     Statistical Mechanics" (K. Singer, Ed.), Vol. 2, Chap.
     4, Chemical Society, London, 1975.

14. Gibbs, J. W., "Collected Works," Vol. 1, Yale University
     Press, New Haven, 1948.

15. Guggenheim, E. A., "Thermodynamics," North-Holland
     Publishing Co., Amsterdam, 1950.

16. Melrose, J. C., "Thermodynamics Aspects of Capillarity,"
     Ind. Eng. Chem. 60, 53 (1968).

17. de La Place, P. S., "Mecanique celeste," 1806.

18. Buff, F. P., "Some Considerations of Surface Tension,"
     Z. Electrochem. 56, 311 (1952).

19. Reed, T. M. and Gubbins, K. E., "Applied Statistical
     Mechanics," McGraw-Hill, New York, 1973.

20. Gray, C. G. and Gubbins, K. E., "Theory of Surface Tension
     for Molecular Fluids," Molec. Phys. 30, 179 (1975).

156   K. E. GUBBINS AND J. M. HAILE

21.   Percus, J. K., "The Pair Distribution Function in Classical Statistical Mechanics," in "The Equilibrium Theory of Classical Fluids" (H. L. Frisch and J. L. Lebowitz, Eds.), Section II, p. 33. Benjamin, New York, 1964.

22.   Lovett, R. A., DeHaven, P. W., Vieceli, J. J., Jr. and Buff, F. P., "Generalized van der Waals Theories for Surface Tension and Interfacial Width," J. Chem. Phys. 58, 1880 (1973).

23.   Triezenberg, D. G. and Zwanzig, R., "Fluctuation Theory of Surface Tension," Phys. Rev. Letters 28, 1183 (1972).

24.   Lovett, R. A., Mou, C. Y. and Buff, F. P., J. Chem. Phys. 65, 570 (1976).

25.   Wertheim, M., preprint (submitted to J. Chem. Phys.) (1976).

26.   Cahn, J. W. and Hilliard, J. E., "Free Energy of a Nonuniform System. I. Interfacial Free Energy," J. Chem. Phys. 28, 258 (1958).

27.   Fisk, S. and Widom, B., "Structure and Free Energy of the Interface Between Fluid Phases in Equilibrium near the Critical Point," J. Chem. Phys. 50, 3219 (1969).

28.   Widom, B., in "Phase Transitions and Critical Phenomena" (C. Domb and M. S. Green, Eds.), Vol. 2, Chap. 3. Academic Press, New York, 1972.

29.   Bongiorno, V. and Davis, H. T., "Modified van der Waals Theory of Fluid Interfaces," Phys. Rev. A 12, 2213 (1975).

30.   Vargas, A. S., "On the Molecular Theory of Dense Fluids and Fluid Interfaces," Ph.D. Dissertation, University of Minnesota (1976).

31.   Fowler, R. H., "A Tentative Statistical Theory of Macleod's Equation for Surface Tension and the Parachor," Proc. Roy. Soc. A 159, 229 (1937).

32.   Freeman, K.S.C. and McDonald, I. R., "Molecular Theory of Surface Tension," Molec. Phys. 26, 529 (1973).

33.   Chapela, G. A., Saville, G. and Rowlinson, J. S., "Computer Simulation of the Gas/Liquid Surface," Faraday Disc. Chem. Soc. 59, 22 (1975).

34.   Miyazaki, J., Barker, J. A. and Pound, G. M., "A New Monte Carlo Method for Calculating Surface Tension," J. Chem. Phys. 64, 3364 (1976).

35.   Present, R. D. and Shih, C. C., "Nonadditivity Extension of Kirkwood-Buff Surface Tension Formula," J. Chem. Phys. 64, 2262 (1976).

36.   Berry, M. V., Durrans, R. F. and Evans, R., "The Calculation of Surface Tension for Simple Liquids," J. Phys. A 5, 166 (1972).

37.   Fitts, D. D., "Theoretical Determination of Surface
      Thickness of Liquid He^4 and He^3," Physica 42, 205
      (1969).

38.   Haile, J. M., Gubbins, K. E. and Gray, C. G., "Theory
      of Surface Tension for Molecular Liquids, II. Per-
      turbation Theory Calculations," J. Chem. Phys. 64,
      2569 (1976).

39.   Toxvaerd, S., "Statistical Mechanical and Quasithermo-
      dynamic Calculations of Surface Densities and Surface
      Tension," Molec. Phys. 26, 91 (1973).

40.   Salter, S. J. and Davis, H. T., "Statistical Mechanical
      Calculations of the Surface Tension of Fluids," J.
      Chem. Phys. 63, 3295 (1975).

41.   Toxvaerd, S., "Hydrostatic Equilibrium in Fluid Inter-
      faces," J. Chem. Phys. 64, 2863 (1976).

42.   Pressing, J. and Mayer, J. E., "Surface Tension and
      Interfacial Density Profile of Fluids near the
      Critical Point," J. Chem. Phys. 59, 2711 (1973).

43.   Hill, T. L., "Statistical Thermodynamics of the Transi-
      tion Region between Two Phases. II. One Component
      System with a Plane Interface," J. Chem. Phys. 20,
      141 (1952).

44.   Kirkwood, J. G., "Statistical Mechanics of Fluid Mix-
      tures," J. Chem. Phys. 3, 300 (1935).

45.   Plesner, I. W. and Platz, O., "Statistical-Mechanical
      Calculation of Surface Properties of Simple Liquids
      and Liquid Mixtures. I. Pure Liquids," J. Chem.
      Phys. 48, 5361 (1968).

46.   Plesner, I. W., Platz, O. and Christiansen, S. E.,
      "Statistical-Mechanical Calculation of Surface Prop-
      erties of Simple Liquids and Liquid Mixtures. II.
      Mixtures," J. Chem. Phys. 48, 5364 (1968).

47.   Smith, W. R., "Perturbation Theory in Classical Sta-
      tistical Mechanics of Fluids," in "Statistical
      Mechanics" (K. Singer, Ed.), Vol. 1, Chap. 2.
      Chemical Society, London, 1973.

48.   Toxvaerd, S., "Perturbation Theory for Nonuniform
      Fluids: Surface Tension," J. Chem. Phys. 55, 3116
      (1971).

49.   Lee, J. K., Barker, J. A. and Pound, G. M., "Surface
      Structure and Surface Tension: Perturbation Theory
      and Monte Carlo Calculation," J. Chem. Phys. 60,
      1976 (1974).

50.   Abraham, F. F., "A Theory for the Thermodynamics and
      Structure of Nonuniform Systems with Applications to
      the Liquid-Vapor Interface and Spinodal Decomposition,"
      J. Chem. Phys. 63, 157 (1975).

51.   Barker, J. A. and Henderson, D., "Perturbation Theory and Equation of State for Fluids, II. A Successful Theory of Liquids," J. Chem. Phys. 47, 4714 (1967).

52.   Abraham, F. F., "A Theory for the Thermodynamics of Nonuniform Mixtures, II," J. Chem. Phys. 63, 1316 (1975).

53.   Barker, J. A., Fisher, R. A. and Watts, R. O., "Liquid Argon: Monte Carlo and Molecular Dynamics Calculations," Molec. Phys. 21, 657 (1971).

54.   McDonald, I. R. and Gubbins, K. E., "Surface Tension of Polar Liquids," J. Chem. Phys. 66, 364 (1976).

55.   Haile, J. M., Gubbins, K. E. and Gray, C. G., "Vapor-Liquid Interfacial Density-Orientation Profiles for Fluids with Anisotropic Potentials," J. Chem. Phys. 64, 1852 (1976).

56.   Reid, R. C. and Sherwood, T. K., "The Properties of Gases and Liquids," 2nd edition, Chap. 8. McGraw-Hill, New York, 1966.

57.   Brock, J. R. and Bird, R. B., "Surface Tension and the Principle of Corresponding States," A.I.Ch.E. J. 1, 174 (1955).

58.   Patterson, D. and Rastogi, A. K., "The Surface Tension of Polyatomic Liquids and the Principle of Corresponding States," J. Phys. Chem. 74, 1067 (1970).

59.   Prigogine, I. (with A. Bellemans and V. Mathot), "The Molecular Theory of Solutions," Chap. 17. North-Holland Publishing Co., Amsterdam, 1957.

60.   Gubbins, K. E., "Perturbation Methods for Calculating Properties of Liquid Mixtures," A.I.Ch.E. J. 19, 684 (1973).

61.   Egelstaff, P. A. and Widom, B., "Liquid Surface Tension near the Triple Point," J. Chem. Phys. 53, 2667 (1970).

62.   Mayer, S. W., "A Molecular Parameter Relationship Between Surface Tension and Liquid Compressibility," J. Phys. Chem. 67, 2160 (1963).

63.   Present, R. D., "On the Product of Surface Tension and Compressibility of Liquids," J. Chem. Phys. 61, 4267 (1974).

64.   Gubbins, K. E. and O'Connell, J. P., "Isothermal Compressibility and Partial Molal Volume for Polyatomic Liquids," J. Chem. Phys. 60, 3449 (1974).

65.   Sprow, F. B. and Prausnitz, J. M., "Vapor-Liquid Equilibria and Surface Tensions for the Nitrogen-Argon-Methane System at 90.67K," Cryogenics 6, 338 (1966).

66.   Sprow, F. B. and Prausnitz, J. M., "Surface Tensions of Simple Liquid Mixtures," Trans. Faraday Soc. 62, 1105 (1966).

67. Hildebrand, J. H., Prausnitz, J. M. and Scott, R. L.,
     "Regular and Related Solutions," Van Nostrand Reinhold
     Co., New York, 1970.

68. Reiss, H., Frisch, H. L., Helfand, E. and Lebowitz, J. L.,
     "Aspects of the Statistical Thermodynamics of Real
     Fluids," J. Chem. Phys. 32, 119 (1960).

69. Reiss, H., "Scaled Particle Methods in Fluids," Adv. Chem.
     Phys. 9, 1 (1965).

70. Lebowitz, J. L., Helfand, E. and Praestgaard, E., "Scaled
     Particle Theory of Fluid Mixtures," J. Chem. Phys. 43,
     774 (1965).

71. Mayer, S. W., "Calculation of Metal Surface Tensions.
     Ionic-Salt and Monatomic Models for Liquid Metals,"
     J. Chem. Phys. 35, 1513 (1961).

72. Mayer, S. W., "Dependence of Surface Tension on Temper-
     ature," J. Chem. Phys. 38, 1803 (1963).

73. Mayer, S. W., "Interrelationships among Thermal Expan-
     sivities, Surface Tensions, and Compressibilities for
     Molten Salts," J. Chem. Phys. 40, 2429 (1964).

74. Reiss, H. and Mayer, S. W., "Theory of Surface Tension
     of Molten Salts," J. Chem. Phys. 34, 2001 (1961).

75. Gambill, W. R., "Surface and Interfacial Tensions,"
     Chem. Eng. 65(9), 143 (1958).

76. Good, R. J. and Elbing, E., "Generalization of Theory
     for Estimation of Interfacial Energies," Ind. Eng.
     Chem. 62, 55 (1970).

77. Abraham, F. F., Schreiber, D. E. and Barker, J. A., "On
     the Structure of a Free Surface of a Lennard-Jones
     Fluid: A Monte Carlo Calculation," J. Chem. Phys.
     62, 1958 (1975).

78. Liu, K. S., "Phase Separation of Lennard-Jones Systems:
     A Film in Equilibrium with Vapor," J. Chem. Phys. 60,
     4226 (1974).

79. Opitz, A.C.L., "Molecular Dynamics Investigation of a
     Free Surface of Liquid Argon," Phys. Letters 47A, 439
     (1974).

80. Welsh, W. J. and Fitts, D. D., unpublished calculations
     (1976); see also, Welsh, W. J., Ph.D. Dissertation,
     University of Pennsylvania (1975).

81. Buff, F. P., Lovett, R. A. and Stillinger, F. H., Jr.,
     "Interfacial Density Profile for Fluids in the
     Critical Region," Phys. Rev. Letters 15, 621 (1965).

82. Stansfield, D., "The Surface Tensions of Liquid Argon
     and Nitrogen," Proc. Phys. Soc. (London) 72, 854
     (1958).

83. Ananth, M. S., Gubbins, K. E. and Gray, C. G., "Perturba-
     tion Theory for Equilibrium Properties of Molecular
     Fluids," Molec. Phys. 28, 1005 (1975).

INTERFACIAL RHEOLOGICAL PROPERTIES OF
FLUID INTERFACES CONTAINING SURFACTANTS

D. T. Wasan and V. Mohan
Department of Chemical Engineering
Illinois Institute of Technology

I. ABSTRACT

The present chapter focuses on the dynamic effects of surface-active agents and polymer additives on the shear and dilatational rheological properties of liquid-liquid and liquid-gas interfaces. An incentive for this study stems from the potential importance of these properties in secondary and tertiary oil recovery and other fluid phase separation processes. New data on interfacial shear and dilational viscosities and elasticities for water soluble chemical additives are reviewed with particular emphasis on correlating these measurements with emulsion stability, foam stability and interfacial mass transport. The effects of surface aging and the non-Newtonian behavior of aqueous surfactant solutions are discussed. It is concluded that in a commercial surfactant system, the concentrations at which both the interfacial viscosities and interfacial tension are minimized cannot be identified by measurements of the interfacial tension alone. These observations clearly point out the importance of interfacial rheological properties in tailoring surfactant and polymer slugs in improved oil recovery.

II. SCOPE

Fluid-fluid interfaces containing surfactants and macromolecules often exhibit interfacial rheological behavior different from that in the bulk. The orientation of these molecules, molecular complexing or structural transformations at the fluid-fluid interface can result in peculiar rheological behavior. Interfacial rheological behavior of fluid-fluid interfaces containing chemical additives is often shear, area and age dependent. The earlier observations of interfacial rheological properties such as the interfacial shear and dilational viscosities and elasticities were qualitative in nature, as no accurate measurement techniques were available to determine their absolute values. However,

in recent years good instrumentation has been developed for
this purpose. It is our main objective to review recent
data on interfacial rheological properties for surfactant
systems with particular emphasis on their role in practical
applications such as secondary and tertiary oil recovery,
emulsion stability, foam stability and interfacial mass
transfer. Four different crude oils were investigated and
aqueous solutions of two different petroleum sulfonates
were used in the present study. The general significance
of interfacial rheological properties in fluid phase separation processes is highlighted.

III. CONCLUSIONS AND SIGNIFICANCE

A number of surfactant formulations used previously in
laboratory and/or field tests and which are reported to yield
good oil recovery exhibit low interfacial viscosities. New
data on interfacial viscosities of crude oil-brine solution
containing petroleum sulfonates show that the concentrations
at which both the interfacial viscosity and interfacial tension are minimized cannot be identified by measurements of
the interfacial tension alone. The effect of the addition
of a surfactant to the brine solution is to decrease its
interfacial tension against the oil phase and produce low
interfacial viscosities. However, the interfacial viscosity
may decrease or increase upon the continued addition of the
surfactant. The crude oil-brine solution exhibits considerable aging effect at its interface; on the other hand, when
brine solution containing a petroleum sulfonate is contacted
against crude oil, the surface aging effect is greatly
diminished. A generalized correlation of the surface shear
viscosity data for the mixed surfactant systems has been
prepared which is a plot of reduced surface shear viscosity
versus the product of shear rate and relaxation time. However, more data are needed to verify the general applicability
of the proposed correlation. The need for a unified molecular approach for predicting interfacial viscosities from
the generalized correlation is pointed out.

The present state-of-art reveals that surfactant selection for an improved oil recovery process is made on the
basis of ultralow interfacial tension between the crude and
the aqueous phase. However, such ultralow values of interfacial tensions also cause immense emulsification problems.
The coalescence promotability of these surfactants have
been studied and emulsion stability tests have been conducted.
These tests clearly show that for oil emulsions in brine
solutions without petroleum sulfonates, the phase separation
is poor, i.e., these emulsions are more stable. However,
with the addition of surfactants to the brine solution, the

separation efficiency is improved. An attempt has been made
to correlate interfacial rheological parameters to the emulsion stability. The addition of a cosurfactant or co-solvent
is found to lower the interfacial viscosity and decrease the
emulsion stability, thus promoting coalescence rates. These
data throw a new light on the possible role of interfacial
rheological properties of surfactant and co-surfactant formulations in promoting coalescence of oil ganglia in porous
media and thereby enhancing the oil bank formation.

Preliminary data have been reported relating foam
stability to the interfacial shear and dilational viscosities and elasticities. However, more extensive data are
needed relating the surface properties of the foam to its
ability to increase oil displacement efficiency.

The need for data relating interfacial rheological properties to the interfacial surfactant transport is pointed
out, as these data will help us firstly to understand the
mechanism of oil-water interface movement through a porous
medium, and secondly, to screen surfactants, co-solvents or
demulsifiers for emulsification/demulsification in oil recovery.

IV.   INTRODUCTION

Surfactants in solution tend to accumulate and adsorb
at interfaces between their solution and adjacent phases.
The orientation of these molecules as well as molecular
interaction and molecular packing result in an interfacial
behavior different from that in the bulk phases. In some
instances, molecular complexing or structural transformations
like the formation of a liquid crystalline phase can occur
resulting in peculiar rheological behavior. Specifically,
such accumulation of surfactants at fluid-fluid interfaces
results in interfacial forces which influence the geometry
of the phase boundary with consequent effect on interfacial
flow and area.

An analysis of the rheological properties, as opposed
to the static properties, such as interfacial tension, is
essential in describing the behavior of fluid interfaces
where interfacial motion is involved. Interfacial rheological behavior of fluid-fluid interfaces containing chemical
additives is often dependent on shear, structure and age.
For simple Newtonian fluid interfaces the significant
rheological properties for interfacial motion are the interfacial shear and dilational viscosities. As the molecular
weight of the adsorbate increases, the interface may exhibit
viscoelastic behavior or display rigidity.

An interface, in general, may be viewed as a viscoelastic region, so that a combination of two types of coefficients is required to represent a real interface. Goodrich (1) suggested five independent pairs of elasticity and viscosity coefficients to describe the capacitive and the dissipative response of a non-isotropic axially symmetric monolayer to various possible motions. As shown in Figure 1, these motions are (a) vertical shear, (b) lateral shear, (c) lateral compression, (d) horizontal shear, and (e) vertical compression. Two additional pairs of viscoelastic constants were introduced by Eliassen (2): bending and torsional viscoelasticity coefficients (Figure 1,f and g).

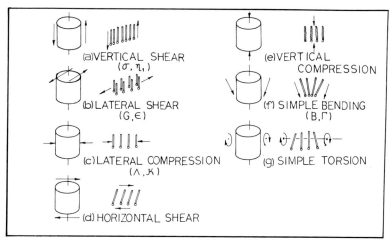

Fig. 1. Possible motions in the interfacial plane

V. DYNAMICS OF FLUID-FLUID INTERFACES

The constitutive equation describes the relation between stress and the deformation or the rate-of-deformation. Boussinesq (3) was the first to propose a two-dimensional analog of the three-dimensional Newtonian fluid to explain the retardation in the terminal velocities of drops and bubbles. Boussinesq expressed his equations in a special coordinate system coinciding with the principal rate-of-strain axes. Oldroyd (4) extended Boussinesq's equations to arbitrary coordinates in a stationary system. These equations were further generalized by Scriven (5), Eliassen (2) and Slattery (6) to an interface undergoing continuous change in shape and extent.

Under the assumptions of negligible mass of the interface and negligible mass transfer across it, the interfacial momentum balance becomes a force balance involving interfacial stress:

$$(\hat{\tau}^{ij} - \tau^{ij})n_1 - (\hat{p} - p)n^i = (t_\alpha^i \cdot T^{\alpha\beta})_{,\beta} \qquad (1)$$

In simple terms, Equation (1) states that the viscous forces due to the two bulk phases ($\hat{\tau}^{ij}$ and τ^{ij}), the pressure forces ($\hat{p}$ and p), and a force due to the interface ($T^{\alpha\beta}$) must be in balance. The term n^i refers to the unit normal to the interface and t_α^i is the hybrid tensor.

For a _purely viscous_ Newtonian interface $T^{\alpha\beta}$ is related to the surface rate-of-strain tensor, $S_{\alpha\beta}$, by the interfacial constitutive equation:

$$T^{\alpha\beta} = \sigma a^{\alpha\beta} + (\kappa-\varepsilon) a^{\delta\nu} a^{\alpha\beta} S_{\delta\nu} + 2\varepsilon a^{\alpha\lambda} a^{\beta\mu} S_{\lambda\mu} \qquad (2)$$

where

σ = surface tension, dyne/cm
κ = surface dilatational viscosity, dyne sec/cm = surface poise (s.p.)
ε = surface shear viscosity (s.p.)
$a^{\alpha\beta}$ = surface metric tensor; α,β = 1,2

The surface rate-of-strain tensor is related to the velocity (v_i) by:

$$S_{\alpha\beta} = 1/2(t_\alpha^i v_{i,\beta} + t_\beta^i v_{i,\alpha}) \qquad (3)$$

The constitutive equation (2) will serve as the basis for measurement of both the interfacial shear and dilational parameters. In essence one hopes to determine with a sur-face viscometer both the interfacial stress, and the shear or dilational strain, and then compute the surface viscosity.

For a _purely elastic_ (Hookian solid) surface, the inter-facial stress tensor $T^{\alpha\beta}$ can be related (4) to the surface strain tensor, $d^{\alpha\beta}$, by:

$$T^{\alpha\beta} = [\sigma + (\Lambda - G) d_\mu^\mu] a^{\alpha\beta} + 2G d^{\alpha\beta} \qquad (4)$$

where

$T^{\alpha\beta}$ = interfacial stress tensor
Λ = interfacial dilatational elasticity, dyne/cm
G = interfacial shear elasticity, dyne/cm
$d^{\alpha\beta}$ = shear strain tensor

When a surface is viscoelastic, i.e., exhibits both viscous as well as elastic character at the interface, a suitable model can be assumed for the combination of Equa-tions (2) and (4) as is done in the case of bulk rheology. The simplest of these models are the Voigt model, the Maxwell model, and combinations thereof.

With this general discussion of interfacial rheological properties, we shall now turn our attention specifically to the measurements of interfacial shear and dilational properties of both gas-liquid and liquid-liquid interfaces containing surface active agents and other chemical additives. This is followed by the presentation of results and discussion.

VI. MEASUREMENTS OF INTERFACIAL RHEOLOGICAL PROPERTIES

The magnitude of interfacial shear and dilational properties for a particular system depend so much on the method of measurement that one should be very cautious in comparing interfacial viscosities and elasticities measured with different devices.

It is not our intention to review in this paper the various methods used for surface rheological measurements. Some recent reviews on this subject are already available (7,92,8). We shall now briefly describe the equipment used in our laboratory for the measurement of interfacial shear and dilational viscosities and elasticities of fluid interfaces containing adsorbed surfactants and polymeric films.

A. Interfacial Shear Viscosity

A deep-channel viscous traction interfacial viscometer (Figure 2) has been in use in our laboratory over the past ten years (9-16) for determining interfacial shear viscosities at liquid-gas as well as liquid-liquid interfaces. A description of this instrument is given by Gupta and Wasan (17).

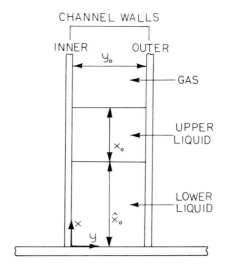

Fig. 2. Cartesian representation of flow in the viscometer

The flow in the viscous traction surface viscometer has been solved for the liquid-gas system by Burton and Mannheimer (18), Mannheimer and Schechter (19) and Pintar, Israel and Wasan (20). The surface shear viscosity (ε) of a Newtonian surface fluid is given by

$$\pi E = \frac{\pi \varepsilon}{\eta y_o} = \frac{V_c^*}{V_c} - 1 = \frac{4}{\pi \, V_c \, \cosh(\pi D)} - 1 \tag{5}$$

where η is the viscosity of the fluid, y_o the channel gap, D the dimensionless depth ($= x_o/y_o$), V_c the dimensionless surface centerline velocity (v_c/v_b) and v_b the velocity of the floor. The quantity V_c^* represents the value of V_c when the interface does not contain the contaminant or for a system of negligible surface or interfacial viscosity.

The method described above which utilizes a small particle at the fluid-fluid interface to determine the interfacial velocity is not particularly suited to a liquid-liquid system where the top liquid is opaque, as is the case with the crude oil-aqueous system. The interface is not viscible through the crude oil phase. This difficulty has been overcome by Wasan, Gupta and Vora (21) who presented a modified method which relates the interfacial shear viscosity at a liquid-liquid interface to the centerline velocity at the liquid-gas surface in a liquid-gas and a liquid-liquid-gas experiment. The dependability of this technique has been recently established in our laboratory by Vora and Wasan (14).

Using a superscript ($\hat{\ }$) to denote quantities associated with the lower liquid and the interface, the results of our analysis can be summarized as

$$\pi \hat{E} = \frac{\pi \hat{\varepsilon}}{\hat{\eta} y_o} = \frac{4}{\pi \hat{V}_c \, \cosh(\pi \hat{D})} - 1 - \frac{\eta}{\hat{\eta}} \frac{[\sinh(\pi D) + \pi E \, \cosh(\pi D)]}{[\cosh(\pi D) + \pi E \, \sinh(\pi D)]} \tag{6}$$

or

$$\pi \hat{E} = \frac{4}{\pi V_c \, \cosh(\pi \hat{D})[\cosh(\pi D) + \pi E \, \sinh(\pi D)]} - 1$$
$$- \frac{\eta}{\hat{\eta}} \frac{[\sinh(\pi D) + \pi E \, \cosh(\pi D)]}{[\cosh(\pi D) + \pi E \, \sinh(\pi D)]} \tag{7}$$

When the dimensionless depth of both the top and bottom liquids (D and $\hat{D}$) are greater than $2/\pi$, Equations (6) and (7) simplify to yield

$$\pi \hat{E} = \frac{4}{\pi \hat{V}_c \, \cosh(\pi \hat{D})} - 1 - \frac{\eta}{\hat{\eta}} \tag{8}$$

or,

$$\pi\hat{E} = \frac{4}{V_c \cosh(\pi\hat{D}) \cosh(\pi D)(1+\pi E)} - 1 - \frac{\eta}{\hat{\eta}} \qquad (9)$$

If the top liquid is opaque, a separate top liquid-gas experiment is run to yield πE using Equation (5). The liquid-liquid-gas experiment yields the value of V_c and using Equation (7) or (9), the value of the interfacial shear viscosity $(\hat{\epsilon})$ at the liquid-liquid interface can be determined. If the top liquid is transparent, Equation (6) or (8) may be conveniently used to evaluate $\hat{\epsilon}$ from the direct measurement of the centerline velocity, $(\hat{V}_c)$, at the liquid-liquid interface.

B. Interfacial Shear Viscoelasticity

The deep channel surface viscometer described earlier can be used to obtain data on surface shear viscoelasticity parameters using the stress relaxation technique developed by Mannheimer and Schechter (19). This technique involves starting the turntable from rest, allowing the dish to rotate at constant speed and then suddenly stopping the dish. The snap-back of the particle at the surface is recorded. Recently, we have used this method to determine the shear viscoelasticities of aqueous solutions of high molecular weight polymer systems. The details of these measurements are reported by Mohan, Malviya and Wasan (22). The two-parameter Maxwell model was employed to represent surface behavior.

We have now developed a suitable theoretical analysis for determining the interfacial shear viscoelasticity at liquid-liquid interfaces containing chemical additives. The results of this new analysis as well as some experimental data will be presented at the International Conference on Colloids and Surfaces to be held in Puerto Rico in June (23).

C. Interfacial Dilational Viscosity and Elasticity

A longitudinal wave technique combined with tracer-particle measurements has recently been developed by Maru and Wasan (8) and used to measure both surface dilational viscosity and elasticity. In this technique, a longitudinal wave disturbance is imparted to an interface by means of a sinusoidally oscillating barrier, and the resulting response of the interface is measured in terms of amplitude and time-lag of a tracer-particle motion relative to the imposed disturbance. The details of the method are available elsewhere (8). Appropriate analysis of hydrodynamics and mass transfer interactions for liquid-gas interfaces have been developed and tested.

We have also recently examined the technique of the motion of single drops through a continuous immiscible liquid medium as a method to evaluate the interfacial dilational viscosity at liquid-liquid interfaces. A theoretical investigation (24) has revealed that the terminal velocity of drops in the Stoke's regime is dependent on the dilational viscosity alone. This technique has been recently employed by Agrawal in our laboratory to measure surface dilational viscosities. However, it is our contention that the interpretation of measured quantities is difficult in this type of experiment and that the new longitudinal wave technique based on the motion of the tracer-particle on the interface is the simpler and more viable method for the determination of both dilational viscosity and elasticity.

VII. MEASUREMENTS OF OTHER PROPERTIES

The surface tension or interfacial tension is obtained by using a standard Cenco du Nouy tensiometer or the Wilhelmy plate method in conjunction with a Cahn Electrobalance and a recorder for values of surface or interfacial tension above 1 dyne/cm. Ultra low interfacial tension values are determined using the spinning drop instrument. This instrument works on the principle of deformation of a spinning drop. It is similar in design to the instrument used by Cayias, Schechter and Wade (25). Our instrument is equipped with close temperature controls. Some of our most recent results on low interfacial tensions in chemically-enhanced oil recovery systems were presented at the Petroleum Chemistry Division Symposium at the American Chemical Society's Centennial meeting in New York (26).

The densities of the aqueous and oil phases are determined using the pycnometer. The viscosities of the bulk phases are measured with a standard Cannon-Fenske capillary viscometer.

In all runs, the aqueous and oil phases were pre-equilibrated.

VIII. EXPERIMENTAL RESULTS AND DISCUSSION

The purpose of this chapter is to review new data on interfacial shear and dilational viscosities and elasticities and interfacial tension of fluid-fluid interfaces involved in chemically-enhanced oil recovery systems. The effects of surface aging on surface rheological parameters and non-Newtonian surface behavior of aqueous surfactant solutions are discussed. A preliminary study involving the effects of a cosurfactant or co-solubilizer and polymer-surfactant interaction on interfacial viscosity is reported. The relationship between interfacial rheological properties and foam and

emulsion stability, and interfacial mass transfer is highlighted with particular emphasis on screening of surfactants
and chemical additives for enhanced oil recovery.

A. Interfacial Viscosity, Interfacial Tension and Oil
    Recovery Tests

       During secondary and tertiary recovery processes, oil
and water come in contact with each other. Surfactants,
naturally occurring as well as commercial additives, adsorb at oil-water interfaces and alter the interfacial
properties. In a typical tertiary flooding process, a
slug of surfactant or a combination of surfactants and cosurfactants or co-solubilizer serves as the oil displacing
agent. In addition, a mobility slug of water solution containing polymer (thickened water) is injected to protect the
slug from water invasion. Finally, drive water is used to
push the surfactant and polymer slugs through the reservoirs.

       Work has been in progress in our laboratory to determine
interfacial viscosities and interfacial tensions of oil-water
systems containing surfactants and polymer additives used by
several investigators in laboratory and/or field tests.

       Table I gives results on interfacial shear viscosities
and interfacial tensions for those selected surfactant systems which exhibit ordinary or low values of interfacial
tensions but do not exhibit ultralow values. In the present
investigation, water-soltrol 130 (odorless mineral spirits,
b.p. 365-405°F) system is used. Interfacial viscosities
were determined using the centerline velocity in the viscous
traction viscometer and Equations (6) and (8).

       Triton X-100 (isooctyl phenoxy polyethoxy ethanol), a
water soluble surfactant, has been used by Dunning et al.
(27), Kennedy and Guerrero (28), Dunning and Johansen (29),
Taber (30,31), Mungan (32), and Dullien et al. (33).
Pluronic L62 (polyoxy propylene-polyoxy ethylene type) has
been used by Dunning and Johansen (29) and Taber (30).
Span 20 (sorbitan monolaurate) has been used by Dunning and
Johansen (29) and Onyx-ol WW (a fatty acid alkanolamide) has
been used by Dullien et al. (33) and Batra (34). All of
these nonionic additives which are found by us to give low
interfacial viscosities ($\leq 10^{-4}$ s.p.) have been already
reported in the literature to yield good oil recovery.
In contrast, a surfactant, Igepal CO-430, reported to give
poor recovery was also studied. While 0.1% Igepal CO-430 in
water-soltrol 130 system exhibited an interfacial tension of
7.56 dyne/cm, the interfacial viscosity was measured to be
about 7.5×10^{-4} s.p. which shows that the interfacial viscosity has increased by almost an order of magnitude.

TABLE I

INTERFACIAL VISCOSITY AND INTERFACIAL
TENSION DATA FOR NON-IONIC SURFACTANTS

Water-Soltrol 130 System

Surfactant	Concentration Wt %	Interfacial Tension, dyne/cm	Interfacial Viscosity s.p.
Triton X-100 (2.5% NaCl in water)	0.1%	2.63	$\lesssim 10^{-4}$
Pluronic L62	0.1%	7.77	$\lesssim 10^{-4}$
Span 20	0.01%	--	$\lesssim 10^{-4}$
Onyx-ol WW (2% NaCl in water)	0.1%	0.3	$\lesssim 10^{-4}$
Igepal CO-430	0.1%	7.56	7.5×10^{-4}

Petroleum sulphonates have recently been used extensively in the laboratory and/or field tests since they give
ultralow interfacial tensions ($\sim 10^{-2} - 10^{-4}$ dynes/cm). In
the present study we have determined interfacial properties
of four different crude oils from South Texas, Oklahoma,
Middle East and Gach Saran covering a range of bulk viscosi-
ties--4.2 to 20 c.p. Petronates used in tertiary oil recovery, namely, Petronate TRS 10-80 (80% active) with an
equivalent molecular weight of 425 and Stepan 107 (63% active)
and Stepan 420 (60% active) with equivalent weights of 425
and 420 respectively were added to brine and studied in the
present work. Table II presents the effect of surfactant
additives on interfacial tension and interfacial shear
viscosities for a Middle East crude oil-brine (1% NaCl) system. The interfacial tension against brine is 16.8 dynes/cm
and the addition of Petronate TRS 10-80 reduces it to a very
low value of 3.4×10^{-2} dynes/cm. The interfacial tension
was determined using the spinning drop technique described
earlier in this chapter. The interfacial shear viscosity
at liquid-liquid interfaces was determined by making velocity measurements at liquid-gas interfaces alone (since the
lighter crude oil was opaque) and using Equations (5) and (7).
The interfacial shear viscosity of crude oil-brine (1% NaCl)
was 7.2×10^{-2} s.p. Addition of 0.2% petroleum sulfonate

TABLE II

INTERFACIAL TENSION AND INTERFACIAL VISCOSITY FOR
MIDDLE EAST CRUDE OIL-BRINE SYSTEM

Aqueous Phase	σ, $\dfrac{dyne}{cm}$	$\hat{\epsilon}$, s.p.
Brine Solution (1% NaCl)	16.8	7.2×10^{-2}
0.2% Stepan 107[*] in Brine	3.8×10^{-2}	3.3×10^{-3}
0.2% Witco TRS 10-80[*] in Brine	3.4×10^{-2}	1.2×10^{-2}
0.3% Witco TRS 10-80[*]	2.0×10^{-2}	8.8×10^{-3}

[*]Percent Surfactant on total basis.

(TRS 10-80 obtained from Witco Chemical Company) reduced the
interfacial shear viscosity to 1.2×10^{-2} s.p. for an outer
dish speed of about 1.5 R.P.M. Increasing the concentration
of sulfonate to 0.3% further reduced the interfacial viscosity to 8.8×10^{-3} s.p. at the same dish speed.
The high viscosity of interfacial films in oil-brine
systems containing a mixture of natural surfactants in the
oil phase has been attributed to the presence of polar
asphaltenes and resins (35-38). The addition of commercial
surfactants to a brine solution can displace some of the
natural surfactants from the crude oil-brine interface and
reduce its interfacial viscosity. An increase in concentration of sulfonate further reduces the interfacial viscosity,
since more natural surfactants are displaced from the interface. Such a behavior of decrease in interface viscosity
as a result of increase in concentration of some surfactants
in a mixed surfactant system has been observed by Gupta and
Wasan (17).
It is seen from Table II that while 0.2% Stepan 107 (a
petroleum sulfonate obtained from Stepan Chemical Co.) in 1%
brine solution reduces the interfacial tension against Middle
East crude as effectively as does 0.2% Witco TRS 10-80 in
brine, the former is more effective in reducing the interfacial shear viscosity. A 0.2% concentration of Stepan 107
reduces the interfacial viscosity to 3.3×10^{-3} s.p., a value
hardly attained even by the addition of 0.3% Witco TRS 10-80
(8.8×10^{-3} s.p.). This indicates the effectiveness of Stepan
107 surfactant in enhancing the interfacial mobility for
Middle East crude oil against 1% brine solution.

It should be noted that recent core flooding tests using fresh 1" x 1" x 12" berea sandstone cores with 5% active sulfonate (Stepan 107) and using a 0.05 pore volume of surfactant slug resulted in higher oil recovery than with Witco TRS 10-80 (39).

Table III compares interfacial tension and interfacial viscosity data for four different crude oil-brine (1% NaCl) systems with and without a sulfonate (0.2% Stepan 107). It is evident from this comparison that more viscous the crude, higher is the interfacial tension and interfacial viscosity. The effect of the addition of a surfactant to the brine solution is to decrease the interfacial tension and the interfacial viscosity.

It should be pointed out that in recent laboratory testing studies using berea cores and various crude oils, it has been observed that the use of a 3% Stepan sulfonate resulted in a 64.2% oil recovery with the Oklahoma crude (6.2 c.p.) as compared to the 52.7% recovery with the Salem-Illinois crude (9 c.p.). These tests were carried out with a 0.05 pore volume of Petrostep 420--Stepan 107 type (39).

We varied the concentration of a petroleum sulfonate to determine its effect on interfacial properties. Figure 3 depicts the variation of both the interfacial tension and interfacial viscosity with a sulfonate surfactant (Stepan 107) in a 1% NaCl solution-Oklahoma crude oil system. The interfacial tension decreases progressively upon increasing the surfactant concentration, while the interfacial shear viscosity decreases first and then increases upon continued addition of the surfactant. Recently, Gupta and Wasan (17) have shown that the surface shear viscosity may both decrease and increase upon the continued addition of another surfactant to a mixed surfactant system. Furthermore, in a commercial system (composed of various surfactants), the concentrations at which both the interfacial shear viscosity and the interfacial tension are minimized cannot be identified by measurements of the interfacial tension alone. These observations clearly point the importance of interfacial viscosity in addition to interfacial tension for tailoring surfactant slugs in improved oil recovery.

We were also interested in determining the effect of salt concentration on the interfacial viscosity and interfacial tension. Figure 4 shows these data for the Gach Saran crude-1% sulfonate (Stepan 107). It is evident that the interfacial viscosity is maximum at 1% salt concentration. These data are reported at the interfacial age of one hour.

TABLE III

COMPARATIVE INTERFACIAL TENSION AND INTERFACIAL VISCOSITY PROPERTIES
OF VARIOUS CRUDE OILS-BRINE SYSTEMS AND STEPAN SULFONATE (107)

Crude Oil	Bulk Viscosity c.p.	1% NaCl		0.2% Stepan 107 with 1% NaCl	
		σ dyne/cm	$\hat{\varepsilon}^{@}$ s.p.	σ dyne/cm	$\hat{\varepsilon}^{@}$ s.p.
South Texas	4.2	2.3	4.2×10^{-3}	3.8×10^{-3}	$\sim 10^{-4}$
Oklahoma	6.2	12.9	1.6×10^{-2}	6.5×10^{-1}	8.6×10^{-3}
Middle East	8.2	16.8	7.2×10^{-2}	3.8×10^{-2}	3.3×10^{-3}
Gach Saran	20	28.5	4×10^{-1}*	~ 5#	2×10^{-1}*

@ surface age of 1 hour

* surface age of 3 hours

interpolated

174

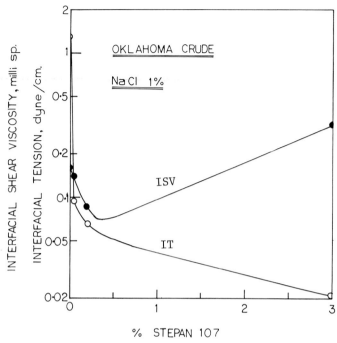

Fig. 3.   *Interfacial properties for Oklahoma Crude against Brine (1% NaCl) containing surfactant Stepan 107*

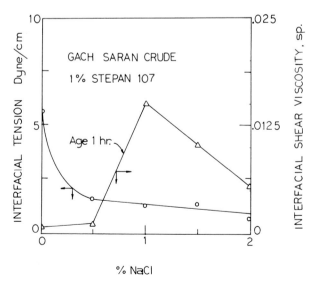

Fig. 4.   *Interfacial properties for Gach Saran Crude against 1% Stepan 107 solution containing different amounts of sodium chloride*

Although, no data are available to-date relating interfacial viscosity at various salt concentrations with the oil
displacement efficiency in a surfactant-enhanced oil recovery
process, Shah and his co-workers (40) have recently conducted
experiments in a secondary recovery process using sandpacks
and measured the percent oil in place recovered when brine
of various NaCl concentrations was pumped through sandpacks.
They found that at 0% NaCl the minimum of oil was recovered
where the interfacial viscosity is maximum, and the maximum
of oil was recovered at 3.5% NaCl where the interfacial
viscosity is minimum. These results confirm the important
role of the interfacial viscosity in oil displacement.

Slattery (41) has theoretically analyzed displacement
of residual oil using an integral momentum balance for the
flow of oil and water in a pore. He has related the displacement velocity in the oil phase to the interfacial
viscosities, interfacial tension, pressure drop across the
pore, pore size and bulk viscosity of the oil as follows:

$$V_o = \frac{A - N_\sigma G}{LD + N_{\kappa+\varepsilon} \dfrac{H}{L}} \tag{10}$$

where V_o = dimensionless displacement velocity in the oil
      phase

  L = ratio of pore length to pore radius

  $N_\sigma = \dfrac{\sigma}{|\Delta P| R}$, dimensionless

  $N_{\kappa+\varepsilon} = \dfrac{\kappa+\varepsilon}{R\eta_o}$, dimensionless

  σ = interfacial tension, dyne/cm
  ΔP = pressure drop over the pore, dyne/cm^2
  R = pore radius, cm
  κ = interfacial dilational viscosity, s.p.
  ε = interfacial shear viscosity, s.p.
  η_o = bulk viscosity of oil, poise

and A,D,G and H are dimensionless groups assumed to be of
order unity.

In order to assess the effect of a change in interfacial
viscosity for the crude oil-water system (Middle East Crude,
Table II) the numerator of Equation (10) A-N_σG may be treated
as constant. Furthermore, for a typical oil bearing rock
structure, R = 5 x 10^{-5} cm and L = 2. Thus the dimensionless
group involving the interfacial viscosities can be estimated.
When ε decreases from 7.2 x 10^{-2} to 8.8 x 10^{-3} s.p. by the
addition of petroleum sulfonate (0.3%), then $N_{\kappa+\varepsilon}$ decreases

from 3.3×10^4 to 4.1×10^3 (assuming κ and ϵ are of similar order). By substituting these values in Equation (10), it is clear that a reduction in the interfacial viscosity by a factor of 0.88 results in an increase in the dimensionless displacement velocity by a factor of 7.2, i.e., velocity is now 8.2 times the original velocity (without surfactant) thereby enhancing oil recovery.

Slattery has concluded from his analysis that the effect of a reduction in the interfacial viscosities upon the efficiency of residual oil recovery is comparable to the effect of a reduction in the interfacial tension when the dimensionless interfacial tension group is sufficiently small. Also, for an interfacial tension less than the critical value, the efficiency of residual oil recovery will increase as the water-oil interfacial viscosities are decreased.

B.   Surface Aging Phenomena
We have observed that the surface viscosity of adsorbed surfactant films changes due to the rapid buildup of a viscous interface (42). This brings out the importance of aging effects and the complexity of the phenomena when a number of natural surfactants are present in the crude oil. Recently, Cash et al. (43) reported data on interfacial tensions of aqueous solutions of petroleum sulfonates against hydrocarbons. The aqueous solutions of petroleum sulfonates were allowed to age for several months. It was found that the interfacial tensions differed substantially with samples of different ages. All the interfacial tensions were measured by Cash et al. without pre-equilibrating the two phases. We are concerned in this chapter with the age dependent interfacial properties of both natural and commercial surfactants when an aqueous surfactant solution is freshly contacted with an oil, each of which has been pre-equilibrated with the other.

Gladden and Neustadter (44) attributed the high stability of Iranian Heavy Emulsion to the rapid buildup of a viscous interface. They also reported that the higher stability of emulsions of the Iranian Crude over Kuwait Crude was due to more significant surface aging effects which lead to increased interfacial viscosity in the Iranian Crude emulsions.

We have taken data on the aging of the liquid-liquid interface in the Gach Saran Crude-brine system with and without the petroleum sulfonate (Stepan 107). Figures 5 and 6 display our measurements for interfacial shear viscosities as a function of the interfacial age which commences when the pre-equilibrated aqueous phase and the oil phase are contacted in the viscous traction interfacial

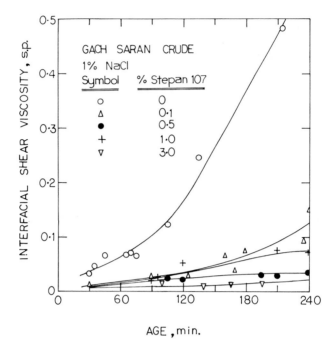

Fig. 5. *Interfacial aging behavior in Gach Saran Crude--*
1% NaCl + Stepan 107 systems

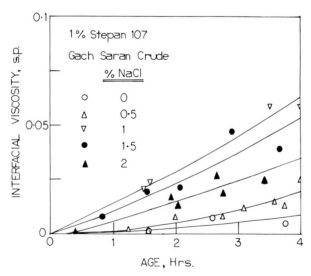

Fig. 6. *Interfacial aging behavior in Gach Saran Crude--*
1% Stepan 107 + NaCl systems

viscometer. The values of the interfacial viscosities for
these pre-equilibrated systems are calculated by extending
our recent analysis for aging liquid-gas interfaces (42).
This extension is a subject of another publication (45).
Work is in progress in our laboratory for non-pre-equilibrated
systems and for systems which have been pre-equilibrated for
various times.

     Figure 5 shows the aging-interfacial viscosity data as
a function of the petroleum sulfonate concentration. These
data are taken with 1% NaCl in the aqueous phase. In the
absence of any commercial sulfonate, the interfacial viscosity of the Gach Saran Crude-brine solution increases very
rapidly over a period of two to three hours and the interface becomes highly viscous. However, the effect of the
addition of petroleum sulfonate is to decrease this rise
in interfacial viscosity with increasing age of the interface. Furthermore, the data indicate that at a given age
there is no systematic trend in interfacial viscosity behavior with increasing concentration of the sulfonate. This
is especially true if the oil phase is a crude oil. Figure
6 shows the variation of interfacial viscosity with aging
of the interface as a function of the varying concentration
of sodium chloride in the aqueous solution for a fixed concentration of the petroleum sulfonate (1%). As also pointed
out earlier, these data show that at a given age of the
phase interface, a 1% sulfonate solution with 1% NaCl gives
the highest value of interfacial viscosity. However, there
is no simple correlation between these trends with varying
amounts of salt and with varying amounts of petroleum sulfonate for a real crude oil. Further measurements are
needed to establish definitive trends and determine the
mechanism of aging process. These data would also give us
insight into the structural transformations in the interfacial layers and their influence on the oil displacement
process.

C.   Non-Newtonian Behavior
     The rheological measurements at fluid-fluid interfaces
containing mixed surfactant systems are further complicated
by the fact that these systems often display shear-dependent
behavior. Some attempts have been made to interpret the
shear-dependent surface flow behavior in a viscous traction
surface viscometer (19,20,46). By analogy with the bulk
rheology of polymer solutions, Schechter et al. (47) and we
(14) have attempted to correlate the non-linear surface shear
flow behavior of commercial surfactants using the molecular
approaches similar to those of Bueche (48), Takemura (49) and
Graessley (50). A generalized correlation of our surface

shear viscosity data for the two different concentrations of
mixed surfactant system consisting of an aqueous solution of
Duponol RA--an anionic detergent and carboxy methyl cellulose--
a polymeric additive is shown in Figure 7. This is a plot
of reduced surface shear viscosity ($\frac{\varepsilon}{\varepsilon_o}$ = viscosity at any
shear rate/viscosity at zero shear rate) versus the product
of shear rate and relaxation time ($\dot{\gamma}\tau_i$). Figure 8 presents
data of all other investigators for spread and adsorbed surfactant films including our own data on mixed solutions of
lauryl alcohol-sodium lauryl sulfate. These data are taken
at steady state. In Figures 7 and 8, we can see that in the
limiting case $\dot{\gamma}\tau_i \to 0$, $\frac{\varepsilon}{\varepsilon_o} \to 1$, which depicts the Newtonian
surface fluid behavior at very low shear rates. For higher
values of $\dot{\gamma}\tau_i$, $\varepsilon/\varepsilon_o$ seems to be proportional to $(\dot{\gamma}\tau)^{-0.486}$
for the systems depicted in Figure 7 and to $(\dot{\gamma}\tau)^{-0.761}$ for
the systems displaying high surface viscosity (Figure 8).
If we recall the bulk rheological theory of Graessley, it
was shown that the slope of the reduced viscosity plot is
$-.75$ for concentrated polymer solutions. On the other hand,
Bueche showed that for dilute polymer solutions the slope
of the reduced viscosity plot is -0.5. Our data shown in
Figure 7 for the low surface viscosity gives the slope of
-0.486. The close agreement between these asymptotic slopes
for the reduced viscosity plots for the bulk and for the
surface flow may be somewhat fortuitous and this concept of
analogy needs to be verified further. However, it is important to point out that this reduced plot for the surface
shear flow behavior may be quite useful in predicting non-
Newtonian surface shear viscosity in the absence of any
good surface flow data and for correlating of data. The
need for such a unified approach for predicting surface
viscosity from a generalized correlation is self-evident.

Work is in progress in our laboratory to determine the
interfacial shear viscosities--shear dependence of mixed
surfactant systems used in enhanced oil recovery processes.
Measurements need to be made at various ages of the interface. This presents a more complex problem in that the
non-Newtonian behavior of the interfacial viscosity becomes
more difficult to assess.

D.   Interfacial Viscoelasticity

The viscoelastic behavior of fluid-fluid interfaces
containing surfactants is much less studied than the purely
viscous resistance, even though the remarkable stability of
emulsions and foams has been attributed to the interfacial
viscoelasticity. In recent years, attempts have been made
by Mannheimer and Schechter (19), Kott et al. (51) and us

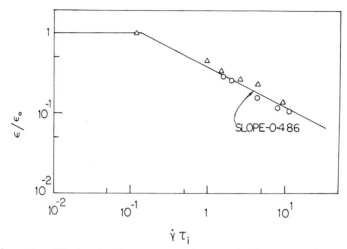

Fig. 7.   *Plot of $\varepsilon/\varepsilon_o$ versus $\dot\gamma\tau_i$:   0.1% Duponol RA + 1% CMC, O; 1% Duponol RA + 1% CMC, $\triangle$*

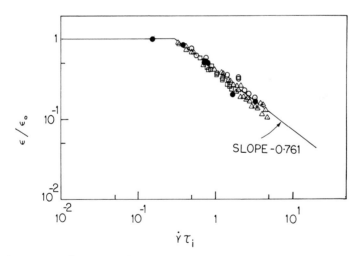

Fig. 8.   *Plot of $\varepsilon/\varepsilon_o$ versus $\dot\gamma\tau_i$:   Joly (94), Mannheimer and Schechter (19), $\triangle$; Suzuki (95), ● ; Vora and Wasan (14), O*

(22,92) to use the stress relaxation technique in conjunction with the deep-channel viscous traction surface visco-meter to study the shear viscoelasticities at both spread and adsorbed surfactant systems. In a forthcoming publica-tion, Mohan, Malviya and Wasan (22) report surface shear viscoelasticity data on aqueous solutions of commercial surfactants, polymers and polymer plus surfactants. They

employed the two-parameter Maxwell model to interpret their
data. This analysis for the liquid-gas interface has been
most recently extended by Mohan and Wasan (23) to the determination of the interfacial shear viscoelasticity phenomenon
at the liquid-liquid interfaces. A partially hydrolyzed
polyacrylamide solution against Soltrol 130 showed finite
shear viscosity as well as shear elasticity. These data are
reported elsewhere (23). The method used for measurement of
the shear viscoelasticity at liquid-liquid interfaces is
similar to that used by Wasan et al. (21) for the interfacial
shear viscosity measurements in a liquid-liquid-gas system.
Work is in progress in our laboratory to assess the effect
of the addition of both petroleum sulfonates and polymer
additives on the interfacial viscoelastic properties.

E.   Interfacial Dilational Viscosity and Elasticity
     We are in the process of adapting our surface longi-
tudinal wave apparatus which employs the motion of a tracer
particle on the surface, for the measurement of both interfacial dilational viscosity and elasticity in a liquid-liquid
system. A preliminary set of experiments has been conducted
with the Gach Saran Crude-brine (1% NaCl) system. Figure 9
shows our preliminary data plotted as the wave amplitude (in
terms of dimensionless amplitude ratio) as measured by the
particle motion and the phase lag ϕ between the barrier and
the particle motion, as a function of the dimensionless
distance from the barrier, X/L. These data are taken at a
barrier speed of 0.261 radians/sec with the barrier amplitude
of 0.23 cm and a period of 24 seconds. The half length of
trough, L equals 30 cm. It is seen from this figure that the
wave amplitude is decreased and the phase lag is increased
due to the addition of petroleum sulfonate (1%) in the Gach
Saran Crude-brine system. Although, we have not completed
our analysis of these data to calculate both the dilational
viscosity and elasticity, these raw data appear to indicate
that the values of these properties will be decreased due to
the presence of petroleum sulfonates.

F.   Emulsion Stability
     In a typical oil recovery process, oil and water are
both forced to the production well either by natural forces
(primary recovery) or by driving forces (secondary and
tertiary recovery processes). During their sojourn, ample
opportunity is given to the oil and water to yield an emulsion even if they had not already been in an emulsion form.
Release of dissolved gas further aids in the emulsification
process. Furthermore, emulsification can occur when the
oil-water mixture passes through perforations in the drill

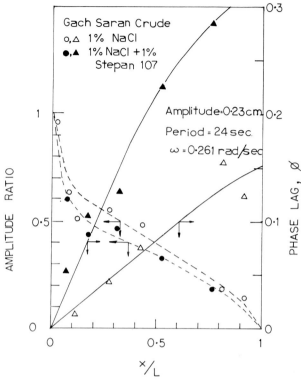

*Fig. 9.   Effect of 1% Stepan 107 on amplitude ratio and
phase lag in longitudinal wave experiments*

pipe, and through chokes, valves and other portions of the
piping system. Also, the presence of naturally occurring
emulsifiers in the crude oil renders highly stable emulsions.
These emulsion stabilizers are complex, polynuclear hydrocarbons and vary from one production site to another. Other
hydrodynamic factors for emulsification include channeling
and breakthrough, when concentrated portions of surfactants
used in a tertiary oil recovery process reach and mix with
the driving fluid causing severe emulsion problems. The
selection of a demulsifier to handle these situations is
difficult, though it is well known that certain simple
materials such as soaps and surfactants would resolve these
emulsions. In some situations a surfactant flood will
solubilize some of the oil underground and disperse the rest
which will later coalesce to form a continuous oil bank which
can be driven to the production well.

The present state-of-art reveals that surfactant selection for a tertiary oil recovery process is made on the basis of ultralow interfacial tension between the crude and the aqueous phase. However, such ultralow values of interfacial tension also cause immense emulsification problems.

From the point of view of interfacial rheology, the remarkable stability of emulsions has been attributed to the interfacial viscoelasticity. It has been shown that the requirements for stability are the presence of a film of high viscosity, the elasticity being of less importance, and a film of considerable thickness. It has also been shown that the rates of coalescence are very sensitive to pH and that the peak in the plots of interfacial shear viscosity and shear elasticity against pH coincides with the maximum in the coalescence times. The coalescence times are reported to be substantially lower in acid media than in alkaline surroundings.

Schulman and Cockbain (52) reported an increase in the stability of emulsions in the presence of lauryl alcohol. This was attributed to the presence of complexes at the oil-water interface which increased the strength of the film against rupture (53,54). A number of investigators (55-61) report an increase in the stability of emulsions due to films of high viscosity. Typically Cumper and Alexander (60) found that a high interfacial viscosity correlated well with drop stability in oil-water systems stabilized by proteins. However, there is some evidence of a stable emulsion in a low interfacial viscosity system (62).

Vold and Mittal (63) reported enhanced stability when the surfactant was less soluble in the bulk. Specifically, lauryl alcohol produced more stable emulsions in Nujol than in olive oil due to the lesser solubility in the former. Our own studies (64) on the behavior of interfacial films indicate that significant shear viscosities are exhibited at the interface when the surfactant is of low solubility in the bulk. Thus, it appears that the results of Vold and Mittal may also be interpreted to mean that an increased stability occurs in a system of larger interfacial rigidity.

The high stability of emulsions of sea water in fuel oil has been attributed to very stable asphaltic films. Petrov et al. (37) and Levchenko et al. (38) ascribed the stability of oil-water emulsions to asphaltenes or silica gel resins depending on the particular crude. Berridge et al. (35) found that the emulsion stability was related to the amounts of asphaltene and vanadium. Fiocco (65) remarked that, with film forming materials, there is little tendency for droplets smaller than 100 microns to coalesce. He added that when a demulsifier is present at the interface, the interfacial viscosity remains low and coalescence is favored.

It has been noted that the stabilizing agents also exhibit time dependent behavior at the interface (65,44).

In his review on "Emulsion Stability and Demulsification," Carroll (66) covers a number of interesting aspects of emulsion stability and, in particular, refers to the work of Sonntag and Strenge who discuss methods of replacing a surfactant at the interface by large amounts of a weakly surface active material like silicones. Measurements of interfacial shear viscosity indicate that, while the interface exhibits aging effects, in the presence of a demulsifier the viscosity remains low.

We have indicated earlier that a reduction in interfacial shear and dilational viscosities may enhance the displacement velocity of an oil-water interface. A second effect of a reduction in interfacial rigidity is to enhance the film thinning rate. Good (67) developed a model for gradient coalescence and showed that the velocity of approach of a drop to an interface increases as the dimensionless interfacial viscosity group $(\kappa+\varepsilon)/R\hat{\eta}$ decreases. Our data on the interfacial shear viscosities (Table III) reveal that the addition of a surfactant Stepan 107 reduces the interfacial viscosity of different crudes against brine. This suggests that petroleum sulfonate formulations in certain concentration ranges may favor coalescence process.

To study the coalescence promotability of these surfactants, we conducted emulsion stability tests in the following manner. Emulsions were generated by agitating 20 ml of the oil phase with 80 ml of the aqueous phase containing chemical additives, in a magnetic stirrer at constant setting for ten minutes. The phase separation of the emulsion so formed is observed in a measuring cylinder by following the progress of the settling front.

The bulk of the aqueous phase separates rapidly, 1-2 minutes in the present study. The oil droplets flocculate during this period. The aqueous phase then drains from between the deformed oil droplets until the surface film ruptures resulting in coalescence (Figure 10). The 'coalescence front' starts from the top and moves downward. The setting of oil droplets continues to occur at the 'settling front' which moves upward. In crude oil systems, the coalescence front cannot be observed due to the opacity of the oil phase. Therefore in all the runs reported here, the progress of phase separation was measured in terms of the position of the settling front.

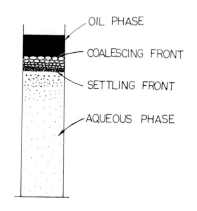

Fig. 10.   Schematic of phase separation

Figure 11 shows the phase separation of Middle East Crude-brine (1% NaCl) emulsions with and without surfactant additives. As can be seen, the emulsions formed without surfactants in the aqueous phase are quite stable and only 83.8% separation occurs. The oil phase sticks as patches on the walls of the measuring cylinder and are not released. With a 0.2% Witco TRS 10-80, the phase separation is improved. The oil patches were still to be observed on the cylinder walls but when the surfactant was present, these oil patches flowed upward as small rivulets. When 0.2% Stepan 107 was added to the brine, the phase separation was dramatically improved. Complete phase separation occurred in 15 minutes. This was expected since Stepan 107 is more effective than TRS 10-80 in reducing the interfacial shear viscosity.

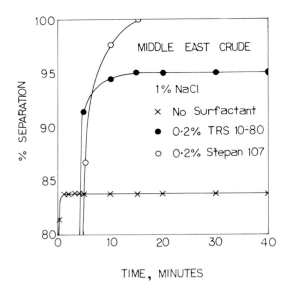

Fig. 11.   Phase Separation of Middle East crude emulsions--Effect of surfactants TRS 10-80 and Stepan 107

This preliminary observation was followed by a more extensive data on phase separation of a Gach Saran Crude-brine emulsion with varying amounts of Stepan 107 and NaCl. Figure 12 shows a photograph of the phase separation with and without the addition of the surfactant Stepan 107. Figure 12a reveals the fact that for emulsions in 1% NaCl without surfactants, oil patches are observed on the cylinder walls and that separation is poor. A dramatic improvement in phase separation is observed (Figure 12b) by the addition of 0.05% Stepan 107. Figure 13 shows the variation of phase separation and interfacial shear viscosity with the percentage of surfactant added (1% NaCl). The interfacial shear viscosity first decreased and then increased before falling again. The inverse trend, namely, an initial rise followed by a decrease and a final rise was observed in the phase separation. However, it is to be noted that the position of the first maximum in phase separation and the first minimum in interfacial viscosity do not correspond possibly due to other factors such as an initial drastic reduction in interfacial tension.

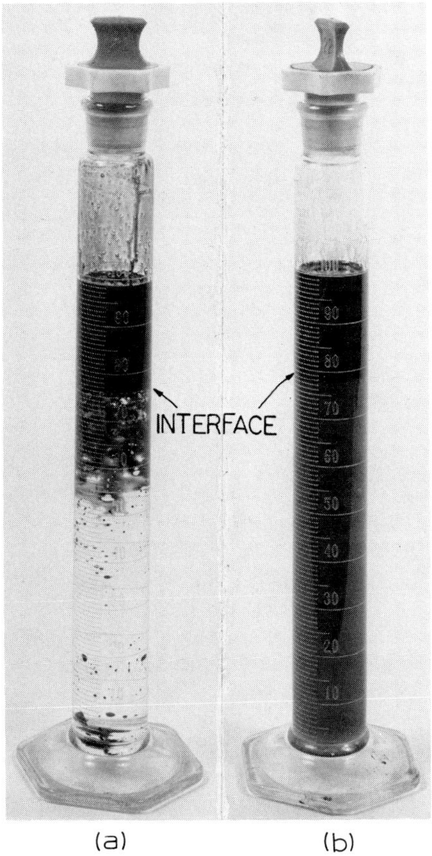

(a)   (b)

Fig. 12. Photograph of the phase separation in Gach Saran Crude emulsions: (a) 1% NaCl (b) 1% NaCl + 0.05% Stepan 107

Figure 14 shows a similar plot of percent separation, interfacial viscosity and interfacial tension for varying salt concentrations. The trends between phase separation and interfacial shear viscosity correspond except in the low concentration range of NaCl. This is possibly due to electrical repulsion between droplets in the absence of NaCl.

A cosurfactant is commonly added to a surfactant to (i) aid in solubilizing the surfactant and (ii) reduce surfactant loss by adsorption.

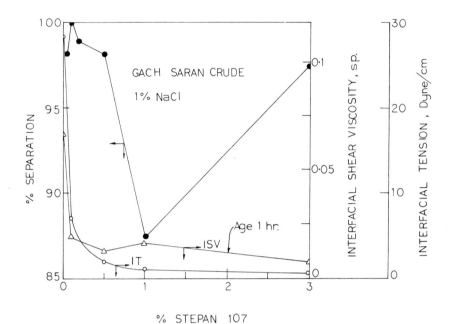

Fig. 13.   *Phase separation at 10 minutes and interfacial properties in Gach Saran Crude - 1% NaCl + Stepan 107*

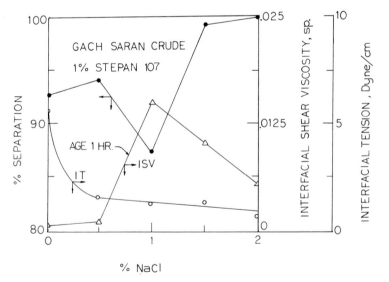

Fig. 14.   *Phase separation at 10 minutes and interfacial properties in Gach Saran Crude - 1% Stepan 107 + NaCl systems*

Furthermore, a cosurfactant also enhances the viscosity of the surfactant slug thus providing mobility control. We explored the effect of a typical cosurfactant on the coalescence phenomenon. Table IV shows the effect of 1% 1-Hexanol added to 1% NaCl + 1% Stepan 107 on the stability of a Gach Saran crude emulsion. The addition of the cosurfactant lowers both the interfacial tension and the interfacial shear viscosity, and also increases the phase separation. These data throw a new light on the possible role of a cosurfactant in promoting coalescence of oil ganglia and thereby enhancing the oil bank formation.

It is recognized that the above discussion presents an oversimplified interpretation of data. In reality, the composition of the aqueous phase has a wider role in that all interfacial and bulk properties are altered. Furthermore, the data on interfacial shear viscosity reported above corresponds to a surface age of one hour. It is to be noted from the aging behavior reported in Figure 5 that when surfactants are present in sufficient amounts, the interfacial shear viscosity is significantly lowered and aging effects are decreased.

G. Foam Stability
      The use of foam as an effective mobility control agent between the micellar solutions and the drive fluid in miscible oil displacement processes has been recently recognized (68). The use of stable foam to improve displacement efficiency in a gas-displacing oil process, to restrict the flow of undesired fluids and improve the sweep efficiency of fluid injection processes in heterogeneous porous media has been well demonstrated (69,70). Although the mechanism for foam stabilization is complex, several investigators have noted the effect of interfacial rheological properties such as surface viscosities on the stability of foams (71-74).

Aqueous solutions of pure sodium lauryl sulfate yield foams of very low stability because they exhibit a very low surface viscosity and a high rate of drainage of solution in foam lamellae. When lauryl alcohol, lauric acid or cetyl alcohol is added to aqueous solutions of sodium lauryl sulfate, a correlation between surface shear viscosity and foam lifetime could be established. Systems of fatty acids--fatty alcohols (decanoic acid-decanol, octanoic acid-octanol) exhibit foam stability at a molar ratio of 1:3 and 9:1, respectively. At the same molar ratio in these systems, the rate of drainage was minimum and the surface shear viscosity was maximum (75).

Mahajan and Wasan (64) determined the surface concentration in mixed surfactant systems such as sodium lauryl sulfate-lauryl alcohol or cetyl alcohol and found that the surface

TABLE IV

EFFECT OF COSURFACTANT ON STABILITY OF
GACH SARAN CRUDE EMULSIONS

Aqueous Phase	Interfacial Tension (dyne/cm)	Interfacial Viscosity (sp)	% Separation at 10 Minutes
1% NaCl + 1% Stepan 107	1.34	0.015	87.5
1% NaCl + 1% Stepan 107 + 1% 1-Hexanol	0.19	$\sim 10^{-4}$	93.8

density of the molecules of the sparingly soluble component
such as lauryl alcohol indicated by surface area in $\overset{\circ}{A}^2$/mole-
cule increases substantially near the CMC (critical micelle
concentration). This leads to a closer packing of the
molecules, a more densely packed film, and, hence, a higher
surface shear viscosity at these concentrations. As the
concentration is increased beyond CMC, the surface viscosity
is decreased. Kanda and Schechter (76) in a recent attempt
to relate surfactant properties of the foam to fluid displacement efficiency, studied a number of surfactant systems
including a commercial grade of sodium lauryl sulfate (which
normally contains lauryl alcohol as its chief impurity), and
found that increasing the surfactant concentration had an increasing effect on the displacement efficiency. The displacement efficiency was found to be a maximum at the CMC
for the sodium lauryl sulfate system. These authors concluded that if displacement efficiency is the practical goal,
then a large surface viscosity is desirable.

In the above discussion, the role of surface dilational
viscosity and elasticity in foam stabilization has not been
delineated. This is because little meaningful research has
been reported to-date to establish the effect of surface
dilational property on foam behavior. The recent efforts of
Lucassen and Giles (77), Van den Tempel and his co-workers
(78) and Prins and Van Voorst Vader (79) in this regard are
especially noteworthy.

Research work in this application area is under way in
our laboratory (80). A preliminary set of experiments have
been completed using the longitudinal wave technique with
the tracer-particle method to determine the surface dilational viscosity and elasticity of surfactant systems.
Figure 15 shows our preliminary data plotted as the amplitude
ratio and phase lag as a function of the distance from the
moving barrier. These data are shown for two different concentrations of purified sodium lauryl sulfate. It is seen
that wave amplitude at 500 ppm of sodium lauryl sulfate is
smaller than that at 1000 ppm and the phase lag is larger
at 500 ppm. Using the analysis recently developed by Maru
and Wasan (8), we have computed both the net surface dilational viscosity and elasticity for these systems. These
results are tabulated in Table V. It is seen from this
table that both the net surface dilational viscosity and
elasticity increase as the concentration of the surfactant
increases, although the increase in the surface dilational
viscosity is much more pronounced. However, more accurate
data are needed to draw significant conclusions from these
measurements and many facets of these properties measurements remain areas of research.

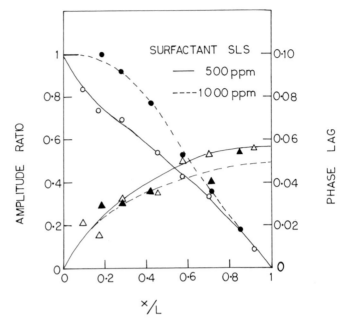

*Fig. 15. Effect of sodium lauryl sulfate on amplitude ratio
and phase lag in longitudinal wave experiments*

TABLE V

DILATIONAL VISCOELASTIC PARAMETERS
===
(ω = 0.261 rad/sec)

SYSTEM	E_n^* dyne/cm	$\eta_n^@$ gm/sec
SLS 500 ppm	0.400	0.202×10^{-2}
SLS 1000 ppm	1.180	0.455

$^*E_n$ = Net dilational elasticity

$^@\eta_n$ = Net dilational viscosity

H.   Interfacial Mass Transfer

It has been suggested that emulsification is largely due to an interfacial instability caused by interfacial mass transport of surfactants. Cayias et al. (25) have discussed spontaneous emulsification as a possible mechanism for enhanced oil recovery. Furthermore, the film between the oil drops drains more slowly when the surfactant is present in the film than when it is present in the drop, so that the mass transport of surfactants to the drop becomes an important factor in screening surfactants, co-solvents or demulsifiers in emulsification/demulsification in oil recovery. Furthermore, during the movement of a water-oil interface through a porous medium, the interface deforms and the surface active agents are redistributed. This may result in a considerable interfacial activity. This redistribution of material at the oil-brine interface may be the key in understanding the "jump" phenomenon in oil displacement process.

A well-planned experimental program is in progress in our laboratory which addresses to this important problem of correlating interfacial surfactant transport with interfacial rheological properties of oil-water interfaces.

A mass transfer cell, a modified version of the Lewis Cell (81), and similar to the one employed by Austin and Sawistowski (82) is used. The cell is designed to yield a well-defined interface and uniform concentrations in the bulk of the liquid phases. The bulk phases are stirred by independently adjustable stirrers, and the interface is held symmetrical with respect to the two stirrers. Samples can be withdrawn from each of the bulk phases. These samples are analyzed for their composition by suitable techniques such as refractive index, electrical conductivity, titration, gas chromatography or spectrophotometry.

We have carried out preliminary experiments to determine interfacial transport rates of solutes in the presence of monolayers of Bovine Serum Albumin (BSA) deposited on aqueous substrates. The system used is benzene-water containing isopropanol in the aqueous phase which is transferred to the organic phase. This system has been chosen because Davies and Mayers (83) have used the same system so that we could check our mass transfer data against theirs. We have also measured the interfacial shear viscosity of the benzene-water system containing BSA using our viscous traction interfacial viscometer and Equations (6) and (8). Figure 16 displays our data for interfacial mass transfer coefficient and interfacial viscosity as a function of surface concentration of the surfactant. It is seen from this figure that by the presence of the surfactant film, the mass transfer

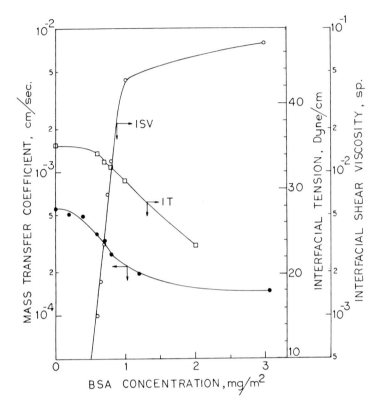

Fig. 16.   *Variation of mass transfer coefficient and interfacial shear viscosity with the BSA concentration for the benzene-water + isopropanol system*

coefficient has been reduced and the interfacial shear viscosity has been increased from a value of about 10^{-4} to 10^{-1} s.p. The interfacial viscosity is very low ($\sim 10^{-4}$) for concentrations of BSA less than 0.5 mg/m^2, and there is a sharp rise in the interfacial viscosity for BSA concentration ranging from 0.5 to 0.8 mg/m^2. It should be remarked that although our mass transfer data agreed well with those reported by Davies and Mayers (83), our interfacial shear viscosity values differ from theirs. As pointed out by Burton and Mannheimer (18), the method used by Davies and Mayers has several drawbacks and, consequently, their values for interfacial viscosity are not very accurate.

Figure 17 shows a plot of interfacial resistance to mass transfer versus interfacial shear viscosity for the benzene-water-isopropanol system containing BSA. The interfacial

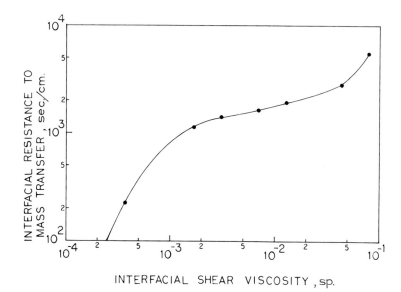

*Fig. 17.   Interfacial resistance versus interfacial shear
viscosity for the benzene-water + isopropanol
system containing BSA*

resistance values are calculated from the mass transfer data.
     We are attempting to develop a simple model for cor-
relating the interfacial diffusional transport to the interfacial shear and dilational viscosities and elasticities.
Various theoretical and semitheoretical equations are available to relate bulk shear viscosity of the solvent and the
dimensions of the solute to its diffusivity and, in turn, to
the diffusional resistance of the bulk to solute mass transfer. However, no theory is available to calculate interfacial resistance from easily measured interfacial properties, like the interfacial ·shear viscosity. It would
therefore seem important to have some molecular interpretation of the interfacial properties of specific surfactant
systems to establish this correlation.

IX.   CONCLUDING REMARKS
     The earlier observations of interfacial rheological
properties like the surface shear viscosity of fluid-fluid
interfaces containing surfactants and polymer additives were
qualitative in nature, as no accurate measurement techniques
were available to determine their absolute values. Consequently, these properties were not recognized as significant
parameters in controlling dynamic rate processes in industrial systems. In addition, much of the attention was

focused on deposited monolayers or single surfactant systems,
the behavior of which differs from the complex phenomena
observed in adsorbed systems as well as mixed or commercial
surfactant systems where molecular complexing, molecular
interactions or structural transformations occur. Furthermore, surface shear viscosity measurements alone were employed to establish or explain many complex phenomena
observed in industrial processes without due regard to the
role of other important interfacial properties such as
dilational properties. For example, a number of attempts
have been made to explain foam drainage with the help of
surface shear viscosity alone (84-89,70,90) despite the
predominant dilational strains present during drainage.

Recently, Maru and Wasan (8) pointed out that interfacial dilational properties have often been ignored or
misunderstood. One main reason for misinterpretation of the
dilational properties is the confusion added by the mass-
exchange effects accompanying dilational strains. This difficulty is not encountered when only shear strains (such as
in the deep channel viscous traction surface shear viscometer)
are present. Levich (90) questioned the validity of the dilational viscosity as a true interfacial property in Boussinesq's
constitutive equations. He claimed that dilational viscosity
in Boussinesq's equation for terminal velocity of drops and
bubbles can be interpreted solely in terms of mass transfer at
the surface and between the surface and the bulk. His claim
implies a total denial of any intrinsic interfacial rheological properties.

In spite of this aura of misunderstanding and confusion,
several investigators have recognized the importance of interfacial shear and dilational viscosities and elasticities with
the consequence that some good instrumentation has been
developed for their measurements. As we have pointed out
earlier, we are using some of the most viable techniques in
our laboratory for making measurements of interfacial rheological properties of fluid-fluid interfaces containing
mixed or commercial surfactants and polymeric materials,
which may very well shed some light on complex mechanisms
in several areas of industrial processes such as foaming
and antifoaming, emulsification and demulsification, mass
transfer, and fluid phase separation processes. In particular, our research program at this stage of development is
directed towards an important aspect of improved oil recovery by surfactant and polymer flooding, in which we seek
to correlate interfacial rheology of oil-water interfaces
with both coalescence rates of oil drops in flow through
porous media and emulsion stability. We have presented here
only a tentative discussion of the data as in some instances
only preliminary data have been gathered on interfacial

rheological properties. Therefore, more extensive data need
to be taken and a refined interpretation of the data needs
to be sought to give a definitive insight into the mechanisms
of oil drop displacement and of oil bank formation. Furthermore, future effort also needs to be directed toward the
molecular interpretation of interfacial properties if our
overall effort is to result in an advance in fundamental
knowledge.

X.   ACKNOWLEDGEMENT
     This work has been supported by the NSF Grant No. GK-
43135. Many of the ideas presented here are direct results
of the efforts of many of the present as well as former
students in our research laboratory and we are grateful to
them for their contributions.

XI.   LITERATURE CITED
1.   Goodrich, F. C., "On the Damping of Water Waves by Mono-
        molecular Films," J. Phys. Chem. 66, 1858 (1962).
2.   Eliassen, J. D., "Interfacial Mechanics," Ph.D. Thesis,
        Univ. of Minnesota, Minneapolis (1963).
3.   Boussinesq, J., "The Existence of Surface Viscosity in
        a Thin Layer Separating One Liquid from Another Con-
        tinuous Fluid," Ann. de Chemie Physique 29, 349; 357;
        364 (1913).
4.   Oldroyd, J. G., "The Effect of Interfacial Stabilizing
        Films on the Elastic and Viscous Properties of Emul-
        sions," Proc. Royal Soc. (London) A232, 567 (1955).
5.   Scriven, L. E., "Dynamics of Fluid Interface," Chem. Eng.
        Sci. 12, 98 (1960).
6.   Slattery, J. C., "Surfaces I," Chem. Eng. Sci. 19, 379
        (1964); "Surfaces II," ibid. 453.
7.   "Rheological Properties of Monomolecular Films. Part I:
        Basic Concepts and Experimental Methods," Surface
        Colloid Sci. Vol. 5 (E. Matijevic, Ed.), Wiley-
        Interscience, New York (1972).
8.   Maru, H. C. and Wasan, D. T., "The Interfacial Visco-
        elastic Properties from Longitudinal Wave Measure-
        ments," Manuscript in preparation (1976).
9.   Pintar, A. J., "The Measurement of Surface Viscosity,"
        Ph.D. Thesis, Illinois Institute of Technology,
        Chicago (1968).
10. Israel, A. B., "Surface Viscosity Measurements of Sur-
        factant Solutions," M.S. Thesis, Illinois Institute
        of Technology, Chicago (1968).
11. Gupta, L., "The Role of Curved Interfaces in Surface
        Viscometric Measurements," M.S. Thesis, Illinois
        Institute of Technology, Chicago (1970).

12. "Interfacial Shear Viscosity of Films Adsorbed from Surfactant Solutions," Ph.D. Thesis, Illinois Institute of Technology, Chicago (1972).

13. Wasan, D. T., "Interfacial Phenomena at Fluid Interfaces: Interfacial Viscosity and Emulsion Stability," Report, API Research Project No. 133 (Mar., 1972).

14. Vora, M. K. and Wasan, D. T., "Interfacial Shear Viscosity of Adsorbed Films," Manuscript in preparation (1976).

15. Shah, S. T. and Wasan, D. T., "Rheological and Thermodynamical Aspects of Fluid Interfaces," Manuscript in preparation (1976).

16. Goyal, A. and Wasan, D. T., "Interfacial Shear Viscosity of Surfactants and Polymer Solutions used in Tertiary Oil Recovery," Manuscript in preparation (1976).

17. Gupta, L. and Wasan, D. T., "Surface Shear Viscosity and Related Properties of Adsorbed Surfactant Films," Ind. Eng. Chem. Fundam. 13, 26 (1974).

18. Burton, R. A. and Mannheimer, R. J., "An Analysis and Apparatus for Surface Rheological Measurements," Ordered Fluids and Liquid Crystals, Advances in Chem. Ser. 3, 315 (1967).

19. Mannheimer, R. J. and Schechter, R. S., "An Improved Apparatus and Analysis for Surface Rheological Measurements," J. Colloid Interface Sci. 32, 195 (1970).

20. Pintar, A. J., Israel, A. B. and Wasan, D. T., "Interfacial Shear Viscosity Phenomena in Solutions of Macromolecules," J. Colloid Interface Sci. 37, 52 (1971).

21. Wasan, D. T., Gupta, L. and Vora, M. K., "Interfacial Shear Viscosity at Fluid-Fluid Interfaces," AIChE J. 17, 1287 (1971).

22. Mohan, V., Malviya, B. K. and Wasan, D. T., "Interfacial Viscoelastic Properties of Adsorbed Surfactant and Polymeric Films at Fluid Interfaces," Can. J. Chem. Eng. in press (1976).

23. Mohan, V. and Wasan, D. T., "Interfacial Viscoelastic Properties of Surfactant Films at Liquid-Liquid Interfaces," Paper to be presented at the International Conference on Colloids and Surfaces, San Juan, Puerto Rico (June 21-25, 1976).

24. Agrawal, S. S., "Measurement of Interfacial Dilational Viscosity of Surfactant Films by the Motion of Fluid Globules," M.S. Thesis, Illinois Institute of Technology, Chicago (1975).

25. Cayias, J. L., Schechter, R. S. and Wade, W. H., "The Utilization of Petroleum Sulfonates for Producing Low Interfacial Tensions Between Hydrocarbons and Water," Private communications; see also "Adsorption at Interfaces," ACS Symp. Ser. No. 8, 234 (1975).

26. Wasan, D. T. and Mohan, V., "Interfacial Rheology in Chemically-Enhanced Oil Recovery Systems," Paper presented at the Symp. on Advances in Petroleum Recovery, ACS, N.Y. (Apr. 4-9, 1976).

27. Dunning, H. N., Gustafson, H. J. and Johansen, R. J., Displacement of Petroleum from Sand Surfaces by Solutions of Polyoxyethylated Detergents," Ind. Eng. Chem. 46(3), 591 (1954).

28. Kennedy, H. T. and Guerrero, G. T., "The Effect of Surface and Interfacial Tensions on the Recovery of Oil by Waterflooding," Trans. AIME 201, 124 (1954).

29. Dunning, H. N. and Johansen, R. J., "Laboratory Evaluation of Water Additives for Petroleum Displacement," Bureau of Mines, Report of Investigations 5352 (July, 1957).

30. Taber, J. J., "The Injection of Detergent Slugs in Water Floods," Petrol. Trans. AIME 213, 186 (Mar., 1958).

31. _____, "Dynamic and Static Forces Required to Remove a Discontinuous Oil Phase from Porous Media Containing both Oil and Water," Soc. Petrol. Engrs. J. 3 (Mar., 1969).

32. Mungan, N., "Role of Wettability and Interfacial Tension in Waterflooding," SPE 705, Paper presented at Soc. Petrol. Engrs., AIME, Dallas, Texas (Oct. 6-9, 1963).

33. Dullien, F.A.L., Dhawan, G. K., Gurak, N. and Babjakt, L., "A Relationship Between Pore Structure and Residual Oil Saturation in Tertiary Surfactant Floods," Soc. Petrol. Engrs. J. 289 (Aug., 1972).

34. Batra, V. K., Ph.D. Thesis, Univ. of Waterloo (1973).

35. Berridge, S. A., Thew, M. T. and Loriston-Clarke, A. G., "The Formation and Stability of Emulsions of Water in Crude Petroleum and Similar Stocks," J. Inst. Petrol. 54(539), 333 (1968).

36. Strassner, J. E., "Effect of pH on Interfacial Films and Stability of Crude Oil-Water Emulsions," J. Petrol. Tech. 303 (Mar., 1968).

37. Petrov, A. A., Pozdnyzhev, G. N. and Borizov, Khimya i Tekhnologiya Topliv i Masel 11 (Mar., 1969).

38. Levchenko, D. N., Khydyakova, A. D. and Ratich, L. I., Khimya i Tekhnologiya Topliv i Masel 21 (Oct., 1970).

39. Knaggs, E. A., "The Role of the Independent Surfactant Manufacturer in Tertiary Oil Recovery," Paper presented at ACS Marketing Symp., New York, N.Y. (Apr. 8, 1976).

40. Shah, D. O., Walker, R. D., Gordon, R. J., Gubbins, K. E., O'Connell, J. P., Schweyer, H. E. and Tham, M. K., "Research on Chemical Oil Recovery Systems," First Semi-annual Report, NSF(RANN) (Feb. 15, 1976).

41. Slattery, J. C., "Interfacial Effects in the Entrapment and Displacement of Residual Oil," AIChE J. 20(6), 1145 (1974).

42. Mohan, V., Gupta, L. and Wasan, D. T., "Effect of Aging on Surface Shear Viscosity of Surfactant Films," J. Colloid Interface Sci. in press (1976).

43. Cash, R. L., Cayias, J. L., Hayes, M., MacAlister, D. J., Schares, T., Wade, W. H. and Schechter, R. S., "Surfactant Aging: A Possible Detriment to Tertiary Oil Recovery," paper presented (1973).

44. Gladden, G. P. and Neustadter, E. L., "Oil/Water Interfacial Viscosity and Oil Emulsion Stability," J. Inst. Petrol. 58(564), 351 (1972).

45. Mohan, V., Vora, M. K. and Wasan, D. T., "Shear Viscosity of Aging Liquid-Liquid Interfaces," Manuscript in preparation (1976).

46. Hegde, M. G. and Slattery, J. C., "Studying Non-Linear Surface Behavior with the Deep Channel Surface Viscometer," J. Colloid Inter. Sci. 35, 593 (1971).

47. Schechter, R. S., Kott, A. T. and Kanda, M., "Surface Elasticity and Viscosity of Monolayers," Paper presented at the 65th Annual AIChE Meeting, New York (Nov. 26-30, 1972).

48. Bueche, F., "Influence of Rate of Shear on the Apparent Viscosity of: A--Dilute Polymer Solutions, and B--Bulk Polymer," J. Chem. Phys. 22, 1570 (1954).

49. Takemura, T., "Influence of Rate of Shear on the Apparent Viscosity of Dilute Polymer Solutions," J. Polymer Sci. 27, 549 (1958).

50. Graessley, W. W., "Viscosity of Entangling Polydisperse Polymers," J. Chem. Phys. 47, 1942 (1967).

51. Kott, A. T., Gardner, J. W., Schechter, R. S. and De Groot, W., "The Elasticity of Pulmonary Lung Surfactant," J. Colloid Inter. Sci. 47(1), 265 (1974).

52. Schulman, J. H. and Cockbain, E. G., "Molecular Interactions at Oil/Water Interfaces. Part I Molecular Complex Formation and Stability of Oil in Water Emulsions," Trans. Faraday Soc. 36, 651 (1940).

53. Kung, H. C. and Goddard, E. D., "Studies of Molecular Association in Pairs of Long-Chain Compounds by Differential Thermal Analysis. I Lauryl and Myrsityl Alcohols and Sulfates," J. Phys. Chem. 67, 1965 (1963).

54. _____, "Molecular Association in Pairs of Long-Chain Compounds. II Alkyl Alcohols and Sulfates," J. Phys. Chem. 68, 3465 (1964).

55. Mysels, K. J., Shinoda, K. and Frankel, S., "Soap Films," Pergamon Press, N.Y. (1959).

56. Sherman, P., "Studies in Water-in-Oil Emulsions III. The Properties of Interfacial Films of Sorbitan Sesquioleate," J. Colloid Sci. 8, 35 (1953).

57. Alexander, A. E., "Emulsions and Emulsification," J. Royal Inst. Chem. 4, 221 (1948).

58. Adam, N. K., "The Physics and Chemistry of Surfaces," Oxford Univ. Press (1948).

59. Lawrence, A.C.S., "Emulsions and Films: Symposium at Sheffield," Nature, 170(4319), 232 (1952).

60. Cumper, C.W.N. and Alexander, A. E., "The Surface Chemistry of Proteins," Trans. Faraday Soc. 46, 235 (1950).

61. Becher, P., "Theoretical Aspects of Emulsification, A Background for Cosmetic Formulation," Amer. Perfumer Cosmet. 7(7), 21 (1962).

62. Blakey, B. C. and Lawrence, A.C.S., "The Surface and Interfacial Viscosity of Soap Solutions," Disc. Faraday Soc. 18, 268 (1954).

63. Vold, R. D. and Mittal, K. C., "The Effect of Lauryl Alcohol on the Stability of Oil-in-Water Emulsions," J. Colloid Interface Sci. 38(2), 451 (1972).

64. Mahajan, V. V. and Wasan, D. T., "Adsorption of Surfactants at Fluid Surfaces," Manuscript in preparation (1976).

65. Fiocco, R. J., "Separation by Chemical Demulsification," Paper presented at the 68th National AIChE Meeting, Houston, Texas (Mar. 4, 1971).

66. Carroll, B. J., "Emulsion Stability and Demulsification: A Literature Review," Project Coordinator, Lucassen, J. PPS 73 1226, BD/SCC (July 30, 1973).

67. Good, P. A., "Gradient Coalescence: Thin Liquid Film Mechanics and Stability," Ph.D. Thesis, Univ. of Minnesota, Minneapolis (1974).

68. Froning, H. R., Raza, S. H. and Askew, W. S., "Method of Mobility Control in Miscible Displacement Process," U.S. Patent No. 3,722,590 (Mar. 27, 1973).

69. Raza, S. H., "Foam in Porous Media: Characteristics and Potential Applications," Soc. Petrol. Engrs. J. 328 (Dec., 1970).

70. Minnsieux, L., "Oil Displacement by Foams in Relation to their Physical Properties in Porous Media," J. Petrol. Tech. 100 (Jan., 1974).

71. Ivanov, I. B. and Dimitrov, D. S., "Hydrodynamics of Thin Liquid Films: Effect of Surface Viscosity on Thinning and Rupture of Foam Films," Colloid and Polymer Sci. 252, 982 (1974).

72. Bikerman, J. J., "Foams," Springer-Verlag (1973).

73. Kanner, B. and Glass, J. E., "Surface Viscosity and Elasticity," Ind. Eng. Chem. 61(5), 31 (1969).

74. _____, "Surface Viscosity," Recent Progress in Surface Science (J. F. Danielli, K.G.A. Pankhurst and A. C. Riddiford, Eds.), Vol. 1, Academic Press, New York, 1964.

75. Djabbarah, N. F., Wasan, D. T. and Shah, D. O., "The Role of Surface Viscosity in Foam Drainage," Paper presented at the 80th National AIChE Meeting, Boston, Mass. (Sept. 7-10, 1975).

76. Kanda, M. and Schechter, R. S., "The Insitu Production of Foams in Porous Media," Private communication (1975).

77. Lucassen, J. and Giles, D., "Dynamic Surface Properties of Nonionic Surfactant Solutions," J. Chem. Soc., Faraday Trans. I 71, 217 (1975).

78. Van den Tempel, M., Private communication (1974).

79. Prins, A. and Van Voorst Vader, F., "Relation Between Film Draining Behavior and Dynamic Surface Properties of Aqueous Dodecyl Sulfate Solutions," Chem. Phys. Chem. Anwendungstech. Grenzflaechenaktiven Stoffe, Ber. Int. Kongr., 6th, year 1972 (1973).

80. Djabbarah, N. F. and Wasan, D. T., "Foam Stability in Mixed Surfactant Systems," Manuscript in preparation (1976).

81. Lewis, J. B., "The Mechanism of Mass Transfer of Solute Across Liquid-Liquid Interfaces, Part I. Determination of Individual Transfer Coefficients: for Binary Systems," Chem. Eng. Sci. 3, 248 (1954); ___, "Part II. The Transfer of Organic Solute Between Solvent and Aqueous Phase," ibid., 260.

82. Austin, L. J. and Sawistowski, H., "Mass Transfer in a Liquid-Liquid Stirred Cell," Inst. Chem. Engrs. Symp. Ser. 26, 3 (1967).

83. Davies, J. T. and Mayers, G.R.A., "The Effect of Interfacial Films on Mass Transfer Rates in Liquid-Liquid Extraction," Chem. Eng. Sci. 16, 55 (1961).

84. Trapeznikov, A. and Rebinder, R., "Stabilizing Effect of Adsorption Layers and their Mechanical Properties," Comp. Rend. Acad. Sci. URSS, 18, 427 (1938).

85. Brown, A. G., Thuman, W. C. and McBain, J. W., "The Surface Viscosity of Detergent Solutions as a Factor in Foam Stability," J. Colloid Sci. 8, 491 (1953).

86. Davies, J. T., "A Study of Foam Stabilizers Using a New (Viscous Traction) Surface Viscometer," Second International Congress on Surface Activity (J. H. Schulman, Ed.), 1, 220 (1957).

87. Ross, J., "Transition Temperature of Monolayers, Surface Viscosity of Dilute Aqueous Solutions of Lauryl Alcohol in Sodium Lauryl Sulfate," J. Phys. Chem. 62, 531 (1958).

88. Sonntag, H., "Eigenschaften Grenzflaechenaktive Stoffe an den Flüssig-Flüssig Phasen grenze," Z. Physik. Chem. 225, 284 (1964).

89. Leonard, R. A. and Lemlich, R., "A Study of Interstitial Liquid Flow in Foam. Part I: Theoretical Model and Application to Foam Formation," AIChE J. 11, 18 (1965).

90. Levich, V. G., "Physicochemical Hydrodynamics," Prentice Hall, Englewood Cliffs, N.J. (1962).

91. Shah, D. O., Djabbarah, N. F. and Wasan, D. T., "Surface Viscosity and Foam Drainage," Manuscript in preparation (1976).

92. Malviya, B. K., Wasan, D. T. and Mohan, V., "Interfacial Viscoelastic Properties of Adsorbed Surfactant and Polymeric Films," Paper presented at the 25th Canadian Chemical Engineering Conference, Montreal, Canada (Nov. 2-5, 1975).

93. _____, "Instrumentation for Surface Shear Viscosity," Paper presented at the 65th Annual AIChE Meeting, New York (Nov. 26-30, 1972).

94. Joly, M., "General Theory of the Structure, Transformations and Mechanical Properties of Monolayers," J. Colloid Sci. 5, 49 (1950).

95. Suzuki, A., "A Study of Non-Newtonian Surface Viscosity," Kolloid Z. und Z fur Polymere 250, 360; 365 (1972).

PHYSICO-CHEMICAL ASPECTS OF ADSORPTION
AT SOLID/LIQUID INTERFACES

I. Basic Principles

P. Somasundaran and H. S. Hanna
Columbia University

I. ABSTRACT

Adsorption of surfactants on solids from aqueous solutions is of interest for tertiary oil recovery by micellar flooding since it can be a major cause for surfactant retention during the flooding. Molecular and ionic forces such as electrostatic attraction, covalent bonding, nonpolar bonding, and molecule-ion complex formation that contribute towards the adsorption process are discussed in this paper with typical examples. Effects of factors that can markedly affect adsorption process, such as solution pH, temperature, chain length of the surfactant, chemical state and structure of the surfactant, nature and concentration of dissolved species including polymers and pretreatment of mineral surfaces, are illustrated with the help of examples.

II. INTRODUCTION

Adsorption of surfactants on minerals is an important phenomenon both in tertiary oil recovery using micellar flooding and mineral froth flotation. Research in the mineral processing area during the last fifty years has yielded valuable information on the adsorption characteristics of minerals and mechanisms governing the adsorption. In addition, the importance of adsorption in controlling other interfacial processes such as dyeing, bloodclotting, lubrication, sizing, chromatography, etc., has also led to research that has yielded useful information on a variety of adsorption processes. Past studies on adsorption in various systems have been discussed in a number of publications (1-16). In this paper, physico-chemical aspects of surfactant adsorption on minerals will be reviewed, and effects of relevant factors such as solution pH, temperature, ionic strength, etc., will be discussed with the help of examples.

III. ADSORPTION MECHANISMS

Adsorption of surfactants on solids is the result of favorable interactions of the surfactant species or its various complexes with the chemical species on or near the solid surface. The extent of adsorption will be determined also by the nature of interaction forces between solvent species and surfactant species as well as the solid surface. It is the presence of solvent species that makes the phenomenon of adsorption from solution more complicated than that of adsorption of gases on solids. It is to be noted that the possibility of the presence of a third phase can further increase the complexity of the systems such as that under condition here. The adsorption at the solid/aqueous solution (fluid-1) interface is related under equilibrium conditions to that at solid/gas or solid/fluid-2 and fluid-1/fluid-2 interfaces (17,18). For example, adsorption in a solid/solution/oil system can be related by the following expression:

$$\Gamma_{SL_1} = \Gamma_{SL_2} + \Gamma_{L_1L_2} \cos \theta + \frac{\gamma_{L_1L_2} \sin \theta}{RT} \cdot \frac{d\theta}{d\ln c} \qquad (1)$$

Interestingly, it was found using a semi-empirical approach for the quartz-amine system at natural pH that the surfactant adsorption density at the liquid/gas interface is several times higher than that at the solid/liquid interface (17). Furthermore, surfactant migration at the solution/gas interface can be significantly faster than that from the bulk aqueous solution to an interface (19). These relationships are to be taken into consideration while interpreting interfacial processes that take place in systems containing more than two phases.

In the case of adsorption from solution, a number of interactive forces, individually or in combination with each other, can be responsible for it. The major forces that can contribute to the adsorption process include electrostatic attraction, covalent bonding, hydrogen bonding or nonpolar bonding between the surfactant and interfacial species, van der Waals or steric interactions among the adsorbed species, and solvation or desolvation of adsorbate and adsorbent species.

The concentration in moles/ml. of counter-ions in the interfacial region can be given by the following expression:

$$c_s = c_b \exp \left[\frac{- \Delta G^o_{B \to S}}{RT} \right] \qquad (2)$$

To express the adsorption density in terms of the more meaningful unit of surface area, the right-hand side of Equation 2 is multiplied by the thickness of the adsorbed layer to yield the expression:

$$\Gamma_s = 2rc_b \exp \left[\frac{- \Delta G^o_{ads}}{RT} \right] \qquad (3)$$

ΔG^o_{ads}, the driving force for adsorption, will be the sum of a number of contributing forces (electrostatic attraction, covalent bonding, cohesive chain-chain interaction among long chain surfactant species, non-polar interaction between chains and hydrophobic sites on the solid, hydrogen bonding, solvation or desolvation of species due to the adsorption process) and can be written as (20,8):

$$\Delta G^o_{ads} = \Delta G^o_{elec} + \Delta G^o_{cov} + \Delta G^o_{c-c} + \Delta G^o_{c-s} + \Delta G^o_h + \Delta G^o_{solv} \qquad (4)$$

For each system one or more of the above terms can be contributing depending on the mineral, surfactant type and concentration, pH, temperature, nature and concentration of dissolved species, etc. For non-metallic minerals, electrostatic and lateral interaction forces are considered to be major factors determining adsorption of surfactants. On the other hand, for salt-type minerals such as calcite and sulfides, such as pyrite, the chemical term can become significant. Spectroscopic techniques have been used in the past to identify the nature of bonding between the mineral surface and adsorbate species (21-24). These studies cannot, however, be considered to provide a definite confirmation on the bonding states since the samples are usually subjected to various treatments for the spectroscopic analysis itself. Chemical states identified after such treatments cannot be considered to be identical to the actual state of the chemical species adsorbed on the mineral. Indirect evidence can, however, be obtained from measurements of zeta potential, wetting characteristics, or the density of packing in relation to the size and shape of the adsorbed molecules.

In a number of cases adsorption has been reported to take place in amounts higher than that required for the formation of a monolayer (25-28). The adsorption mechanism for the uptake of a second or a third layer can indeed be of a different nature from that for the uptake of the first layer.

There are also a number of reported cases where adsorption isotherms have been found to pass through a maximum particularly near the critical micelle concentration of the surfactant (15,29,30). Adsorption in this range can involve additional factors such as micellar exclusion from the interfacial region, competition between various monomeric and micellar species and even phase separation and precipitation.

A. Electrostatic Factors

For certain systems in which the mineral and the surfactant are oppositely charged, and for systems in which micellar exclusion phenomenon (29) is significant, forces of adsorption due to electrostatic attraction or repulsion between the charged mineral surface and the charged surfactant species can play a governing role. This mechanism is ordinarily studied without much difficulty with the help of data obtained from electrokinetic and titration measurements. Correlation of the adsorption of the surfactant under various conditions with the zeta potential of the mineral as well as changes in it owing to the adsorption has yielded information on the role of electrostatic adsorption in mineral systems.

The generation of surface charge on the mineral particles is considered to be either due to preferential dissolution or due to hydrolysis of surface species followed by pH-dependent dissociation of surface hydroxyl groups (31-34). Alternate mechanisms that have been proposed include dissolution of lattice ions followed by hydrolysis in the bulk and subsequent adsorption of the complexes (31,34,35). The latter two mechanisms are, however, thermodynamically indistinguishable from each other and yield identical final results.

For such oxides as silica and alumina, the hydrolysis of the surface species followed by pH-dependent dissociation is considered to be a major governing mechanism:

$$— M(H_2O)^+_{surface} \underset{H^+}{\overset{OH^-}{\rightleftarrows}} — MOH_{surface} \overset{OH^-}{\underset{}{\rightleftarrows}} — MO^-_{surface} + H_2O \quad (5)$$

It can be seen by examining this equation that the surface would be positively charged under low pH conditions and negatively charged under high pH conditions. The pH at which the net charge of the surface is zero is called the point of zero charge (pzc). The ions, such as H^+ and OH^- in the case of oxides, that determine surface charge are called potential determining ions. In contrast to oxides, simple salts such as AgI are considered to get charged due to preferential dissolution of lattice ions. Thus for AgI, concentration of silver or iodide ions are potential determining.

For salt-type minerals such as calcite and apatite, the preferential hydrolysis of the surface species, and

preferential dissolution of ions which is often accompanied
by reactions with the solution constituents and possible uptake of the solid, have been proposed to be the major controlling mechanisms. Calcite, for example, can undergo the
following reactions upon contact with water and generate a
number of complexes (34):

$$CaCO_{3(s)} \quad \rightleftarrows \quad CaCO_{3(aq)} \qquad K_1 = 10^{-5.09} \qquad (6)$$

$$CaCO_{3(aq)} \quad \rightleftarrows \quad Ca^{++} + CO_3^{--} \qquad K_2 = 10^{-3.25} \qquad (7)$$

$$CO_3^{--} + H_2O \quad \rightleftarrows \quad HCO_3^- + OH^- \qquad K_3 = 10^{-3.67} \qquad (8)$$

$$HCO_3^- + H_2O \quad \rightleftarrows \quad H_2CO_3 + OH^- \qquad K_4 = 10^{-7.65} \qquad (9)$$

$$H_2CO_3 \quad \rightleftarrows \quad CO_{2(g)} + H_2O \qquad K_5 = 10^{1.47} \qquad (10)$$

$$Ca^{++} + HCO_3^- \quad \rightleftarrows \quad CaHCO_3^+ \qquad K_6 = 10^{0.82} \qquad (11)$$

$$CaHCO_3^+ \quad \rightleftarrows \quad H^+ + CaCO_{3(aq)} \qquad K_7 = 10^{-7.90} \qquad (12)$$

$$Ca^{++} + OH^- \quad \rightleftarrows \quad CaOH^+ \qquad K_8 = 10^{1.40} \qquad (13)$$

$$CaOH^+ + OH^- \quad \rightleftarrows \quad Ca(OH)_{2(aq)} \qquad K_9 = 10^{1.37} \qquad (14)$$

$$Ca(OH)_{2(aq)} \quad \rightleftarrows \quad Ca(OH)_{2(s)} \qquad K_{10} = 10^{2.45} \qquad (15)$$

It can be seen from these equations that when calcite approaches equilibrium with water at high pH values, an excess
of negative HCO_3^- and CO_3^{--} will exist, whereas at low pH values
an excess of positive Ca^{++}, $CaHCO_3^+$ and $CaOH^+$ species will occur.
These ionic species may be produced at the solid/solution
interface or may form in solution and subsequently adsorb on
the mineral in amounts proportional to their concentration in
solution. In either case, the net result will be a positive
charge on the surface at low pH values and a negative charge
at high pH values. The potential determining ions in this
case are, in addition to Ca^{++} and CO_3^{--}, also OH^-, H^+ and HCO_3^-.
Since the activities of these species can be calculated as a
function of pH from the available thermodynamic data, the point
of zero charge of the solid can be obtained by estimating the
pH at which the total activity of the negative ions is equal to
that of the positive ions, assuming a close correspondence
between the solid and the point of zero charge of the solution.
For calcite a value of 8.2 was obtained in close agreement with
that obtained from the measurement of solution pH changes
caused by the addition of calcite as well as that estimated
from solubility measurements (see Figure 1). It is to be

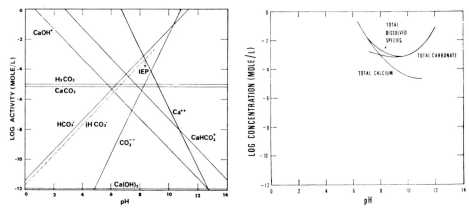

Fig. 1a. *Determination of point of zero charge and potential-determining ions for calcite-aqueous solution-air system using thermodynamic data (34)*

Fig. 1b. *Determination of point of zero charge from experimental measurements of total calcium and total carbonate concentrations in aqueous solutions in equilibrium with calcite and air (34)*

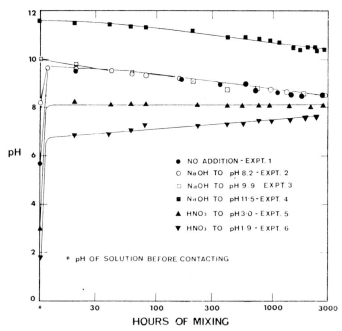

Fig. 1c. *Determination of point of zero charge from the measurement of solution pH change (34)*

pointed out that the measured isoelectric point[*] has been observed to undergo time-dependent changes during equilibration.
Furthermore, the values obtained are dependent also on the
history of the mineral as well as the pretreatment it receives.
Thus, whereas an isoelectric point of pH 5.2 was obtained for
a fluor apatite sample leached with nitric acid, a much lower
value of about 3.2 was obtained for the same sample treated
with hydrochloric acid (36).

Clay minerals which have layered structures consisting
of sheets of SiO_4 tetrahedra and sheets of AlO_6 octahedra
linked with each other by means of shared oxygen ions are
negatively charged under most natural conditions mainly due
to the substitution, for example, of Al^{3+} for Si^{4+} in the
silica tetrahedra. This charge is internal to the structure
and is not dependent on solution concentrations (7). The
edges of the clay particles will, on the other hand, exhibit
pH-dependent charge characteristics due to hydroxylation and
ionization of the broken Si-O and Al-O bonds at the edges.
The point of zero charge of the clay is, thus, determined by
the algebraic sum of face and edge charges. It is to be noted
that at the point of zero charge both the sides and faces will
be charged and thus possess adsorptive properties that other
minerals might not possess at their points of zero charge.

In contrast to clays, non-clay silicates such as
chrysotile possess points of zero charge that are determined
by the mineralogical composition of the surface exposed to
the solution; this composition can be different from that
expected from bulk composition (7,35). Thus, chrysotile
possesses a point of zero charge that is more alkaline than
that expected on the basis of the bulk chemical composition
since the surface exposed to solution is richer in magnesium.

Point of zero charge of minerals is an important experimentally accessible interfacial property since the adsorption
of various charged species is related to the location of the
point of zero charge with respect to the concentrations of the
potential determining ions in the solution under consideration.

---

[*]Condition under which the zeta potential, as obtained
from various electrokinetic measurements, is zero is called
the isoelectric point (iep). It need not be identical to the
point of zero charge, particularly in the presence of specifically adsorbing species such as polyvalent or surfactant ions.
While surface charge or surface potential of minerals cannot
be directly measured accurately, changes in the zeta potential
(ζ)-by definition, the potential at the plane of shear-due to
the addition of various reagents can be monitored using electrokinetic methods. Such changes in zeta potential can be
helpful for establishing the reasons for adsorption.

Typical pzc values are given in Table I. It is important to note that the values obtained are affected by the presence of impurities, previous history including pretreatments, method of storing and ageing, and the extent of ageing (36–37). Variations in the source or method of preparation, including mechanical treatments and washing and drying, and the presence of surface defects and of adsorbed and structural impurities, also produce significant changes in the pzc (36). It was already mentioned that the type of acid used for cleaning minerals can also cause considerable effects. In fact, it has been found recently that the pzc of quartz can be raised from below 2 to as high as 6 by leaching it in hydrofluoric acid solution. Upon ageing the HF-leached quartz in water, the pzc can be brought to its original value but only over a period of several days. Washing the minerals with hot solutions also produces similar long-term effects (38).

TABLE I

POINT OF ZERO CHARGE OR ISOELECTRIC
POINT OF VARIOUS TYPICAL MINERALS (6)

Minerals	pzc or iep
Quartz, SiO_2	pH 2–3.7
Rutile, TiO_2	pH 6.0
Corundum, Al_2O_3	pH 9.0
Magnesia, MgO	pH 12.0
Fluorapatite (natural), $Ca_5(PO_4)_3$ (F,OH)	pH 6
Fluorapatite (synthetic)	pCa 4.4, pF 4.6, $pHPO_4$ 5.22
Hydroxyapatite, $Ca_5(PO_4)_3(OH)$, Synthetic	pH 7–7.15, $pHPO_4$ 4.19–4.48
Calcite, $CaCO_3$	pH 9.5
Barite, $BaSO_4$	pBa 6.7
Silver iodide, AgI	pAg 5.6
Silver sulfide, Ag_2S	pAg 10.2

The dependence of adsorption on the electrical nature of the interface has been clearly shown for several mineral/surfactant systems. Adsorption of sodium dodecylsulfonate on alumina is given in Figure 2 as a function of pH (39). Electrophoretic mobility of the alumina particles in the presence of sodium dodecylsulfonate is also given in this figure. The point of zero charge of this alumina has been determined to be 9.1. It can be seen from the figure that only below the point of zero charge, when the alumina is positively charged, does the sulfonate adsorb to any measurable extent. Adsorption of the sulfonate anions below pH 7 is sufficiently high to counteract even the increase in number of positive sites owing to the decrease in solution pH. This aspect will be discussed in detail later. Results obtained for the flotation of calcite with anionic dodecylsulfate below the pzc and with cationic dodecylammonium salt above the pzc have confirmed the strong dependence of adsorption on the electrical condition of the interface (34). The correlation is presented in Figure 3 at two concentrations of the surfactants. Similarly good correlations have been found for adsorption of surfactants and flotation of monazite (40), zircon (41),

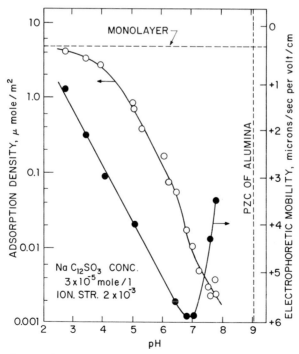

Fig. 2. Adsorption of dodecylsulfonate on alumina as function of pH (39) from J. Phys. Chem., Courtesy of Am. Chem. Soc.

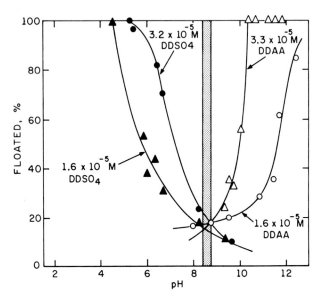

*Fig. 3. Flotation of calcite with dodecylammonium acetate
(DDA) and sodium dodecylsulfate (DDSO₄) solutions
(34)*

corundum (42,43), quartz (44), fluorite (45), magnetite (46),
apatite (17), calcite (34), tricalciumphosphate (47), and
calcium carbonate (48). The correlation obtained for the adsorption of quarternary ammonium salts on tricalciumphosphate
and calcium carbonate with the zeta potential and wetting
properties of these solids is particularly interesting. It
can be seen from Figure 4 that the adsorption density of
cetyltrimethylammonium bromide (CTAB) on tricalciumphosphate
is about four times that on calcium carbonate. A close-packed
monolayer with a molecular area of 51 $\overset{\circ}{A}^2$/CTAB molecule was
found for the case of $Ca_3(PO_4)_2$ but not for $CaCO_3$. This was
attributed to the lesser number of negative sites on $CaCO_3$
than on $Ca_3(PO_4)_2$. Moreover, it can be seen from Figure 4
that the adsorption and desorption isotherms do coincide with
each other, indicating total reversibility of adsorption.
These observations support the above interpretations based
on the hypothesis that adsorption mechanisms for these systems
are mainly due to electrostatic interaction (49).
Inorganic Electrolytes. If the driving force is mainly electrostatic attraction, an increase in concentration of inorganic
electrolytes can affect the adsorption of surfactant owing to
alterations that they bring about in the interfacial electrical properties. A significant increase in ionic strength will

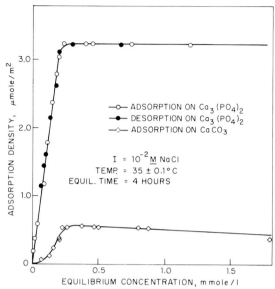

Fig. 4.   Adsorption isotherm of cetyltrimethylammonium bromide (CTAB) on tricalciumphosphate and calcium carbonate (49)

normally decrease the surfactant adsorption on solids owing to competition for adsorption sites by inorganic ions that are charged similarly to the surfactant ion. Thus flotation (which is a measure of the surfactant adsorption) of quartz at pH 6 using dodecylammonium chloride can be depressed to a measurable extent by the addition of 10^{-4} mole/1 KNO_3 (see Figure 5) (50). In this case the adsorption of dodecyl-ammonium ions on quartz is reduced by the competing potassium ions. This can be viewed as resulting from a reduction in electrical potential, but specific counter ion effects can also be encountered. Electrokinetic experimemts (51) have shown that if the added salt contains multivalent ions their adsorption can, in some cases, be sufficiently strong to produce a reversal of the zeta potential. A more marked depression, which has been obtained for quartz flotation using calcium nitrate (50), results from the stronger tendency of bivalent ions over monovalent ions to adsorb and compete with collector ions. It is of interest that, if the bivalent ions are of opposite charge to that of the particle and the surfactant ions, activation of flotation can occur due to charge reversal of the particle caused by the adsorbed bi-valent ions (43). In this connection it might be mentioned that alkaline earth ions such as calcium have been reported to function most effectively in the pH range where they are in hydrolyzed soluble form (52-55).

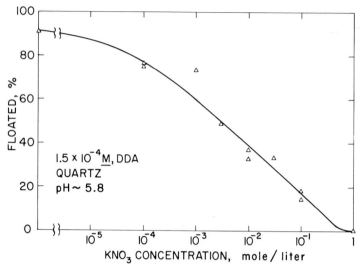

Fig. 5.   Flotation of quartz at pH 5.8 using dodecylammonium
acetate (DDA) as a function of KNO₃ concentration
(50)

The effect of inorganic salts on the adsorption of cetyltrimethylammonium bromide on $Ca_3(PO_4)_2$ and $CaCO_3$ is shown in
Figures 6a and 6b. It is interesting to note that while the
introduction of SO_4^{--} ions does increase the adsorption of the
cationic surfactant on $CaCO_3$, it has no such effect for the
adsorption on $Ca_3(PO_4)_2$ (48,49). Apparently the uptake of
SO_4^{--} ions by $CaCO_3$ has caused an increase in the number of
negative sites on the mineral surface, but unable to do so
on $Ca_3(PO_4)_2$ since, as indicated earlier, this mineral has a
surface that is sufficiently saturated with negative sites to
yield a closely packed monolayer adsorption even without the
addition of SO_4^{--} into the solution. Introduction of Mg^{++} ions
into the solution, on the other hand, is found to reduce the
adsorption of cationic surfactants on both the minerals.

B.   Chemical Forces
Surfactants such as fatty acids and sulfonates have been
proposed by a number of investigators to adsorb on minerals
such as calcite, fluorite, apatite and barite. Chemisorption
was first proposed by French et al. (21) for oleate/fluorite
systems on the basis of the infrared results. Subsequently,
Peck and Wadsworth (22) reported chemisorption of oleate on
calcite and barite in addition to fluorite. The infrared
spectra obtained by Peck and Wadsworth for fluorite before
and after contact with oleic acid are given in Figure 7, along
with those for oleic acid and calcium oleate. Comparison of

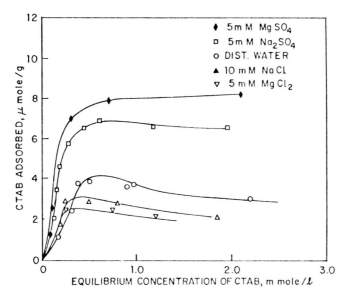

Fig. 6a. *Effect of inorganic electrolytes on CTAB adsorption on CaCO₃ (48)*

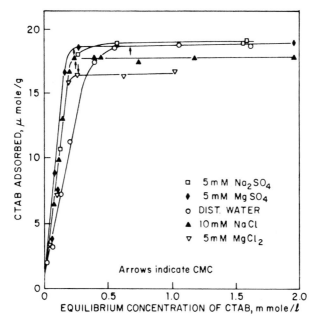

Fig. 6b. *Effect of inorganic electrolytes on CTAB adsorption on tricalcium phosphate (47)*

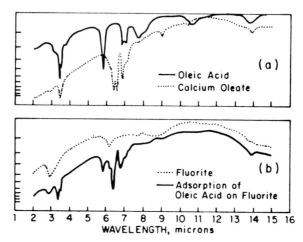

Fig. 7.   Infrared spectra of a) oleic acid and calcium oleate
b) fluorite and oleic acid adsorbed on fluorite (22)

the spectra led the authors to propose the presence of both
physically adsorbed and chemically adsorbed oleate on the
mineral.   Chemisorption has been proposed to take place by
an ion exchange process in which the surfactant anions replace an equivalent amount of the lattice ions such as F^- to
form a surface layer of the alkaline earth oleate, whereas
physical adsorption of oleic acid or oleate occurs on account
of the van der Waals and Coulombic forces.   The above mechanism has been confirmed by the data obtained by Bahr et al.
(56), and Bilsing (57) for stoichiometric release of fluoride
ions by oleate ions during their adsorption on fluorite.
Similar mechanisms resulting in the formation of salts have
also been suggested by Shergold (24) for the adsorption of
dodecylsulfate on fluorite (26) for that of dodecylsulfate
on barite, and Fuerstenau and Miller (58) for that of alkylsulfonates on calcite.

On the other hand, oleate has been suggested to adsorb
on calcite and apatite electrostatically below their points
of zero charge and chemically above them (34,59).   Results
of oleate adsorption and zeta potential changes supporting
such a mechanism are given in Figures 8 and 9.   Adsorption
of oleate results at the pH value of 9.6 for the present
system due to its chemical interaction with calcium species.
The sharp increase in the slope of the adsorption curve above
3×10^{-5} mole/1 can be attributed to precipitation of calcium
oleate since, with calcium present at a concentration of
$\sim 1.5 \times 10^{-4}$ mole/1, the solubility limit is exceeded for
calcium oleate above about 5×10^{-5} mole/1 of oleate.
Oleate adsorption at concentrations below 10^{-6} mole/1 and

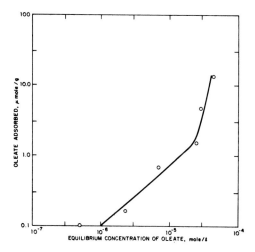

Fig. 8.   Adsorption isotherm of potassium oleate on calcite at a natural pH of 9.6 (85)

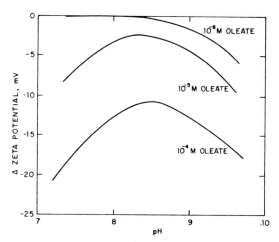

Fig. 9.   Change in zeta potential of calcite particles as function of pH at constant ionic strength (1 mM KNO_3) (85)

at pH values below the pzc of calcite (pH 8.2) can be considered as mainly due to electrostatic attraction between the
negative oleate ions and positive surface sites since under
these conditions there is very little change in zeta potential
due to oleate addition, as long as the ionic strength is maintained at a constant value.

   Change in zeta potential and even reversal of it occurs
at higher concentrations owing to the two-dimensional aggregation that will be discussed in detail elsewhere. Above the

pzc, the zeta potential of calcite is seen to change continuously with oleate concentration. Some type of specific adsorption is suggested here since simple electrostatic adsorption does not produce such changes in zeta potential under constant ionic strength conditions. In fact, the increase in calcium solubility observed at pH 12.1 upon the addition of oleate indicates chemical binding of calcium with the oleate in solution. It is likely that similar interactions leading to chemisorption of oleate on calcite take place with the surface calcium species also.

Salt formation between the lattice cations and alkylsulfates and sulfonates has been proposed, as mentioned earlier, by a number of investigators, as a mechanism of surfactant adsorption on salt-type minerals. Hanna (60,49) observed the adsorption density of Aerosol OT (alkylsulfosuccinate), sodium dodecylebenzinesulfonate and sodium dodecylesulfate on calcite, precipitated $CaCO_3$ and $Ca_3(PO_4)_2$ to correspond to that required for the formation of a bilayer on the surface of these minerals. The nature of adsorption isotherms given in Figures 10 and 11 suggested salt formation between the surfactant anions and the surface calcium species (chemisorption) to form the first layer followed possibly by interchain cohesion (physical adsorption) to form the second layer.

The above mechanism is supported by the observation that the contact angle and flotation response obtained were maximum at concentrations corresponding to the completion of the first layer. The absence of a definite plateau at monolayer formation was attributed to the surface heterogeneity of the mineral samples involved.

C. Chain-Chain Interactions

Adsorption isotherms for certain surfactants such as alkylsulfonate on alumina have been found to undergo a marked increase in slope around a particular surfactant concentration (39) (see Figure 12). This has been attributed to the two-dimensional lateral aggregation between the adsorbed long chain surfactant species above a given adsorption density. Below this point in Region 1 in Figure 12, adsorption is proposed to take place individually due to the electrostatic attraction between the charged mineral surface and the oppositely charged surfactant species (see Figure 13). Upon further increasing the surfactant concentration two-dimensional aggregates called hemi-micelles form (Region 2). Under these conditions, adsorption is favored, in addition to electrostatic forces, by the favorable energetics of removal of the alkyl chains from an aqueous environment. In Region 3, electrostatic forces work against adsorption but adsorption continues to increase with bulk surfactant concentration due to

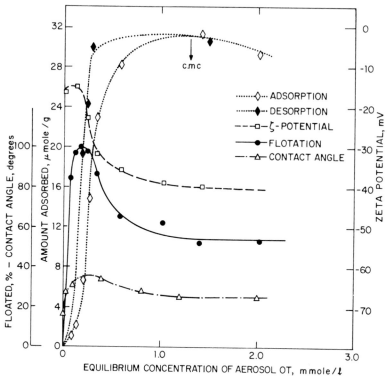

Fig. 10. Adsorption, zeta potential, contact angle and flotation of calcium carbonate by Aerosol OT at constant ionic strength (10 mM NaCl) (28)

the aggregation at the surface. The hemi-micellization phenomenon is analogous to the three-dimensional micellization in the bulk. The energy gained by two-dimensional aggregation has been estimated to be about 1kT for quartz/dodecyleamine and alumina/dodecylesulfonate systems. This energy, as shown in the schematic diagram in Figure 14 for free energy for transfer of CH_2 groups between various environments, is close to but less than that involved in micellization. Hemi-micelles possibly contain more water in their environment than micelles since possibly a smaller fraction of the CH_2 groups are removed from the aqueous environment in the case of the former. Other thermodynamic quantities, heat and entropy changes, involved in the association of the adsorbed surfactants have been calculated for the quartz/amine solution system (61). It is interesting to note that the above association resulted in a net increase in the entropy of the system presumably due to the decrease upon association in the ordering of the water molecules that were

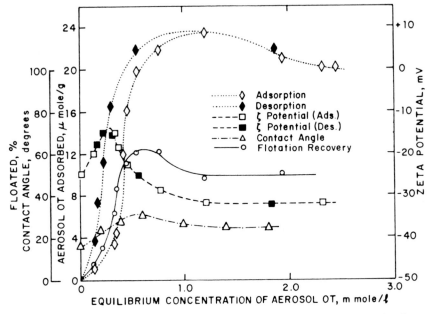

Fig. 11.   *Adsorption, zeta potential, contact angle and flotation of tricalcium phosphate by Aerosol OT at constant ionic strength (10 mM NaCl) (47)*

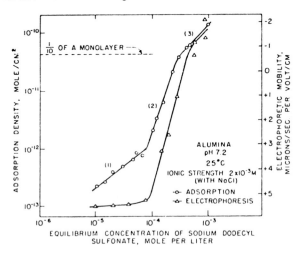

Fig. 12.   *Adsorption density of dodecylsulfonate ions on
alumina and the electrophoretic mobility of
alumina as a function of the concentration of
sodium dodecylsulfonate at pH 7.2 and at constant
temperature and ionic strength (39) from* J. Phys.
Chem., *Courtesy of Am. Chem. Soc.*

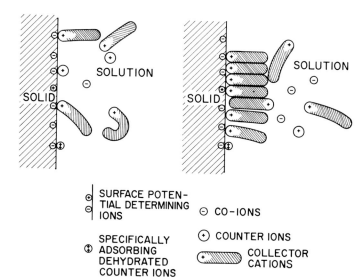

Fig. 13.   Schematic representation of a long chain anionic surfactant (a) individually at low concentrations and (b) with lateral association between chains at higher concentrations. (6) from AIChE Symp. Ser., Courtesy of Am. Inst. Chem. Eng.

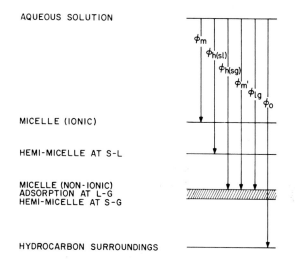

Fig. 14.   Schematic diagram for free energy of transfer of $-CH_2-$ groups from aqueous solution to various environments (91)

originally around isolated surfactant chains. It is also to
be noted that the two-dimensional aggregation is dependent
upon solution pH and hence surface potential of the particles,
temperature, chain length, ionic strength, and the chemical
state and structure of the surfactant (32,39,61,62).

D. Hydrogen Bonding
     Hydrogen bonding between surfactant species and the min-
eral surface species has been proposed for a number of systems,
particularly those containing hydroxyl, phenolic, carboxylic
and amine groups. As mentioned earlier, adsorption of oleic
acid on fluorite and HF-treated surface of beryl has been
attributed by Peck and Wadsworth (22) to hydrogen bonding
between the oxygen of the surfactant functional group and the
fluoride at the solid surface. According to Parks' (7) review of Giles' work, phenols form strong hydrogen bonds and
possibly adsorb by such bonding on alumina and silica as well
as certain textile substrates. In this regard, the mechanism
suggested by Sorensen (63) based on the similarity in geometry
between the mineral crystal and the collector is noteworthy.

E.   Structural Compatibility
     Sorensen suggested that in the case of anionic flotation
of simple salts such as fluorite, hydrogen bonding between the
oxygen of the collector and fluoride species is active and
that it is assisted by the electron resonance of the polar
groups, the structure of which must be compatible with the
geometry of the mineral crystal. Other systems where the
structural compatibility has been taken into consideration
are those of soluble salts. Fuerstenau and Fuerstenau (64)
proposed that adsorption in the case of salts such as sylvite
is governed by a matching of size of the functional group of
the surfactant with that of the lattice ion of the solid that
carried the same charge as the hydrophilic part of the surfactant. Thus, aminium ion adsorbs on sylvite (KCl) but not on
halite (NaCl) due to comparable sizes of the aminium ion and
K^+. This theory is, however, unable to explain why, for
example, the anionic alkylsulfate should distinguish between
KCl and NaCl. The sulfate adsorbs on the former but not on
the latter.

F. Hydration and Solubility Factors
     According to a theory proposed by Rogers and Schulman
(65), adsorption on soluble salts is governed by the hydration properties of the solid; the one with the largest
negative heat of solution being a better adsorbent than the
others. This theory also fails to provide adequate explanation as to why a particular mineral such as KCl will be

floated by certain collectors as alkyl sulfates, but not by
carboxylates or phosphates even though the authors do provide
alternate explanations for certain cases. Correlation that
exists in several cases between the adsorption and insolubility
of compounds formed between the collector and the chemical
constituents of the solid suggest another mechanism dependent
essentially on what could be considered as the precipitation
of such compounds on solids (10).

G. Hydrophobic Bonding

A mechanism that could be of particular importance for
oil wettable sand, coal and other naturally hydrophobic minerals is the hydrophobic bonding between surfactant molecules
adsorbed flat on the mineral, and hydrophobic sites on the
solid. Such adsorption can also take place on other types of
minerals that are originally hydrophilic but have acquired
some hydrophobicity owing to reactions with organic as well as
inorganic compounds in solution. This type of bonding has
also often been called van der Waals bonding which depends
essentially upon the polarizability of the various species
involved. Energy involved in this type of bonding is additive,
and therefore the adsorption due to it is expected to increase
with increasing size of the adsorbate. Morgan and Stumm (66)
in fact observed this to be true; whereas polystyrene sulfonate
adsorbed strongly on silica, toluene sulfonate did not.

IV. ADSORPTION KINETICS

Kinetics of adsorption as well as desorption should be
of importance in tertiary oil recovery since the process as
a whole is dynamic in character than several other interfacial
processes. To the authors' knowledge kinetics of sorption has
not, however, been examined in the past studies with reservoir
rock minerals. A few systems studied in other areas include
that of dodecylbenzenesulfonate on earth sediments (67),
sodium dodecylsulfonate on alumina (68), oleate on hematite
(69), and alkylbenzenesulfonate on activated carbon (70), and
clay minerals (71). For the case of earth sediments (non-
porous sands), equilibration times ranging from 3 to 5 days
were reported by Tallmadge and Tan (67). This is a much
slower system than those encountered normally.

In contrast to the above, adsorption of dodecylsulfonate
on non-porous alumina was observed by Somasundaran (68) to be
complete in less than one hour (see Figure 15). Similar
results were also reported by Hanna (47) for the adsorption
of quaternary ammonium salts on calcium carbonate and tricalcium phosphate. Kinetics of adsorption of oleate on
hematite was, on the other hand, found to be very much
dependent on the temperature and pH of the solution (72).

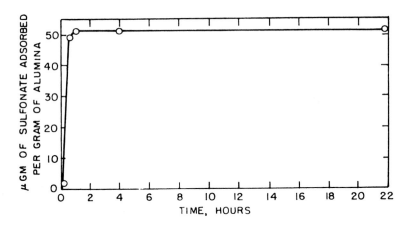

Fig. 15. Adsorption of sodium dodecylsulfonate on alumina as a function of time (68), Courtesy of University of California at Berkeley

It was found that at pH 4.8 and 25°C, the adsorption was very slow and an equilibrium adsorption density was not obtained even after four hours (see Figure 16). An increase in pH or temperature improved the kinetics of adsorption significantly. In fact, at pH 8, the adsorption was complete in less than 5 minutes at both the temperatures (25°C and 75°C) investigated. It is to be noted that we have observed the adsorption at the solution/air interface also, as measured by the decrease in surface tension, to be governed by similar kinetics. A most interesting observation was that the adsorption rate was maximum at a pH value of about 8 where the acid-soap complex formation, as mentioned earlier, is also predominant (see Figure 17) (19). Both at lower and higher pH values the adsorption was found to take place at the air/solution interface at a much slower rate.

V.   ADSORPTION NEAR AND ABOVE THE CRITICAL MICELLE
      CONCENTRATIONS
       In contrast to the adsorption from dilute solutions, that from micellar solutions has received only limited attention to date. Adsorption isotherms in this concentration range have been reported to exhibit shapes that have not been encountered elsewhere. For example, Trushenski et al. (73) reported the existence of a depletion maximum followed by a minimum in the micellar concentration range. The system reported is that of Berea sandstone/Mahogany petroleum sulfonate/isopropyl alcohol micellar fluid at 110°F. The presence of a depletion minimum is most interesting from an industrial viewpoint, since operation at surfactant concentrations corresponding to its occurrence could lead to

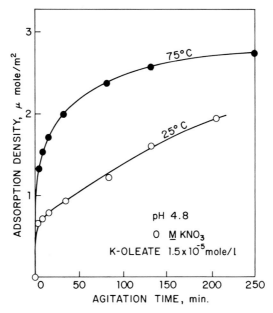

Fig. 16a. *Diagram illustrating the effect of temperature on the kinetics of oleate adsorption on hematite at pH 4.8 (19)*

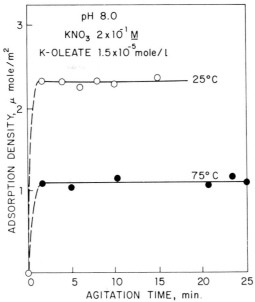

Fig. 16b. *Diagram illustrating the effect of temperature on the kinetics of oleate adsorption on hematite at pH 8 and constant ionic strength (2 x 10^{-1} $\underline{M}$ KNO₃) (19)*

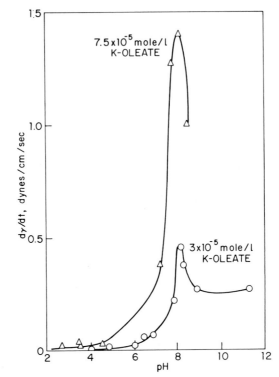

Fig. 17. Diagram illustrating dependence of surface tension decay rate on pH and oleate concentration at 25°C (19)

a minimum cost increment for the surfactant usage. The presence of adsorption maxima has been reported by Sexsmith and White (74) for cetyltrimethylammonium bromide on viscose monofil oxycellulose and cotton, by Gotshal et al. (75) for cetyltrimethylammonium bromide on viscose monofil, by Fava and Eyring (76) for sodium dodecylbenzenesulfonate on cotton, Mukerjee and Anavil (29) for sulfonates on bioglass and by Hanna (47) for Aerosol OT on precipitated $CaCO_3$ and $Ca_3(PO_4)_2 nH_2O$ and for sodium oleate on $Ca_3(PO_4)_2$. A few typical isotherms are reproduced in Figures 18 to 21. These reports have shown that even for the same surfactant one does not obtain an adsorption maximum or minimum for every solid substrate. In other words, the properties of the surface is one of the major factors governing adsorption above critical micelle concentration.

There are several suggested reasons for the presence of an adsorption maximum, but none of them appears to be sufficiently substantiated to be considered as a confirmed mechanism for surfactant adsorption near and above critical

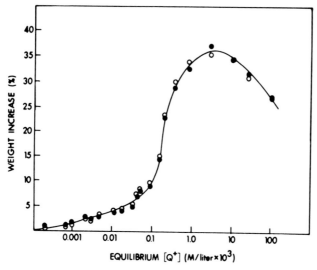

Fig. 18.   Increase in weight of oxycellulose due to adsorption
from CTAB solutions. ● Direct observations; ○ calculated from change of titer (74)

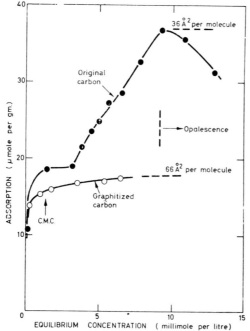

Fig. 19.   Adsorption isotherms of Aerosol OT on carbon black
("Sterling MT") and graphitized Sterling MT from
solutions containing 0.01M NaCl (30), from J. Chem.
Soc., Courtesy of Chemical Soc. (London)

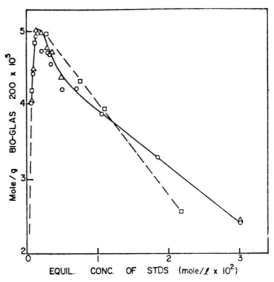

Fig. 20. *Adsorption isotherm of sodium tetradecylsulfate Bio-Glas at high concentrations.* ○ = 4 days, 30°C; △ = 6 days, 30°C; □ = 7 days, 35°C (29), from ACS Symp. Ser., *Courtesy of Am. Chem. Soc.*

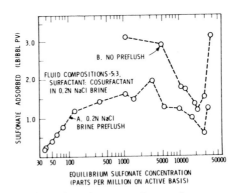

Fig. 21. *Adsorption isotherms of a mahogany petroleum sulfonate on Berea sandstone at 100°F (73), from J. Pet. Eng., Courtesy of SPE/AIME*

micelle concentration. Interpretation of the experimental
adsorption data has often been made difficult due to the
possibilities of precipitation accompanying the adsorption.
Even more serious is the possibility of the existence of
experimental artifacts owing to the surfactant entrapment in
pores which might not be removed during the final steps of
adsorption tests. Yet another source for experimental error
is the possibility of removing the adsorbed surfactant itself
during the washing of the core that is normally conducted to
remove the bulk surfactant that is not adsorbed. Two possible reasons for the presence of an adsorption maxima,
which are also not thermodynamically unacceptable are: 1)
exclusion of micelles from the near surface region of the
solid due to higher electrostatic repulsion between the
particles and micelles than that between the former and
singly charged surfactant monomers, and 2) alteration in
solid properties such as effective surface area, owing to
change in particle morphology upon excessive surfactant adsorption that can cause repulsion between various parts
(such as platelets of clay) of the particle.

Sexsmith and White (77) have described the existence of
maximum in adsorption isotherm by considering micelle formation as a simple association reaction without activity coefficient changes, rather than a phase change. Indeed,
recent careful experiments by Elworthy and Mysels (78) have
given evidence against consideration of micelles as a separate
phase. They observed a definite, even though small, decrease
in surface tension with surfactant concentration above the
critical micelle concentration. The suggested increase in
monomer activity is not compatible with the hypothesis of a
phase separation. The law of mass action was applied by
Sexsmith and White to calculate the concentration of monomer
species in solution for the case of micelles with different
ratios of number of surfactant species to number of counter
ions. For a micelle, $Q_n X_m$, that contains n number of surfactant species and m number of counter ions, the concentration of the monomer surfactant species, c_Q, would be governed
by the mass balance expression:

$$c = nC_M + c_Q = mc_M + c_X \qquad (16)$$

The first and second derivatives of the equation for concentration of the monomer with respect to the total surfactant
concentrations suggest that there will be a maximum in c_Q at
the concentration determined by the expression, provided
$n \geq m \geq 2$

$$c_Q = \left[\frac{(m-1)^{m-1}}{m^m (n-m)K_m} \right]^{\frac{1}{n+m-1}} \tag{17}$$

The effects of variation in n and m on the concentration of the monomer are shown in Figures 22 and 23. Clearly such existence of a maximum in monomer concentration can be expected to produce a maximum in adsorption also. A theoretical adsorption isotherm constructed by Sexsmith and White on the basis of the above hypothesis is reproduced in Figure 24. The isotherms for both the adsorption of the surfactant ion and the inorganic ion are clearly characterized by the presence of maxima. White (15) has discussed the thermodynamic limitations of the problem of existence of a true maximum. Uptake of the cationic surfactants on cellulose substrates has been interpreted by assuming ion-pair absorption that is dependent on cation exchange. Existence of a maximum for the adsorption of anionic surfactants on cellulose leaves such an assumption invalid, since there are no anion exchange sites here. Other explanations for the existence of a maximum adsorption include that of Moilliet et al. (79) and Hanna (80) based on the presence of impurities that would relieve the thermodynamic restrictions, and that of Kitchener (14) based on the release of supersaturation at high surfactant concentrations. It is very clear from an examination of this problem that it requires additional careful experimental and theoretical studies.

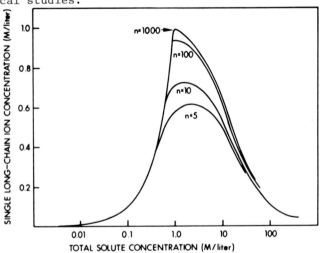

Fig. 22.   Concentration of long chain monomer ions as a function of total solute concentration-effect of n (K = 10; m/n = 0.9) (77)

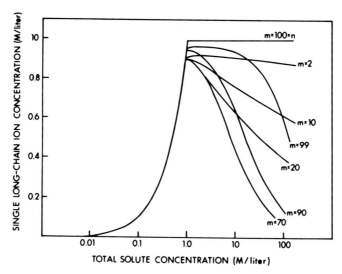

Fig. 23. Concentration of long chain monomer ion as a function of total solute concentration-effect of m at constant $\underline{n}$ (K = 10; $\underline{n}$ = 100) (77)

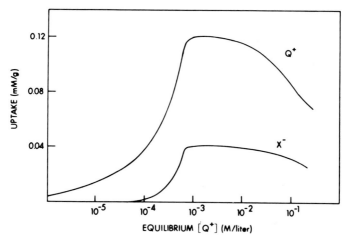

Fig. 24. Hypothetical isotherms for the absorption of QX by a cellulosic substrate (77)

VI. EFFECTS OF VARIABLES

A. Solution pH

The effect of solution pH is shown in Figure 25 for the case of dodecylsulfonate adsorption on alumina (39). At lower pH values the adsorption density required for lateral interaction is attained at a lower surfactant concentration.

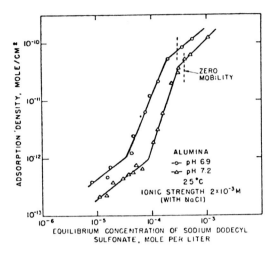

Fig. 25. *Diagram illustrating the effect of pH on the adsorption of dodecylsulfonate on alumina (39) from J. Phys. Chem., Courtesy of Am. Chem. Soc.*

This is clearly due to the larger electrostatic attractive forces resulting from a larger surface potential. It must be noted that pH can also influence adsorption through its control of the hydrolysis of certain surfactants. This is discussed in detail elsewhere.

B. Temperature

The effect of solution temperature on adsorption in general is determined by the type of adsorption. Whereas any increase in temperature is expected to decrease the extent of physical adsorption, the reverse is found to be generally true for the case of chemisorption. Thus, adsorption of dodecylsulfonate on alumina is found to decrease with an increase in temperature (see Figure 26) (39). Changes in the heat and entropy of adsorption and association of long chain surfactants at the alumina aqueous solution interface calculated from these data agreed with the postulate of two-dimensional aggregation at the solid/liquid interface (61).

The effect of temperature on the adsorption of oleate on hematite, which is considered to involve chemisorption, is illustrated in Figure 27. It can be seen that in this case the surfactant adsorption increases, but only at low ionic strength conditions. Above an ionic strength of about 2×10^{-3} N, adsorption was found to decrease markedly with increase of temperature. This interesting ionic strength-temperature interaction is discussed in other publications (19).

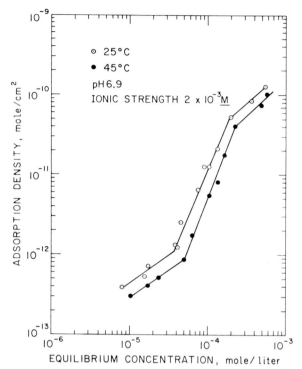

Fig. 26. *Diagram illustrating the effect of temperature on*
the adsorption of dodecylsulfonate on alumina (39)
from J. Phys. Chem., Courtesy of Am. Chem. Soc.

C.   Chain Length of Surfactant
      Increase in the length of the non-polar part of a sur-
factant generally causes an increase in adsorption owing to
increased lateral interaction between chains. The presence
of an environment with an effectively lower dielectric constant at the interface would also promote adsorption of a
long chain material there. Results in Figure 28 for the
quartz flotation using alkyl ammonium acetates of varying
chain length clearly indicate an increase in adsorption with
an increase in chain length (81). Adsorption measurements
by Wakamatsu and Fuerstenau (82) of sulfonates of varying
chain length also show the dependence of adsorption and
lateral chain-chain interaction on the length of the hydrocarbon chain.

D.   Chemical Structure of the Surfactant
      Structural differences due to, for example, polar sub-
stitution or chain branching that hinder lateral interaction,
can decrease the adsorption. The effect of the position of

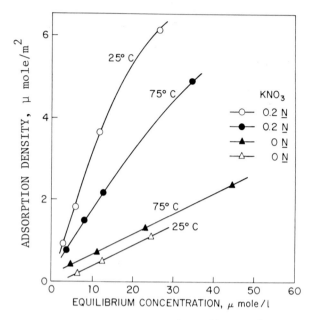

*Fig. 27. Diagram illustrating the effect of temperature on
the adsorption of potassium oleate on hematite (19)*

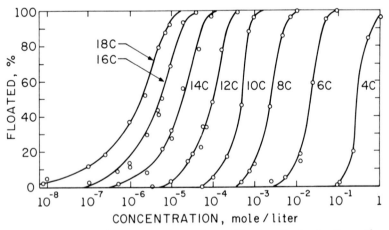

*Fig. 28. Diagram illustrating the effect of chain length of
alkylammonium acetate on the flotation of quartz
(81) from Trans. AIME, Courtesy of Am. Inst. Min.
Met. Engrs.*

the functional group in the hydrocarbon chain was studied by
Dick et al. (83) for the alumina/alkyl benzene sulfonate
system. Adsorption of the sulfonate was found to progressively decrease as the functional group was shifted from the
end to the center position. On the other hand, substitution
of hydrogen in CH_2 or CH_3 groups of the surfactant with
fluorine can be expected to increase the surface activity and
adsorption. This is clearly indicated in Figure 29 where
flotation of alumina, using perfluorocarboxylic acid is compared with that obtained using the corresponding carboxylic
acid (62).

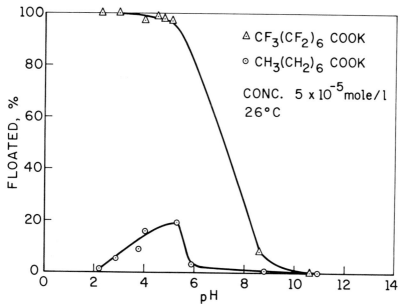

Fig. 29. *Comparison of the flotation of alumina using perfluorocarboxylic acid with that using the corresponding carboxylic acid (62) from Trans. IMM
(London), Courtesy of Institute of Mining and
Metallurgy (London)*

E. Chemical State of the Surfactant (Complex Formation)
    Chemical state of the surfactant is influenced by varia-
tion in solution conditions such as pH, and this can have a
significant effect on adsorption. Surface activity of various
soaps and amines is in fact influenced considerably by pH
since they can form ionomolecular surfactant complexes (84).
Flotation of hematite by oleic acid, for example, was found to
be highest in the pH range where acid-soap complex formation
is expected (72). Evidence of high surface activity of this
complex was obtained by surface tension measurements (shown

in Figure 30) of oleate solutions. Similarly, maximum flotation of quartz with alkylamine observed around pH 10.2 can
also be explained on the basis of the formation of ion-dipole
complexes (84). The pH of maximum flotation does coincide
also with the pH at which maximum lowering of the adhesion
tension of the system and of the surface tension of amine
solutions occur.

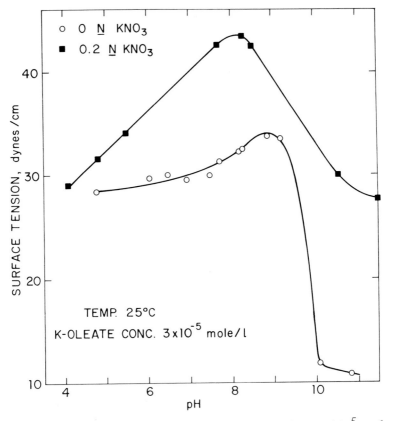

Fig. 30. *The equilibrium surface pressure of 3 x 10⁻⁵ mole/1
potassium oleate solution as a function of pH and
ionic strength (72) from AIChE Symp. Ser., Courtesy
of Am. Inst. of Chemical Engineers*

F.   Polymeric Reagents
     The effect of presence of polymers such as starch in
solution on surfactant adsorption has been investigated only
to a limited extent. In our study on calcite/oleate/starch
system, it was observed that the oleate adsorption on calcite
was enhanced by the addition of starch and vice versa (see
Figures 31 and 32) (85,86). It is to be noted that even

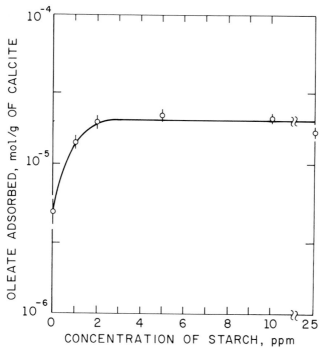

Fig. 31. Adsorption of oleate on calcite as a function of
          starch concentration at a natural pH of 9.6-9.8
          (85)

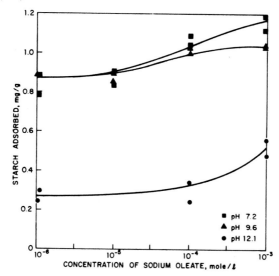

Fig. 32. Adsorption of starch on calcite as a function of
          oleate concentration at various pH values (85)

though the particles adsorbed more surfactant in the presence
of starch, according to flotation results shown in Figure 33,
they had become less hydrophobic under these conditions.
This interesting effect was ascribed to helical structure that
starch assumes in the presence of hydrophobic species or under
alkaline conditions with a hydrophilic exterior and a hydrophobic interior. Mutual enhancement of adsorption was possible
under such conditions owing to the formation of a helical
clathrate with the hydrophobic oleate held inside the hydrophobic starch interior. Similar results have also been
reported for the coadsorption of tannin and oleic acid on
fluorite, barite and calcite where each reagent was found to
mutually enhance the adsorption of the other (87). Enhancement of polymer adsorption by surfactant was also observed
by Tadros (88) for silica/polyvinyl alcohol/sodium dodecylbenzenesulfonate (see Figures 34 and 35) and silica/polyvinyl
alcohol/cetyltrimethylammonium bromide systems. It might be
noted that the addition of quebracho did not enhance the
adsorption of oleic acid on the above minerals (89).

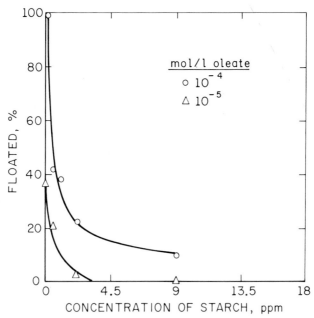

Fig. 33. *Flotation of calcite using sodium oleate as a function of starch concentration (85)*

VII. INTERACTIONS BETWEEN VARIABLES

Under most practical conditions, one can expect a variation in more than one of the above variables. Even though
rarely done, it is therefore important to conduct adsorption

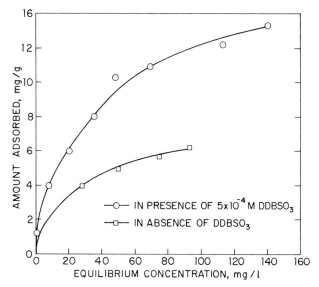

Fig. 34. *Adsorption isotherm for polyvinyl alcohol on silica at different sodium dodecylbenzenesulfonate concentrations (88)*

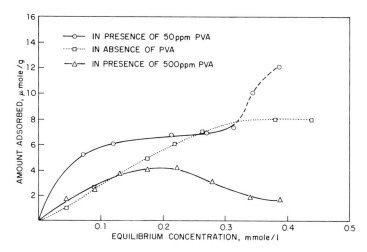

Fig. 35. *Adsorption isotherm for sodium dodecylbenzene-sulfonate on silica at different polyvinyl alcohol concentrations (88)*

studies as a function of several variables at a time. Use of multifactorial experimental designs and computer techniques in conducting multivariate adsorption tests was demonstrated recently for the calcite/oleate/starch system (90). Example

of a three-dimensional response map drawn by a plotter in the
system is given in Figure 36. The interaction between the
oleate and the starch variables can be easily read from this
plot.

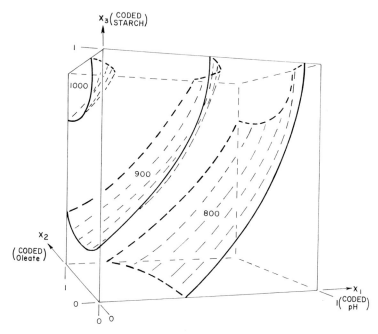

*Fig. 36. Response maps for the adsorption of starch (ppm per
gram on calcite as a function of the three variables,
i.e., pH, oleate concentration and starch concentration (90) from Canad. Min. Met. Bulletin, Courtesy
of Canad. Inst. Min. Met.*

VIII. CLOSING REMARKS

Mechanisms of adsorption on minerals in solutions have
been discussed along with the effect of variables such as pH,
salinity, temperature, chemical structure including the chain
length of the surfactant and polymer concentration. Both the
amount of surfactant adsorbed and the related kinetics are
complex functions of the above variables as well as the type
and the pretreatment history of the mineral. In addition,
the effect of interactions between variables on the adsorption
process can also be significant. The number of factors involved and possible interactions between them under reservoir
conditions is indeed of a higher magnitude than those in the
simple adsorption systems normally investigated by surface
and colloid chemists. The need for extensive basic research
with well characterized mineral/surfactant system becomes

very clear upon reviewing various phenomena discussed here.

IX. ACKNOWLEDGMENTS
     National Science Foundation (ENG-76-08756) and Amoco
Production Company are gratefully acknowledged for their
support of this work.

X. LIST OF SYMBOLS

Ads.   Adsorption
Des.   Desorption
c_b The bulk concentration of the adsorbate in moles/ml.
c_s The interfacial concentration of the adsorbate in
        moles/ml.
c_Q Concentration of the monomer surfactant species
c_M Concentration of the micelle species
c_X Concentration of counter ions
ΔG_{solv} The standard free energy of solvation of the adsorbate
        species or any species displaced from the surface due
        to the adsorption process
ΔG_h^o The standard free energy of hydrogen bonding
ΔG_{ads}^o The standard free energy of adsorption
ΔG_{elec}^o The standard free energy of electrostatic interaction
ΔG_{c-c}^o The standard free energy of non-polar cohesive
        interaction among long chain surfactant species
ΔG_{cov}^o The standard free energy of covalent bonding
ΔG_{c-s}^o The standard free energy of non-polar interaction
        between the hydrocarbon chains and hydrophobic
        sites on the solid
$\Delta G_{B \to S}^o$ The standard free energy for transfer of surfactant
        from the bulk to the interfacial region
$K_{1,2,3..}$ Equilibrium constants
   etc.
K_m Equilibrium constant of the micelle association
m   Number of counter ions in a micelle
n   Number of surfactant species in a micelle
iep   Conditions under which the zeta potential as measured
        by electrokinetic methods is zero
pzc   Conditions under which the net surface charge is zero
ϕ_o Energy involved in the transfer of a $-CH_2$-group from
        aqueous environment to hydrocarbon environment
$\phi_{h(sl)}$ Energy involved for $-CH_2$-group in the process of
        hemi-micelle formation at the solid/liquid interface
$\phi_{h(sg)}$ Energy involved for $-CH_2$-group in the process of
        hemi-micelle formation at the solid/gas interface
ϕ_{lg} Energy involved in the transfer of a $-CH_2$-group from
        aqueous environment to the air/liquid interface

ϕ_m Energy involved in the transfer of a $-CH_2-$group from aqueous solution into a charged micelle

$\phi_{m'}$ Energy involved in the transfer of a $-CH_2-$group from aqueous solution into a non-ionic micelle

r The effective radius of the adsorbed ion

R The universal gas constant

T The absolute temperature

Γ_s Adsorption density in moles/cm^2

Γ_{SL_1} Adsorption at the solid/aqueous solution interface

Γ_{SL_2} Adsorption at the solid/oil interface

$\Gamma_{L_1L_2}$ Adsorption at the aqueous solution/oil interface

θ Contact angle measured through L_1

$\gamma_{L_1L_2}$ Aqueous solution/oil interfacial tension

Q_n Number of surfactant species in a micelle

X_m Number of counter ions in a micelle

ζ Zeta potential in mV

XI.   LITERATURE CITED

1.   Somasundaran, P. and Grieves, R. B., ed., <u>Advances in Interfacial Phenomena of Particulate Solution/Gas Systems</u>, AIChE Symp. Ser., Vol. 71, No. 150 (1975).

2.   Goddard, E. D. and Somasundaran, P., "Adsorption of Surfactants on Mineral Solids," to be published in <u>Croatica Chem. Acta</u> (1976).

3.   Hanna, H. S. and Somasundaran, P., "Flotation of Salt-Type Minerals," in <u>Flotation</u>, A. M. Gaudin Memorial Volume (M. C. Fuerstenau, Ed.), Vol. 1, Chap. 8, pp. 197-272, AIME, 1976.

4.   Mittal, K. L., ed., <u>Adsorption at Interfaces</u>, Amer. Chem. Soc., Symp. Ser. 8 (1975).

5.   Gould, R. F., ed., <u>Adsorption from Aqueous Solution</u>, Adv. Chem. Ser. 79, Am. Chem. Society (1968).

6.   Somasundaran, P., "Interfacial Chemistry of Particulate Flotation," in <u>Advances in Interfacial Phenomena of Particulate Solution/Gas Systems</u> (P. Somasundaran and R. B. Grieves, Eds.), AIChE Symp. Ser., Vol. 71, No. 150, 1-15 (1975).

7.   Parks, G. A., "Adsorption in the Marine Environment," in <u>Chemical Oceanography</u>, 2nd ed. (Riley and Skirrow, Eds.), pp. 241-308, Academic Press, New York, 1975.

8.   Fuerstenau, D. W., "The Adsorption of Surfactants at Solid/Water Interfaces," in <u>The Chemistry of Bio-surfaces</u> (M. L. Hair, Ed.), Vol. 1, pp. 143-176, Marcel Dekker, Inc., New York, 1971.

9.   Healy, T. W., "Principles of Adsorption of Organics at Solid-Solution Interfaces," J. Macromol. Sci. Chem. Vol. A8, No. 3, 603-619 (1974).

10. Du Rietz, C., "Chemisorption of Collectors in Flotation," 11th Int'l Mineral Processing Congress, Cagliari, Paper 13, 29 pp. (1975).

11. Lipatov, Yuis and Sergeeva, L. M., Adsorption of Polymers, John Wiley, New York, 1974.

12. Ponec, V., Knor, Z. and Cerny, S., Adsorption of Solids, CRC Press, 1974.

13. Kipling, J. J., Adsorption from Solutions of Non-Electrolytes, Academic Press, London, 1965.

14. Kitchener, J. A., "Mechanisms of Adsorption from Aqueous Solutions: Some Basic Problems," J. Photographic Sci. Vol. 13, 152-158 (1965).

15. White, H. J., Jr., "Adsorption of Cationic Surfactants by Cellulosic Substrates," in Cationic Surfactants (E. Jungermann, Ed.), Chap. 9, 311-341. Marcel Dekker, New York, 1970.

16. Ginn, M. E., "Adsorption of Cationic Surfactants on Minerals and Miscellaneous Solid Substrates," in Cationic Surfactants (E. Jungermann, Ed.), Chap. 10, 11, 341-385. Marcel Dekker, New York, 1970.

17. Somasundaran, P., "Zeta Potential of Apatite in Aqueous Solutions and its Change During Equilibration," J. Colloid Interf. Sci. Vol. 27, No. 4, 659-666 (1968).

18. Smolders, C. A., "Contact Angles: Wetting and De-Wetting of Mercury, Part II. Theory of Wetting," Rec. Trav. Chim. Vol. 80, No. 7, 8, 650-720 (1961).

19. Kulkarni, R. D. and Somasundaran, P., to be published (1976).

20. Haydon, D. A. and Taylor, F. H., "Adsorption of Na Octyl and Decyl Sulfates and Octyltrimethylammonium Bromide at the Decane-water Interface," Proc. 3rd Int'l Cong. Surf. Active Sub. Vol. 1, 157 (1960).

21. French, R. O. et al., "Applications of Infrared Spectroscopy to Studies in Surface Chemistry," J. Phys. Chem. Vol. 58, 805-810 (1954).

22. Peck, A. S. and Wadsworth, M. E., "Infrared Studies of the Effect of Fluoride, Sulfate and Chloride on Chemisorption of Oleate on Fluorite and Barite," Proc. 7th Int'l Mineral Processing Congress (N. Arbiter, Ed.), pp. 259-267. Gordon and Breach, 1965.

23. Lovell, V. M., Goold, L. A. and Finkelstein, N. P., "I. R. Studies of the Adsorption of Oleate Species on Calcium Fluoride," Int'l J. Mineral Processing Vol. 1, 183-192 (1974).

24. Shergold, H. L., "Infrared Study of Adsorption of Sodium Dodeculsulfate by CaF_2," Transactions Instn. Min. Metall. Sec. C, Vol. 81, No. 3, C148-156 (1972).

25. Razouk, R. I., Saleeb, F. Z. and Hanna, H. S., "Wettability of Phosphate Ore Constituents in Certain Cationic and Anionic Surfactants," Proceedings 5th Int'l Congress on Surface Active Substances, Barcelona, 1968, Ediciones Unidas S. A., Vol. 2, 695-700, 1969.

26. Cumming, B. D. and Schulman, J. H., "Two-Layer Adsorption of Dodecylsulfate on Barium Sulfate," Australian J. Chem. Vol. 12, 413-423 (1959).

27. Schulman, J. H. and Smith, T. D., "Selective Flotation of Metals and Minerals," Recent Development in Mineral Dressing, Instn. Ming. Metall., London, 393-413 (1953).

28. Hanna, H. S., "Adsorption of Anionic Surfactants on Precipitated $CaCO_3$ and Calcite," Paper presented at 4th Arab Chem. Conference, National Research Center, Cairo (1975).

29. Mukerjee, P. and Anavil, A., "Adsorption of Ionic Surfactants to Porous Glass: The Exclusion of Micelles and Other Solutes from Adsorbed Layers and the Problem of Adsorption Maxima," in Adsorption at Interfaces (K. L. Mittal, Ed.), ACS Symp. Ser. 8, 107-128 (1975).

30. Saleeb, F. Z. and Kitchener, J. A., "The Effect of Graphitization on the Adsorption of Surfactants by Carbon Black," J. Chem. Soc. Vol. 167, 911 (1965).

31. Parks, G. A., "The Isoelectric Points of Solid Oxides, Solid Hydroxides, and Aqueous Hydroxo Complex Systems," Chemical Reviews 65, 177-198 (1965).

32. Somasundaran, P., Healy, T. W. and Fuerstenau, D. W., "Surfactant Adsorption at the Solid/Liquid Interface; Dependence of Mechanism on Chain Length," J. Phys. Chem. Vol. 68, No. 12, 3562-3566 (1964).

33. Yopps, J. and Fuerstenau, D. W., "The Zero Point of Charge of Alpha-Alumina," J. Coll. Sci. Vol. 19, 61-71 (1964).

34. Somasundaran, P. and Agar, G. E., "The Zero Point of Charge of Calcite," J. Colloid Interface Sci. Vol. 24, 433-440 (1967).

35. Parks, G. A., "Aqueous Surface Chemistry of Oxides and Complex Oxide Minerals," Advances in Chemistry Series Vol. 6, No. 67, 121-160 (1967).

36. Somasundaran, P., "Pretreatment of Mineral Surfaces and its Effect on their Properties," in Clean Surfaces, Their Preparation and Characterization for Interfacial Studies, pp. 285-306. Marcel Dekker, New York, 1972.

37. Kulkarni, R. D. and Somasundaran, P., "The Effect of
     Ageing on the Electrokinetic Properties of Quartz in
     Aqueous Solutions," in Oxide-Electrolyte Interfaces
     Am. Electrochem. Soc. 31-44 (1972a).
38. Kulkarni, R. D. and Somasundaran, P., "Pretreatment
     Effects on the Electrokinetic Properties of Quartz,"
     164th ACS Meeting, New York (1972b).
39. Somasundaran, P. and Fuerstenau, D. W., "Mechanism of
     Alkylsulfonate Adsorption of the Alumina-water Inter-
     face," J. Phys. Chem. Vol. 70, 90 (1966).
40. Choi, H. S. and Whang, K. U., "Mechanism of Collector
     Adsorption on Monazite," Korean Chem. Soc. Vol. 7,
     p. 91 (1963a).
41. Choi, H. S. and Whang, K. U., "Surface Properties and
     Floatability of Zircon," Trans. Can. Inst. Min. Met.
     Vol. 66, 242-244 (1963b).
42. Modi, H. J. and Fuerstenau, D. W., "Streaming Potential
     Studies on Corundum in Aqueous Solutions of Inorganic
     Electrolytes," J. Phys. Chem. Vol. 61, 640-643 (1957).
43. Modi, H. J. and Fuerstenau, D. W., "The Flotation of
     Corundum--an Electrochemical Interpretation," Trans.
     AIME Vol. 217, 381-387 (1960).
44. Iwasaki, I., Cooke, S.R.B. and Colombo, A. F., "Flotation
     Characteristics of Goethite," U.S. Bureau of Mines,
     R.I. 5593 (1960).
45. Choi, H. S., "Correlation of Electrochemical Phenomena
     at Fluorite/Solution Interface with Floatability of
     Fluorite," Canadian Metallurgical Quarterly, Ottawa,
     Vol. 2, No. 4, 410-414 (1963).
46. Iwasaki, I., Cooke, S.R.B. and Kim, Y. S., "Some Sur-
     face Properties and Flotation Characteristics of
     Magnetite," Trans. AIME Vol. 223, 113-120 (1962).
47. Hanna, H. S., "Contribution to the Flotation of Phosphate
     Ores," Ph.D. Thesis, Ain Shams Univ., Cairo (1968).
48. Saleeb, F. Z. and Hanna, H. S., "Correlation of Adsorp-
     tion, Zeta Potential, Contact Angle and Flotation
     Behaviour of Calcium Carbonate," J. Chem. U.A.R.
     (Egypt) Vol. 12, No. 2, 229-236 (1969a).
49. Hanna, H. S., "Relation Between Crystal Lattice Struc-
     ture and the Adsorption Behavior of Sparingly Soluble
     Salts," Paper No. 237, Sec. 51c, presented at the VIIth
     Int'l Congr. Surface Active Sub., Moscow (1976).
50. Somasundaran, P., "The Cationic Depression of Amine
     Flotation of Quartz," Trans. AIME Vol. 255, 64-68
     (1974).
51. Aplan, F. F. and Fuerstenau, D. W., "Principles of Non-
     metallic Mineral Flotation," in Froth Flotation
     (D. W. Fuerstenau, Ed.), AIME, 50th Ann. Vol., 170-214
     (1962).

52. Fuerstenau, M. C., Rice, D. A., Somasundaran, P. and Fuerstenau, D. W., "Metal Ion Hydrolysis and Surface Charge in Beryl Flotation," Trans. I.M.M. (London) Vol. 73, 381 (1965).

53. Fuerstenau, M. C., Martin, C. C. and Bhapu, R. B., "The Role of Metal Ion Hydrolysis in Sulfonate Flotation of Quartz," Trans. AIME Vol. 226, 449 (1963).

54. Healy, T. W., James, R. O. and Cooper, R., "The Adsorption of Aqueous Co(II) at Silica/Water Interface, in Adsorption from Aqueous Solution," in Advances in Chem. Series , No. 79, pp. 62-81. Am. Chem. Soc., Washington, D.C., 1968.

55. Saleeb, F. Z. and Hanna, H. S., "Flotation of Calcite and Quartz with Anionic Collectors, The Depressing and Activating Action of Polyvalent Ions," J. Chem. U.A.R. (Egypt) Vol. 12, No. 2, 237-244 (1969b).

56. Bahr, A., Clement, M. and Surmatz, H., "On the Effect of Inorganic and Organic Substances on the Flotation of some Non-Sulfide Minerals by Using Fatty Acid-type Collectors," 8th Int'l Mineral Processing Congress, Leningrad, paper S-11, 12 pp. (1968).

57. Bilsing, U., "The Mutual Interaction of the Minerals during Flotation, for Example the Flotation of CaF_2 and $BaSO_4$," Dissertation, Bergakademie Frieberg (1969) (Ger. Text).

58. Fuerstenau, M. C. and Miller, D. J., "The Role of the Hydrocarbon Chain in Anionic Flotation of Calcite," Trans. AIME Vol. 238, No. 2, 153-160 (1967).

59. Somasundaran, P. and Agar, G. E., "Further Streaming Potential Studies on Apatite in Inorganic Electrolytes," Trans. SME/AIME Vol. 252, 348-352 (1972).

60. Hanna, H. S., "Role of Cationic Collectors on Selective Flotation of Phosphate Ore Constituents," Powder Technology Vol. 12, No. 1, 57-64 (Jul./Aug. 1975).

61. Somasundaran, P. and Fuerstenau, D. W., "The Heat and Entropy of Adsorption of Long Chain Surfactants on Alumina in Aqueous Solutions," Trans. AIME Vol. 252, 275-279 (1972).

62. Somasundaran, P. and Kulkarni, R. D., "The Effects of Chain Length of Perfluoro Surfactants as Collectors," Trans. Inst. Min. Met. 82 (802), C164-C167 (1973).

63. Sorensen, E., "On the Adsorption of Some Anionic Collectors on Fluoride Minerals," J. Colloid and Interface Sci. Vol. 45, No. 3, 601-607 (1973).

64. Fuerstenau, D. W. and Fuerstenau, M. C., "Ionic Size in Flotation Collection of Alkali Halides," Trans. AIME Vol. 204, 302-307 (1956).

65. Rogers, J. and Schulman, J. H., "A Mechanism of the Selective Flotation of Soluble Salts in Saturated Solutions," Proc. 2nd Int'l Congr. Surface Activity Vol. III, Butterworths, London, 243-251 and Discussion, 1957.

66. Morgan, J. J. and Stumm, W., "Colloid-Chemical Properties of Mn-Dioxide," J. Coll. Interf. Sci. Vol. 19, No. 4, 347-359 (1964).

67. Tallmadge, J. A. and Tan, J. K., "Adsorption of Dodecyl-benzenesulfonate on Earth Sediments. Liquid-Solid Phase Equilibrium," J. Chem. Eng. Data Vol. II, 125-128 (1966).

68. Somasundaran, P., "The Effect of van der Waals Interaction between Hydrocarbon Chains on Solid-Liquid Interfacial Properties," Ph.D. Thesis, University of California, Berkeley (1964).

69. Kulkarni, R. D. and Somasundaran, P., "Oleate Adsorption at Hematite/Solution Interface and its Role in Flotation," AIME Annual Meeting, New York (February 1975).

70. Weber, W. J., Jr. and Morris, J. C., "Kinetics of Adsorption on Carbon from Solution," J. Sanitary Engng. Div., ASCE, Vol. 89, No. SA2, Proc. Paper 3483, 31-59 (1963) or Vol. 90, 79 (1964).

71. Wayman, C., "Surfactant Sorption on Heteroionic Clay Minerals," Proc. Int'l Clay Conference Vol. 1, Pergamon Press, 329-341 (1964).

72. Kulkarni, R. D. and Somasundaran, P., "Kinetics of Oleate Adsorption at the Liquid/Air Interface and its Role in Hematite Flotation," in Advances in Interfacial Phenomena of Particulate Solution/Gas Systems (P. Somasundaran and R. B. Grieves, Eds.), AIChE Symp. Ser., Vol. 150, 124-133 (1975).

73. Trushenski, S. P., Dauben, D. L. and Parrish, R. D., "Micellar Flooding-Fluid Propagation, Interaction and Mobility," Soc. Pet. Eng., Paper 4582, presented at 48th Ann. Meeting SPE, Las Vegas (Sept. 1973).

74. Sexsmith, F. H. and White, H. J., Jr., "The Absorption of Cationic Surfactants by Cellulosic Materials. I. The Uptake of Cation and Anion by a Variety of Substrates," J. Coll. Interf. Sci. Vol. 14, 598-618 (1959).

75. Gotshal, Y., Rebenfeld, L. and White, H. J., Jr., "The Absorption of Cationic Surfactants by Cellulosic Materials. II. The Effects of Esterification of the Carboxyl Groups in the Cellulosic Substrates," J. Coll. Interf. Sci. Vol. 14, 619-629 (1959).

250   P. SOMASUNDARAN AND H. S. HANNA

76.   Fava, A. and Eyring, H., "Equilibrium and Kinetics of
      Detergent Adsorption - A Generalized Equilibrium
      Theory," J. Phys. Chem. Vol. 60, 890-898 (1956).
77.   Sexsmith, F. H. and White, H. J., Jr., "The Absorption
      of Cationic Surfactants by Cellulosic Materials. III.
      A Theoretical Model for the Absorption Process and a
      Discussion of Maxima in Absorption Isotherms for
      Surfactants," J. Coll. Interf. Sci. Vol. 14, 630-639
      (1959).
78.   Elworthy, P. H. and Mysels, K. J., "The Surface Tension
      of Sodium Dodecylsulfate Solutions and the Phase
      Separation Model of Micelle Formation," J. Coll.
      Interf. Sci. Vol. 21, 331-347 (1966).
79.   Moilliet, J. L., Collie, B. and Black, W., Surface
      Activity, 2nd ed., Van Nostrand, Princeton, 1961.
80.   Hanna, H. S., "Flotation Behavior of Spar Minerals.
      Study of the Influence of Unavoidable Ions on the
      Depressing Action of Tannins and Starches during
      Fatty Acid Flotation of Calcite, Fluorite and Barite,"
      Dr. Eng. Dissertation, Bergakademie Freiberg, Freiberg,
      1971 (Ger. Text).
81.   Fuerstenau, D. W., Healy, T. W. and Somasundaran, P.,
      "The Role of Hydrocarbon Chain of Alkyl Collectors
      in Flotation," Trans. AIME Vol. 229, 321 (1964).
82.   Wakamatsu, T. and Fuerstenau, D. W., "The Effect of
      Hydrocarbon Chain Length on the Adsorption of Sul-
      fonate at the Solid/Water Interface," Advances in
      Chem. Series No. 79, Chap. 13, ACS, 161-172 (1968).
83.   Dick, S. G., Fuerstenau, D. W. and Healy, T. W., "Ad-
      sorption of Alkylbenzenesulfonate (A.B.S.) Surfactants
      at Alumina-Water Interface," J. Colloid and Interface
      Sci. Vol. 37, No. 3, 595-601 (1971).
84.   Somasundaran, P., "Role of Ionomolecular Surfactant
      Complexes in Flotation," Int'l J. Mineral Processing
      Vol. 3, 6 pp. (1976).
85.   Somasundaran, P., "Adsorption of Starch and Oleate and
      Interaction between them on Calcite in Aqueous Solu-
      tions," J. Coll. Interf. Sci. Vol. 31, No. 4, 557-565
      (1969).
86.   Hanna, H. S., "Adsorption of Some Starches on Particles
      of Spar Minerals," 2nd Cairo Solid State Conference,
      in Recent Advances in Science and Technology (A. Bishay,
      Ed.), Vol. 1, pp. 365-374, Plenum Press, 1974.
87.   Hanna, H. S., "Adsorption of Tannin on Spar Minerals,"
      Paper presented at 3rd Arab. Chem. Conference,
      National Research Center, Cairo (1972).

88.   Tadros, Th. F., "The Interaction of Cetyltrimethyl-
       ammonium Bromide and Sodium Dodecylbenzenesulfonate
       with Polyvinyl Alcohol. Adsorption of the Polymer-
       Surfactant Complexes on Silica," J. Coll. Interf. Sci.
       Vol. 46, 528-540 (1974).

89.   Iskra, J., Gutierrez, C. and Kitchener, J. A., "Influence
       of Quebracho on Flotation of Fluorite, Calcite, Hematite
       and Quartz with Oleate as Collector," Trans. IMM (London)
       Sec. C Vol. 82, C73-78 (1973).

90.   Somasundaran, P. and Prickett, G. O., "A Response Sur-
       face and Computer Methodology for Adsorption and Dis-
       solution Studies," Canadian Met. Bull. Vol. 66, 92-96
       (1973).

91.   Lin, J. J. and Somasundaran, P., "Free Energy Transfer
       of Surface-Active Agents Between Various Colloidal
       Interfacial States," J. Colloid Interf. Sci. Vol. 37,
       No. 4, 731 (1971).

PHYSICO-CHEMICAL ASPECTS OF ADSORPTION
AT SOLID/LIQUID INTERFACES

II.   Mahogany Sulfonate/Berea Sandstone, Kaolinite

H. S. Hanna and P. Somasundaran
Columbia University

I.   ABSTRACT

Adsorption isotherms recently obtained for the Mahogany sulfonate/Berea sandstone and pure alkyl sulfonate/kaolinite systems have provided insight into the role of factors that control the nature of adsorption in these systems. The appearance of maxima and minima in the adsorption isotherms is found to be dependent primarily on the type of surfactant and on inorganic dissolved species that are present. Impurities as well as oils in the surfactant are also found to produce significant effects on the adsorption.

II.   INTRODUCTION

In a study of the adsorption of oil soluble Mahogany petroleum sulfonate on Berea sandstone cores, Trushenski et al. (1) observed an adsorption maximum near the critical micelle concentration followed by a minimum in the concentration region where the fluid changes from turbid to transparent amber in appearance (20,000-30,000 ppm). They also observed preflushing of the core with a 0.2N NaCl solution to reduce the uptake of the surfactant by the core. Bae et al. (2) also observed the prewetting of the silica gel they used with a 40 weight percent brine solution to reduce the adsorption from an oil-external microemulsion (as opposed to the water-external microemulsion studied by Trushenski et al. (1)) by almost one-half. These workers found using both dynamic and static tests, the sulfonate adsorption from the oil-external microemulsion to be higher than that from the water-external microemulsion and aqueous solutions. The results of Bae et al. (2) and Bae and Patrick (3) for adsorption from aqueous solutions also indicated the presence of an adsorption maximum, in general agreement with those reported by Trushenski et al.

Other noteworthy studies of adsorption on reservoir rock minerals include that of Gale and Sandvik (4) for petroleum sulfonate fractions on calcium montmorillonite, kaolinite and

Berea sandstones, Hill et al. (5) for petroleum sulfonate on
Berea sandstone in the presence of poly-phosphates and Hurd
(6) for petroleum sulfonates on reservoir sandstone in the
presence of carbonates and phosphates. Gale and Sandvik (4)
observed preferential adsorption of the high equivalent weight
(500 or more) sulfonates on montmorillonite. A similar effect
of molecular weight of the surfactant has also been reported
by Hurd (6) for petroleum sulfonate in reservoir sandstones,
Somasundaran et al. (7) and Fuerstenau et al. (8) for alkyl-
amines on quartz, Somasundaran and Kulkarni (9) for perfluoro-
carboxylates on alumina, Wakamatsu and Fuerstenau (10) for
sodium alkylsulfonates on alumina, Weber and Morris (11) for
alkylbenzenesulfonates on activated carbon. The above observations are of particular interest since it suggests undesirable
changes in composition of the slug as it moves through a reservoir. Hill et al. (5) also observed the adsorption of petroleum sulfonate to be reduced in the presence of tripolyphosphate and a Dow pusher polymer.

     It is the objective of the present investigation to study
the adsorption of relevant surfactants on reservoir rock minerals under various experimental conditions and thereby to
understand the mechanisms involved in this interfacial process.
In this paper, the results obtained for adsorption of sulfonates on Berea sandstone are presented along with some data
for adsorption on kaolinite for the purposes of comparison.

III. EXPERIMENTAL

A. Substrates
     Materials used in this study as adsorbents include Berea
sandstone, and kaolinite.
1. *Berea Sandstone (BS)*
     Two crushed samples of Berea sandstone supplied by Amoco
Production Company were used. The size analysis and X-ray
diffraction analysis provided for the sample are given in
Table I. The above samples were prepared for adsorption
studies by dry grinding in a planetary porcelain ball mill.
Size analysis of the initial crushed and final ground products are given in Table II.
2. *Kaolinite*
     Kaolinite used was from a well-crystallized sample with
a surface area of 9.816 m^2/g purchased from the clay repository
at the University of Missouri.

B. Surfactants and Chemicals
     Sodium dodecylbenzenesulfonate (SDDBS) specified to be
95-99% and sodium dodecylsulfonate (SDDS) specified to be
99+% pure were purchased from K and K Laboratories and Aldrich
Chemical Company respectively. Surface tension versus

TABLE I

SIZE ANALYSIS AND X-RAY DIFFRACTION DATA FOR BEREA SANDSTONE SAMPLES (12)

Sample No.	Size Analysis		X-ray Diffraction Data, Percentages						
	Size Fraction, Mesh	Wt. %	Quartz	Feldspar	Kaolinite	Illite	Calcite	Dolomite	Siderite
B.S.-1	-25 + 50	40.0							
	-50 + 100	25.0							
	-100 + 200	25.0							
	-200 (PAN)	10.0							
	Total	100.0	80	1	7	4	1	4	3
B.S.-2	-25 + 50	6.8	86		6	2		6	Tr.
	-50 + 100	79.7	93		5			2	
	-100 + 200	10.0	67	7	16	5		5	
	-200 (PAN)	3.5	41	8	26	9		16	Tr.
	Total	100.0	80	4	8	6		2	

TABLE II

SIZE ANALYSIS AND SURFACE AREA OF B.S.-1 AND
B.S.-2 BEFORE AND AFTER GRINDING

Sample	B.S.-1		B.S.-2	
Size Fraction, mesh	Crushed	Ground	Crushed	Ground
+100	65.0	1.0	86.5	1.0
-100 + 200	25.0	61.6	10.0	68.0
-200 (PAN)	10.0	37.4	3.5	31.0
Total	100.0	100.0	100.0	100.0
Surface Area, m^2/g.	1.224	1.673	1.102	1.526

concentration curve obtained for SDDS showed it to be surface
chemically pure. Dodecylbenzenesulfonate, however, appeared
to contain some surface active impurity.

A sample of Mahogany AA sulfonate (MS-AA) was provided
by Amoco Production Company. This sample contained 37% by
weight of unsulfonated oil. The oil-free active sulfonate
fraction was separated into three fractions (MS-1, MS-2, MS-3)
using a chromatographic separation technique (13). Average
equivalent weights of the active sulfonate determined using
the two phase mixed indicator method (14) were found to be
457, 424 and 460 respectively. Surface tension versus concentration curves of these fractions did not show any minimum;
this indicates that these fractions contain surfactants that
were close in molecular weight (see Figure 1).

The inorganic salts used to adjust the ionic strength and
pH were of A.R. grade. Triple distilled water was used for
all the tests.

C. Experimental Procedure

Desired amounts of solids (5 to 20 grams) were agitated
with the required volume of surfactant solutions for known
times in pyrex vials partially immersed in a water bath or
placed in an incubator maintained at the desired temperature.
At the end of a test, a sample of the supernatant solution
was centrifuged at 1500 G for 20 minutes, the liquid above
the mineral layer was thoroughly mixed and the supernatant
was analyzed for the residual concentration of the sulfonate.
From the difference between initial and final values, adsorption of the surfactant was calculated. For evaluating

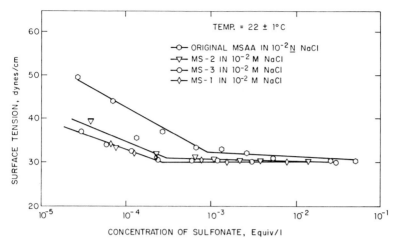

Fig. 1. Surface tension versus concentration curves for various Mahogany sulfonate fractions, MS1, MS2, and MS3, and the MSAA bulk sample

retention instead of adsorption, the same procedure was used but without any mixing. Samples for the determination of the residual concentration were taken in this case from the upper region of the supernatant. The concentration of the surfactant was determined by a two phase titration technique using a mixed indicator (dimidium bromide + disulphine blue (14-16)). It might be noted that even after mixing the supernatant, some of the precipitate trapped inside the centrifuged mineral bed might not be mixed into the supernatant. This effect, if present, will cause apparently higher adsorption values. The approach to equilibrium adsorption at the mineral/solution interface is given in Figures 2 and 3 for dodecylsulfonate/ kaolinite system and Mahogany sulfonate/Berea sandstone systems respectively. For sodium dodecylbenzenesulfonate/ montmorillonite system, a similar trend has been observed by Wayman et al. (17).

The surface tension was determined using a Wilhelmy plate (Pt) set up in combination with a microbalance connected to a recorder (18,19).

IV. RESULTS AND DISCUSSION

Adsorption of Mahogany sulfonate and dodecylbenzenesulfonate was determined on crushed and ground samples of Berea sandstone. The isotherms obtained are given in Figures 4 and 5.

All of the isotherms exhibit an adsorption maximum, with the main difference being that the maximum adsorption on the ground sample is about two times that on the original crushed

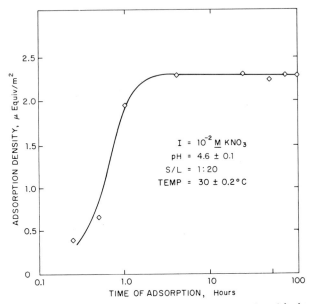

Fig. 2.   *Adsorption of dodecylsulfonate on kaolinite as a function of time*

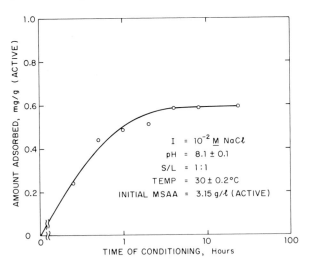

Fig. 3.   *Adsorption of Mahogany sulfonate AA on Berea sandstone as a function of time*

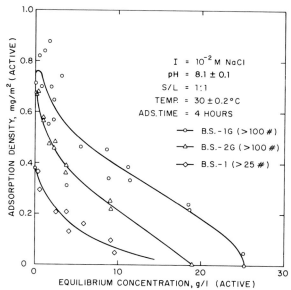

Fig. 4. *Adsorption isotherms of Mahogany sulfonate AA on two*
ground Berea sandstone samples (B.S.-1G and B.S.-2G)
and one crushed Berea sandstone sample (B.S.-1)

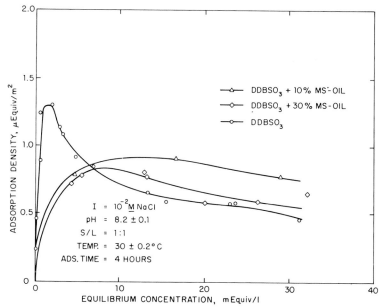

Fig. 5. *Adsorption isotherms of dodecylbenzenesulfonate on*
Berea sandstone in the presence and in the absence
of oil extracted from Mahogany sulfonate AA sample

sample on a surface area basis and three to four times on a weight basis. This suggests an increase in the number of sites that are suitable for the adsorption of the Mahogany sulfonate. The increase in adsorption may be due to an increase in the surface area of the sample upon grinding and possibly increased surface contamination of the harder quartz particles with softer minor constituents such as clay, carbonate or even possibly alumina from the grinding medium.

The lower adsorption capacity of the second ground sample, B.S.-2G, than that of the first ground sample, B.S.-1G, is attributed to its lower content of clay and carbonate minerals (see Table I). Quartz by itself is not expected to adsorb significant quantities of anionic sulfonates in the absence of specifically adsorbing multivalent metal cations, such as calcium which admittedly will be present in the system studied here.

The isotherm obtained for the Mahogany sulfonate on kaolinite in the presence of 10^{-2} $\underline{M}$ NaCl is given in Figure 6. An adsorption maximum is clearly exhibited by the Mahogany sulfonate/kaolinite system. However, the limiting adsorption density obtained in this case does not tend to zero like in the case for Berea sandstone.

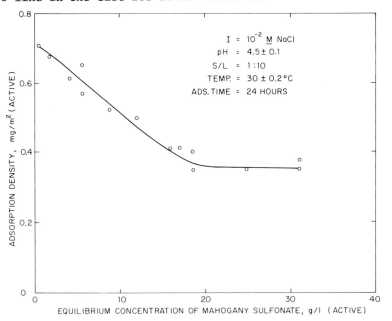

Fig. 6. Adsorption isotherm of Mahogany sulfonate AA on kaolinite

The adsorption isotherms obtained for sodium dodecyl-benzenesulfonate and sodium dodecylsulfonate on kaolinite are given in Figures 7 and 8 respectively. The isotherm obtained for the SDDS is characterized by a moderate increase in adsorption with increase in sulfonate concentration initially, an apparently limiting value being approached around the critical micelle concentration (CMC) of the sulfonate. Increase of the ionic strength caused enhanced adsorption, even though the general nature of the adsorption isotherm was retained. This result is in general agreement with that obtained in the past for adsorption of sulfonates on oxide minerals.

The scattering of the data in these cases was considerably less than that obtained for the sample B.S.-1G in Figure 4. This comparison shows that sampling of the mineral is not a major cause for the scattering since the reproducibility for the case of dodecylbenzenesulfonate is excellent. Sampling of the supernatants can still be one of the factors responsible for the scatter, particularly if phase separation is involved.

The marked difference in shape between the isotherms for Mahogany sulfonate and dodecylsulfonate on the same mineral is to be noted. The above results clearly show that basic differences in the adsorption behavior can occur due to differences in the nature of the surfactant used. While it is necessary to conduct basic studies on well-characterized systems to elucidate the mechanism involved, it can be stated at this point that the structure of the surfactant including the presence of aryl groups as well as the presence in the surfactant system of co-surfactants or even impurities can be responsible for the observed effects. In fact, the method of calculation of adsorption density itself can cause marked changes in the shape of the isotherm obtained. This is clearly seen in Figure 9 which shows isotherms obtained with and without taking the activity of the sulfonate into consideration.

One major difference between the above sulfonates is the presence of unsulfonated oil in the Mahogany sulfonate. To determine the role of this oil in determining the shape of the isotherm, adsorption of dodecylbenzenesulfonate was determined in the presence of various amounts of oil. The results obtained are shown in Figure 5. It can be seen that even though the general shape of the isotherms remained unaltered, addition of the oil did produce a measurable change in the amount of the sulfonate adsorbed. This observation is in agreement with that obtained for the kaolinite/sulfonate system (Figure 7). Non-polar compounds such as long chain hydrocarbons, alcohols and polymers have been reported in

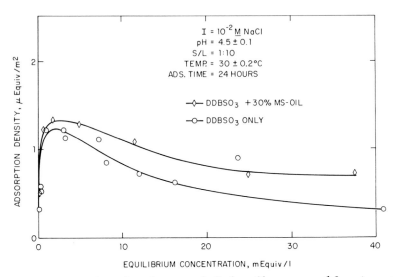

Fig. 7. *Adsorption isotherms of dodecylbenzenesulfonate on*
kaolinite, in the presence and in the absence of
oil extracted from Mahogany sulfonate AA sample

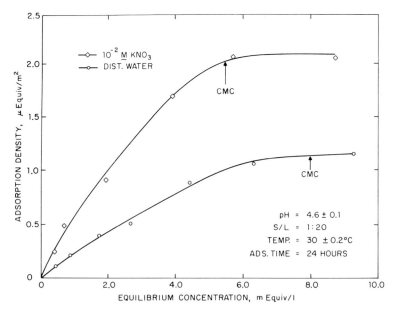

Fig. 8. *Adsorption isotherms of dodecylsulfonate on kaolinite*

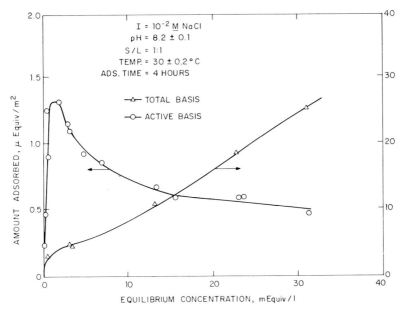

*Fig. 9.   Effect of consideration of the activity of the
surfactant on the shape of the adsorption isotherm of Berea sandstone/Mahogany sulfonate
system*

the past to enhance or depress adsorption of surfactants
depending, among other things, on the chain length of the
non-polar compound. Increased adsorption results if the
surfactant is present partly in the neutral molecular form
or is replaced partly by an alcohol of approximately the
same chain length. On the other hand, if the co-surfactants
present are short chain alcohols, adsorption is depressed.
The nature of the chain length distribution or chemical composition of the unsulfonated oil is, however, unknown to us
and further analysis of the results is hampered by this fact.

Adsorption isotherms obtained for the three oil-free
Mahogany sulfonate fractions (MS-1, MS-2, MS-3) on Berea
sandstone at constant ionic strength of 10^{-2} $\underline{M}$ NaCl are given
in Figure 10 along with that obtained for the original Maho-
gany sulfonate containing ~ 37% unsulfonated oil. It can be
seen that the oil-free Mahogany sulfonate/Berea sandstone
systems also exhibit adsorption maxima. Moreover, at concentrations above about 10 g/l negative adsorption was obtained
for these systems. Such negative adsorption of anionic
surfactants on clay minerals has been attributed in the
past to the higher affinity of clay to water than that to
the surfactant. Other interfacial phenomena such as micellar

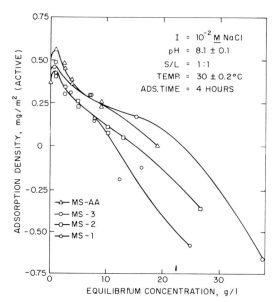

Fig. 10. Adsorption isotherms of Mahogany sulfonate oil-free fractions compared with that of the original MS AA on Berea sandstone

exclusion could also possibly be responsible for the observed negative adsorption. Surfactant concentration of zero adsorption is found to increase in the order MS-1 < MS-2 < MS-AA < MS-3. This order apparently follows that of the increase in their solubility in water rather than the estimated equivalent weight of the fractions.

A. Phase Separation in the MS-AA/B.S. Systems

The phase separation and/or precipitation is found to play an important role in the Mahogany sulfonate/Berea sandstone system as suggested by the following observations:

1) In contrast to the MS-AA/B.S. system, all the aqueous solutions of the oil-free Mahogany sulfonate fractions were turbid even at concentrations as low as 1 g/l. Clear solutions of MS-AA could be obtained at all concentrations up to 100 g/l in water or 10^{-2} M NaCl solutions.

2) Results for the retention of sulfonate (obtained from the difference between the concentration of the sulfonate in the uppermost layer of the supernatant and the initial concentration) provided indications for the existence of sulfonate precipitation or phase separation. The isotherms shown in Figure 11 exhibit a shallow minimum followed by a sharp increase in the high concentration range. The limiting bulk residual concentration may correspond to

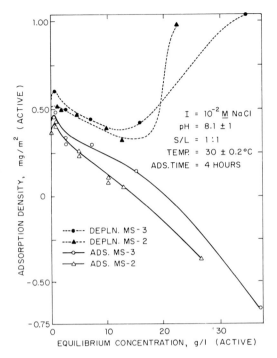

Fig. 11. Depletion isotherms of Mahogany sulfonate fractions on Berea sandstone (B.S.-2G)

the solubility limit of salts such as calcium and magnesium sulfonates.

B. Effect of Inorganic Salts on Mahogany Sulfonate Adsorption

Effect of inorganic salts such as NaCl, Na_2SO_4 and Na_2HPO_4, which contain ions of different charges, on the adsorption of Mahogany sulfonate on Berea sandstone was investigated. Adsorption isotherms obtained at different concentrations of NaCl are given in Figure 12. Addition of NaCl to the system is found to both increase the adsorption density and alter the shape of the isotherms in a very interesting manner. In the presence of 10^{-2} M NaCl the adsorption density (0.66 mg/m²) is about twice as that adsorbed in water (0.28 mg/m²) at maximum adsorption. The adsorption maximum is found to exist below a NaCl concentration of 2.5 x 10^{-2} M. At a concentration of 2.5 x 10^{-2} M NaCl, the isotherm exhibits a maximum followed by a shallow minimum as the concentration of the sulfonate is increased. Increase in NaCl concentration to 5 x 10^{-2} M causes disappearance of the maximum and the minimum. The isotherm is now a typical smooth H-type curve (20). A further increase in the salt concentration

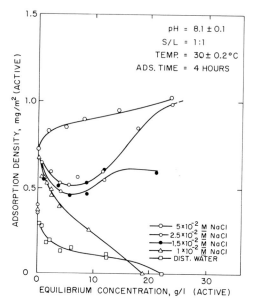

Fig. 12.   Effect of NaCl additions on the adsorption of Mahogany sulfonate AA on Berea sandstone (B.S.-2G)

merely increases the sulfonate adsorption without any additional effect on the shape of the isotherm.

Inorganic electrolytes are known to increase adsorption
of a surfactant at both the solid/liquid and the solution/air
interfaces. Such increase in surface activity is also the
cause for a decrease in critical micelle concentration upon
addition of salts. In addition, adsorption of anionic
surfactants on negatively charged solids can increase upon
the addition of salts presumably due to the reduced electrostatic repulsion between the surfactant ions or micelles and
the solid particles. It is, however, difficult to attribute
the drastic change in shape of this adsorption isotherm to
such effects. The concept of micellar exclusion phenomena
can also hardly be considered to cause such changes since
the change in thickness of the electrical double layer owing
to a five-fold increase in salt concentration is, even though
not insignificant, not as drastic.

Addition of Na_2SO_4 to the adsorption system also increases the maximum adsorption density from 0.28 mg/m^2 in
water to 0.48 mg/m^2 and 0.84 mg/m^2 in 5 x 10^{-3} $\underline{M}$ and
5 x 10^{-2} $\underline{M}$ Na_2SO_4 solutions respectively (see Figure 13).
These values are, however, lower than those obtained for
systems containing NaCl at similar ionic strength conditions.
Also, unlike that of NaCl, addition of Na_2SO_4 (up to 5 x 10^{-2}
$\underline{M}$) did not produce any significant change in the shape of the

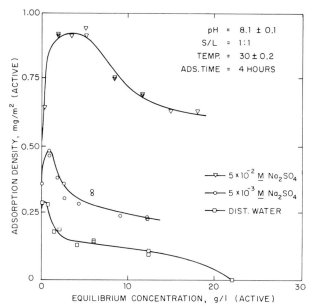

Fig. 13. *Effect of Na_2SO_4 additions on the adsorption of Mahogany sulfonate AA on Berea sandstone (B.S.-2G)*

isotherm. An adsorption maximum was always observed. It is to be noted that this maximum shifted to higher sulfonate concentrations with increase in concentration of the added sulfate. The increase in adsorption due to the addition of Na_2SO_4 can be attributed, as in the case for NaCl, to reduced electrostatic repulsion between the various negatively charged entities in the interfacial region. The lower adsorption in Na_2SO_4 solutions compared with that in NaCl solutions can be ascribed to the bivalent sulfate species that will compete more effectively with negatively charged sulfonate groups of the surfactant for adsorption sites. In addition, any specific adsorption that is normally associated with such bivalent species will increase the negative charge of the mineral and thereby reduce adsorption of anionic surfactants.

Addition of Na_2HPO_4 is found to depress the adsorption of Mahogany sulfonate under all conditions (see Figure 14). Adsorption is even lower than that obtained in water. Adsorption maxima, even though not as marked as those obtained in water, are obtained in Na_2HPO_4 solutions. The observed decrease in adsorption is in accord with the explanations given earlier for the SO_4^{--} effect. The polyvalent phosphate species will in a similar manner provide an increased competition to the sulfonates as well as an increase in the negative charge of the mineral both of which will in turn cause a depression of the sulfonate adsorption.

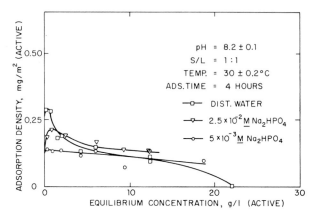

Fig. 14.   Effect of Na_2HPO_4 addition on the adsorption of
           Mahogany sulfonate AA on Berea sandstone (B.S.-2G)

In summary, effect of addition of inorganic electrolytes
on sulfonate adsorption can be either to enhance it or depress
it depending on the charge of the added species and their
concentration. Monovalent sodium ions appear to enhance adsorption of the sulfonate by acting as counter ions and
possibly by salting-out effects. In contrast to this, anions
are seen to compete with the sulfonate groups of the surfactant and decrease its adsorption on the mineral. The effect
of anions was found to increase in the order $Cl^- < SO_4^{--} <$
phosphate (mostly in HPO_4^{--} form in the pH region under
consideration). It is to be noted that inorganic electrolytes
will also change the ionomolecular composition of the adsorption system by changing the solubility of the constituent
minerals as well as the distribution of the surface active
constituents in the oil phase, aqueous phase and the interfacial region. Adsorption of the surfactant may be more
sensitive to such changes that have been shown in the past
to affect adsorption mechanisms greatly and thereby the
shape of the isotherm.

The significance of the effect of inorganic electrolytes
cannot be overemphasized since reservoir systems can be expected to contain various dissolved species depending on the
mineralogical composition of the rock.

C.   Effect of Solid/Liquid Ratio

Adsorption isotherms obtained for Mahogany sulfonate AA
on Berea sandstone ·at different solid/liquid ratios are given
in Figure 15. The ratio did not have any effect on the shape
of the isotherm itself. However, the amount adsorbed is increased at high surfactant concentrations with an increase

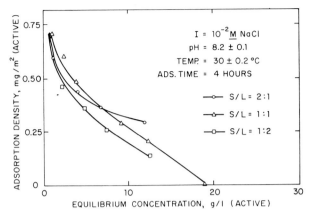

Fig. 15. *Effect of solid/liquid ratio on the adsorption of Mahogany sulfonate AA on Berea sandstone (B.S.-2G)*

of the amount of solid, there being no definite trend at low surfactant concentrations.

Ideally, for a well-characterized system under equilibrium conditions there should be no effect of solid/liquid ratio on adsorption isotherm provided there are no such effects as micellar exclusion. The micellar exclusion, if present, could contribute towards an increase in adsorption with increase in solid/liquid ratio if the solid content is sufficiently high so that the double layers of individual particles will begin interacting.

D. Effect of Temperature on Mahogany Sulfonate Adsorption

The effect of solution temperature is illustrated in Figure 16. Adsorption of the sulfonate increases with increase in solution temperature. The original shape of the isotherm is, however, maintained in that the maximum is present at all temperatures.

The increase in adsorption with temperature indicates possible chemical interaction between the sulfonate and the surface species such as calcium. It is to be noted, however, that for the present system such an effect could also be the result of variations with temperature in the distribution of the sulfonate between the aqueous phase and oil phase as well as the changes in such characteristics as concentration of micelles and/or microemulsion in the system.

E. Effect of pH

Experiments were conducted on Berea sandstone/Mahogany sulfonate and Kaolinite/dodecylsulfonate systems to determine the effect of solution pH on adsorption. For the former system at a constant ionic strength of 10^{-2} M NaCl, the

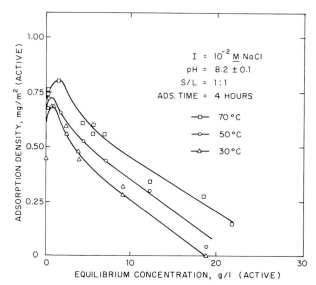

Fig. 16.   *Effect of temperature on the adsorption of Mahogany*
          sulfonate AA on Berea sandstone (B.S.-2G)

adsorption density were found to be 0.66 and 0.4 mg/m^2 for the
initial pH conditions of 5 and 11, respectively, and the corresponding final pH values were not much different from each
other (8.3 and 8.8).   Results obtained for the kaolinite/
dodecylsulfonate system are given in Figure 17.   It can be
seen that the adsorption of sulfonate on kaolinite decreases

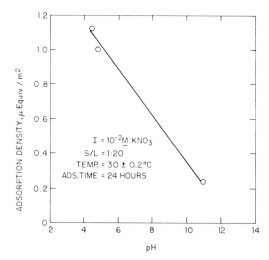

Fig. 17.   *Effect of pH on the adsorption of dodecylsulfonate*
          in 10^{-2} M KNO$_3$ on kaolinite

with increase in pH in this case also. These observations are
in agreement with what would be expected from a consideration
of the fact that the mineral will become increasingly negatively charged with an increase in pH and thereby possibly
retard the adsorption of an anionic surfactant such as
sulfonate on it.

It is to be noted that natural pH values obtained with
Berea sandstone (pH 8.2 ± 0.1) were higher than those obtained
with clay (pH 4.5 ± 0.1) owing to the presence of carbonate
minerals associated with the former. Minerals with significant buffering capacity can be expected to influence the
adsorption, in addition to their direct role, also by changing
the ionomolecular composition of the solution. Decreased adsorption of sulfonate on Berea sandstone samples compared
with that on kaolinite under natural pH conditions is attributed to the above pH change.

V.   SUMMARY AND CONCLUSIONS

Adsorption isotherms for sodium dodecylsulfonate, sodium
dodecylbenzenesulfonate and Mahogany sulfonate, on Berea sandstone and kaolinite were determined by measuring the change
in concentration of the sulfonate in the solution upon contacting with the minerals. Effects on adsorption of the
presence of NaCl, Na_2SO_4, Na_2HPO_4 and varying amounts of unsulfonated oil fractions of the Mahogany sulfonate AA, were
investigated. Also, effects of solid to liquid ratio,
temperature, solution pH and pregrinding of the rock were
examined for selected systems. Major findings are summarized
below:

1)   The nature of adsorption isotherm obtained is markedly
dependent upon the type of sulfonate used. Adsorption maxima
were obtained for the Mahogany sulfonate and dodecylbenzenesulfonate on both kaolinite and Berea sandstone. Isotherms of
the dodecylsulfonate exhibited only positive slopes in the
concentration range studied.

2)   An increase in ionic strength due to the addition of
NaCl increased the adsorption in all cases. Adsorption in the
presence of KNO_3 was higher than that in the presence of NaCl
for the dodecylsulfonate/kaolinite system.

3)   Adsorption maximum was sensitive to the amount of
NaCl added. At lower NaCl concentrations the maximum existed
for Mahogany sulfonate AA/Berea sandstone system; at an intermediate concentration of 2.5 x 10^{-2} M NaCl, the isotherm exhibited a maximum followed by a shallow minimum; at still
higher concentrations the adsorption maximum was not present.

4) Na_2SO_4 addition also caused an increase in adsorption, but the effect of this uni-bivalent salt was less than
that caused by the addition of the uni-univalent salt NaCl.

5) Addition of Na_2HPO_4 (the uni-<u>trivalent</u> salt) was found to depress the adsorption of the sulfonate under all conditions.

6) Addition of unsulfonated oil, separated from the original Mahogany sulfonate AA, altered the adsorption of the oil-free sulfonate fractions as well as that of dodecyl-benzenesulfonate. Possible role of the unsulfonated oil in causing the existence of the adsorption maximum is discussed.

7) The three oil-free sulfonate fractions obtained using chromatographic separation also yielded adsorption maxima and, at higher concentrations, even negative adsorption.

8) The effects of phase separation on sulfonate retention (in contrast to adsorption) and dependence of the extent of phase separation on salinity and unsulfonated oil concentration are discussed.

9) The effect of pH increase for the dodecylsulfonate/kaolinite system was to decrease the sulfonate adsorption with increase in pH.

10) Solid to liquid ratios tested did not have an effect on the shape of the isotherm and the presence of the maximum itself. The amount adsorbed was higher at higher solid to liquid ratio but only at higher sulfonate concentrations.

11) Increase of temperature of the Mahogany sulfonate AA/Berea sandstone system increased the adsorption. Adsorption maximum was present at all temperatures.

An analysis of the above observations has been presented in this paper. As the type of adsorption isotherms obtained is found to depend significantly on the type of sulfonate, mineral constituents, inorganic electrolytes and possibly even various organic fractions present, it becomes evident that further studies with well-characterized systems will be of substantial help in elucidating the mechanisms involved in determining the adsorption or retention of surfactants in reservoir rocks.

VI. ACKNOWLEDGMENTS
     Amoco Production Company and National Science Foundation (ENG-76-08756) are acknowledged for support of this research and the former for providing us with the necessary technical information and various samples.

VII. LITERATURE CITED
1.   Trushenski, S. P., Dauben, D. L. and Parrish, D. R., "Micellar Flooding--Fluid Propagation, Interaction and Mobility," SPE Paper 4582 presented at 48th Annual SPE Meeting, Las Vegas (1973).

2. Bae, J. H., Petrick, C. B. and Ehrlich, R., "A Comparative
    Evaluation of Microemulsions and Aqueous Systems," SPE
    Paper 4749 presented at Improved Oil Recovery Symposium
    of SPE, Tulsa (1974).
3. Bae, J. H. and Petrick, C. B., "Adsorption of Petroleum
    Sulfonates in Berea Cores," SPE Paper 5819 presented
    at Improved Oil Recovery Symposium of SPE, Tulsa (1976).
4. Gale, W. W. and Sandvik, E. I., "Tertiary Surfactant
    Flooding: Petroleum Sulfonate Composition--Efficiency
    Studies," Soc. Pet. Eng. J. Vol. 13, 191-199 (1973).
5. Hill, H. J., Reisberg, J. and Stegemeier, G. L., "Aqueous
    Surfactant Systems for Oil Recovery," J. Petroleum
    Technology Vol. 25, 186-194 (1973).
6. Hurd, B. G., "Adsorption and Transport of Chemical Species
    in Laboratory Surfactant Waterflooding Experiments," SPE
    Paper 5818 presented at Improved Oil Recovery Symposium,
    Tulsa (1976).
7. Somasundaran, P., Healy, T. W. and Fuerstenau, D. W.,
    "Surfactant Adsorption at the Solid-Liquid Interface--
    Dependence of Mechanism on Chain Length," J. Phys.
    Chem. Vol. 68, 3562-3566 (1964).
8. Fuerstenau, D. W., Healy, T. W. and Somasundaran, P.,
    "The Role of Hydrocarbon Chain of Alkyl Collectors
    in Flotation," Trans. AIME Vol. 229, 321-325 (1964).
9. Somasundaran, P. and Kulkarni, R. D., "Effect of Chain
    Length of Perfluoro Surfactants as Collectors," Trans.
    IMM (London) Vol. 82, C164-C167 (1973).
10. Wakamatsu, T. and Fuerstenau, D. W., "The Effect of Hydro-
    carbon Chain Length on the Adsorption of Sulfonate at
    the Solid/Water Interface," in Adsorption from Aqueous
    Solution, No. 79, pp. 161-172. American Chemical
    Society, 1968.
11. Weber, W. J., Jr. and Morris, J. C., "Kinetics of Adsorp-
    tion on Carbon from Solution," J. Sanit. Eng. Div.
    Proc. ASCE Vol. 90, 70 (1964).
12. Froning, H. R., Personal Communication, May 1975.
13. ASTM Method D2548-69, "Analysis of Oil-Soluble Sodium
    Petroleum Sulfonates by Liquid Chromatography," Am.
    Nat. Standard 211, 278 (1970).
14. Powers, G. W., "The Volumetric Determination of Organic
    Sulfonates or Sulfates by the Double Indicator Method,"
    Amoco Method C-225 (1970).
15. Reid, V. W., Longman, G. F. and Heinerth, E., "Determina-
    tion of Anionic-Active Detergents by Two-Phase Titra-
    tion," Tenside Vol. 4, No. 9, 292-3-4 (1967).
16. Ibid., Vol. 5, No. 3-4, 90-96 (1968).

17.   Wayman, C. H., Robertson, J. B. and Page, H. G., "Adsorp-
      tion of the Surfactant ABS[35] on Montmorillonite," U.S.
      Geol. Survey, Paper 475-B, Article 59, B213-216 (1963).
18.   Somasundaran, P. and Moudgil, D. M., "The Effect of
      Dissolved Hydrocarbons in Surfactant Solutions on
      Froth Flotation of Minerals," J. Coll. Interf. Sci.
      Vol. 47, 290-299 (1974).
19.   Somasundaran, P., Danitz, M. and Mysels, K. J., "A New
      Apparatus for Measurements of Dynamic Interfacial
      Properties," J. Coll. Interf. Sci. Vol. 48, 410-416
      (1974).
20.   Giles, C. H. et al., "Adsorption: XI A System of Clas-
      sification of Solution Adsorption Isotherms, etc.,"
      J. Chem. Soc. 3973 (1960).

THE ADSORPTION LOSSES OF SURFACTANTS
IN TERTIARY RECOVERY SYSTEMS

E. W. Malmberg and L. Smith
Suntech, Incorporated

I. ABSTRACT

The retention of sulfonates from chemical slugs during surfactant flooding is a determinant in process performance and economy. The factors affecting adsorption on reservoir mineral can be investigated on small cores by determining the saturation adsorption of sulfonate from formulations under reservoir conditions but in the absence of oil. The retention in recovery experiments is much less than the saturation adsorption, because residual oil is present, partial-pore-volume slugs are used, and polymer solutions are injected for mobility control. By taking into account the effects of these conditions, the retention of sulfonates in a recovery experiment can be estimated from the saturation adsorption under the same conditions and on the same core material.

II. SCOPE

Retention of sulfonates by reservoir mineral depletes their concentration in chemical slugs injected for recovery of residual oil. The extent of this loss of chemical and loss of process performance and lifetime is a determinant in the economic feasibility of surfactant flooding.

The retention of surfactants by core materials is a combination of two effects. Surfactant retained by physical or mechanical trapping in the pore spaces can be flushed from the core by flow of brine. Surfactant which cannot be removed by brine flushing, but is eluted by isopropyl alcohol is considered to be adsorbed on the mineral surface (1). It has been suggested that some sulfonate may be desorbed in the brine flush (2); however, for the purposes of this paper, the above definitions are adopted. Furthermore, while total sulfonate retained is dependent on flooding rate (2), saturation adsorption is not rate dependent.

The adsorption of sulfonates from surfactant formulations injected into small cores under the conditions of a proposed application demonstrates the effects of temperature, salinity, mineralogy, and formulation composition. However, these

saturation adsorptions are not expected to be the same as retention of sulfonates in a recovery experiment in which a partial pore volume slug is injected, residual oil saturation is
present, and the slug is followed by a polymer solution. Data
demonstrating these effects on retention of sulfonates from
turbid or birefringent formulations are presented and discussed.

III.   CONCLUSIONS AND SIGNIFICANCE

It is concluded that saturation adsorption can be correlated to retention in recovery experiments, and, moreover,
by taking into account the effects of partial pore volume of
surfactant, oil and polymer solution on retention, the retention can be approximated from saturation adsorption on the
same core material.

IV.   BACKGROUND

The adsorption of surface active agents on reservoir mineral has long been considered an impediment to their effective
use as oil recovery agents. Early investigators who advocated
injecting a continual waterflbod containing a low concentration
of detergent recognized that because of adsorption the processes would be economic failures (3-5). Rather than continual injection, Preston and Calhoun (6) suggested the use of a
partial pore volume of chemical solution followed by water to
decrease the amount of chemical required. Adsorption of the
chemical was a major parameter considered by Johnson (7) in
describing methods to evaluate the technical and economic
feasibility of this process. Taber (8) showed adsorption was
decreased and chromatographic transport of nonionic detergents
improved when the concentrations were increased.

Subsequent developments in surfactant flooding have
yielded a variety of chemical systems which generally contain
petroleum sulfonates as the surface active agents, and retention of the sulfonates by core material has been included in
the parameters investigated in characterizing these processes.
In some cases, factors contributing to adsorption of petroleum
sulfonates have been explored separately from recovery experiments. Among variables affecting the adsorption and retention
of petroleum sulfonates are:

- (a) Equivalent weight and equivalent weight distribution
       (9-11).
- (b) Mineralogy (12,9,13) and surface area (11).
- (c) Salinity and divalent cation content of the water in
       place and in the formulation (14,15,11).
- (d) Concentration of the petroleum sulfonates (9,1,16,2)
       and the ratio of coagent to sulfonate (14).
- (e) Physical characteristics of the formulation such as
       whether water or oil external microemulsions or
       aqueous solutions (17) and slug deterioration (1).

(f) Water or oil wetness of the core material (18) and
whether or not oil is present (14).

Several methods of screening surfactant formulations for
adsorption and retention of sulfonates have been described.
By the definitions adopted herein, the method of Gale and
Sandvik (9) and that of Smith et al. (19) measure only adsorption. The procedures of Trushenski (16) and Bae (17)
measure adsorption and physical entrapment. These procedures
approximate to varying degrees the circumstances of a recovery
in application. Trushenski (16) has reported agreement between
adsorption of sulfonates determined in a screening test and
retention observed in recovery experiments. A correlation
between retention observed in a pilot test (20) and laboratory
results (15) has been reported.

The screening method described by us (19) was designed to
measure adsorption of sulfonate and sulfonate surfactants on
core materials under the conditions of their proposed applications, but in the absence of oil. A small core, equilibrated
to flow of synthetic reservoir brine at reservoir temperature,
was injected with surfactant until the effluent concentration
equaled the injected concentration. The mechanically retained
surfactant was removed by flushing with brine, and the adsorbed
surfactant was eluted with isopropyl alcohol. The surfactant
eluted by the isopropyl alcohol represents saturation-level
adsorption in the core material under the experimental conditions. The method lends itself to investigation of temperature, salinity, mineralogy, and variations in formulation as
factors affecting adsorption. Excellent material balances
between injected and produced surfactant were obtained from
sample analyses by the modified Epton titration.

A similar procedure of flushing cores with brine and
isopropyl alcohol to measure sulfonates retained during recovery experiments has been used by us and has been described by
Healy et al. (1). Generally, the retention of sulfonates in
recovery experiments is much less than the adsorption in the
screening method. Some of the effects contributing to this
difference have been investigated.

V. EXPERIMENTAL PROCEDURES AND RESULTS

In the following experiments, the modified Epton titration (19) was used for sulfonate analyses. Unless stated,
Berea sandstone, approximately 500 md was used and flow rates
were 1 ft/day.

A. Adsorption on Different Core Materials

The relative effects of mineralogy on adsorption were
investigated for Formulation B, Table I. The cores used were
Berea sandstone, Seeligson, Cottage Grove, and South Texas

TABLE I

BRINE AND FORMULATION COMPOSITIONS

==
==

Synthetic Seeligson Brine

Sodium	9,945 ppm
Calcium	1,263
Magnesium	65
Sulfate	14
Chloride	17,715
Bicarbonate	55
Total	29,057 ppm

Formulation A

TRS-10	2.5 percent
NEODOL 25-3S	1.5
HEXYL CARBITOL	1.0
in the above brine.	

Formulation B

3% active synthetic sulfonate
3% alcohol coagent
    in 1% sodium chloride

Formulation C

1.8% active synthetic sulfonate
1.9% alcohol coagent
    in 1% sodium chloride

Formulation D

1.8% active petroleum sulfonate
1.0% alcohol coagent
    in 1.3% brine

---

reservoir core A. the mineralogy for these cores is listed in Table II. The cores were approximately 2.5 cm in diameter and 5.0 cm long. The brine used was one-half the concentration of the brine described in Table I. The temperature was 71°C. The experimental procedure was (a) injection of the formulation at 4 ft/day into cores equilibrated to brine flow at temperature until the effluent concentration equaled the injected concentration, (b) brine flush until sulfonate could not be measured in the effluent, and (c) isopropyl flush to elute the adsorbed sulfonate (19). The data are presented

TABLE II

CLAY MINERALOGY OF CORE MATERIALS

	Mont.	Mixed Layer	Illite	Kaolinite
Berea Sandstone	0.10	0.20	1.24	7.38
Seeligson	6.65	0.88	1.01	0.82
South Texas Core A	8.52	0	0.04	0.70
South Texas Core B	4.07	2.20	0.26	1.10
Cottage Grove	1.54	0.57	1.71	9.47

in Table III and shown as a function of montmorillonite content in Figure 2.

TABLE III

SATURATION ADSORPTION ON SMALL CORES

Mineral	Adsorption, lbs/acre-ft.
Berea	9,000
Seeligson	11,600
South Texas Core A	18,000
Cottage Grove	12,000

B. Adsorption from Partial-Pore-Volume Slugs

To demonstrate adsorption of surfactants from partial-pore-volume slugs, Formulation A, Table I (21), was injected into Berea sandstone cores 26 cm long and 5.1 cm in diameter previously equilibrated to flow of 2.9 percent brine, Table II, at 71°C. The cores contained no oil. After the surfactant slug, brine was injected until no surfactant could be measured in the produced fluid. The adsorbed surfactants were flushed from the cores with isopropyl alcohol. The adsorption data are presented in Table IV as total anionic surfactant and in Figure 1 as a function of slug size. In each case, the eluted surfactant contained sulfonate and sulfate in approximately the same ratio as Formulation A. The saturation adsorption for Formulation A on Berea sandstone is also listed in Table IV.

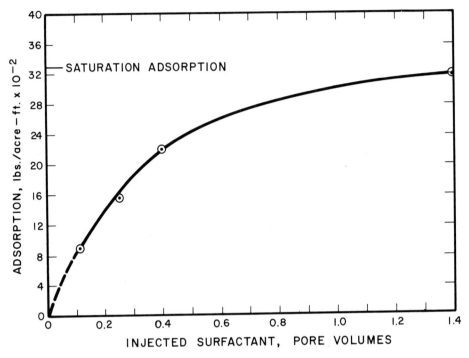

FIGURE 1.   ADSORPTION   OF   FORMULATION   'A' AS   A
FUNCTION   OF   SLUG   SIZE.

C. Adsorption in the Presence of Oil and Polymer
     The effects of residual oil and polymer solutions present
in oil recovery experiments are not accounted for in the adsorption tests described above. To examine these effects on
adsorption, approximately 0.1 pore volume slugs of Formulation
B, 3% active synthetic sulfonate and 3% alcohol, were injected
into 3.8 cm x 3.8 cm x 122 cm Berea sandstone cores equilibrated to flow of 1% sodium chloride, 1 ft/day, at 52°C. In
experiment (a) the injected slug was followed by brine, two
pore volumes, and the core was flushed with isopropyl alcohol.
In experiment (b), the injected surfactant was followed by one
pore volume of 500 ppm Dow PUSHER 700 in fresh water and one
pore volume of 1% sodium chloride brine. The core was flushed
with isopropyl alcohol. In a recovery experiment (c), a core
was oil flooded to residual water with a South Texas crude oil
and water flooded to residual oil, 35% pore volume, with brine.
The surfactant was followed by one pore volume of polymer
solution, one pore volume of brine, and flushed with isopropyl
alcohol. The adsorption data for these experiments are summarized in Table V.

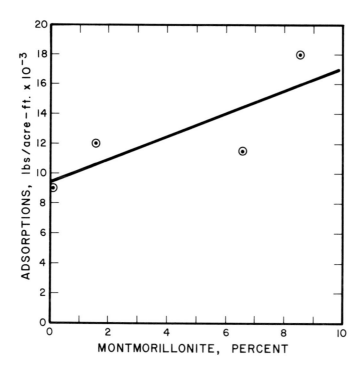

FIGURE 2. SATURATION ADSORPTION ON
CORE MATERIAL AS A FUNCTION OF
MONTMORILLONITE CONTENT.

D. Retention in Recovery Experiments with Increasing
   Slug Size
      The retention of sulfonates from various sizes of surf-
actant slugs in recovery experiments was examined by injecting increasing amounts of Formulation D, Table I, into Berea
sandstone cores, 122 x 3.8 x 3.8 cm, containing residual
saturation of a South Texas crude oil at 52°C. Formulation
D is 1.8 percent active petroleum sulfonate with 1 percent
alcohol coagent in 1.3 percent brine. The retention was the
difference between injected and produced sulfonate. The
amounts of sulfonate injected and retained and the oil recovery are summarized in Table VI.

E. Retention in Recovery Experiments on Different Core
   Materials
      The effect of mineralogy on retention of sulfonate is
illustrated by recovery experiments on composite cores of
Berea sandstone and Seeligson reservoir core plugs, on
Cottage Grove sandstone and unconsolidated South Texas

TABLE IV

ADSORPTION OF SURFACTANTS FROM PARTIAL-PORE-VOLUME
SLUGS BEREA SANDSTONE; FORMULATION A

		Injected Surfactant	Adsorbed Surfactant		
	pore volume	meq/g-rock x 10^3	meq/g-rock x 10^3	lbs/acre-ft	percent of injected
(a)	0.12	0.485	0.357	860	74
(b)	0.25	1.04	0.648	1570	62
(c)	0.40	1.66	0.903	2200	54
(d)	1.4	5.81	1.32	3200	44
(e)		Saturation	1.36	3300	

Core B. The mineralogy is described in Table II. The composite cores consisted of ten sections, 2.5 x 7.5 cm. The
Cottage Grove core was 3.81 x 64 cm and the unconsolidated
core material was packed in a 2.5 x 30.5 cm tube. The cores
were evacuated and flooded with synthetic Seeligson brine,
Table I, and stabilized to flow at 71°C. The cores were oil
flooded to residual water with Seeligson crude oil and waterflooded to residual oil. Approximately 0.1 pore volume of
Formulation C, 1.8 percent active synthetic sulfonate and
one percent alcohol coagent in one percent sodium chloride
solution, was injected into each core. The surfactant was
followed by a 500 ppm solution of Dow PUSHER 700 in fresh
water. The retention data reported in Table VII were obtained by subtracting produced sulfonate from the measured
quantity injected. The retention is presented as a function
of montmorillonite content in Figure 3.

VI. DISCUSSION
There are numerous factors which determine the adsorption
and retention of sulfonates. Adsorption increases with increasing average equivalent weight of the sulfonate (9-11) and
with increasing salinity or divalent cation concentration in
the formulation or in connate brine (14,15,11). Adsorption
decreases with increasing temperature. Greater retention
from oil-external microemulsions than from water-external
has been reported by Bae (17); that this comparison may be
very complex, however, is shown by the evidence that retention is dependent upon the stage of slug degradation (1).

TABLE V

THE EFFECTS OF OIL AND POLYMER ON ADSORPTION
OF SULFONATE FROM FORMULATION B

| | | Surfactant Injected | Surfactant Adsorbed | | Concentration Produced |
		meq/g-rock x 10⁴	meq/g-rock x 10⁴	percent of injected	Maximum C/Co
(a)	Surfactant Slug	7.53	7.00	93	0.003
(b)	Surfactant Slug; Polymer	7.40	3.14	42	0.08
(c)	Residual Oil; Surfactant Slug; Polymer	7.50	2.70	36	0.44
		91% recovery of residual oil			
(d)	Saturation Adsorption		30.0		1.0

283

TABLE VI

THE EFFECT OF SLUG SIZE ON RETENTION OF SULFONATES IN
RECOVERY EXPERIMENTS: FORMULATION D AND SOUTH TEXAS
CRUDE OIL IN BEREA SANDSTONE

Injected Sulfonate meq/g-rock x 10^4	Retained Sulfonate meq/g-rock x 10^4	Oil Recovery percent of residual
1.8	0.7	40
2.6	1.5	52
5.9	1.5	95
6.0	1.8	98
saturation	11.1	

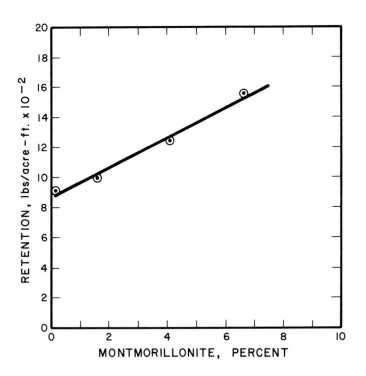

FIGURE 3. RETENTION IN RECOVERY
EXPERIMENTS AS A FUNCTION OF
MONTMORILLONITE CONTENT.

TABLE VII

RETENTION OF SULFONATE IN RECOVERY EXPERIMENTS
ON COMPOSITE CORES

	Retention	lbs/acre-ft.	Recovery of Residual Oil
Berea Sandstone	3.69×10^{-4} meq/g-rock	900	99%
Seeligson Core Plugs	6.85×10^{-4} meq/g-rock	1540	51%
South Texas Core B	5.50×10^{-4} meq/g-rock	1240	82%
Cottage Grove	4.45×10^{-4} meq/g-rock	1000	42%

The adsorption of sulfonates in oil-wet cores has been found to be greater than in water-wet cores (18). Adsorption and retention are also dependent upon the sulfonate concentration (9,1) and the ratio of coagent to sulfonate (14). A minimum in retention with increasing sulfonate concentration has been reported (17,2,16) and has been described as "retrograde adsorption."

Bae (2) has reported that retention, measured by his procedure, is flow rate dependent. Adsorption, as defined herein, is rate independent, but it is not surprising that mechanical and physical retention are dependent upon flow rate.

In addition to the factors cited above, mineralogy, interactions with oil and polymer, and the quantity of sulfonate injected influence the quantity of sulfonate retained. While the effects of sulfonate average equivalent weight, formulation characteristics, salinity, temperature, and core mineralogy and wetness can be effectively investigated in small cores by the method of Smith et al. (19), the correlation of saturation adsorption to the retention observed in recovery experiments requires that the effects of oil, polymer, and flow of partial-pore-volume slugs be considered.

A.   The Effect of Partial-Pore-Volume Slugs on Adsorption

The flow of several pore volumes of surfactant through a core is required before equilibrium or saturation adsorption is achieved (19,2,16). Therefore, the adsorption of sulfonate from a partial-pore-volume slug would be expected to be less than the saturation value. In the experiments summarized in Table IV, the adsorbed sulfonate and sulfate increase when increasing amounts of Formulation A were injected into cores. The percentage of injected sulfonate adsorbed decreased from 74 percent at 0.1 pore volume to 44 percent at 1.4 pore volume. The saturation value for Formulation A under the experimental conditions was approached at approximately 1.5 pore volumes of injected slug, Figure 1.

It is noteworthy that the ratio of sulfate to sulfonate eluted from the cores by isopropyl alcohol in the course of the experiments was approximately the same ratio as of Formulation A. Fractionation or separation of these two anionic surfactants during the adsorption process by virtue of their different brine solubilities might have been expected (9,11), but did not occur.

B.   The Effects of Oil and Polymer on Retention

The series of experiments summarized in Table V, illustrated the effects of residual oil and succeeding polymer solution on retention of sulfonate from Formulation B.

Injection of a polymer solution behind a slug of Formulation B reduced the amount of sulfonate retained. An increase in the amount of sulfonate eluted by IPA experiments (a) and (b), Table V, was observed. An additional small decrease in retention may have occurred because the polymer solution was in fresh water rather than one percent sodium chloride. In the recovery experiment, (c), the maximum sulfonate concentration in the produced fluids was increased again and a further decrease in adsorption resulted. The recovery was 91 percent of the residual oil. The amount of sulfonate retained per gram of rock was about one-tenth of the saturation adsorption, (d).

C. The Effect of Slug Size on Retention and Recovery

As would be expected, the retention of sulfonates in recovery experiments is also dependent upon the slug size. In the recovery experiments in Table VI, the retention was observed to increase with the quantity of sulfonate injected. The oil recovery in Table VI suggests that a minimum concentration of Formulation D is required; and, when retention depletes the sulfonate in the surfactant slug, recovery decreases.

Similar effects have been observed by other investigators. Lower sulfonate retention in cores containing oil than in cores without oil has been reported by Gilliland and Conley (14). Healy et al. (1) have reported that an increase in injected slug, 0.05 to 0.20 pore volume, in recovery experiments of the same length resulted in a 30 to 50 percent increase in sulfonate retained. Increasing the core length, 4 ft. to 16 ft., increased retention about 30 percent for the same fractional pore volume slug. It was concluded surfactant retention increased with increasing slug deterioration (1).

D. The Effect of Mineralogy on Adsorption and Retention

The quantities of sulfonate adsorbed in saturation experiments, Table III, correlate with clay content in the core material, Table II. This correlation is shown in Figure 2. No pure silica sandstone was available for measurement for loss in a clay-free system. The data in Tables II, III, and VII show that as kaolinite content decreases and montmorillonite increases, surfactant loss increases. This result is in general agreement with the losses reported by Gale and Sandvik (9) on the pure clay mineral. The adsorption of sulfonates on clays, has been reported previously (13,9,12). A very similar correlation, Figure 3, was found between core mineralogy, Table II, and retention of sulfonate in recovery experiments, Table VII. In Figure 3, the sulfonate retained from 0.1 pore volume of surfactant during recovery experiments was approximately one-tenth of the saturation adsorption in Figure 2. The oil recovery again decreases with increasing retention. In some reservoirs, retention will result in relatively greater slug depletion than in Berea sandstone.

E. Correlation of Saturation Adsorption with Retention

Saturation adsorption data can be obtained on small cores which can be used repeatedly (19). Since reservoir materials are frequently of limited availability, projecting the retention in a recovery experiment from saturation adsorption data obtained on small cores would be desirable.

If one assumes the relationship between saturation adsorption and retention in recovery experiments is the same in Berea sandstone and Seeligson core material, the saturation adsorptions on Berea and Seeligson cores, Table III, and the retention in a recovery experiment on Berea cores, Table VII, can be used to estimate retention in a recovery experiment on Seeligson cores. The retention during the recovery experiment on Berea sandstone was 900 lbs/acre-ft., one-tenth the saturation adsorption, 9,000 lbs/acre-ft. In a recovery experiment with the same size slug on Seeligson cores, the retention was then projected to be 1,200 lbs/acre-ft., one-tenth of the saturation adsorption on Seeligson cores. The observed retention with a slightly larger slug was 1,520 lbs/acre-ft.

Similarly, estimates of retention in recovery experiments can be made by adjusting saturation adsorption data with the effects of a partial-pore-volume slug, residual oil and succeeding polymer solution. The retention from approximately 0.1 pore volume of surfactant was 0.26 of the saturation adsorption, (a) and (e) in Table IV. The retention from 0.1 pore volume of surfactant in the presence of residual oil and succeeded by polymer solution was 0.39 the value obtained in the absence of oil and polymer, (a) and (e), Table V. These adjustments on 11,600 lbs/acre-ft., the saturation adsorption on Seeligson cores, also yield 1,200 lbs/acre-ft. estimated retention in a recovery experiment on Seeligson cores.

From the data in Table III, the estimated retention for a recovery experiment on Cottage Grove sandstone was 1,200 lbs/acre-ft. by the above procedure, and the observed retention was 1,000 lbs/acre-ft., Table VII. Since mineralogically different South Texas cores were used for the adsorption and recovery experiments, a direct comparison cannot be made between the estimate, 1,800 lbs/acre-ft., from Table III, and the observed retention, 1,240 lbs/acre-ft., in Table VII.

VII. CONCLUSION

Saturation adsorption experiments on small cores can be useful in studying the effects of temperature, salinity, mineralogy, and formulation characteristics on the adsorption of sulfonates by core materials. The relationship between saturation adsorption from turbid or birefringent surfactant dispersions on Berea sandstone and reservoir cores and the retention of sulfonate in a recovery experiment on Berea sandstone can be used to estimate retention in recovery experiments on reservoir cores under the same conditions. Similarly, the adsorption on a reservoir core can be adjusted by the mitigating effects of a partial-pore-volume of surfactant, residual oil, and succeeding polymer solution to yield an estimate.

Limited availability of reservoir core materials often makes numerous recovery experiments impractical. Estimates of retention from saturation adsorption data affords needed information for comparison of surfactant formulations. Examples of the interrelationship of loss to the formation, surfactant slug size, and percent oil recovery have been included.

VIII. ACKNOWLEDGEMENT

The authors wish to express their appreciation to Sun Oil Company for permission to publish this paper, to R. E. Pinchera, Sam Slovak, and Sara Fowler for their laboratory work, and to W. M. Hensel and Richard Stallings for core and mineralogical analyses.

IX. LITERATURE CITED

1. Healy, R. N., Reed, R. L. and Carpenter, C. W., "A Laboratory Study of Microemulsion Flooding," Soc. Pet. Eng. J. 15, 87-103 (February, 1975).

2. Bae, J. H. and Petrick, C. B., "Adsorption of Petroleum Sulfonates in Berea Cores," SPE Number 5819, presented at the Symposium on Improved Oil Recovery, Tulsa, Oklahoma, March 22-24, 1976.

3. Kennedy, H. T. and Guerrero, G. T., "The Effect of Surface and Interfacial Tensions on the Recovery of Oil by Water Flooding," Trans. AIME 201, 124 (1954).

4. Dunning, N. H. and Hsiso, L., "Experiments with Detergents as Water Flooding Additives," Prod. Monthly 18, 24 (1954).

5. Ojeda, E., Preston, F. W. and Calhoun, J. C., Jr., "Correlation of Oil Residuals Following Surfactant Floods," Prod. Monthly 18, 20 (1953).

6. Preston, F. W. and Calhoun, J. C., Jr., "Applications of Chromatography to Petroleum Production Research," Prod. Monthly 16, 22 (1952).

7. Johnson, C. E., Jr., "Evaluation of Surfactants for Oil Field Flooding," Jour. Am. Oil Chem. Soc. 34, 209-214 (1957).

8. Taber, J. J., "The Injection of Detergent Slugs in Water Floods," Pet. Trans. AIME 213, 186-192 (1958).

9. Gale, W. W. and Sandvik, E. I., "Tertiary Surfactant Flooding: Petroleum Sulfonate Composition--Efficacy Studies," Soc. Pet. Eng. J. 13, 191-199 (1973).

10. Fuerstenau, D. W. and Wakamatsu, T., "The Effect of Hydrocarbon Chain Length on the Adsorption of Sulfonates at the Solid/Water Interface," in Adsorption from Aqueous Solutions, Advances in Chemistry Series Number 79 (Robert F. Gould, ed.), pp. 161-172, American Chemical Society, 1968.

11. Hurd, B. G., "Adsorption and Transport of Chemical Species in Laboratory Surfactant Waterflooding Experiments," SPE Number 5818, presented at the Symposium on Improved Oil Recovery, Tulsa, Oklahoma, March 22-24, 1976.

12. Hower, W. F., "Adsorption of Surfactants on Montmorillonite," in Clays and Clay Minerals 18, 97-105 (1970).

13. Froning, H. R. and Treiber, L. E., "Development and Selection of Chemical Systems for Miscible Waterflooding," SPE Number 5816, presented at the Symposium on Improved Oil Recovery, Tulsa, Oklahoma, March 22-24, 1976.

14. Gilliland, H. E. and Conley, R. F., "Pilot Flood Mobilizes Residual Oil," Oil and Gas Journal 43-48 (1976).

15. Hill, H. J., Reisberg, J. and Stegemeier, G. L., "Aqueous Surfactant Systems for Oil Recovery," J. Pet. Tech. 186-194 (February, 1973).

16. Trushenski, S. P., Dauben, D. L. and Parrish, D. R., "Micellar Flooding--Fluid Propagation, Interaction, and Mobility," Soc. Pet. Eng. J. 14, 633-642 (Dec., 1974); Trans. AIME, Vol. 257.

17. Bae, J. H., Petrick, C. B. and Ehrlich, R., "A Comparative Evaluation of Microemulsions and Aqueous Surfactant Systems," SPE Number 4749, presented at the Symposium on Improved Oil Recovery, Tulsa, Oklahoma, April 22-24, 1974.

18. Boneau, D. F. and Clampitt, R. L., "A Surfactant System for the Oil-Wet Sandstone of the North Burbank Unit," SPE Number 5820, presented at the Symposium on Improved Oil Recovery, Tulsa, Oklahoma, March 22-24, 1976.

19. Smith, L., Malmberg, E. W., Kelley, H. W. and Fowler, S., "The Quantitative Analysis of Adsorbed Petroleum Sulfonates," SPE 5369, SPE 45th Annual California Regional Meeting, Ventura, California, April 2-4, 1975.

20. French, M. S., Keys, G. W., Stegemeier, G. L., Ueber,
     R. C., Abrams, A. and Hill, H. J., "Field Test of an
     Aqueous Surfactant System for Oil Recovery, Benton
     Field, Illinois," J. Pet. Tech. 195-204 (February,
     1973).
21. Malmberg, E. W., Wilchester, H. L., Shepard, J. C.,
     Schultze, E. F., Parmley, J. B. and Dycus, D. W.,
     "Laboratory Studies on Oil Recovery with Aqueous
     Dispersions of Oil-Soluble Sulfonates," SPE Number
     4742, presented at the Symposium on Improved Oil
     Recovery, Tulsa, Oklahoma, April 22-24, 1974.

THE STRUCTURE, FORMATION AND PHASE-INVERSION
OF MICROEMULSIONS

D. O. Shah, V. K. Bansal, K. Chan, and W. C. Hsieh
University of Florida

I. ABSTRACT

This paper reviews the studies on the structure, formation and phase-inversion of microemulsions by previous investigators as well as that carried out in the present investigators' laboratory. The historical development in the field of microemulsions is reviewed. The differences between micellar solutions and microemulsion systems are discussed. Molecular aggregation and the formation of various liquid crystalline phases in surfactant solutions are summarized. Various theories for the formation of microemulsions and the role of electrical double layer in causing interfacial instability are discussed. The differences between microemulsions and cosolubilized systems are mentioned and the experimental approaches to distinguish between them are discussed. The experimental investigation on the formation of various liquid crystalline phases in the phase-inversion region of microemulsions is described. This review paper is an attempt to present the current state of knowledge of the structure, formation and phase-inversion of microemulsions.

II. CONCLUSIONS AND SIGNIFICANCE

Microemulsions are isotropic, clear or translucent, thermodynamically stable oil/water/emulsifiers dispersions. The droplet diameter in microemulsions ranges from 100–1000Å. The droplets are stabilized by a mixed interfacial film of surfactant and alcohol. Penetration of the interfacial film with oil as well as the interaction of water with the polar group are essential for the formation of microemulsions. The structure of microemulsions consists of close-packed spheres of water-in-oil or oil-in-water. The gap between the surfaces of adjacent droplets is smaller compared to the diameter of the droplets. The structure and the length of alcohol can influence the phase continuity of microemulsions. The shorter chain-length alcohols (C_3 to C_5) tend to make water-external microemulsions whereas the higher chain-length alcohols (C_6 to C_{10}) tend to form oil-external microemulsions. The structure of oil also influences the phase continuity in microemulsions. It can also be said that the relative wettability of the interfacial film with oil and water determines the phase continuity (i.e., the microemulsion will be oil-external or water-external).

It is true that microemulsions can be considered as swollen micelles. However, all micellar solutions cannot be swollen to the extent of microemulsions unless the specific structural requirements and conditions are satisfied. Based on our investigations using various physical techniques, we have proposed that isotropic, stable, clear dispersions can be one of the three types; micellar solutions, microemulsions, or co-solubilized systems. The co-solubilized systems resemble molecular solutions of all four components. However, the possibility of the presence of very small aggregates of water or emulsifiers is not ruled out. The co-solubilized systems exhibit very low electrical resistance even when relatively small amounts of water are added to the system.

The electrical double layer at the oil/water interface can also influence the formation and phase continuity of microemulsions. When water:oil ratio is increased, the water-in-oil microemulsion can be converted into oil-in-water microemulsion. The system will undergo a phase-inversion region which consists of liquid crystalline phases. The cylindrical and lamellar structures of water/oil/emulsifiers are formed in this region. The orientation of lamellar structure with respect to one another can strikingly influence the rheological properties of this system. The ion-dipole association between adjacent droplets in oil-in-water microemulsion can also produce high viscosity gel state for the dispersions. The presence of salt can influence the interdroplet association in microemulsions.

In several microemulsion systems, an oil-in-water microemulsion inverts to water-in-oil type upon increasing salinity. At an intermediate salinity (optimal salinity) a middle phase microemulsion is formed. The middle phase microemulsion is in equilibrium with excess oil and brine. It is proposed that the addition of salt can decrease the interdroplet repulsion and hence produce a close-packed state for the oil-in-water microemulsion which subsequently results in a phase separation and formation of the middle phase. The need for further research in this area is stressed.

III.   HISTORICAL DEVELOPMENTS

In 1940, Schulman and his co-workers published three papers on molecular interactions at oil/water interfaces

which reported the results of their studies on molecular complex formation and the stability of oil-in-water or water-in-
oil emulsions, phase-inversion and interfacial tension
measurements at the oil/water interfaces (Schulman and
Cockbain, 1940a,b; Alexander and Schulman, 1940). From their
studies on the properties of emulsions, they concluded that
the reactions occurring at the oil/water interface were
closely analogous to the corresponding reactions at an air/
water interface. The state of the interfacial film as observed at the air/water interface, i.e., solid, viscous or
liquid, is reflected in the nature of the resulting emulsion
being respectively a grease, viscous or liquid emulsion.
These papers represented one of the first attempts to correlate the properties of the interfacial film with emulsion
characteristics. Hoar and Schulman, in 1943, published a
paper entitled "Transparent Water-in-Oil Dispersion: the
Oleopathic Hydro-Micelle." They used the term "oleopathic
hydro-micelle" for transparent oil-water dispersions which
are now commonly referred to as "microemulsions." At that
time such oil-continuous systems were known as "soluble-oils"
and the essential conditions for their formation were (a)
high soap/water ratio, and (b) the presence of a short-chain
alcohol, fatty acid, alkyl amine or other nonionized amphipathic substance in a mole fraction approximately equal to
that of the soap. They proposed a method to calculate the
radius, r, of water droplets in such systems as follows:

$$r = \frac{3(\text{volume of water})}{\text{area of oil/water interface}} \tag{1}$$

The radius of the water droplet plus the thickness of the
interfacial film of surfactant and alcohol yields the total
radius for the dispersed droplets. They also conjectured that
the spontaneous emulsification of the added water in such systems is made possible by the very low interfacial tension.
They called such water droplets "inverted" micelles which is
the analogue of the hydrophilic "swollen" soap micelle containing oil. Using electrical conductivity, Schulman and
McRoberts (1946) investigated the effect of chain-length and
structure of alcohol and oil on the formation of transparent
water/oil dispersions (solubilized oil). They reported that
ethyl-, propyl-, butyl- and amyl-alcohols yield electrically
conducting systems with benzene over a wide concentration
range of the soap. Such systems become electrically non-
conducting very sharply for hexyl alcohol and for all higher
alcohols, i.e., inversion of the continuous phase takes place
between n-pentanol and n-hexanol. Cyclohexanol and phenol
both give a water-continuous system, but m-cresol produces

an oil-continuous system. They suggested that the alcohol molecules must associate with the soap molecules at the interface and change the wettability of the interface by the oil and water according to the hydrophilic-lipophilic balance of the interfacial film. They used the concept of increasing disorder in the interfacial film due to the addition of alcohol to explain the formation of microemulsions. The crystalline arrangement of the soap molecules in the interfacial film is somewhat broken down by the penetration of alcohol molecules which would create a certain disorder in the lattice. This disorder will permit a liquid or viscous interfacial film to form. The liquid film could thus expand and contract and permit the formation of droplets containing oil or water according to the wettability of the interfacial film.

Schulman, McRoberts, and Riley (1948) used low angle x-ray measurements to confirm that the diameter of droplets calculated using Equation (1) was indeed in agreement with that calculated from x-ray investigation. They assumed an area per pair of soap-alcohol molecules of 60 $\overset{\circ}{A}{}^2$ and the thickness of the interfacial monolayer 24 $\overset{\circ}{A}$. Diameters of the droplets up to 600 $\overset{\circ}{A}$ were measured in such systems. Riley (1949) also reported x-ray diffraction investigations on the structure of transparent oil/water emulsions. It was clearly shown that the Bragg spacing increases with decreasing amounts of soap or with increasing amounts of oil in the oil-continuous system. Schulman and Riley (1948) reported their extensive x-ray investigation on 30 transparent oil/water dispersions, both oil-continuous as well as water-continuous types. They showed that these systems consist of close-packed uniform water spheres in oil or close-packed oil spheres in water. By changing the soap and/or alcohol concentration, the diameter of the spheres can be varied from 100 $\overset{\circ}{A}$ to 550 $\overset{\circ}{A}$. Schulman and Friend (1949) used light scattering measurements to determine the size of droplets in the transparent oil-water dispersions which they had previously studied using the x-ray diffraction technique. The diameter of droplets determined from light scattering measurements agreed reasonably well with the values obtained from x-ray diffraction as well as those calculated using Equation (1). The results showed that as the concentration of surfactant (potassium oleate) increases, the droplet size decreases in both oil- and water-external microemulsions. For a specific concentration of potassium oleate, an increase in the amount of water decreases the size of water droplets in oil. The analysis of data obtained using x-ray diffraction and light scattering also showed that for the diameters of 100, 200 and 400 Å of water droplets obtained using decreasing quantities of potassium oleate, the distance between the surfaces of adjacent droplets was 10, 64 and 185 Å respectively for a phase volume ratio of 5.0. The phase volume ratio g for water-in-oil dispersions is defined as follows:

$$g = \frac{\text{volume of intermicellar liquid}}{\text{volume of intermicellar liquid}} = \frac{\text{volume of oil}}{\text{volume of water}} \quad (2)$$

Therefore, g = 5.0 indicates the oil/water ratio of 5/1.

Schulman, Matalon, and Cohen (1951) studied systems of oil/water/nonionic (ethoxylated) surfactants and reported the formation of various structures (e.g., lamellar, cylindrical, or spherical) depending upon the distribution of the length of hydrocarbon chains and ethylene oxide groups in the surfactant. When all the surfactant molecules had the identical chain-length and the same number of ethylene oxide groups, the oil/water/surfactant system easily formed lamellar structures. A mixture of nonionic surfactants with the same hydrocarbon chain length but unequal ethylene oxide part produced cylindrical structures. A mixture of nonionic surfactants in which both the hydrocarbon as well as the ethylene oxide parts were of unequal length produced isotropic fluids of low viscosity.

Bowcott and Schulman (1955) proposed a mechanism for the control of droplet size and phase continuity in transparent oil-water dispersions stabilized by soap and alcohol. Accordingly, coarse emulsions of nonpolar oils and water stabilized by a soap such as potassium oleate may be titrated to transparent isotropic fluids by the addition of long chain alcohols. These systems can be prepared when the aqueous concentration of the soap varies between approximately 10 and 40%. One of the phases, either oil or water, is dispersed in the form of spherical droplets of about 100 to 600 Å in diameter, depending upon the concentration of soap in the system. Each droplet is surrounded by a mixed monolayer of soap and alcohol which is in a liquid condensed state. The diameter of the droplets is less than a quarter wavelength of visible light and hence the systems are transparent. The mixed interfacial film around microemulsion droplet is considered as a distinct interphase having two interfaces, one between the oil and hydrocarbon chains of the interfacial film and the other between the water and polar groups of the interfacial film. It was proposed that these two interfaces have distinct interfacial tensions, and that the sum of these interfacial tensions determines the curvature and phase-continuity of the microemulsion.

The x-ray studies of soap solutions by various investigators (Kiessig, 1941; Hughes <u>et</u> <u>al</u>., 1945; Stauff, 1939)
have shown that soap micelles in concentrated solutions have
a definite lamellar structure which can be swollen, within
very restricted limits, by oil and can be diluted by water
above the Krafft point. However, when soap micelles are
penetrated by unionized amphipathic molecules such as aliphatic alcohols, they are able to swell almost unlimitedly
both with oil and water. From the results of monolayer
penetration experiments at the air/water interface (Goddard
and Schulman, 1953), it is evident that the alcohol molecules
can penetrate between the soap molecules and disorder the
regular condensed 2-dimensional packing in the micelles to
produce a liquid interphase; an effect which can also be
brought about at appropriate temperatures by using a soap
with a sufficiently large head group such as ethanol amine
oleate (Pink, 1946). This permits surface tension forces to
act at the liquid interface and produce a curvature according
to the difference in tension between the hydrocarbon part of
the interphase and the oil phase, and the polar part of the
interphase and the water phase. Calculations of droplet
diameters, assuming they are spherical, agree well with x-ray
and light scattering data (Schulman and Riley, 1948; Schulman
and Friend, 1949). When the interphase is more readily wetted
by oil than by water, then oil is the continuous phase. Thus,
the longer the alcohol chain, the greater the hydrophobic
nature of the mixed film so that water is dispersed into
droplets and a water-in-oil (w/o) microemulsion is formed.
According to this concept, the interphase presents a smaller
surface area to the water and a larger external area to the
oil. Decreasing the chain length of alcohol makes the interphase less hydrophobic which causes the reverse state and the
system is an oil-in-water (o/w) microemulsion. Schulman and
McRoberts (1946) demonstrated the influence of the length of
the alcohol molecule on the continuous phase of microemulsions
containing equal quantities of benzene and water. A balance
between the hydrophobic and hydrophilic nature of the interphase can be considered in terms of the two interfacial
tensions; the one between the interfacial monolayer and oil,
$\gamma_{m/o}$, and the other between the monolayer and water, $\gamma_{m/w}$.
If the monolayer is liquid, then the phase continuity will
be determined by the curvature of the interphase brought
about by these two forces. When $\gamma_{m/o}$ is less than $\gamma_{m/w}$ the
system is w/o and when $\gamma_{m/o}$ is greater than $\gamma_{m/w}$ it is o/w
since the greatest surface tension will produce the smaller
surface area, i.e., the inner surface of the droplet. Thus,
to change a microemulsion from an oil-continuous to a water-

continuous system it is necessary to increase $\gamma_{m/o}$ relative to $\gamma_{m/w}$ and this effect is achieved by decreasing the chain length of the alcohol to produce a more hydrophilic inter-facial monolayer. Another means of bringing about this change is to use an oil having a larger interfacial tension with respect to water. The interfacial tension between the higher paraffins, e.g., Nujol, and water is about 50 dynes/cm, compared with 35 dynes/cm for benzene and water. Therefore, hexanol can produce a dispersion of water in benzene, whereas under the same conditions, it produces an oil-in-water micro-emulsion for Nujol/water system.

The conditions governing the phase continuity of the transparent systems are the same as those which apply to the coarser emulsions (Schulman and Cockbain, 1940). These con-ditions can be stated by the following general rules.
1)   If an interfacial film is ionized, the system will be o/w, since the charges will by mutual repulsion orientate the film molecules with their ionized groups outward.
2)   If the stabilizing agent is unionized and preferably soluble in oil, the system will be w/o.
3)   If the stabilizing agent is unionized and preferably soluble in water, the system will be o/w.
4)   Addition of salt to remove the diffuse layer and surface charge in rule (1) to produce ion pairs will make the agent behave as in (2) or (3).

The same factors also apply to emulsions stabilized by fine solid powders (Schulman and Leja, 1954). Solid surfaces having contact angles of water smaller than 90° yield o/w emulsions, whereas those having contact angles above 90° form w/o emulsions. Transparent alcohol-soap systems are able to form oil-continuous emulsions only because the aqueous con-centration of the soap is high enough to ensure that the soap is in the form of an undissociated ion-pair. Addition of such a system to a large volume of water produces a coarse o/w emul-sion, the ionization of the soap as well as a change in the phase volume ratio causing the transition from micro to macro-emulsion. The size of the droplets in microemulsions is deter-mined by both the volume of the dispersed phase and the inter-facial area. The latter is of course determined by the amount of soap and alcohol present in the system.

IV.   THE STRUCTURE OF THE INTERPHASE IN MICROEMULSION SYSTEMS
Bowcott and Schulman (1955) also noted that although the most convenient method of preparation of a transparent dis-persion was titration of the coarse emulsion of oil, water and soap with alcohol, it was not essential to add the com-ponents in any particular order, nor is vigorous shaking

essential as when preparing coarse or macroemulsions. This
fact indicates that the phases are in equilibrium with each
other (see also McBain and McBain, 1936). Therefore, one
can imagine that the alcohol is distributed between three
phases, namely, oil, interphase, and water, whereas soap
(e.g., potassium oleate) is entirely in the interphase.
It is possible to make this last assumption since the concentrations of soap used are so high that the concentration
of ionized soap molecules in water is negligible. This assumption is limited and is not valid where there is insufficient water present to provide a large enough interfacial
area for entire soap to be at the interface; from previous
work it appears that 100 Å diameter water droplets is about
the smallest size occurring in these systems.

Since the area of an oleate molecule at an oil/water
interface is about 30 $Å^2$ while that for a straight chain
alcohol is about 20 $Å^2$, a 1:1 soap/alcohol ratio in the
interface should account for an area/oleate molecule of
approximately 50 $Å^2$. However, determinations of droplet
diameter have agreed most favorably with calculated values
if an area of 70 $Å^2$ is assumed for complexes with both
straight chain and cyclic alcohols, suggesting that the
alcohol to soap ratio in the interface may be nearer 2:1.
An analysis of systems stabilized with straight chain
alcohols shows that the alcohol:soap molecular ratio at the
interface may vary from 1:1 to about 3:1, depending on both
the amount of water and soap dispersed and the chain length
of the alcohol.

It was concluded that there is also an upper limit to
the amount of water which the systems are able to dissolve.
To dilute the system it is necessary to maintain a constant
alcohol to oil ratio in the continuous phase, i.e.,

$$\frac{n_a^o}{n_o} = k \qquad (3)$$

where n_a^o is the number of moles of alcohol in oil phase o and
n_o is the number of moles of oil. The total alcohol n_a is

$$n_a = n_a^o + n_a^w + n_a^i \qquad (4)$$

where n_a^w and n_a^i are the number of moles of alcohol in water
and in the interface respectively. On substituting Equation
(3) in (4) we get,

$$n_a = k \, n_o + (n_a^w + n_a^i) \qquad (5)$$

Equation (5) suggests that the total moles of alcohol is linearly proportional to the total moles of oil. The slope of the plot gives the value of k and the intercept the value of $n_a^w + n_a^i$. The constitution of the dispersed and continuous phases may be determined simply by plotting the total alcohol needed to form a clear system against the amount of oil, keeping the amounts of water, n_w, and soap, n_s, constant. The slope (k) and the intercept at $n_o = 0$ indicates the amount of alcohol in the water swollen micelle, $(n_a^w + n_a^i)$, both at the interface and dissolved in the internal aqueous phase. It is not possible to separate n_a^w from n_a^i so that two values for n_a^i have been determined: 1) assuming $n_a^w = 0$, i.e., the alcohol is insoluble in water, and 2) assuming that alcohol dissolves up to its saturation value in water. In the cases examined, these assumptions gave values for the molar ratio of alcohol to soap at the interface differing by no more than 1.3% at the most. For the system benzene-water-potassium oleate-hexanol, Bowcott and Schulman (1955) showed that the alcohol to soap ratio at the interface varied from 2:1 to about 3.2:1. It was difficult to produce the transparent dispersion above a molar ratio of water to soap of about 75. As compared to n-hexanol, n-heptanol and n-octanol were able to solubilize a maximum of 55 and 45 moles of water per mole of soap respectively. Between 12° and 35°C there was no temperature effect on the slopes (k) for these systems. For the hexanol system, 1 ml of oleic acid can dissolve a maximum of 4 ml of water (molar ratio of water to soap equal to 70) and the maximum droplet size is about 220 Å.

At a water to soap molar ratio of 35, the diameter of water droplets increases from 170 Å to 220 Å as the chain length of alcohol increases from n-hexanol to n-decanol. Since the amount of water is constant, this implies that a longer chain length of alcohol produces fewer droplets of larger size. By using n-pentanol it is possible to produce transparent dispersions for water to soap molar ratios up to 80, and with p-methylcyclohexanol, where the ring structure prevents tight adlineation of the soap and alcohol, this value can be taken up to 160. This implies that potassium oleate and p-methylcyclohexanol system can dissolve 9.15 ml of water per ml of oleic acid.

V. EXPERIMENTAL STUDIES ON MICROEMULSION SYSTEMS

Bowcott and Schulman (1955) determined the sedimentation velocity of droplets in the transparent dispersions using ultracentrifugation. The system they studied had the following ratios of components $n_w/n_s = 70$, $n_b/n_s = 140.5$ and $n_a/n_s = 26.7$ where n_w, n_s, n_b and n_a are respectively the number of moles of water, soap, benzene and alcohol (n-hexanol). Using a centrifugal field of about 130,000g at 21°C ± 1° the sedimentation constant, S, was found to be 22.4 ± 1 (x 10^{-13}).

They also studied the transparent systems of benzene/water/potassium oleate with n-butanol and n-pentanol. Again a direct proportionality was shown to exist between the total alcohol and benzene in the systems. The solubility of these alcohols in water may make it difficult to interpret the results. Both of these systems (containing butanol or pentanol) have high electrical conductivities. The fact that they can be diluted with the nonpolar oil and yet have a high conductivity suggests that the continuous phase, although it is predominantly benzene, may in fact be a four component molecular solution. It is important to realize that in this work, Bowcott and Schulman (1955) as well as Schulman and Riley (1948) alluded to the fact that there are so-called anomalous systems which exhibit very high electrical conductivity although the amount of water may be relatively small in the systems. It is this type of systems which Shah (1974,1976) has recently termed as cosolubilized systems based on high resolution NMR, electrical conductivity and other physical measurements. Bowcott and Schulman (1955) also investigated a system containing 70% by volume of the dispersed phase and demonstrated that it did not show any sedimentation peak but exhibited an almost stationary structure expected for a close-packed system.

Schulman, Stoeckenius and Prince (1959) used the term "microemulsion" to describe such transparent oil/water dispersions. Using electronmicroscopy, they established that such dispersions consist of uniform spherical droplets of either oil or water dispersed in the appropriate continuous phase and are therefore in fact microemulsions. They further pointed out that the necessary degree of disorder in the films can be achieved in several ways: (a) penetration of a mixed interfacial film, consisting of a complex of a soap or detergent and an amphiphile, by a nonpolar hydrocarbon from the oil phase. It has been demonstrated that such penetration can occur when the association between at least one component of the complex and the hydrocarbon is strong; (b) use of large positive counterions to make the resultant soap molecules asymmetric and thus create disorder among the associating species in the film; and (c) penetration of a monolayer composed of asymmetric molecules by a molecular species from the oil phase. From experimental studies it was shown that association between the molecules of the interfacial film and the hydrocarbon in the dispersed or continuous phase is very important in the formation of microemulsions.

From measurements of surface pressure, Zisman (1941) showed
that for alcohol monolayers at the oil/water interface there
is a 1:1 molar association between the alcohol and oil mole-
cules since the area per alcohol molecule was about 44-50 $\overset{\circ}{A}^2$.
On the other hand, films of fatty acids at the benzene/water
interface gave the correct surface area/fatty acid molecule
(about 20-25 $\overset{\circ}{A}^2$) showing that the benzene could be ejected
readily from the alkyl hydrocarbon chains in the monolayer
at the benzene/water interface. Therefore, using the same
alcohol and surfactant components, they showed that some oils
produce microemulsions, whereas others do not, thus establishing the importance of the interaction between the oil molecules and interfacial film. Where disorder is required, 2
amino-2-methyl-1-propanol (AMP) works much better than potassium ions. Increasing the temperature also helps to produce
disorder in such systems. Schulman et al. (1959) further
noted that during the phase inversion, upon increasing the
amount of water in the system, such dispersions first became
strongly viscoelastic and anisotropic, but then reverted back
to an isotropic fluid system to form an oil-in-water microemulsion. The two types of microemulsions formed, water-in-
oil and oil-in-water, were isotropic and of low viscosity.
They claimed that when the size of microemulsion droplets is
in the range of 1200 $\overset{\circ}{A}$ diameter, a considerable scatter in
size is observed in the electronmicrographs. In the range
of 450 $\overset{\circ}{A}$ diameter, the droplets are uniform such that
hexagonal packing is nearly visible in the electronmicrographs. Droplets of 250 $\overset{\circ}{A}$ diameter can be seen clearly by
the osmic acid staining technique at magnifications of 40,000
or 80,000. These authors succeeded in developing a technique
to obtain the electronmicrographs of these systems without
microtoming or shadow casting.

Stoeckenius, Schulman and Prince (1960) reported an im-
pressive study on the structure of liquid crystalline phases
and microemulsions using electronmicroscopy. The combination
of electronmicroscopy and x-ray investigation of the myelinic
structures of soap, above the Krafft point, and phospholipid
has given an indication that the structures are liquid crystals in which the hydrocarbon chains are in the fluid state
and the polar groups are arranged in a crystal order. This
has been inferred from the diffused nature of the side spacing
and high order spacing of the main layer structure unit in the
x-ray pictures. If further disorder is placed in these molecular liquid crystal arrays, the systems will become completely
liquid and surface tension forces will take over at the interfaces between the monolayer and the liquid phases surrounding
them. This results in the monolayer breaking up into spherical

droplets. One phase or another is occluded inside the molecular array according to the tensions on each side of the monolayer.

Droplets of diameter less than 300 Å did not give enough contrast in the electronmicroscope to show them clearly. Hence, they used the negative staining method for electronmicroscopy of microemulsions consisting of oil droplets smaller than 300 Å diameter. In this technique the material is dried on the grid from a suspension in 1% phosphotungstate. This forms a very dense film in which the much less dense oil droplets of diameter less than 300 Å are embedded. Where the thickness of the phosphotungstate film is of the same order of magnitude as the droplet diameter, the droplets can therefore be clearly seen standing out brightly against a dark background. Using this technique they obtained electronmicrographs of droplets as small as 75 Å, 150 Å, and 200 Å in diameter. In the ultracentrifuge, the two sizes appeared as two sedimentation zones. The 75 Å droplets approximate to the dimensions of a swollen micelle and agreed with the low angle x-ray pictures taken of this system. The 150 Å diameter droplets were referred to as a microemulsion.

In summary, these authors seem to differentiate between swollen micelles and microemulsions. The investigation carried out by Schulman and his collaborators in the period 1940-1960 developed the picture of a microemulsion system as a highly concentrated, transparent oil/water dispersion which is isotropic, clear, and has low viscosity. In microemulsions the gap between surfaces of adjacent droplets is much less than the diameter of the droplets. Hence, microemulsions appear to be systems of close-packed spheres in a continuous medium of oil or water. They also suggested that in water-in-oil type microemulsions, there is no surface charge associated with the interfacial film and hence the electrical double layer effect is negligible. However, in oil-in-water type microemulsions, they proposed that the droplets have a surface charge and hence the electrical double layer could contribute to the formation and stability of the microemulsion. In the 1960 publication, they proposed that the phase continuity may be controlled by the surface charge at the oil/water interface. This is important from an electrostatic point of view, as the presence of charge on the polar group will cause mutual repulsion and hence polar groups will tend to go away from one another and hence the interface will develop a curvature such that the area in the polar group region will be greater than that at the terminal end of their hydrocarbon chains. Hence, the dispersions will form oil-in-water type microemulsions. In the absence of any surface charge, however,

the polar groups can be packed together and hence the hydrocarbon chains can spread apart due to the penetration of oil
molecules into the interfacial film, leading to the formation
of water-in-oil microemulsions.

VI.   MOLECULAR AGGREGATES IN SURFACTANT SOLUTIONS

When a surfactant is dissolved in water, it tends to adsorb at the air/water interface. The adsorption of surfactant
at the interface results in a greater concentration at the
interface as compared to that in the bulk solution. Above a
critical concentration, which depends upon the structure of
surfactant molecules as well as physicochemical conditions,
the surfactant molecules form aggregates called micelles
(Figure 1A). This characteristic concentration is called the

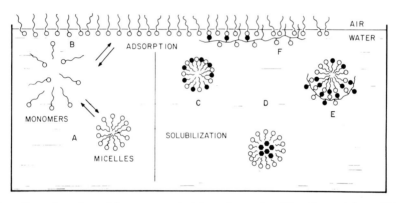

ADSORPTION, MICELLE FORMATION, SOLUBILIZATION
AND INTERACTIONS AT THE MICELLE SURFACE

Fig. 1.   A schematic presentation for micelle formation (A),
          adsorption (B), mixed micelle formation (C), solu-
          bilization of oil in micelles (D), polymer-micelle
          interaction (E), and surfactant-polymer mixed film
          at interfaces (F) in surfactant solutions

critical micelle concentration (CMC). Specifically the CMC
represents a narrow range of surfactant concentration (Preston,
1948). In general, micelles are spherical aggregates of surfactant molecules containing 20 to 100 molecules per micelle.
The formation of micelles in aqueous solution creates local
nonpolar environments within the aqueous phase. Any oil
soluble material such as dyes, pigments or nonpolar oils can
dissolve within the micelles (Figure 1D). Using ionic and
nonionic surfactants, one can produce mixed micelles which
are often larger in size and in the number of molecules per

micelle (Figure 1C).   If a surfactant solution contains a
surface active polymer, then a mixed adsorbed film of polymer
and surfactant can occur at the interface (Figure 1F).   The
polymer-surfactant interaction can also occur at the micellar
surface (Figure 1E).   The solubilization of oil within micelles can also occur when such micellar solutions are in
contact with crude oil (e.g., injection of micellar solutions
in the oil fields).

     Surfactant molecules can be considered as building blocks.
One can make various association structures by increasing the
concentration of surfactant in water and adjusting physico-
chemical conditions.   Figure 2 schematically shows various

STRUCTURE FORMATION IN SURFACTANT SOLUTION

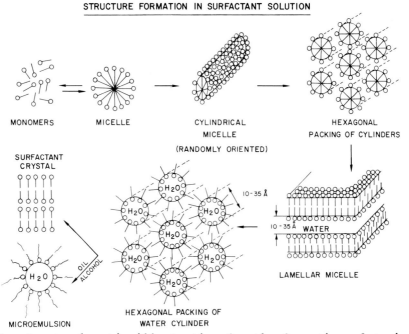

Fig. 2.   A schematic illustration for the formation of various
          structures in surfactant solution upon increasing the
          concentration of surfactant

structures that are likely to form in the surfactant solutions
upon increasing the concentration of surfactant.   The spherical micelles become cylindrical ones.   Upon further increasing the surfactant concentration, there is a hexagonal packing
of surfactant cylinders.   If the concentration is further increased, the lamellar structure is formed.   Upon further
addition of surfactant, the lamellar structure is converted
to a hexagonal packing of water cylinders.   Upon addition of

oil and a short-chain alcohol, one can convert such water
cylinders into water-in-oil microemulsions. It is possible
to induce a transition from one structure to another by
changing the physicochemical conditions such as temperature,
pH and addition of mono- or divalent cations to the surfactant
solution. It should be emphasized that the scheme shown in
Figure 2 is a general scheme and a surfactant may skip several
phases depending upon its structure and the physicochemical
conditions.

VII. MICROEMULSIONS VS. MICELLAR SOLUTIONS

There is some disagreement, confusion and controversy
for describing such isotropic, clear, low viscosity oil-
water-emulsifiers systems as "microemulsions" (Adamson, 1969;
Tosch et al., 1969; Prince, 1975). We believe that this is
due to a lack of complete characterization of the molecular
association and aggregation in such systems. For the sake
of convenience, we have classified various dispersed systems
as shown in Figure 3. Solutions of inorganic salts such as

DIMENSIONS OF DISPERSED PHASES IN VARIOUS SYSTEMS

*Fig. 3.   A schematic presentation of dimensions of dispersed
phases in various systems*

sodium chloride or compounds such as sugar form true molecular
solutions in which the dispersed phase (solute) has dimensions
up to 10 Å. The dimensions of the dispersed phase in micellar solutions may range from 20 to 100 Å depending upon the
length of the surfactant molecule and the aggregation number
in micelles. The dispersed phase in microemulsions may have
a diameter in the range of 100-1000 Å. In colloidal dispersions of polymers, polysaccharides, or polymer latex, the
dimensions of dispersed phase can be in the range of 0.1 to
1 micron. In macroemulsions, the dimension of the dispersed
phase may be in the range of 1 to 10 microns. The classifica-
tion of dispersed systems shown in Figure 3 is arbitrary and
is given only to snow the relative order of the size of dispersed phase in these systems. It is obvious that there will
be considerable overlap of dimension for two adjacent systems.

It has been proposed that microemulsions are colloidal
solutions (Shinoda and Friberg, 1975). Unfortunately, the
word "colloidal" has been used in literature to denote systems having the dispersed phase of 100 Å to more than 1 micron

in dimension. It has been proposed also that microemulsions should be referred to as "swollen micelles" or "micellar emulsions" (Adamson, 1969; Ahmad et al., 1974). Here, we believe that it is true that microemulsions are equivalent to swollen micelles but it is also true that all micelles cannot be swollen to the limit of microemulsions. In other words, we believe that one can produce microemulsions only under very specific structural requirements for the inter- facial film. Schulman et al. (1959) have shown that there are strict requirements for the formation of microemulsions such as the penetration of hydrocarbon into the interfacial film of surfactant and alcohol. They varied the chain length of surfactant, alcohol, and oil and showed that microemulsions form only when there is penetration of oil into the inter- facial film. In order to emphasize the intricate balance of interfacial forces, it is desirable to label such systems as microemulsions. The following are our thoughts on specific differences and similarities between microemulsions and micellar solutions.

Let us first consider the aqueous micellar solutions and oil-in-water microemulsions. In both cases the continuous phase is water. In aqueous surfactant solutions, the solu- bilization of nonpolar oils, dyes, etc. occurs upon micelle formation. McBain and Hutchinson (1955) have reported exten- sively on solubilization and related phenomena in micellar solutions. It is evident from the solubilization properties of micellar solutions of various surfactants reported by these authors that the molar ratio of solubilized oil/ surfactant rarely exceeds the value 1.5. Only in some unusual cases this ratio was found to be close to 2.0. Therefore, it appears that in most of the conventional micellar solutions, a maximum of two oil molecules/ surfactant molecule can be solubilized. In contrast, many oil-in-water microemulsion systems reported by Schulman and co-workers can solubilize anywhere from 10 to 25 oil molecules/surfactant molecule.

If we consider the solubilization of water in oil con- taining surfactant, we find that in the case of reversed micelles the number of water molecules solubilized per sur- factant molecule is mostly below 10 and in some cases up to 30 (Palit and Venkateswarlu, 1951), whereas the water-in-oil microemulsions studied by Schulman and others can solubilize as much as 75 to 150 water molecules per surfactant molecule. It is interesting that Frank and Zografi (1969) reported that in water/Aerosol OT/n-octane system at 25°C there is a tran- sition from colorless to turbid to blue translucent and back again to turbid as the amount of water is increased. The

system remains colorless when moles of water/mole of Aerosol
OT is increased from 1 to 20. However, the second blue translucent region occurs when the number of water molecules per
Aerosol OT molecule is in the range of 140 to 160. We believe
that the first colorless region represents the reverse micelles in n-octane whereas the second blue translucent region
represents microemulsion region.

   Schulman and Riley (1948) have shown that microemulsions
resemble close-packed spheres and the gap between the surfaces
of adjacent spheres is much smaller than the diameter of the
spheres. We believe that the reverse micellar solutions are
considerably dilute systems and interparticle distance or the
gap between adjacent micelles presumably is much greater than
that in microemulsions. The close packing of spherical droplets in microemulsions appears to be required for the stability of the microemulsion system. And it is for this reason
that one cannot dilute a microemulsion indefinitely. We have
observed that many water-in-oil microemulsions when diluted
either by oil or by a mixture of oil and alcohol eventually
separates into two phases. This happens because of the increase in the interparticle distance and a transition occurs
from a microemulsion to a reverse micellar solution state.
In contrast, the reverse micellar solution can be diluted
considerably by oil without causing a phase separation.

   From the observation that most of the microemulsions are
produced using a combination of surfactant and a short chain
alcohol, we believe that the entropic contribution may play
a significant role in the stabilization of microemulsion
systems. As pointed out by Schulman and co-workers, the interfacial disorder is an essential condition for the formation of microemulsions, and, therefore, surfactants and
alcohols of greatly dissimilar chain lengths are commonly
used in producing microemulsions. For the formation of
reverse micellar solutions neither the chain length requirement nor the presence of alcohol is necessary. However, we
would like to mention that in some exceptional cases such as
Aerosol OT, one can produce water-in-oil microemulsion using
surfactant alone because of its unusual molecular shape. Figure 4 schematically shows the water-in-oil and oil-in-water
microemulsions stabilized by surfactant and alcohol molecules.

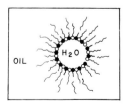

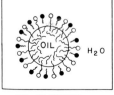

*Fig. 4. A schematic il-
lustration of the struc-
ture of surfactant-alcohol
film in water-in-oil and
oil-in-water microemulsions*

WATER-IN-OIL   OIL-IN-WATER

VIII.   THEORIES OF MICROEMULSION FORMATION

During the past three decades various investigators have
proposed several molecular mechanisms for the formation of
microemulsions. As mentioned in the previous sections,
Schulman and his collaborators visualized the mechanism of
formation of microemulsions in terms of spreading pressure
of the mixed interfacial film at the oil-water interface and
subsequent penetration of oil molecules into the interfacial
film. To account for the spontaneous uptake of water in oil-
continuous or oil in water-continuous microemulsions, they
postulated that the transient interfacial tension has to be
negative for the spontaneous uptake of water or oil in microemulsions. Accordingly, one phase breaks up into the maximum
number of droplets whose diameter depends upon the interfacial
area produced by the surface-active molecules. The transient
negative interfacial tension (i.e., the spontaneous tendency
of the interface to expand) produced by the mixing of the
components will, at equilibrium, become zero and the dispersion, and not separation, will be the equilibrium condition.
Schulman's explanation of transient negative interfacial
tension has been misunderstood and misquoted by several investigators who conceived that even at equilibrium microemulsions have negative interfacial tension. In fact,
Schulman and his collaborators have specifically mentioned
that the negative interfacial tension is a transient phenomenon and that at equilibrium the oil-water interface in a
microemulsion has either zero or a very small positive interfacial tension. Possible ways of producing negative
interfacial tension and hence microemulsions have been indicated in the previous publication by Schulman et al. (1959).

Davies and Haydon (1957) have extended the concept of
transient negative interfacial tension from their experiments
on microemulsion formation. They described an experiment by
Ilkovic (1932) in which a negative potential was applied to
a mercury drop in an aqueous solution of a quaternary ammonium
compound. The interfacial tension can be reduced to zero such
that the mercury drop begins to disintegrate into a brown
cloud of colloidal particles in the solution. At -8 V/cm
applied potential, the spontaneous emulsification of mercury
was very striking. The spontaneous emulsification was also
observed when a detergent solution was brought in contact
with an oil containing cetyl alcohol or cholesterol. For an
oil containing a specific amount of alcohol, the interfacial
tension decreases as the concentration of the surfactant increases in the aqueous solution. Upon extrapolation, this
plot indicates a zero interfacial tension at a specific
surfactant concentration. For all surfactant concentrations

above this value, the extrapolated interfacial tension is negative. The spontaneous emulsification was observed for surfactant concentrations which exhibited negative values for interfacial tension upon extrapolation. These results were used as an indication that for spontaneous emulsification the dynamic interfacial tension may reach transient negative values. However, this does not indicate that at equilibrium the dispersed droplets will have a negative interfacial tension. At equilibrium it is expected that the droplets may exhibit a small positive interfacial tension value in the range of 10^{-2} to 10^{-4} dynes/cm.

Schulman et al. (1959) further proposed that the phase continuity can be controlled readily by controlling the surface charge. If the concentration of the counterions for the ionic surface-active agent is high and the diffuse electrical double layer at the interface is compressed, water-in-oil microemulsions are formed. If the concentration of the counterions is reduced sufficiently to produce a surface charge at the oil/water interface, the emulsion presumably inverts to an oil-in-water type microemulsion. They also proposed that if the droplets are spherical, the resulting microemulsion is isotropic, exhibits Newtonian flow behavior with one diffused band in x-ray diffraction pattern. If the dispersions consist of cylindrical structures, they are optically anisotropic and non-Newtonian and show two diffused x-ray bands giving dimensions related by a factor of $\sqrt{3}$.

Prince (1967,1969) has further extended the concept of molecular interaction at the oil/water interface for the formation of microemulsions. In a subsequent paper, Prince (1975) discussed the differences between microemulsions and micellar systems.

Adamson (1969) proposed a model for water-in-oil microemulsion in terms of a balance between Donnan osmotic pressure due to the higher total ionic concentration in the water drops in oil and the Laplace pressure associated with the interfacial tension at the oil/water interface. The microemulsion phase can exist in equilibrium with an essentially noncolloidal aqueous second phase provided there is an added electrolyte distributed between the droplet's aqueous interior and the external aqueous medium. Both aqueous media contain some alcohol and the total ionic concentration inside the aqueous droplet exceeds that in the external aqueous phase. Miller and Scriven (1970a,1970b) proposed the role of the electrical double layer in causing interfacial instability. They divided the total interfacial tension into two components as follows:

$$\gamma_T = \gamma_p - \gamma_{dl} \qquad (6)$$

where γ_T is the total interfacial tension which is the excess tangential stress over the entire region between the homogeneous bulk fluids including the diffuse double layer, γ_p is the phase interface tension which is that part of the excess tangential stress which does not arise in the region of the diffuse double layer and $-\gamma_{dl}$ is the tension of the diffuse layer region. This equation suggests that when γ_{dl} exceeds γ_p, the total interfacial tension γ_T becomes negative. For a plane interface the destabilizing effect of a diffuse layer is primarily that of a negative contribution to interfacial tension. Levine and Robinson (1972) modified the model proposed by Adamson for water-in-oil microemulsions to allow for the diffuse electrical double layer in the interior of aqueous droplet. These authors derived a relation governing the equilibrium of the droplet for a 1-1 electrolyte, which was based on a balance between the surface tension of the film at the boundary in its uncharged state, the osmotic pressure due to the ion concentration, and the Maxwell electrostatic stress associated with the electric field in the internal diffuse layer.

Recently Ruckenstein and Chi (1975) have discussed the stability of microemulsions and the size of droplets in stable microemulsions using a thermodynamic approach. These investigators' approach accounts for the free energy of the electrical double layer, the Van der Waals and electrical double layer interaction potentials among the droplets as well as for the entropy of formation of the microemulsion. This treatment can predict the occurrence of phase inversion. Accordingly, the free energy change, ΔG_m, consists of changes in the interfacial free energy, ΔG_1, the interaction energy among the droplets, ΔG_2, and the effect caused by the entropy of dispersion, ΔG_3 ($= T\Delta S_m$). The competition among the various terms contributing to the free energy change determines whether or not microemulsions can form. When the free energy change ΔG_m is negative, spontaneous formation of microemulsion occurs. The variation of ΔG_m with the radius of the droplet, R, at constant values of water/oil ratios can be determined using $\Delta G_m(R) = \Delta G_1 + \Delta G_2 - T\Delta S_m$. R^* is that value of R which leads to a minimum in ΔG_m, and is the most stable droplet size for a given volume fraction of the dispersed phase. R^* may be obtained from

$$\left(\frac{d\Delta G_m}{dR} \right)_{R = R^*} = 0 \qquad (7)$$

if the condition, $\left(\dfrac{d^2\Delta G_m}{dR^2}\right)_{R=R^*} > 0$ is satisfied.

In order to determine which type of microemulsion is more stable and to predict the possible occurrence of phase inversion, the values of ΔG_m^* for both types of microemulsion have to be compared at the same composition. The one having the more negative value of ΔG_m^* at that particular composition of the mixture will be favored. Phase inversion occurs at that volume fraction for which the values of ΔG_m^* for both kinds of microemulsion are the same.

Gerbacia and Rosano (1973) measured the interfacial tension at the oil/water interface while alcohol was injected into one of the phases. They found that the interfacial tension could be temporarily lowered to zero while the alcohol diffuses through the interface. They proposed that the diffusion of surface-active molecules across the interface is an important requirement for the formation of microemulsions. They further claimed that, in contrast to the commonly accepted notion, the formation of microemulsions is not independent of the order in which the components are added. They reported two systems in which the order in which the components were mixed determined the formation of microemulsions. In a recent review article, Shinoda and Friberg (1975) have summarized their extensive studies on the formation of microemulsions using nonionic surfactants. They proposed the following conditions to produce microemulsions with minimum amount of surfactants:
1)   Microemulsions should be produced near or at the phase inversion temperature (PIT) or HLB temperature for a given nonionic surfactant, since the solubilization of water (or oil) in a nonaqueous (or aqueous) solution of nonionic surfactant shows a maximum at this temperature.
2)   The mixing ratio of surfactants should be such that it produces an optimum HLB value for the mixture.
3)   The closer the phase inversion temperatures of two surfactants, the larger the solubilization, hence the minimum amount of the nonionic surfactants is required.
4)   The larger the size of the nonionic surfactant, the larger the solubilization of oil and water.

Robbins (1976) has recently proposed a theory for the phase behavior of microemulsions. According to this theory, the hydrophilic heads and lipophilic chains of the interphase are treated as independent interfaces, water interacting with the heads and oil with the chains. Direction and degree of curvature are imposed by a lateral stress gradient in the interface, resulting from differences in interaction on

either side of the interphase. This stress gradient is expressed in terms of physically measurable quantities, namely,
surfactant molecular volume, interfacial tension and compressibility. For ethoxylated surfactants, increasing temperature,
salt concentration and oil aromaticity result in increased oil
uptake and decreased water uptake. Decreasing head/chain
volume and compressibility ratio have the same effect.

IX. MICROEMULSIONS VS. COSOLUBILIZATION

Using a combination of physical techniques such as electrical resistance, high resolution NMR (220 Mc), spin-spin
relaxation time (T_2), and viscosity measurements, Shah et al.
(1976) have shown that two isotropic clear systems with identical compositions, except that one contains n-pentanol and
the other n-hexanol, are structurally quite dissimilar systems. Figure 5 shows the optical appearance and electrical

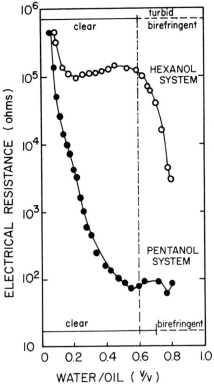

Fig. 5. *Optical appearance and
electrical resistance of hexadecane-water-potassium oleate-
alcohol (hexanol or pentanol)
system as a function of water-
oil ratio*

resistance of dispersions containing n-hexanol or n-pentanol
as the cosurfactant. It is evident that both systems remain
isotropic, clear and stable up to a water to oil ratio of 0.6.

Upon further addition of water, both systems become birefringent and translucent. However, the pentanol containing
system was more clear in the birefringent region as compared
to the hexanol system. The electrical resistance of these
two systems showed a striking difference in the region 0.1 to
0.7 water to oil ratio. As the water to oil ratio is increased from 0.2 to 0.6, the hexanol containing dispersions
maintained their electrical resistance at 10^5 ohms. However,
the pentanol containing dispersions exhibited a continuous
decrease in the electrical resistance from 10^5 to about 50
ohms upon addition of water. These results very clearly indicate that although both these systems are isotropic, clear
and stable, their electrical properties are strikingly different, and that the difference of one carbon atom in the
chain-length of the cosurfactant molecule can strikingly influence the electrical resistance of these systems.

Figure 6 shows the chemical shift of water and hydroxylic
protons resonance peak in the high resolution NMR (220 Mc)

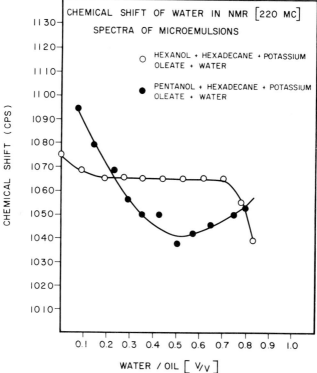

Fig. 6.   High resolution NMR (220 Mc) chemical shift of water
          and hydroxylic protons in isotropic, clear, stable
          dispersions of hexadecane-water-potassium oleate-
          alcohol (hexanol or pentanol)

spectra. As water to oil ratio is increased from 0.1 to 0.7, the chemical shift of these protons remained constant for the hexanol containing systems whereas there was a continuous up-field shift of the resonance peak in the pentanol system. This suggests that the environment of water and hydroxylic protons changed continuously as more water was added to the pentanol containing systems whereas in the hexanol containing systems the environment of these protons remained the same (as shown by the relatively constant chemical shift). It had also been observed that chemical shifts of methylene and methyl protons showed no significant change upon addition of water for both the pentanol and hexanol containing systems.

Figure 7 shows the spin-spin relaxation time (T_2) of these two systems. It is evident that T_2 remained the same

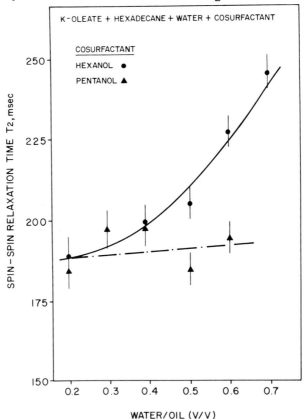

Fig. 7.   *The spin-spin relaxation time (T_2) of isotropic, clear, stable dispersions of hexadecane-water-potassium oleate-alcohol (hexanol or pentanol) as a function of water to oil ratio*

in the pentanol system whereas it increased for the hexanol
system upon addition of water. Using high resolution (100 Mc)
and pulsed NMR spectroscopy, Hansen (1974) investigated the
potassium oleate-n-hexanol-water-benzene system. He observed
that the polar end of the oleate molecules were relatively
immobilized at the aqueous interface, while the terminal
methyl end of the molecule was free to reorient in the benzene
phase. In contrast, hexanol showed no motional restriction
and presumably exchanged rapidly between the interfacial film
and the benzene phase. Figure 8 shows the relative viscosity

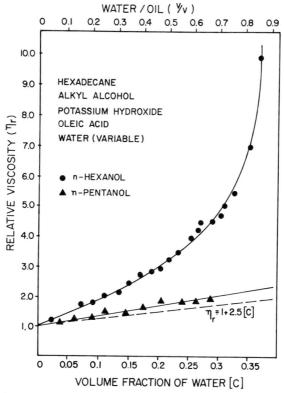

Fig. 8.   *The influence of volume fraction of dispersed water
on the relative viscosity of isotropic, clear,
stable dispersions of hexadecane-water-potassium
oleate-alcohol (hexanol or pentanol)*

plotted against the volume fraction of dispersed water in
hexanol and pentanol systems. The relative viscosity increases much more rapidly in the hexanol containing systems
as compared to pentanol system. Figure 8 illustrates very
clearly the striking effect of the chain length of co-

surfactant molecules on rheological properties of such isotropic, clear and stable dispersions.

Figure 9 shows our proposed structures for the isotropic,
clear, stable dispersions prepared by using n-pentanol or n-
hexanol as cosurfactant. The proposed structure for the pentanol containing system is a <u>cosolubilized system</u> in which
one can visualize the surfactant and the cosurfactant forming
a liquid which can dissolve both oil or water as a molecular
solution, whereas hexanol containing system is a <u>true water-
in-oil microemulsion</u> in which water is present as spherical
droplets. In the cosolubilized system, as one increases the

- ●   WATER
- ⊂▭   HEXANOL OR PENTANOL
- ⊂▭▭   POTASSIUM OLEATE
- ▨▨▨   HEXADECANE

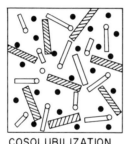
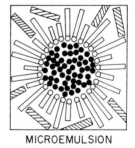

COSOLUBILIZATION   MICROEMULSION

*Fig. 9.   A schematic presentation of cosolubilized systems and
water-in-oil microemulsions. Pentanol produces a cosolubilized system whereas hexanol produces a water-
in-oil microemulsion*

amount of water, the average distance between the water molecules as well as between alcohol molecules would change and
this consequently would influence the hydrogen bonding ability of water and alcohol molecules, which in turn would influence the chemical shift of the resonance peak. Also in
the cosolubilized system as one adds more water, it becomes
more and more electrically conducting and hence exhibits a
continuous decrease in the electrical resistance. However,
in the hexanol containing system, since it contains water
spheres in a continuous oil medium, the addition of water
creates more spheres. The continuous medium is still an oil
phase and hence the electrical resistance is maintained at a
high value (10^5 ohms).

We would like to emphasize that the structures shown in Figure 9 are schematic and should not be taken rigidly. We have not ruled out the possibility that at a higher water content there may be aggregates of water molecules or surfactant and alcohol molecules in the cosolubilized system. However, the structure of the cosolubilized system shown in Figure 9 is consistent with the change in chemical shift as well as the decrease in electrical resistance upon addition of water to such systems.

In summary, from the results presented so far and from the discussion in the previous section we propose that the transparent, isotropic, clear, stable systems prepared from oil/water/emulsifier can be classified into one of three main categories, namely, normal or reverse micelles, water-in-oil or oil-in-water microemulsions, or cosolubilized systems. One can distinguish between these three classes by using a combination of physical techniques to study the properties of such systems.

X. PHASE-INVERSION OF MICROEMULSIONS AND THE FORMATION OF LIQUID-CRYSTALLINE PHASES

It is expected that as the water/oil ratio increases, the water-in-oil type microemulsions may invert to oil-in-water type microemulsions. For the hexadecane-potassium oleate-hexanol-water system, Shah and Hamlin (1971) and Shah et al. (1972) proposed a mechanism of phase-inversion based on electrical, birefringence and high resolution NMR (220 Mc) studies upon increasing the water/oil ratio in the system. Microemulsions were produced by mixing the components in the following proportions: for 10 ml of hexadecane, 4 ml of hexanol and 2 gm of potassium oleate were added. Water was added in small amounts to this mixture, which was then shaken vigorously. Increasing amounts of water were added to form the microemulsion at different water/oil ratios. Birefringence was detected with the aid of two polarizing plastic sheets which were arranged perpendicular to one another. The electrical resistance of the microemulsions was measured by dipping the electrodes, connected to an AC conductivity bridge, into the microemulsions. Nuclear magnetic resonance spectra were obtained after the gradual addition of water to the microemulsion in the sample tube of the NMR spectrometer (Varian HA-220 Mc).

Figure 10 shows the optical and birefringence characteristics, as well as the measured electrical resistance, as a function of water/oil ratio. As the amount of water increases, the microemulsion passes through a clear to turbid to clear region. In contrast to the two clear

RESISTANCE vs. WATER CONTENT FOR MICROEMULSIONS

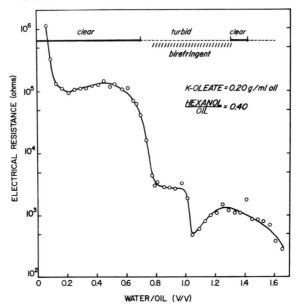

Fig. 10.   The influence of oil-water ratio on the optical
appearance and electrical resistance of hexadecane-
water-potassium oleate-hexanol system

regions, the turbid or translucent region exhibits birefringence. After clarity returns for the second time, further
addition of water causes the dispersion to become opaque,
milky, and nonbirefringent. The variation of the electrical
resistance as a function of water/oil ratio follows a very
unusual pattern as shown in Figure 10. There is no significant change in the resistance values when water/oil ratio is
changed from 0.2 to 0.6. However, there is a sharp decrease
in the resistance at water/oil ratios around 0.7 and again
at 1.0; subsequently, it increases and then decreases.

Figure 11 shows the chemical shift and the bandwidth at
half-height of the major peaks in the NMR spectra of the
microemulsions. From Figure 11 it is evident that in the
birefringent region distinct changes occur in the chemical
shifts and in the broadening of the resonance peaks of water
and hydrocarbon protons. It is also evident that the chemical
shift of water protons is markedly influenced in contrast to
that of methylene or methyl protons. As shown in Figure 10,
the dispersions up to water/oil ratio 0.7 are optically clear
and isotropic. The constant values of electrical resistance
between the oil/water ratios 0.1 to 0.7 suggest that this

NMR DATA vs. WATER CONTENT FOR MICROEMULSIONS

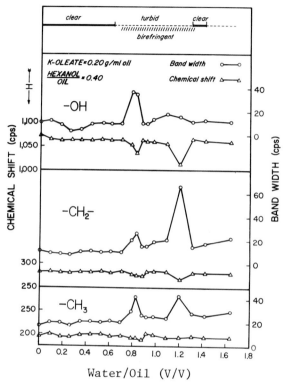

Fig. 11.   *The influence of water to oil ratio on the chemical
shift of hydroxylic and water, methylene, and methyl
protons in hexadecane-water-potassium oleate-hexanol
system.   The upper part of the diagram shows the
optical appearance of the system*

region consists of discrete droplets of water dispersed in a
continuous oil phase, because if this were the region of cosolubilization then electrical resistance would have decreased
continuously with increasing amount of water.   Initially, the
resistance drops from 10^6 to 10^5 ohms as the ratio of water to
oil increases from 0.1 to 0.2; this effect is presumably due
to the molecular solubilization of water in the hexadecane-
hexanol-potassium oleate mixture.   This interpretation is
supported by the observation that the hydroxylic protons of
hexanol show an upfield shift from 1075 to 1965 cycles/sec
with the initial addition of water.   The sharp decrease in
the resistance and the development of birefringence suggest
that there is a transition in the structure of the system

from isotropic to anisotropic liquid–crystalline state. The
NMR data also indicate that water protons exist in two distinct molecular environments in the birefringent region. In
the first environment, the chemical shift of water protons
moves upfield by 25 cps, and in the second environment by 50
cps. Moreover, the bandwidth of resonance peak of water protons is considerably greater in the first as compared to that
in the second environment. The bandwidth is related to molecular mobility or motion. In general, the greater the bandwidth, the smaller the molecular motion. The measurements of
the bandwidth suggest that water molecules are less mobile
in the first environment than in the second (Figure 11). In
contrast to water protons, the bandwidth of methylene ($-CH_2-$)
protons suggests that hydrocarbon chains are less mobile in
the second environment than in the first. These are the
expected characteristics of the system if the first environment consists of water cylinders dispersed in a continuous
oil medium, and the second environment consists of water and
oil lamellae. Water molecules would be less mobile in the
cylinders than in the lamellae because of the restriction
imposed by the cylinder diameter. Moreover, the formation
of these structures would also decrease electrical resistance
since the ions could migrate within water cylinders or lamellae without passing through the oil/water interface. The
formation of these structures can also account for the development of birefringence. The transition from water spheres
in oil to water cylinders to water lamellae to oil droplets
in a continuous water phase represents the mechanism of phase-
inversion in this microemulsion system.

Figure 12 illustrates schematically the mechanism of
phase inversion of this microemulsion based on electrical,
optical, and NMR measurements. From left to right, it shows
water spheres in the continuous oil phase, water cylinders
in the oil, lamellar structure of the surfactants and water
where the oil is solubilized within the surfactant bilayer,
and the water continuous microemulsion, where the oil droplets are stabilized by the potassium oleate-hexanol mixed
film. Increasing temperature can shift the structure from
left to right and addition of cations like Li^+, Na^+, K^+, Ca^{++},
Al^{+++}, etc. can shift the structure from right to left (Shah,
1973). It should be emphasized that the mechanism of phase-
inversion proposed in Figure 12 is valid only for the system
reported here and may not be applicable to other systems.
It is likely that the specific mechanism of phase-inversion
will depend upon the chemical structure of the components,
composition of the system and the physicochemical conditions
such as temperature, etc.

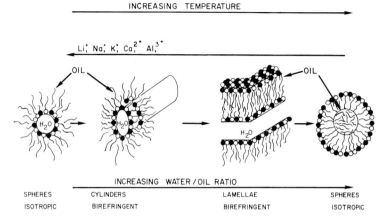

Fig. 12.   A schematic illustration of the mechanism of phase-inversion in the hexadecane-water-potassium oleate-hexanol system of the composition discussed in the text.   Increasing water to oil ratio or temperature shifts the system towards the right, whereas the addition of electrolytes shifts the system towards the left

Falco et al. (1974) studied the effect of water/oil ratio and phase-inversion phenomenon on the viscosity of the microemulsion at different water/hexadecane ratio to correlate the changes in viscosity with the structural changes in the dispersion as determined by electrical resistance, optical birefringence, and NMR spectroscopy. Figure 13 shows the effect of water to hexadecane (oil) ratio on the viscosity of these dispersions. The maximum in the viscosity at a ratio of 1.4 corresponds to the lamellar structures occurring at this ratio (Figure 12). The peak observed between the ratios 2.0 and 3.5 exhibits a very high viscosity and this larger value was quite unexpected. Both these viscosity peaks were observed upon increasing or decreasing the water to oil ratio. The viscosity maximum at the ratio of 1.4 occurred in the turbid region, when the microemulsion undergoes a phase-inversion from water-in-oil to oil-in-water type microemulsions. The change in viscosity observed at the ratio 0.8 corresponds to the formation of water cylinders, and the viscosity peak observed at the ratio 1.4 corresponds to the formation of lamellar structure. The minimum in viscosity at the ratio 2.0 is that of an isotropic, clear, water-continuous microemulsion. The occurrence of the lamellar structure at a higher water to oil ratio (1.4), as indicated by the viscosity measurements, compared to that obtained by NMR technique (1.2) can be explained

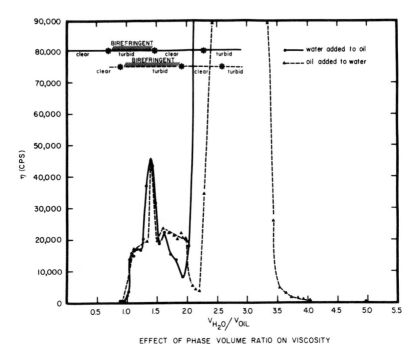

EFFECT OF PHASE VOLUME RATIO ON VISCOSITY

Fig. 13. The influence of water to oil ratio on the viscosity of hexadecane-water-potassium oleate-hexanol system. The continuous line indicates the viscosity upon increasing the water to oil ratio, whereas the broken line represents the viscosity when the water to oil ratio was decreased

on the basis of inherent breakdown of the lamellar structure under shear, and, consequently, a higher water-to-oil ratio is required for the formation and stabilization of the lamellar structure when subjected to shearing forces. An unexpected increase in the viscosity also was observed between the ratios of 2.0 and 3.5. This peak cannot be due to either cylindrical or lamellar structures because it occurs in the nonbirefringent and turbid region. Since the high viscosity region occurs upon addition of water to the oil-in-water microemulsion at a ratio 2.0, the increase in ionization of soap molecules upon increasing the water content may cause interdroplet cross-linking by ion-dipole association between the hydroxyl group of hexanol and the ionized carboxyl group of oleate on the adjacent droplets, resulting in a gel-like high viscosity structure (Figure 14). This interpretation was supported by the observation that the addition of salt strikingly decreased this viscosity peak due to neutralization of carboxylic groups.

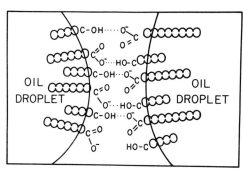

*Fig. 14.   A schematic illustration of ion-dipole association
between adjacent oil droplets proposed to explain
the viscosity peak between 2.5 to 3.5 water/oil
ratio shown in Figure 13*

Shah et al. (1976) have been able to elucidate the structure as well as the mechanism of rheological changes that are
induced in birefringent systems (at water/oil ratio = 1.4)
upon shearing. It was observed that the liquid-crystalline
phase (birefringent region phase) had very unusual rheological
properties. If the sample was kept on a shelf for a day or so
the viscosity decreased strikingly and the sample became very
fluid. However, if such a sample was vigorously shaken for
30 seconds, it became a gel. Figure 15 shows the liquid-
crystalline phase of the n-hexanol containing system at the
water to oil ratio of 1.4 before (Figure 15A) and after
(Figure 15B) shaking the sample tube.

Figures 16 and 17 show the x-ray scattering intensity of
the birefringent phase before and after shaking the sample
tube. The intensity of scattered x-rays is plotted as a function of scattering parameter, S (i.e., $S = 2 \sin \theta/\lambda$, where
θ is the scattering angle and λ the wavelength of the x-rays).
The first-order diffraction maximum at $S = 0.0053$ and the
birefringent properties as well as high resolution NMR characteristics of the sample indicate that this system is in the
form of parallel lamellae. The increased width of the first
order maximum for the shaken sample (Figure 17) can be due to
variation in the separation or disorder of the lamellae. It
is evident from Figures 16 and 17 that the maximum intensity
of scattered x-rays at $S = 0.0053$ decreases sharply after
shaking and the deconvoluted bandwidth increases. The x-ray
scattering data suggest that after shaking the sample tube,
the degree of order decreased in the sample.

Figures 18 and 19 are electronmicrographs of the lamellar
liquid-crystalline structure before and after shaking the
sample tube, obtained using freeze-etching technique. It is

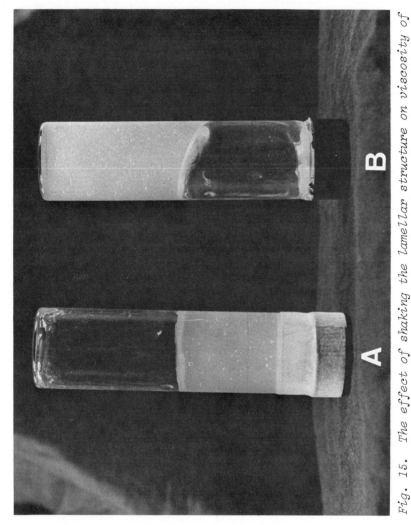

Fig. 15. The effect of shaking the lamellar structure on viscosity of hexadecane-water-potassium oleate-hexanol system. (A) shows the sample tube not shaken whereas (B) shows a sample tube after shaking

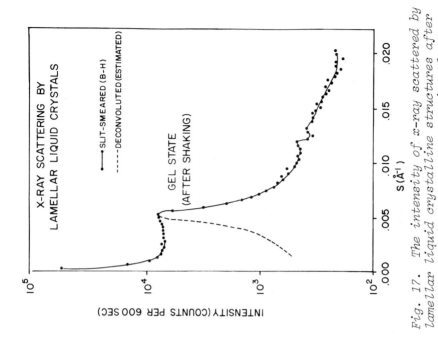

Fig. 17. The intensity of x-ray scattered by lamellar liquid crystalline structures after shaking (hexadecane-water-potassium oleate-hexanol)

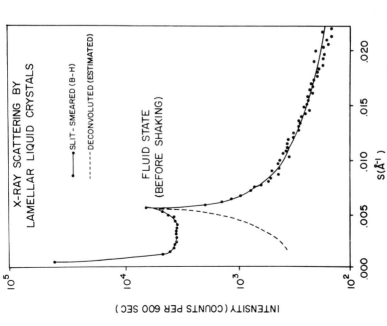

Fig. 16. The intensity of scattered x-ray by lamellar liquid crystalline structures before shaking (hexadecane-water-potassium oleate-hexanol)

*Fig. 18. The freeze-etching electronmicrograph of lamellar
liquid crystalline system before shaking (hexadecane-water-potassium oleate-hexanol system). The
bar at the lower right hand side corner represents
the distance of 0.5 microns*

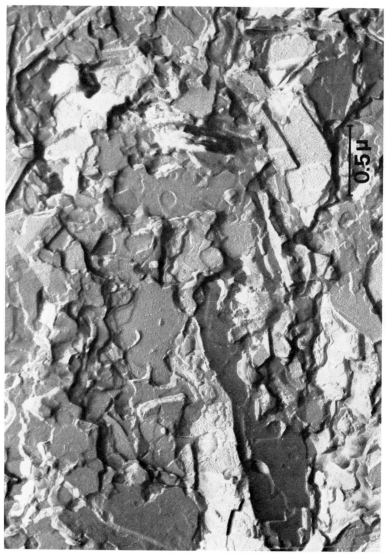

Fig. 19. The freeze-etching electronmicrograph of the lamellar liquid crystalline system after shaking. It clearly shows the random orientation of planes of lamellar structures. The bar on the lower right hand corner of the diagram represents 0.5 microns (hexadecane-water-potassium oleate-hexanol system)

obvious that before shaking the tube, the lamellae orientate
parallel to one another as shown in Figure 18, whereas after
shaking the sample tube, disordering as well as significant
breakdown of the lamellae occur. It is also evident from
Figure 18 that each leaflet is about 80 to 90 Å thick, and
this corresponds to the thickness of a bilayer of surfactant
and cosurfactant molecules swollen with oil. Therefore, it
appears that upon standing on the shelf the lamellae orientate
parallel to one another and therefore flow past one another
during flow thus exhibiting the characteristics of a low
viscosity fluid. However, upon agitation or shaking, the
lamellae become disordered and get entangled with one another,
resulting in a gel-like state (Figure 19).

     Harusawa et al. (1974) have reported their phase-equilibria
studies for the binary and ternary systems of water-pentaoxy-
ethylene dodecylether system and water-dodecane systems containing 10, 20, 30 and 40 wt. % of pentaoxyethylene dodecylether using x-ray diffraction, electrical conductivity,
viscosity and differential scanning calorimetry. They observed a Bragg spacing of 40 to 50 Å in the lamellar liquid
crystalline region of the binary system, which corresponds to
unswollen surfactant bilayers.

XI. PHASE-EQUILIBRIA IN MICROEMULSION SYSTEMS

     In general, microemulsions are composed of three major
components, namely, oil, emulsifiers and water or brine solution. It is therefore both convenient and instructive to
employ a ternary representation for a phase equilibrium study.
If the microemulsion is formed by a mixture of surfactants
then all surfactants including alcohol (cosurfactant) are
taken as one component. A large variety of phases can exist
in equilibrium with each other and these phases can have different microstructures. Figure 20 is a hypothetical phase
diagram (Prince, 1975) in which E (emulsifier) at the apex is
the sum of the soap and alcohol. The macroemulsion region
lies below the microemulsion region at lower emulsifier content. This phase diagram illustrates the formation of various
phases when the system is diluted by water.

     Phase diagram studies of microemulsions involved in ter-
tiary oil recovery provide a useful method to predict the
formation of various phases in the reservoir as a function of
composition. Although these diagrams may be quite complex
when only oil, water, and surfactant are involved, the addition of an appropriate alcohol as a cosolvent causes simplification in phase behavior. For such simple systems the
ternary diagram can be divided into four regions as shown in
Figure 21 (Healy and Reed, 1974). Every compositional point

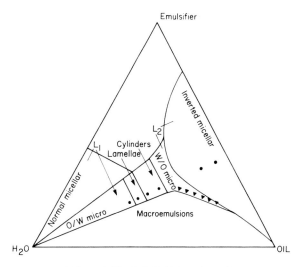

Prince, L.M., J. Colloid Interface Sci.
(1975) 52, 182-187

*Fig. 20. A theoretical phase diagram of water-oil-emulsifier
representing regions of various structures as a
function of composition of the system*

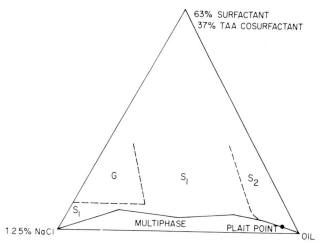

Healy, R.N. and Reed, R.L., SPE J.
(Oct. 1974) 491-501

*Fig. 21. A phase diagram of brine-oil-emulsifier showing
regions of various structures. The transition
from S_2 to S_1 occurs without the system going
through a liquid crystalline region*

within the single phase region above the binodal curve corresponds to a microemulsion or a gel state; however, the
structure of the phase may vary with the location of the
point. Compositional points below the binodal curve correspond to a multiphase region, comprising of microemulsions
of various kinds, excess oil and excess water, and therefore
opaque emulsions of these are observed upon mixing. Spherical
micelles consisting of oil cores in a water continuous medium
are called an S_1 or water-external phase. The inverse of this
is the S_2 or oil-external phase. An intermediate lamellar
structure, which may be a gel or liquid crystal, is called
the G phase. It is evident from the comparison of the two
phase diagrams (Figures 20 and 21) that the position of the
liquid crystalline phase differs depending upon the chemical
composition of the system. In Figure 20 the liquid crystalline phase is in between the oil-in-water and water-in-oil
microemulsions as compared to Figure 21 where the liquid
crystalline region is at the corner of the phase diagram.
It is possible to convert S_2 to S_1 phase without going
through the gel type structure. In recent years phase diagram studies of various nonionic surfactants have been
carried out by various workers (Shinoda and Kunieda, 1973;
Attwood et al., 1974; Friberg et al., 1976). Their emphasis
was basically on the effect of phase inversion temperature
and HLB value of surfactant on the phase behavior. The
effect of important variables such as salt, surfactant and
cosolvents (alcohols) on phase behavior have been studied by
Jones (1975), Lorenz (1974) and Anderson et al. (1976).

     Many oil/water systems containing ionic or nonionic sur-
factants produce a surfactant-rich middle phase when salt
concentration or temperature is changed. Under these conditions, the middle phase or surfactant-rich phase is in equilibrium with excess oil and water (or brine). The surfactant-
rich phase has been referred to as the middle phase microemulsion or the surfactant phase (Reed and Healy, 1976;
Shinoda and Kunieda, 1973). Since the middle phase is in
equilibrium with excess oil and brine, Reed and Healy have
suggested that the middle phase microemulsion is neither
water-external nor oil-external. Friberg et al. (1976) have
proposed a structure consisting of spherical aggregates and
planar structures resembling bicontinuous dispersion of oil
and water for such a middle phase. Scriven (1976) has also
proposed the existence of bicontinuous oil-water structures
in the middle phase. However, in contrast to the bicontinuous
structure, Shah (1976) has proposed that such a middle phase
microemulsion may consist of water-external microemulsion.
It is proposed that upon increasing salt concentration, the

surface charge on micelles decreases due to charge neutralization and the solubilization of oil within micelles increases.
Consequently, the swollen micelles approach one another and
phase separation occurs due to buoyancy effect. However, we
believe a more detailed study is required to establish the
precise structure of the middle phase microemulsions observed
under narrow conditions of salt concentrations or temperature.

XII. ACKNOWLEDGEMENTS

The authors wish to express their sincere appreciation
and thanks to the National Science Foundation-RANN, Energy
Research and Development Administration, and the consortium
of 21 major oil and chemical companies for supporting parts
of the research presented in this review article. The permission given by various authors and publishing companies to
reproduce several diagrams are also gratefully acknowledged.

XIII. LITERATURE CITED

Adamson, A. W., "A Model for Micellar Emulsions," J. Colloid
    Inter. Sci. Vol. 25, No. 2, 261-267 (1969).

Ahmed, S. I., Shinoda, K. and Friberg, S., "Microemulsions and
    Phase Equilibria," J. Colloid Inter. Sci. Vol. 47, No. 1,
    32-37 (1974).

Alexander, A. E. and Schulman, J. H., "Molecular Interaction
    at Oil-Water Interfaces, Part III," Trans. Faraday Soc.
    960-964 (1940).

Anderson, D. R., Bidner, M. S., Davis, H. T., Manning, C. D.
    and Scriven, L. E., "Interfacial Tension and Phase Behavior
    in Surfactant-Brine-Oil Systems," SPE 5811 presented at
    Improved Oil Recovery Symposium, Tulsa, OK, March 22-24
    (1976).

Attwood, D., Currie, L.R.J. and Elworthy, P. H., "Studies of
    Solubilized Micellar Solutions," J. Colloid Inter. Sci.
    Vol. 40, No. 2, 249-254 (1954).

Bowcott, J. E. and Schulman, J. H., "Emulsions," Zeitschrift
    für Elektrochemie Berichte der Bunsengesellschaft für
    Physikalische Chemie Band 59, Heft 4, 283-290 (1955).

Davis, J. T. and Haydon, D. A., "Spontaneous Emulsification,"
    The Proceedings of the Second International Congress of
    Surface Activity, Vol. I, 417-425 (1957).

Falco, J. W., Walker, R. D., Jr., and Shah, D. O., "Effect of
    Phase Volume Ratio and Phase-Inversion on Viscosity of Micro-
    emulsion and Liquid Crystals," AIChE J. Vol. 20, No. 3, 510-
    514 (1974).

Frank, S. S. and Zografi, G., "Determination of Micellar
    Weights for Di-Alkyl Sodium Sulfosuccinates in Anhydrous
    and Hydrous Hydrocarbon Solutions," J. Pharma. Sci. Vol. 58,
    No. 8, 993-997 (1969).

Friberg, S., Lapczynska, J. and Gillberg, G., "Microemulsions Containing Nonionic Surfactants: The Importance of the PIT Value," J. Colloid Inter. Sci. Vol. 56, No. 1, 19–32 (1976).

Garbacia, W. and Rosano, H. L., "Microemulsion: Formation and Stabilization," J. Colloid Inter. Sci. Vol. 44, No. 2, 242–248 (1973).

Goddard, E. D. and Schulman, J. H., "Molecular Interaction in Monolayers, I. Complex Formation," J. Colloid Sci. Vol. 8, 309–328 (1953).

Hansen, J. R., "High Resolution and Pulsed NMR Studies of Microemulsion," J. Phys. Chem. Vol. 78, No. 3, 256–261 (1974).

Harasawa, F., Nakamura, S. and Mitsui, T., "Phase Equilibria in the Water-Dodecane-Pentaoxylethylene Dodecyl Ether System," Colloid and Polymer Sci. Vol. 252, 613–619 (1974).

Healy, R. N. and Reed, R. L., "Physicochemical Aspects of Microemulsion Flooding," Soc. Pet. Eng. J. Vol. 14, 441–501 (1974).

Hoar, T. P. and Schulman, J. H., "Transparent Water-in-Oil Dispersions: The Oleopathic Hydro Micelle," Nature Vol. 152, 102–103 (1943).

Hughes, E. W., Sawyor, W. M. and Vinograd, J. R., "X-ray Diffraction Study of Micelle Structure in K-Laurate Solutions," J. Chem. Phys. Vol. 13, 131–132 (1945).

Ilkovic, D., "Polarographic Studies with the Dropping Mercury Kathode-Part XXVIII," Coll.Trav.Chim.Tchecosl. Vol.4,480(1932).

Jones, S. C. and Dreher, K. D., "Surfactants in Micellar Systems Used for Tertiary Oil Recovery," SPE 5566 presented at SPE Meeting, Dallas, TX, Sept. 28-Oct. 2 (1975).

Kiessig, H., "X-ray Investigation of the Structure of Soap Solutions," Kolloid-Z. Vol. 96, 252–255 (1941).

Levine, S. and Robinson, K., "The Role of the Diffuse Layer in Water-in-Oil Microemulsions," J. Phys. Chem. Vol. 76, No. 6, 876–886 (1972).

Lorenz, P. B. and Bayazeed, A. F., "The Effect of Phase Behavior on Recovery Efficiency with Micellar Floods," SPE 4751 presented at Improved Oil Recovery Symposium, Tulsa, OK, April 22-24 (1974).

McBain, J. W. and McBain, M.E.L., "The Spontaneous Stable Formation of Colloids from Crystals or from True Solutions Through the Presence of a Protective Colloid," J. Amer. Chem. Soc. Vol. 58, 2610–2612 (1936).

McBain, M.E.L. and Hutchinson, E., "Solubilization and Related Phenomena," p. 66. Academic Press, Inc., New York, 1955.

Miller, C. A. and Scriven, L. E., "Interfacial Instability Due to Electrical Forces in Double Layers I: General Considerations," J. Colloid Inter. Sci. Vol. 33, No. 3, 360–370 (1970a).

Miller, C. A. and Scriven, L. E., "Interfacial Instability Due to Electrical Forces in Double Layers II: Stability of Interfaces with Diffuse Layer," J. Colloid Inter. Sci. Vol. 33, No. 3, 371-383 (1970b).

Palit, S. R. and Venkateswarlu, V., "Solubilization of Water in Non-polar Solvents by Cationic Detergents," Proceedings of the Royal Society, A Vol. 281, 542 (1951).

Pink, R. C., "Critical Effect of Temperature on the Absorption of Water by Solutions of Ethanolamine Oleate in Benzene," Trans. Faraday Soc. Vol. 42B, 170-173 (1946).

Preston, W. C., "Some Correlating Principles of Detergent Action," J. Phys. Colloid Chem. Vol. 52, 84-97 (1948).

Prince, L. M., "A Theory of Aqueous Emulsions I. Negative Interfacial Tension at the Oil/Water Interface," J. Colloid Inter. Sci. Vol. 23, 165-173 (1967).

Prince, L. M., "A Theory of Aqueous Emulsions II. Mechanism of Film Curvature at the Oil/Water Interface," J. Colloid Inter. Sci. Vol. 29, No. 2, 216-221 (1969).

Prince, L. M., "Microemulsions Versus Micelles," J. Colloid Inter. Sci. Vol. 52, No. 1, 182-188 (1975).

Reed, R. N. and Reed, R. L., "Physico-Chemical Aspects of Microemulsion Flooding," paper presented at Improved Oil Recovery Symposium at 81st National AIChE Meeting, Kansas, MO, April 13-14 (1976).

Riley, D. P., "X-ray Investigations of the Structure of Liquid Disperse Systems," British Science News Vol. 3, No. 25, 7-10 (1949).

Robbins, M. L., "Theory for the Phase Behavior of Microemulsions," SPE 5836 presented at Improved Oil Recovery Symposium, Tulsa, OK, March 22-24 (1976).

Ruckenstein, E. and Chi, J. C., "Stability of Microemulsions," J. Chem. Soc. Faraday Trans. II Vol. 71, 1650-1707 (1975).

Schulman, J. H. and Cockbain, E. G., "Molecular Interactions at Oil/Water Interfaces, Part I," Trans. Faraday Soc. Vol. 36, 651-661 (1940a).

Schulman, J. H. and Cockbain, E. G., "Molecular Interactions at Oil/Water Interfaces, Part II," Trans. Faraday Soc. Vol. 36, 661-668 (1940b).

Schulman, J. H. and McRoberts, T. S., "On the Structure of Transparent Water and Oil Dispersions (Solubilized Oil)," Trans. Faraday Soc. Vol. 42B, 165-170 (1946).

Schulman, J. H. and Riley, D. P., "X-ray Investigation of the Structure of Transparent Oil-Water Disperse Systems," J. Colloid Sci. Vol. 3, 383-405 (1948).

Schulman, J. H., McRoberts, T. S. and Riley, D. P., "X-ray Investigation of Transparent Oil-Water Systems," Proceedings of the Physiology Society, J. Physiology Vol. 107, 15 (1948).

Schulman, J. H. and Friend, J. P., "Light-Scattering Investigation of the Structure of Transparent Oil-Water Disperse Systems II," J. Colloid Sci. Vol. 4, 457-505 (1949).

Schulman, J. H., Matalon, R. and Cohen, M., "X-ray and Optical Properties of Spherical and Cylindrical Aggregates in Long Chain Hydrocarbon Polyethylene Oxide Systems," Faraday Soc. Disc. Vol. 11, 117-121 (1951).

Schulman, J. H., Stoeckenius, W. and Prince, L. M., "Mechanism of Formation and Structure of Microemulsions by Electron Microscopy," J. Phys. Chem. Vol. 63, 1677-1680 (1959).

Schulman, J. H. and Leja, J., "Molecular Interactions at the Solid-Liquid Interface with Special Reference to Flotation and Solid-Particle Stabilized Emulsions," Kolloid-Z. Vol. 136, 107-119 (1954).

Scriven, L. E., "Equilibrium Bicontinuous Structure," Nature Vol. 263, 123-124 (1976).

Shah, D. O., "High Resolution NMR (220 Mc) Studies on the Structure of Water in Microemulsions and Liquid Crystals," Annals of the New York Academy of Sciences Vol. 204, 125-133 (1973).

Shah, D. O. and Hamlin, R. M., Jr., "Structure of Water in Microemulsions: Electrical, Birefringence, and Nuclear Magnetic Resonance Studies," Science Vol. 171, 483-488 (1971).

Shah, D. O., "Research on Chemical Oil Recovery Systems," Third Semi-Annual Report, University of Florida, A-133 (Dec. 1, 1976).

Shah, D. O., Walker, R. D., Jr., Hsieh, W. C., Shah, N. J., Dwivedi, S., Nelander, J., Pepinsky, R. and Deamer, D. W., "Some Structural Aspects of Microemulsions and Cosolubilized Systems," SPE 5815 presented at Improved Oil Recovery Symposium, Tulsa, OK, March 22-24 (1976).

Shah, D. O., Tamjeedi, A., Falco, J. W. and Walker, R. D., Jr., "Interfacial Instability and Spontaneous Formation of Microemulsions," AIChE J. Vol. 18, 1116-1120 (1972).

Shah, D. O., "How to Distinguish Microemulsions from Cosolubilization," 48th National Colloid Symposium, Austin, TX, June 24-26 (1974).

Shinoda, K. and Friberg, S., "Microemulsion: Colloidal Aspects," Advances in Colloid and Inter. Sci. Vol. 4, 281-300 (1960).

Shinoda, K. and Kunieda, H., "Conditions to Produce So-called Microemulsions: Factors to Increase the Mutual Solubility of Oil and Water by Solubilizer," J. Colloid Inter. Sci. Vol. 42, No. 2, 381-387 (1973).

Stauff, J., "Micellar Types of Aqueous Soap Solutions,"
Kolloid-Z. Vol. 89, 224-233 (1939).

Stoeckenius, W., Schulman, J. H. and Prince, L. M., "The
Structure of Myelin Figures and Microemulsions as Observed
with the Electron Microscope," Koll Zeil Band 169, Heft 1-2,
170-180 (1960).

Tosch, W. C., Jones, C. C. and Adamson, A. W., "Distribution
Equilibria in a Micellar Solution System," J. Colloid Inter.
Sci. Vol. 31, No. 3, 287-305 (1969).

Zisman, W. A., "The Spreading of Oils on Water III. Spreading Pressures and the Gibbs Adsorption Relation," J. Chem.
Phys. Vol. 9, 789 (1941).

SOME THERMODYNAMIC ASPECTS AND MODELS OF
MICELLES, MICROEMULSIONS AND LIQUID CRYSTALS

John P. O'Connell and Robert J. Brugman
University of Florida

I.   ABSTRACT

The considerations are reviewed for developing thermodynamic and molecular models for solutions containing high
concentrations of amphiphilic compounds. Discussion includes
the types of molecular interactions present and the influence
of molecular fluctuations with emphasis on their importance
in phase behavior and formation for quantitative correlation.
Rigid-body theories for correlation of the "hydrophobic effect" in micelle formation and microemulsion interactions
are illustrated and correct formulation of counterion binding
is discussed. Also reviewed are recent theories for lyotropic
liquid crystal phase transitions and for microemulsion formation and phase equilibria.

II. SCOPE

Solutions containing amphiphilic compounds have recently
assumed greater importance in practical situations such as in
detergency and chemical formulations for improved oil recovery,
because these substances are capable of radically changing the
phase behavior, interfacial structure, adsorption, and mass
transfer of fluid phases. Most of the phenomena of interest
involve equilibrium properties, such as interfacial tension,
the existence of molecular aggregates, such as micelles, and
the properties of microstructured phases such as microemulsions
and lyotropic liquid crystals. Thus, the thermodynamics of
phase stability and equilibria and pseudoreaction equilibrium
must be applied to these systems. Further, molecular models
should be able to be established to correlate critical micelle
concentrations (CMC), species activities and distribution coefficients, and interfacial tension as functions of temperature, pressure and composition including additional species
such as salts and organic substances.

The purpose of this paper is to discuss some of the fundamental aspects of these systems which have not been considered in the context of recently acquired knowledge. In particular, more accurate experimental data have become available

and quantitative calculations using the statistical thermodynamics of fluid structure have been developed to a high
degree. Several significant theories of micelles and of
microemulsion formation have appeared within the past three
years. Some generalizations of phenomena are illustrated and
the theories are reviewed. We also address some of the questions regarding the true nature of the "hydrophobic effect"
in micelle formation, the binding of counterions to micelles
and the number of phases which can coexist in some oil/water
amphiphile systems.

III.   CONCLUSIONS AND SIGNIFICANCE

The molecular interactions of micellar systems and microstructured phase systems have many similarities. The free
energy of formation of amphiphilic aggregates involve some
energetic effects of repulsion between head groups and between
aggregates, particularly for ionic species, and large entropic
effects. We believe the latter are mostly associated with the
"hydrophobic effect," that is, the coalescence of volume excluded to the solvent when micelles are formed from dissolved
monomers or with the configurational entropy of dispersed
droplets of a microemulsion compared to the bulk phases. In
both cases, it appears that statistical mechanical expressions
for fluids of rigid bodies can quantitatively describe observed
entropic effects. Examples include the carbon number dependence of CMC, which is essentially the same for all ionic surfactants, and the relationship of interfacial tension and size
in microemulsions.

The presence of molecular fluctuations causes well-known
distributions of micelle size. It also explains why the so-
called "phase separation model" violates the phase rule.
They often cannot be ignored in thermodynamic formulations
for micelle formation although for some effects they are less
than the experimental uncertainty. Further, we speculate that
fluctuations of the interfacial area allow the anomalous thermodynamic variance which appears when microstructured phases
involving lamellar forms exist as in some ternary oil/water/
surfactant systems.

The electrostatic interactions in amphiphilic systems may
allow some generalizations for correlating purposes without
determining the precise nature of the system. The dependence
of CMC on added salt concentration and the distribution coefficients and phase uptake in a variety of microemulsion systems appear to be controlled by general electrostatic features
rather than by the particular species involved.

We believe that the present work develops some new understanding and describes some constraints not emphasized by
others which should be met by models and theories of amphiphile

systems and suggests potential methods for generating quantitative expressions for correlating their phenomena.

IV. INTRODUCTION

In recent years there has been a great increase of interest in systems with significant interfacial area and/or concentrations of amphiphiles. Detergent action, mineral flotation, evaporation retardation, chemical tertiary oil recovery, and liquid crystal optical displays are but a few of the practical applications to which attention is being directed. Much of the scientific basis for analysis of these systems lies in thermodynamics beginning with Gibbs (1) and continuing with Defay et al. (2) and Melrose (3,4) as well as the small system treatment by Hill (5) and finally, important statistical mechanical developments by Buff (6,7), Toxvaerd (8) and others as described by Gubbins and Haile (9) elsewhere in this volume. In general, recent research into systems of continuous aqueous or nonaqueous phases containing microstructured dispersed phases enclosed by amphiphiles, e.g., microemulsions and lyotropic liquid crystals has not been analyzed within the context of the thermodynamics of phases. In addition, except for the work of Mukerjee (10), Anacker (11), and Lin and Somasundaran (12) there has not been any recent published attempt to enumerate and collect together the various contributions of thermodynamic models (involving concepts such as free energies of transfer, etc.) for equilibrium phenomena of micelles, microemulsions and liquid crystals and how they have been modelled. It is our purpose to examine both of these situations in order to develop more understanding of the phenomena and lead to more unified models which can describe broader classes of systems than heretofore. The format is not a strict review in the sense that description of the work of some has been minimized. Instead, our discussion attempts to bring out aspects which might provoke further thought and development in the area of improved oil recovery.

V. MODELLING OF AMPHIPHILE SYSTEMS

Since there is a great deal of similarity between the phase equilibrium thermodynamics of micelles and of fluid microstructure phases, and the structural organization of the amphiphiles between both types of lipophilic and hydrophilic regions is also similar, it seems appropriate to attempt to bring molecular models of these two types of systems into a single framework. In fact, it can be argued that micelles are actually only a special case of microstructured fluids. Since there are many more theories for describing micelles than for microemulsions and lyotropic liquid crystals, an attempt at unification might provide insight into the best

advances for the latter systems. (Many of the molecular phenomena are also present in surfactant monolayers but we have
not attempted to include them in detail here.)

Table I and Figure 1 describe our categorization of the
kinds of thermodynamic contributions which must be taken into
account by theories describing concentrated amphiphile systems.
Many of these have been discussed by others (Poland and
Scheraga, 13,14; Debye, 15; Lin and Somasundaran, 12). However, there are some aspects we feel are unique to our description. The breakdown is into overall energetic effects in
various regions of the system and entropic contributions due
to differences in molecular conformations and excluded volume
effects of the dispersed phase in the continuous phase. We
choose the last as a separate category (which has some energy
and large entropy contributions) because it is usually modelled
directly in some way in all theories under names such as the
"hydrophobic" effect. We have indicated electrostatic effects
which would be restricted to ionic amphiphiles. Finally, we
discriminate between those contributions which are associated
with aspects internal to the dispersed phase, including intra-
molecular constraints, those which occur across the interface
of the dispersed and continuous phases, and those which occur
between dispersed phases across the continuous phase (we assume that whatever internal effects occur in the continuous
phase are localized near the interface and thus no category is
set aside for these).

In principle, all of these effects should be taken into
account. However, such a development would lead to models
containing too many parameters and perhaps excessively complicated expressions. Empirically, it has been observed that
the entropy change upon micellization is large and positive
while the enthalpy change is generally small and can be either
positive or negative. This indicates that the major molecular
factors are due to changes in structure and conformation rather
than to intermolecular potential energy changes. The latter
cannot be ignored, particularly for ionic amphiphiles where
significant charge repulsion exists between head groups. Yet
the former dominate and we assert that the major structural
change is not due to condensation of the hydrocarbon chains,
which is an energy effect, or to rearrangement of the water
as a "cage" around a monomer or a micelle. Instead, we consider it due to a change of the volume excluded to the small
water molecules by the amphiphilic species. This is the
"solvaphobic" effect described by Ben-Naim (16) which is extremely large for water because the molecules are very small.
For micellization, the excluded volume which is dispersed with
monomers is coalesced when micelles are present. While it may
seem unusual for coalescence to lead to increased entropy, it

TABLE I

MOLECULAR INTERACTIONS IN SYSTEMS CONTAINING AMPHIPHILES

Interaction/System	Bulk	Mono-layer	Dissolved Molecules	Micellar Aggregate	Anisotropic Phase	Micelle-Solubilized Oil	Microemulsion, Liquid Crystal
Energetic							
Amphiphile-Amphiphile	x	x^a		x^a	x^a	x^a	x^a Internal^b
Amphiphile-Solvent							
Head-Solvent		x	x	x	x	x	x Interfacial
Tail-Solvent			x				
Amphiphile-Oil						x	x Internal^b
Aggregate-Aggregate					x^a	x^a	x^a External
Entropy (Molecular Constraints)							
Amphiphile		x		x	x	x	x Internal^b
Solvent			x	x	x	x	x Interfacial
Excluded Volume of Solvent			x	x	x	x	x Interfacial and/or External
Fluctuations Important				Micelle Number	Radius of Cylinders Thickness of Lamellae	Micelle Number, Oil Concentration	Area of Lamellae Bound Counterions (?)

^{a}Electrostatic effect for ionic amphiphiles

^{b}Effect occurs in interior of aggregates

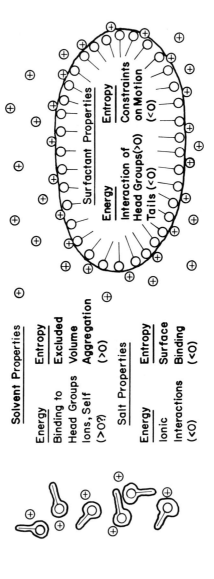

Solvent Entropy Effect Overweighs Surfactant Energy and Salt Entropy Effects

Fig. 1. Contributions to Thermodynamic Properties of Micelle Formation from Various Species

344

must be remembered that the species whose entropy is increased
is the solvent water not the amphiphile; when the excluded
volume is coalesced (note: it is <u>not</u> reduced in value) the
volume available for the water allows for many more configurations of the molecules than when it is dispersed. (We show
calculations below which are qualitatively and quantitatively
consistent with this picture and with correlations for micellization phenomena.)

VI. MICELLIZATION

A. The Effect of Polydispersity

We do not wish to cover all aspects of amphiphilic aggregates in binary systems because many are not relevant to chemical oil recovery systems. However, we do assert that because
of similarity in their molecular effects the important thermodynamic aspects of micelle formation can provide insight into
the analysis of fluid microstructured phase formation and
transitions. In particular, we want to indicate the potential
role of theories for rigid body fluids in developing correlations for the behavior of amphiphile systems. Here we start
with single component micelles, emphasizing the role of ionic
amphiphiles, counterions and the addition of dissolved salts
on the mean aggregate size and the critical micelle concentration. This means we will not here consider mixed systems,
including those with solubilized nonelectrolytes, or the distribution of the number of molecules in a micelle. We prefer
to ignore the effect of polydispersity when possible. (This
is not to say these effects are not of importance, but simply
to limit the scope of this paper.)

First, we need to indicate the limits under which mono-
dispersity is sufficiently accurate. For nonionic amphiphiles,
Hall and Pethica (17) show a relation derived from Hill's small
systems thermodynamics (5) for the Gibbs free energy of formation, Δg^o, of a system of micelles of average size $\bar{N}$, in their
standard state of infinite dilution, from monomeric species,
in their standard state of infinite dilution in a solvent, in
terms of the mole fractions of the monomers, x_1, and of the
micelles, x_m, when ideal solution (infinite dilution) is
assumed

$$\frac{\Delta g^o}{\bar{N}RT} \equiv \frac{\mu_m^o}{\bar{N}RT} - \frac{\mu_1^o}{RT} = \ln x_1 - \frac{1}{N} \ln x_m \qquad (1)$$

At the critical micelle concentration (CMC), relatively abrupt
changes in properties can be observed because the concentration
of monomeric species begins to change very little with the mole
fraction of added amphiphile, x_o

$$x_o \equiv x_1 + \bar{N} x_m \tag{2}$$

In fact, around the CMC value of $\left.\dfrac{\partial x_1}{\partial x_o}\right)_{T,P}$ falls very rapidly from near unity to near zero. The CMC definition of Phillips (18), and explored by Hall (19), is

$$\lim_{x_o \to x_o^+} \left.\frac{\partial^3 \phi_c}{\partial x_o^3}\right)_{T,P} = 0 \tag{3}$$

where x_o^+ is the CMC and ϕ_c is an "ideal colligative property" which depends only on the number of solute species (monomers and micelles). This is inconvenient as Chung and Heilweil (20) show. As an alternative in order to find the value of $\bar{N}$ below which polydispersity is important, we use the expression of Hall and Pethica (17)

$$\lim_{x_o \to x_o^+} \left.\frac{\partial (x_1 + x_m)}{\partial x_o}\right)_{T,P} = 0.5 \tag{4}$$

which is essentially equivalent to Equation (3) for sharp CMC points. Using this relation to obtain

$$x_1^+ = (\overline{N^2} - 2\bar{N}) x_m^+ \tag{5}$$

and the definition of x_o in terms of x_1, $\bar{N}$ and x_m we find

$$x_m^+ = x_o^+ / (\overline{N^2} - \bar{N}) \tag{6}$$

$$x_1^+ = x_o^+ (\overline{N^2} - 2\bar{N}) / (\overline{N^2} - \bar{N}) \tag{7}$$

which then yields

$$\frac{\Delta g^o}{\bar{N}RT} - \ell n\, x_o^+ = \frac{1}{\bar{N}^+} \left[-\ell n\, x_o^+ + \ell n\, \overline{N^2} + \ell n\, (1 - \bar{N}/\overline{N^2}) \right]$$
$$+ \ell n\, [(1 - 2\bar{N}/\overline{N^2}) / (1 - \bar{N}/\overline{N^2})] \tag{8}$$

We wish to assume that the right-hand-side is small enough to neglect, which means its value is approximately 0.05 for 5% error in x_o^+ (a typical experimental uncertainty). By taking

$$1 \gg 1/\bar{N} \approx \overline{\bar{N}/N^2}$$

so that the first order expressions are valid (the latter approximation is conservative), we obtain

$$\frac{\Delta g^o}{\bar{N}RT} - \ell n \ x_o^+ \approx \frac{- \ell n \ x_o^+ + 2\ell n \ \bar{N} - 1}{\bar{N}} \tag{9}$$

The range of $\ell n \ x_o^+$ is -6 to -15 so the value of $\bar{N}$ should be greater than about 300 to 600. Since values of $\bar{N}$ range upward from 10, polydispersity is normally important and thermodynamic theories for $\Delta g^o/\bar{N}RT$ should give an equation with values of N^2 and $\bar{N}$. Further, data analysis to obtain values of Δg^o from CMC values must allow for these terms. For example, where $\bar{N} \approx 25$ and $\ell n \ x_o^+ \approx -6$ the right-hand-side of Equation (9) is 0.46.

The above analysis was developed by Hall and Pethica (17) for nonionic species. To our knowledge, the small systems analysis has not been applied to ionic systems where the effect of counterions on the thermodynamics must be included. As mentioned elsewhere the fluctuation of counterions may in fact be important in the phase behavior of fluid microstructured systems, so it seems appropriate to suggest that expressions for ionic amphiphiles be developed within the framework of the small systems analysis.

The division of the free energy change of micellization into enthalpy (or energy) and entropy portions is accomplished by the relations

$$\frac{\Delta h^o}{\bar{N}R} = \frac{1}{\bar{N}} \frac{\partial \Delta g^o/RT}{\partial 1/T}\bigg)_{P,n} - \Delta g^o/\bar{N}RT \frac{\partial \ell n\bar{N}}{\partial 1/T}\bigg)_{P,n} \tag{10a}$$

$$\approx \left(1 - \frac{1}{\bar{N}}\right) \frac{\partial \ell n \ x_o^+}{\partial 1/T}\bigg)_{P,n} + \frac{1}{\bar{N}} \frac{\partial \ell n\bar{N}}{\partial 1/T} [\ell n \ x_o^+ - 2\ell n \ \bar{N} + 3] \tag{10b}$$

$$\frac{\Delta s^o}{\bar{N}R} = \frac{\Delta h^o - \Delta g^o}{\bar{N}RT} \tag{11a}$$

$$\approx - \ln x_o^+ + \frac{1}{\bar{N}} \left\{ \frac{[\ln x_o^+ + 1 - 2\ln \bar{N}][1 - \frac{\partial \ln \bar{N}}{\partial \ln T}] - 2 \frac{\partial \ln \bar{N}}{\partial \ln T}}{(N-1) \dfrac{\partial \ln x_o^+}{\partial \ln T}} \right\} \tag{11b}$$

The form of Equation (10b) indicates that the last term is small so that the standard state enthalpy change should be close to the temperature derivative of the CMC. Since this is small for many amphiphiles the major contribution to Δg^o is $- T\Delta s^o$. Values for $\Delta g^o/\bar{N}RT$ range from about -6 to -15, but values for $\Delta h^o/\bar{N}RT$ are of the order of zero to one-fourth these values and may be positive or negative. Thus, theories which attempt to correlate the data will usually have terms in $\Delta g^o/\bar{N}RT$ which are constant or with weak temperature dependence. The "hydrophobic" effect is the major one of these.

B.   Thermodynamics of Micellization

At present, the thermodynamics of micellization of ionic systems divides itself into two approaches based on the mass action approach which ignores fluctuations. These are reviewed by Mijnlieff (21). In the first, such workers as Stigter (22-26), Emerson and Holtzer (27-29), and Mukerjee (30) focus on the changes associated with the amphiphilic ions forming an aggregate. Thus, for a singly charged anionic monomer, the reaction is

$$\bar{N} M_1^- \rightleftharpoons M_m^{-\bar{N}} \tag{12}$$

and the thermodynamic formulation is

$$\mu_1 = \mu_1^o + \mu_1^{el} + RT \ln a_1 \tag{13}$$

$$\mu_m = \mu_m^o + \mu_m^{el} + RT \ln a_m \tag{14}$$

$$\mu_m^+ - \bar{N}\mu_1^+ = 0 = (\mu_m^o - \bar{N}\mu_1^o) + (\mu_m^{el} - \bar{N}\mu_1^{el}) + RT (\ln a_m^+ - \bar{N}\ln a_1^+) \tag{15}$$

where μ_1^o is the standard state (infinite dilution) chemical potential of the uncharged monomer, μ_m^o is the standard state chemical potential of the uncharged micellar aggregate, and $(\mu_m^{el} - \bar{N}\mu_1^{el})$ is the chemical potential difference associated with changing the charge on the micelle and the monomers from zero to full value while it is in the presence of the ionic atmosphere of the counterions. Such a change involves the response of the counterions and is sensitive to the detailed molecular structure assumed, as the calculations of Stigter (25,26) and Mukerjee's criticism (30) of Emerson and Holtzer (27-29) show. It is particularly sensitive to the fraction of ions assumed bound to the micelle in the Stern layer as related to electrophoretic and electrical conductance measurements (22). This fraction is apparently of the order of one-half when the micelle is fully charged but how this value depends upon the charging process is unclear. (Stigter (25) does indicate that an increase of bound ions with charging leads to more consistent predictions but the actual form and the value in the limit of an uncharged micelle is not determinable.)

The value of μ_1^{el} is determined from some expression such as that of Debye-Huckel theory leading to

$$\mu_1^{el} = \mu_1^o + RT \ln \gamma_1 \tag{16}$$

The relationship of the counterions to μ_m^{el} is one of equilibrium between those in bulk solution and those in the Stern (bound) layer and the Gouy-Chapman (diffuse) layer (22)

$$\mu_c(\text{solution}) = \mu_c^o(s) + \mu_1^{el} + RT \ln a_c(s)$$

$$= \mu_c(\text{micelle}) = \mu_c^o(m) + \frac{\mu_m^{el}}{\bar{N}} + \frac{RT}{\bar{N}} \ln a_c(m) \tag{17}$$

or

$$\mu_c^o(m) - \mu_c^o(s) + \frac{\mu_m^{el}}{\bar{N}} - \mu_1^{el} = RT \ln a_c(s)/a_c(m) \tag{18}$$

Substituting for $\mu_m^{el} - \bar{N}\mu_1^{el}$ in Equation (15) and combining the standard state chemical potentials together yields

$$\mu_m^o - \bar{N}(\mu_1^o + \mu_c^o) + RT \ln \left\{ \frac{a_m^+ a_c^+(m)}{[(a_1^+ a_c^+(s))]^{\bar{N}}} \right\} = 0$$

Assuming that we can replace a_m^+ and $a_c^+(m)$ by unity (micelles) and a_1^+ and $a_c^+(s)$ by mole fractions (solution), using the definition of Δg^o from Equation (1) where all the species are uncharged, and rearranging gives

$$\frac{\Delta g^o}{\bar{N}RT} = \ell n \ x_1^+ \ x_c^+ \tag{19a}$$

which for no added salt $(x_c = x_1)$ is

$$\frac{\Delta g^o}{\bar{N}RT} = 2 \ \ell n \ x_1^+ \tag{19b}$$

This relation also appears in the work of Shinoda and Hutchinson (31). It is important to note that all standard state chemical potentials and activities given above are for neutral species and the electrostatic effects are taken into account by expressions for activity coefficients. We believe it is correct.

The second approach to the thermodynamic relationships for ionic amphiphiles (32-36,18) writes the reaction for this system as

$$\bar{N} \ M_1^- + \bar{N} \ \alpha \ C^+ \ \rightleftarrows \ M_m^{-\bar{N}(1-\alpha)} \tag{20}$$

where α is the apparent fraction of amphiphiles whose charge is neutralized by bound counterions. The chemical potential relation is then

$$\mu_m^+ - \bar{N}\mu_1^+ - \bar{N}\alpha\mu_c^+ = 0 = \mu_m^o - \bar{N}\mu_1^o - \bar{N}\alpha\mu_c^o + RT[\ell n \ a_m^+ - \bar{N}\ell n \ a_1^+(a_c^+)\alpha] \tag{21}$$

where the standard state is the charged species at unit activity. With the above substitutions,

$$\frac{\Delta g^o}{\bar{N}RT} = \ell n \ x_1^+ \ (x_c^+)^\alpha \tag{22}$$

In these relations, the chemical potentials are for ionic species, a concept which is tenuous since in the definition

$$\mu_i = \partial G/\partial n_i)_{T,P,n_j \neq i}$$

charge neutrality prevents holding all n_j constant while n_i is varied if species i is charged. For the case of no added salt

Equation (22) yields

$$\frac{\Delta g^o}{\bar{N}RT} = (1 + \alpha) \ln x_1^+ \qquad (23)$$

In order for Equation (23) to yield Equation (20), the value of α must be unity.

This conclusion has two important consequences for theoretical analyses. The first is that since α has not been assumed to be unity in the semiempirical expressions and data analysis of Phillips (18), Molyneaux and Rhodes (35) and others, it is not clear what interpretation can be placed on their results. While it could be argued that the value of α is a way of including the deviations from unity of the activities of the charged species involved, this would require the activity coefficients to vary as the mole fraction to the α power. Perhaps this is why most workers obtain $\alpha = 1/2$, the Debye-Huckel dependence.

The second consequence involves the work of Sexsmith and White (32,33) (see also White, 34) which, when assuming $\alpha < 1$, gives a maximum in the monomeric amphiphile concentration. Using Equations (6), (7) and (22) with mole fractions for activities at all concentrations plus

$$x_c = x_o - \alpha(x_o - x_1)$$

yields the relation

$$x_1 = \left(\frac{x_o - x_1}{\bar{N}}\right)^{1/\bar{N}} \frac{1}{K[x_o - \alpha(x_o - x_1)]^\alpha} \qquad (24)$$

where

$$K \equiv \exp [\Delta g^o/\bar{N}RT] \ll 1$$

At small values of x_o, $x_o \approx x_1$, but at larger values of $x_o \gg x_1$ two limiting cases appear when terms involving fluctuations $(1/\bar{N})$ are ignored

$$x_1 \approx 1/x_o^\alpha K(1-\alpha)^\alpha \qquad \alpha < 1 \qquad (25)$$

$$x_1 \approx x_o^{1/2\bar{N}}/K^{1/2} \qquad \alpha = 1 \qquad (26)$$

Equation (25) is chosen by Sexsmith and White (33) which indicates a rapidly decreasing monomeric concentration with total amphiphile while we would argue for Equation (26) which gives a slowly increasing monomer concentration. Experimental evidence is mixed (17). White's concern was with adsorption maxima as a function of increasing total amphiphile

concentration. There are apparently other arguments to explain this, including micelle exclusion from the surface due
to electrostatic repulsion (37) and mixed surfactant solubilization such as gives surface tension minima (2).

C.   Some Theories for Free Energy Changes Upon Micellization
     Before proceeding to describe the theories for calculating
Δg^o, we wish to point out a significant phenomenological observation previously discussed by Lin and Somasundaran (12).
We, using the tables of Mukerjee and Mysels (38), have found
the critical micelle concentration for amphiphiles with paraffinic tails varies with the number of carbons in the following
way:
a)   for all ionics such as sulfates, sulfonates, alkanoates,
trimethyl ammonium chlorides, and pyridinium bromides, each
additional carbon changes $\ln x_o^+$ within experimental error by
a value of -0.69 ($\pm$ 0.02) with negligible effect of temperature
and added salt concentration (Lin and Somasundaran (12) cited
values essentially the same except for the alkyl ammonium
chlorides);
b)   for nonionics such as oxyethylene-3 alcohols, -6 alcohols,
and n- and c-betaines the variation is from -1.09 to -1.28 although it is constant for each compound;
c)   by contrast, Tanford (39) quotes the results of McAuliffe
(40) for each carbon group changing the alkane solubility, $\ln$
x_w, in water at 25°C by -1.49 ($\pm$ .02) (Lin and Somasundaran
quote a value of -1.39). It is not surprising that the nonionics should show some difference of carbon number effect
with head group and/or perhaps micelle number (mean number of
molecules in the micelle). However, it is quite surprising
that these effects do not appear for the ionics. (The fluctuation effects of the micelle number given by using Equation
(9) are within the uncertainty listed since the variation of
$\bar{N}$ and x_o^+ with carbon number compensate.)
     One explanation which can be advanced is that, except for
small differences in potential energy and in conformational
entropy of the hydrocarbon tails in bulk alkanes compared to
micelles of amphiphiles (and even these should vary proportionally to the carbon number), the carbon number dependence
of micelles and alkane solubility should be the same since it
is caused solely by the "hydrophobic" effect. The variation
to be described is that of

$$\frac{\Delta \ln x}{\Delta n_c} = \frac{\Delta(\Delta g^o/\bar{N}(1+\alpha)RT)}{\Delta n_c} \qquad (27)$$

where $\alpha = 0$ for alkanes and nonionics and $\alpha = 1$ for ionics and
x is x_o^+ for micelles and x_w for alkane solubility. Because of

the small effects mentioned above, the value of the right-hand
side will be more negative for alkane solubility than for micelle formation. As cited above, the value for ionics (-0.69)
is slightly less than one-half that for alkane solubility
(-1.49). While we are unsure of why the nonionics do not have
a value equal to twice that for the ionics, it is possible that
the volume excluded to water by nonionic micelles varies with
the nature of the compound due to differences of penetration
of water around the head groups and this causes differences in
the "hydrophobic" effect.

The above observation has been considered by Tanford, in
a series of articles (41-43) in which a theory due to Tartar
(44) is further developed for micelle formation and size distribution. Tanford separates $\Delta g^o/\bar{N}RT$ into a portion linear
in the carbon number, n_c, a portion which depends upon the
area of hydrocarbon core in the micelle, A_{HM}, plus a portion
dependent only on the area per head group A_{RM}. Tanford's
empirical expression for an ideal solution is

$$-\frac{1}{\bar{N}} \ln x_m + \frac{1}{\bar{N}} \ln x_1^+ = \Delta g^o/\bar{N}RT$$
$$= [-k_1 - k_2 n_c + k_3 A_{HM}] + \sum \delta_i / A_{RM} \quad (28)$$

where the constants k_i are positive, the δ_i are constants and
there may be as many as three different terms in the δ_i sum.
The first group of terms of the right-hand side is the same
as the terms $(\mu_m^o/\bar{N} - \mu_1^o)$ in Equation (15) while the sum
apparently is $(\mu_m^{el}/\bar{N} - \mu_1^{el})$. Tanford identifies A_{RM} with that
of an ellipsoid whose minor axis is that of the flexible hydro-
carbon chain plus 0.03 nm. Tanford's values for k_1 and k_2 are
apparently derived empirically for micelles since they are not
the same as those for alkane solubility.

We prefer the "hydrophobic" concept of Ben-Naim (16), who
indicates that the effect arises from the volume of the solution occupied by the hydrocarbon tail which is excluded from
occupancy by the water. The primary effect is then on the
changes in solvent entropy as opposed to energy, as usually
assumed. To illustrate this, we compare Tanford's form with
that which is given for rigid spheres from scaled particle
theory. (We will show below that they can be in close quantitative agreement.) The hydrophobic free energy change
$\Delta\mu_{HS}/RT$ associated with a sphere of diameter σ being inserted
into a solvent of diameter σ_s is essentially given by

$$\frac{\Delta\mu_{HS}}{RT} = -\ln(1-y) + \left(\frac{\sigma}{\sigma_s}\right) \frac{3y}{2(1-y)} \left[2 + \frac{(2+y)}{(1-y)} \left(\frac{\sigma}{\sigma_s}\right)\right] \quad (29)$$

where $y = \frac{\pi \rho}{6} \sigma_s^3$ and ρ is the solvent molecular density. (We have ignored a very small term which varies as the pressure.) At constant temperature this means that $\Delta\mu_{HS}/RT$ is a quadratic in the ratio of solute to solvent diameter.

$$\Delta\mu_{HS}/RT = a + b(\sigma/\sigma_s) + c(\sigma/\sigma_s)^2 \tag{30}$$

The constants are of the order of 0.5 to 5 with c being largest, so that the last term is the dominant one.

To create a micelle of diameter σ_m from $\bar{N}$ monomers of diameter σ_1, the free energy change per monomer will vary as

$$\frac{\Delta g_{HS}^o}{\bar{N}RT} = \left.\frac{\Delta\mu_m}{\bar{N}RT}\right)_{HS} - \left.\frac{\Delta\mu_1}{RT}\right)_{HS}$$

$$= a\left(\frac{1}{\bar{N}} - 1\right) + b\frac{(\sigma_m/\bar{N} - \sigma_1)}{\sigma_s} + c\frac{(\sigma_m^2/\bar{N} - \sigma_1^2)}{\sigma_s^2} \tag{31}$$

For most ionic amphiphiles values of $\bar{N}$ are 25 or greater and (σ_m/σ_s) is of the order of 5. This means that the only terms of significance are

$$\frac{\Delta g_{HS}^o}{\bar{N}RT} \simeq - [a + b(\sigma_1/\sigma_s) + c(\sigma_1/\sigma_s)^2] + c\left(\frac{\sigma_m^2}{\bar{N}\,\sigma_s^2}\right) \tag{32}$$

For this to coincide with Tanford's concept we see that the first bracketed terms must vary linearly in the carbon number n_c, and there must be a direct correspondence between the hydrocarbon water contact area A_{HM} and σ_m^2. This is precisely what is appropriate for the monomer if we are modelling a cylinder of constant radius r whose length ℓ is proportional to n_c as a sphere of equal area. Thus

$$A_{sphere} = \pi\sigma_1^2 = 2\pi\ell r = A_{cylinder} \tag{33}$$

with

$$\ell = c_1 + c_2 n_c \tag{34}$$

so

$$\sigma_1^2 = c_1' + c_2' n_c \tag{35}$$

The ellipsoidal micelle geometry is close to spherical so the relation between A_{HM} and σ_m^2 may be preserved although they would vary differently with $\bar{N}$.

To add further evidence to our assertion, we examine the results of Stigter (25,26) who has developed a very detailed theory for $\left(\dfrac{\mu_m^{el}}{\bar{N}} - \mu_1^{el}\right)$. When these calculated contributions are subtracted from experimental $\ln x_0^+$ values, he finds a correlation with the amphiphile-water contact areas of monomer and micelle of the form

$$\frac{\Delta g^o}{\bar{N}RT} - \frac{\left(\dfrac{\mu_m^{el}}{\bar{N}} - \mu_1^{el}\right)}{RT} = k_1' - k_2' A_1 + k_3' A_m/\bar{N} \tag{36a}$$

$$= -[a' + b'n_c] + c' \sigma_m^2/\bar{N} \tag{36b}$$

where the constants α', a', etc., are all positive. The second form is our rearrangement of Stigter's equation, and is again entirely consistent with our assertion. Thus the form of the hydrophobic effect for micelle formation is accounted for by rigid body effects.

There are two quantitative tests available. The first is that the constants k_1, a, and a' in Equations (28), (32) and (36b) have values of 1.9, 0.56, 0.25, respectively, differences well within the variation allowed by the fit of data to theory. (For example, Tanford (43) acknowledges that the value of 1.9 is probably too high.) In another test, Table II compares results from the hard sphere model with the "residual" free energy change of Stigter. The micelle radius was calculated from Stigter's formulae

$$\sigma_m = 2\ell[(2\bar{N} + n_1)/3n_1]^{1/2} \tag{37}$$

$$\Delta g^o/\bar{N}RT = -0.249 - 9.94\ell + 33.7\ell^2 (2\bar{N} + n_1)/n_1\bar{N} \tag{38}$$

where

$$\ell = 0.1257(n_c - 1) + 0.28 \quad \text{(nm)} \tag{39}$$

Using Equation (29) with $y = 0.4273$, $\sigma_s = 0.29$nm, to yield $a = 0.557$, $b = 2.238$, $c = 4.743$, Table II gives the monomer radii to obtain the values of $\Delta g^o/\bar{N}RT$ using listed values of $\bar{N}$ from Tartar (45) and n_1 from Stigter. The radii are about two-thirds the value of the radii of the spheres whose areas equal the areas of the cylinders assumed by Stigter perhaps

TABLE II

RIGID-SPHERE RADII FOR EXCLUDED VOLUME EFFECT TO CORRELATE
RESIDUAL (NONELECTROSTATIC) FREE ENERGY
OF MICELLIZATION (STIGTER)

n_c	$\bar{N}^a$	n_1^b	σ_m^c, nm	$-\Delta g^o/\bar{N}RT)^d$	σ_1^e, nm
8	23-24	26.8	2.15-2.195	6.43-6.51	0.500-0.501
9	30-31	32.9	2.46-2.50	7.79-7.84	0.520-0.5205
10	36-50	39.4	2.70-3.07	9.01-9.53	0.538-0.532
11	42	46.7	2.915	10.22	0.555
12	40-100	54.5	2.85-4.15	11.09-12.43	0.568-0.559
14	70-92	72	3.77-4.17	14.08-14.51	0.606-0.600
			Average Increment per-CH_2-group		0.017±0.001

^{a}Range reported by Tartar (45)

^{b}Stigter (26)

^{c}Equation (37)

^{d}Equation (38)

^{e}Equation (32)

due to molecular coiling. However, the results do give an essentially constant increment of 0.017nm in σ_1 with varying carbon number.[1] The calculations are not extremely sensitive to $\bar{N}$ so they are not sensitive to the type of head group. Thus, calculations of the nonelectrostatic contributions to micellization from rigid-body volumes excluded to the solvent (water) appear to be consistent with present data and knowledge. In other solvents, the effect will be significantly

---

[1]It may not truly be of significance, but the increment from odd to even n_c (e.g., 9 to 10) is smaller than from even to odd (e.g., 8 to 9) even though the increment from even to odd or odd to odd is constant. This may be due to the effective diameter of the cylinder being smaller when the line of axial symmetry passes through an even number of units than when it passes through an odd number. Such an explanation has been invoked to explain the variation of melting point with carbon number in normal paraffins.

smaller due to σ_s being larger; this may explain why nonaqueous micellization is of considerably less importance and the values of $\bar{N}$ are much smaller (46).

D. The Effect of Added Salt on Micellization

One final aspect of micellization is concerned with the effect of added salt on the CMC of ionic amphiphiles. Examination of the better data reported by Mukerjee and Mysels (38) for systems such as alkyl ammonium chlorides and bromides, alkyl trimethyl ammonium chlorides, sodium alkyl sulfates and alkyl sulfonates, alkyl pyridinium bromides and potassium alkanoates with added salts such as the sodium halides, potassium nitrate and potassium bromide up to 1M, confirm the relation first given by Corrin and Harkins (47), Hobbs (48) and by Shinoda (49) and described in detail by Mijnlieff (21) and Lin and Somasundaran (12). The data can be reproduced to within the estimated experimental error with

$$\ln \frac{x_1^+}{x_1^{+o}} = K' \ln \left| \frac{\overline{(x_1^+ + x_2)}}{x_1^{+o}} \right| \tag{40}$$

where x_1^{+o} is the CMC with no added salt, x_1^+ is the value with added salt of mole fraction x_2 and K' is a constant independent of the salt whose value is $-0.66 \pm .03$ for anionic amphiphiles and $-0.58 \pm .03$ for cationics.[2] It is remarkable that the values of K' are as consistent as they are, though there is a definite difference between the two charge types when the counterion is of single charge. The only data for a multiply charged salt was for sodium sulfate with sodium dodecyl-sulfate where the correlation was within 10% only up to 0.05M salt. In this case, if ionic strength or counterion concentration were used instead of mole fraction salt, the correlation was better but the value of k'' was -0.40 or -0.47. There needs to be more accurate data of this phenomenon to confirm this relation. Mijnlieff writes the reaction for the neutral species ($M_1 \equiv$ Amphiphilic Salt, $S_2 \equiv$ Added Salt, $M_{MQ} \equiv$ Micelle)

$$\bar{N} M_1 + \bar{Q} S_2 \rightleftarrows M_{MQ} \tag{41}$$

---

[2]Lin and Somasundaran (12) report some values of K different than these but they involve a theoretical assumption. In addition, the uncertainty of the data allows the above range of values that would be consistent with experiment.

and the mass action relation for amphiphile (1) and salt (2) as

$$\bar{N} \ \mu_1 + \bar{Q} \ \mu_2 \rightleftarrows \mu_{MQ} \tag{42}$$

Now for an ideal solution where the added salt has a common ion with the amphiphilic salt

$$\mu_{MQ} = \mu_{MQ}^o \tag{43}$$

$$\mu_1 = \mu_1^o + RT \ \ell n \ x_1 x_c \tag{44}$$

$$\mu_2 = \mu_2^o + RT \ \ell n \ x_2 x_c \tag{45}$$

where $x_c = x_1 + x_2$ is the mole fraction of counterion in the system from both a 1-1 amphiphilic salt and a 1-1 added salt. The relations for other salts would be similar in form but more complex in detail. The equilibrium relation is then

$$\frac{\Delta g^o}{\bar{N}RT} = \frac{\mu_{MQ}^o - \bar{N} \ \mu_1^o - \bar{Q} \ \mu_2^o}{\bar{N}RT} = \ell n \ \{x_1^+ \ x_2^{\bar{Q}/\bar{N}} \ (x_1^+ + x_2)^{(1+\bar{Q}/\bar{N})}\} \tag{46}$$

or

$$\ell n \ [x_1^+(x_1^+ + x_2)] + \frac{\bar{Q}}{\bar{N}} \ \ell n \ [x_2(x_1^+ + x_2)] = \frac{\mu_{MQ}^o - \bar{N} \ \mu_1^o - \bar{Q} \ \mu_2^o}{\bar{N}RT} \tag{47}$$

In the limit $x_2 = 0$, $\bar{Q} = 0$

$$\frac{\mu_{MO}^o - \bar{N}^o \mu_1^o}{\bar{N}^o RT} = 2 \ \ell n \ x_1^{+o} \tag{48}$$

where $\bar{N}^o$ is the micelle number in the absence of added salt and the standard state chemical potential of the micelle without salt μ_{MO}^o may differ from that with salt, μ_{MQ}^o. Again, these are neutral species, not charged.

Finally, this may be rearranged to give

$$\ell n \ (x_1^+/x_1^{+o}) = - \ \ell n \left(\frac{x_1^+ + x_2}{x_1^{+o}}\right) + \frac{\frac{\mu_{MQ}^o}{\bar{N}} - \frac{\mu_{MO}^o}{\bar{N}^o} - \frac{\bar{Q}}{\bar{N}} \mu_2^o}{RT}$$

$$- \frac{\bar{Q}}{\bar{N}} \ \ell n \ [x_2(x_1^+ + x_2)] \tag{49}$$

For the correlation of Equation (40) to hold, the form of the standard state chemical potential must be

$$\frac{\dfrac{\mu_{MQ}^{o}}{\bar{N}} - \dfrac{\mu_{MO}^{o}}{\bar{N}^{o}} - \dfrac{\bar{Q}}{\bar{N}}\mu_{2}^{o}}{RT} = \frac{\bar{Q}}{\bar{N}}\,\ln x_{2} + \left|\frac{\bar{Q}}{\bar{N}} + 1 + K'\right|\,\ln\,(x_{1}^{+} + x_{2})$$
$$- (1 + K')\,\ln x_{1}^{+o} \qquad (50)$$

Mijnlieff shows that the reciprocity relation

$$\left.\frac{\partial\mu_{1}}{\partial n_{2}}\right)_{T,P,n_{1}} = \left.\frac{\partial\mu_{2}}{\partial n_{1}}\right)_{T,P,n_{2}} \qquad (51)$$

leads to [with correction of a sign error in his Equation (26)]

$$\frac{\bar{Q}}{\bar{N}} = - \frac{(1+K')}{2 + (1-K')x_{1}^{+}/x_{2}} < 0 \qquad (52)$$

In the limit $x_{2}/x_{1}^{+} \ll 1$

$$\frac{\bar{Q}}{\bar{N}} = - \frac{(1 + K')x_{2}}{(1 - K')} \qquad (53)$$

This equals zero when $x_{2} = 0$. In the limit $x_{2}/x_{1}^{+} \gg 1$, $\frac{\bar{Q}}{\bar{N}} =$ -0.16 for anionics and -0.21 for cationics. The fact that it is constant, but different for the charge types must be of significance.

Finally, after some rearrangement

$$\frac{\dfrac{\mu_{MQ}^{o}}{\bar{N}} - \dfrac{\mu_{MO}^{o}}{\bar{N}^{o}} - \dfrac{\bar{Q}}{\bar{N}}\mu_{2}^{o}}{RT} = (1+K')\left\{\frac{(1-K')x_{1}^{+}/x_{2}}{2 + (1-K')x_{1}^{+}/x_{2}}\,\ln x_{2}\right.$$
$$\left. + [1 + (1-K')x_{1}^{+}/x_{2}]\,\ln\,(1 + x_{1}^{+}/x_{2}) - \ln x_{1}^{+o}\right\}$$
$$(54)$$

Thus, theories for the standard state Gibbs free energy change should be of the above form. Again when $x_{2}/x_{1}^{+} \gg 1$ or high salt concentration, Equation (54) becomes

$$\frac{\dfrac{\mu_{MQ}^{o}}{\bar{N}} - \dfrac{\mu_{MO}^{o}}{\bar{N}^{o}} + \dfrac{(1 + K')\mu_{2}^{o}}{2}}{RT} = - (1 + K')\ln x_{1}^{+o} = \text{constant} \qquad (55)$$

This implies that μ_{MQ}^{o} varies proportionally with $\bar{N}$ since all other terms are constant.

For Stigter's expression in Equation (38), this is true when $2\bar{N} \gg N_1$ although with the values for n_i he quotes ($n_1 = 54.5$ for C_{12}), this does not happen in the range of salt concentrations for which Equation (55) applies. The values of Tanford for the same quantity are considerably less (for C_{12} the value is 20) implying that this theory may need to be modified. For the rigid sphere expressions, Equation (55) implies that σ_m^2 varies essentially with $\bar{N}$. Such a variation can be developed within the framework of Stigter or Tanford with $N_1 \ll \bar{N}$.

We have not evaluated Equation (54) for the rigid sphere theory, mainly because of the probable sensitivity of the results to attempting to model the spherocylindrical monomers and oblate ellipsoidal micelles as spheres. (Accounting for shape awaits the development of a theory for mixtures of rigid bodies of different shapes.) However, we have made calculations of the quantity $\ln x_o^+ - \left| \dfrac{\mu_m^{-el}}{\bar{N}RT} - \dfrac{\mu_1^{el}}{RT} \right|$ for aqueous sodium alkyl sulfate solutions containing sodium chloride using values of $\bar{N}$ from Emerson and Holtzer (28) and electrostatic theories from Emerson and Holtzer (29), Mukerjee (30) and Stigter (25, 26). The results are quite different from each other but a significant conclusion is that the values from Stigter (Equation 38) are close in magnitude and vary precisely the same with added salt as do those from Tanford (41) using the first bracketed terms of Equation (28). Yet, because neither $\bar{N}$ variation appears to be compatible with Equation (55) we are unsure of the proper interpretation of this.

The purpose of the above discussion has been to indicate the electrostatic effects can be separated from "hydrophobic" effects in micellization, and even, such as in added salt phenomena, need not always be dealt with in detail if simple measurements are available. In addition, the "hydrophobic" effects appear to be handled quite well by rigid body theories for the excluded volume effect. We assert that such concepts are likely to be valid in developing theories for microemulsion formation and anisotropic phase transformations. Since this has not been done, however, and little data on well-

defined systems are available, in the next few sections we
merely review the objectives of certain important models and
suggest the directions in which we feel further advances
might lie.

VII. MICROSTRUCTURED PHASES

A. Theory of Amphiphile Microstructure Transitions

While there are many theories of pure component liquid
crystals [e.g., Saupe (50); Wojtowicz (51); Wadati and Isihara
(52); Chandrasakhar and Madhusudana (53)], because these do
not involve the kind of dispersed-phase/continuous-phase structure of interest in oil recovery systems, we focus on those
which occur in only binary and multicomponent systems. An
example is the theory of binary, lyotropic liquid-crystal
phase-transitions by Parsegian (54). The objective was to
calculate the geometry and equilibrium transition from the
fibrous "middle" phase to the lamellar "neat" phase as a
function of concentration and temperature for ionic amphiphiles in water. The basis was the difference of molar Gibbs
free energies of the two structures where the major contributions were the "charging" energy g_c, calculated by using
standard electrostatic formulae for the Poisson-Boltzmann
equation for polyelectrolyte cylinders (55) and planes, and
the "surface" energy, g_s, which is related through a "surface
energy coefficient" to the cylindrical radius a_c or lamellar
thickness a_p and volume fraction of amphiphile ϕ_v (by assuming
the surface area per head group is a constant for each phase
type). Thus, only external energetics and a single term for
all excluded volume effects are included.

An equation for equilibrium is

$$\left.\frac{\partial g}{\partial a}\right)_{T,\phi_v} = 0 \qquad (56)$$

where $g = g_c + g_s$ for both cylinders and planes. This means
that a_p and a_c can be calculated for each geometry at all
compositions. Comparisons with experimental X-ray data of
the variation of microphase geometry for sodium and potassium
normal fatty acid salts $(C_{12}-C_{18})$ were quite good (1-2%). The
values of the "surface energy coefficients" for planar systems
were nearly independent of substance and temperature, although
use of a constant value would lead to errors up to 5% in the
values of a_p and a_c. The g_c values were about one-half the
interfacial tension between paraffins and water, but apparently no physical meaning can be attributed to them. The
transition composition between the two structures was calculated from the difference between their free energies

which changed sign as a function of volume fraction ϕ_v. Figure 2 presents a replotting of the graphs of Parsegian showing the variation of the free energies, the stable structures and the transition region. (The values of g_s in both Parsegian's graphs and ours have arbitrary constants added.) Since the dependence of free energy is linear in volume fraction, no two-phase region appears, though it is observed.

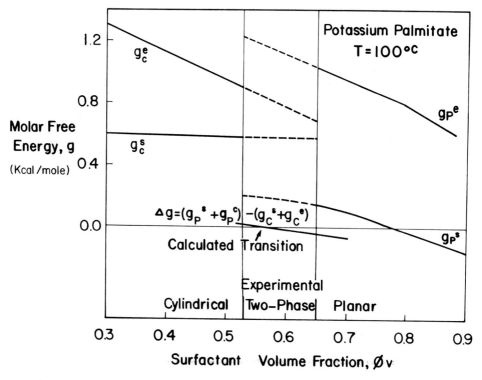

Fig. 2. *Electrostatic and Surface Contributions to Free Energies for Cylindrical and Planar Aqueous Potassium Palmitate Aggregates at 100°C (after Parsegian (54))*

The success of this theory seems to us to be of importance for guidance in prediction of liquid crystal transitions in multicomponent systems, particularly since it is the only one available for such systems. It is simple because it ignores all entropic effects, assumes the internal energetic effects are functions only of geometry, and lumps the external energetic effects and excluded volume effects into a single term which is characterized by one quantity that is insensitive to conditions. Such simplicity is unlikely to

be preserved when the electrolyte concentration is independently varied, but the first order effects of this might be
taken into account by the electrostatic model. While no two-
phase or three-phase regions could be found with this model
as it stands, inclusion of the proper entropic effects may
rectify this situation.

B.   The Phase Rule for Fluid Microstructured Systems

It has apparently always been assumed that a microstructured fluid of a dispersed phase within a continuous phase
such as in microemulsions and lyotropic liquid crystals containing three or more components is a single phase according
to the phase rule. As near as we can tell, no one has actually investigated if this is true. In fact, our discussion
below indicates that this assumption should not be made and
that at least one anomalous degree of freedom exists in some
ionic ternary systems.

The phase rule constrains the number of phases which can
coexist in a system made up of a mixture of components, subject to certain interactions with the surroundings across
boundaries. As Redlich (56) shows, the interactions are those
which arise from a generalized force (pressure, temperature,
surface tension, chemical potential, etc.) acting on a system
through a mode of interaction (mechanical, thermal, electrical, etc.) which changes a generalized coordinate (volume
(times-1), entropy, surface area, number of molecules, etc.).
The number of independent intensive variables to characterize
a system is the minimum number of generalized forces which
must be brought into balance between the surroundings and the
gauges measuring the forces of the system.

To describe an arbitrary state of a system of C components, ϕ bulk phases [regions of homogeneous structure (Defay
et al., 2)], Ψ surface phases and S types of surface, the
total number of variables is the number of orthogonal "work"
modes of interaction (characterized by a generalized force
for each phase), plus the single thermal mode of interaction,
plus the independent mole fractions of each of the components
in each of the bulk and surface phases plus any additional
geometric variables for characterizing mechanical equilibrium,
such as mean radius of curvature of each surface phase. For
nonreacting systems with only mechanical work whose surface
phase areas and number of molecules are variable by an external mode of interaction, this leads to a total of $\phi + \Psi +
1 + (C-1) (\phi + \Psi) + \Psi = C(\phi + \Psi) + \Psi + 1$ variables. However,
there are Ψ relations associated with mechanical equilibrium
between bulk phases across the interfaces, Ψ–S mechanical
equilibrium relations associated with coexistent surface
phases and $C(\phi + \Psi - 1)$ relations associated with equilibrium

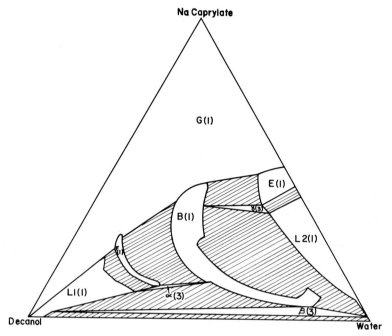

Fig. 3.   *Tertiary Composition Diagram for the Water-Decanol*
Sodium Caprylate at 25°C showing 1, and 3-phase
regions.   (After Ekwall et al. (59); shaded area
indicates 2-phase regions.)

the phase rule, because fluctuations of the number of molecules in a micelle impart an additional degree of freedom
within the context of Hill's small systems analysis (5).
Fluctuations are the likely explanation for any anomalous
variation in multiphase systems.

   The system described above represents the easiest test
for anomalous variance in fluid microstructured systems.
(The same test can be made of the microstructured phases in
a binary system at fixed pressure or temperature but since
they have the dimensions of micelles, fluctuations are assured,
and, in fact, the micelle analysis of Hall and Pethica is appropriate for all aspects of binary systems containing aggregates.) The microscopic phases will contain all three components at different mole fractions. One curved interface
separates the continuous phase from the dispersed spherical
(L1, L2) or fibrous phase (E,F) and the lamellar phase (B)
is made up of a series of alternating phases separated by
a plane interface. Table III shows the variance as calculated from Equation (58) for the various regions of the diagram (fixed T,P)

with respect to transfer of molecules from one phase to another. Thus the net number of independent variables is

$$f = C(\phi + \Psi) + \Psi + 1 - (2\Psi\text{-}S) - C(\phi + \Psi - 1) = 1 + C\text{-}\Psi + S \tag{57}$$

For some interfacial phases out of the total Ψ being
plane, there are up to $(\phi\text{-}1)$ additional constraints which
equate the pressures of the bulk phases. A general formulation for our purposes is that for S_c single-phase curved interfaces out of the total of S $(= \phi - 1 = \Psi)$ single-phase
interfaces. Equation (57) is then

$$f = 2 + C - \phi + S_c \tag{57a}$$

We are particularly interested in the variance of lyotropic liquid crystals as described by Winsor (57,58) and
Ekwall et al. (59) since similar behavior is observed in chemical systems for oil recovery. A typical example which has
been studied extensively is the ternary water-decanol-sodium
caprylate which shows a variety of single, double and triple
phase behavior which, at constant temperature and pressure,
superficially acts according to the variance given by f = 3 -
ϕ where the ϕ phases include molecular solutions, micelles,
isotropic microemulsions or micellar solutions, and lyotropic
liquid crystals of the cubic, fibrous or cylindrical, and
lamellar forms.
     The ternary composition diagrams (at fixed temperature
and pressure) such as Figure 3[3], which Winsor and Ekwall
present have 1) triangular regions where three "phases" of
fixed composition are the vertices, 2) areas bounded by the
compositions of two "phases" whose compositions are connected
by tie lines, and 3) regions of a single "phase" whose composition may be arbitrarily varied. However, as Hall and
Pethica (17) have pointed out for micelles, these "phases"
seem hardly to fit the description of homogeneous structure,
and, furthermore, the last forms have significant surface
areas with high degrees of curvature. The question remains
as to whether an analysis including the interfaces is consistent with the observed behavior. As Hall and Pethica
point out, micelles may not be treated as separate phases in

---

[3]For this analysis we ignore the regions involving some
phases, G, (solid crystalline and hydrated sodium caprylate
with fibrous structure) and "C" for which some controversy of
structure exists, although it appears to behave as the other
microscopic phases do. We accept Winsor's analysis that
regions B and D are the same, at least for our purposes.

TABLE III

VARIANCE OF MULTIPHASE REGIONS OF WATER-
DECANOL-SODIUM CAPRYLATE AT 20°C[a]
(Refer to Figure 3)

Three-microstructured Phases

Region	Phases	ϕ	S_c[b]	f(Eq. 51)[c]	f(exp)[a]
α	L1,B,F	6	2	-1	0
β	L1,L2,B	6	2	-1	0
γ	L2,B,E	6	2	-1	0

Two-microstructured Phases

	L1,B	4	1	0	1
	L2,B	4	1	0	1
	E,B	4	1	0	1
	F,B	4	1	0	1
	L1,L2	4	2	1	1
	L2,E	4	2	1	1
	L1,F	4	2	1	1

One-microstructured Phase

	L1	2	1	2	2
	L2	2	1	2	2
	E	2	1	2	2
	F	2	1	2	2
	B	2	0	1	2

[a] Data of Ekwall (59)

[b] Values of S_c for various phases are as follows: Spheres,
L1(1), L2(1); Lamellae, B(0); Cylinders, E(1), F(1).

[c] Equation (58) $f = 3 - \phi + S_c$

$$f = 3 - \phi + S_c \qquad\qquad (58)$$

All the "three-phase" regions (α, β, γ) yield an f of -1, instead of the observed value of zero except that the single additional degree of freedom brought in by fluctuations precisely accounts for the observed phenomena.[4] Note that Hill's analysis of small systems brings in only one additional degree of freedom, even though there may be several types of fluctuations involved. For the "two-phase" and "one-phase" regions, two values of f are obtained from Equation (58) depending upon whether the lamellar phase (B) is present or not. When it is not, Equation (58) yields the observed value. Otherwise, an additional degree of freedom is required. It would appear that only the lamellar phase yields an anomalous variance which then must be attributed to fluctuations.

It is of interest to speculate about the nature of possible fluctuations in all aspects of the system, particularly in view of the different behavior encountered for systems with and without lamellar phases. In general, fluctuations are "significant" at the 10% level. Fluctuations in size vary inversely as the square root of the number of molecules, so aggregates should contain of the order of 100 or fewer molecules. For this reason, it seems to us unlikely that the fluctuations are those of the size of one or more of the microscopic phases. For example, a sphere of 10 nm radius contains of the order of 10^5 cylindrical molecules of 1 nm length and 0.1 nm radius and a fibrous cross-section or lamellar thickness of 10 nm radius contains of the order of 3000 of the same molecules. Further, fluctuations of the number of surfactant molecules and counterions (as adsorbed species) while greater, are only marginal. For spheres of 10 nm radius with surfactant molecular cross-sectional area of 0.2 nm^2 the number of molecules per sphere is about 6000. The number of surfactant molecules around the perimeter of a cylindrical cross-section of 10 nm radius is of the order of 150. The fraction of adsorbed counterions is of the order of one-half that of the surfactant molecules. Thus, there might be significant fluctuations of charge on cylinders. However, the anomalous variance is found in systems without cylinders.

We believe that the important fluctuations are in the surface area of the lamellar phases due to deformations away

---

[4]This should be contrasted with surfactant systems, containing salts and cosolvents such as alcohols, described by Anderson et al. (60) and Robbins (61), among others, where the actual number of components is four or more, and the variance is at least zero, a possible value. However, it also seems likely that significant fluctuations occur in these systems.

from the time average planar shape. This is the only con-
jecture which is consistent with experiment. The primary
basis for the magnitude of the fluctuations is the work of
Lovett (62). He showed that fluctuations of the planar interface between a liquid and its vapor increase as the critical
temperature is approached and the interfacial tension decreases toward zero. In fact, the square root of the second
moment (standard deviation) is of the order of 1-2 nm even
when the interfacial tension is as high as 10^{-3} N/M (1 dyne/
cm) and becomes 10 nm before it reaches 10^{-6} N/M which apparently is of the order of the tensions in liquid crystal
systems.

While the analysis of Lovett for spheres indicates similar magnitude for curved interfaces, his development assumed
all possible values of all modes of deformations of the
spheres which yields too much freedom and is inconsistent
with the essential incompressibility of the liquid. It would
be of interest to reexamine this analysis to see if the fluctuations of the curved interfaces are computed to be significant (which the present development indicates they are not)
and if they can be experimentally observed by such methods
as inelastic scattering.

C. Theories of Immiscible Microstructured Phase Formation

In these cases, the dispersed phase is generally spherical and of the order of 1-10 nm in radius. The major contributions from the literature are by Reiss (63), Ruckenstein
and Chi (64) and Adamson and coworkers (65-67). Reiss ad-
dressed the general problem of explaining the stability of
very small dispersed phases which are energetically unfavored
because of large surface area but may have lower free energy
because the system entropy is then greater than that for two
bulk liquids. This could hold for either oil-in-water or
water-in-oil systems. Ruckenstein and Chi considered an oil
or water dispersed phase with surface charge, while Adamson
et al. evaluated equilibria between a bulk aqueous phase and
an oil phase having a dispersed aqueous phase containing ionic
amphiphiles and a solvent alcohol. The latter model is appropriate only for the multiphase system and cannot describe
single microstructured phases.

Reiss attempted to determine if the existence of dispersed phases can be justified thermodynamically. The free
energy difference between the dispersed and bulk systems is
calculated from statistical mechanical partition functions
for the molecules in the two cases. The contributions are
divided into 1) a surface free energy part, related to the
size of the droplets and the interfacial tension, 2) a dispersed phase part related to the fluctuation of the droplet

Gibbs dividing surface relative to the droplet center of mass (this subtlety has been pointed out by Reiss (63) in work concerning nucleation theory), and 3) an entropic part due to configuration of the dispersed droplets which is determined from rigid sphere expressions.

The results given by Reiss came from his final formula (which contains several more approximations concerning concentrations, molecular volume, etc.) relating the equilibrium number of droplets in the dispersed phase, n, to the interfacial tension, γ

$$\gamma = \frac{3.74}{n^{2/3}} [1.5 \ln (n) - 1.49] \tag{59}$$

for molecules whose volume is 10^{-1} nm^3. Figure 4 shows the size range for stable droplets as a function of interfacial tension. The important feature is that the interfacial tension is small but positive and is in the range of 10^{-3} - 10^{-4} dynes/cm to obtain dispersed phase droplets of 5-10 nm radius. These are both in the empirically observed range for chemical oil recovery systems. Thus, while ignoring all changes in molecular configurations and energies associated with dispersion, except those which can be lumped into the interfacial tension, the theory contains the essentially excluded volume effect calculated from a simple formula for electrically neutral systems of rigid bodies. As discussed below, this represents a significant improvement in calculational convenience over the complexities of Ruckenstein and Chi (64) to handle the same effect.

This latter model is much more detailed in scope but it does describe more phenomena such as the phase inversion from oil-in-water to water-in-oil systems as a function of increasing moles of oil. The surface contributions to free energy include an "uncharged interfacial free energy" which is left as one parameter, and the Debye-Huckel free energy of formation of surface double layers in ionic systems. The latter relates the droplet size to the surface potential which is set as an additional parameter. The ionic strength is also allowed to be variable. The energetics of the droplets are mainly accounted for by a van der Waals assumption of liquid structure and the double layer repulsion energy formula of Verwey and Overbeek (68) which involve the same quantities as those above. The van der Waals attraction between droplets were examined within the framework of Ninham and coworkers (69) but is ignorable relative to the electrostatic effects. This might not be the case in nonionic systems. The entropic effect is evaluated from an approximate

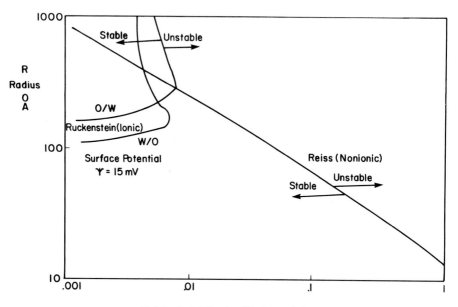

Fig. 4.   *Size - Interfacial Tension Relations for Stable and Unstable Dispersed Fluid Spheres as calculated by Reiss (63) for uncharged species and by Ruckenstein and Chi (64) for charged spheres of fixed surface potential*

geometric probability analysis of droplet configurations. However in contrast to the work of Reiss, only upper and lower bounds to this effect were obtained.

Figure 5 shows the qualitative results of Ruckenstein and Chi (64) for the Gibbs free energy charge of dispersion, Δg_d, as a function of droplet size R, and uncharged surface free energy, f_s. The figure shows three curves depending on the value of f_s. If it is too large, no dispersed phase is stable. If it is very small, a stable dispersion of small droplets can exist. Finally, for an intermediate value, a metastable emulsion of large droplets can exist. These three cases are found in nature, indicating the basic validity of the model.

Quantitatively, the model also predicts the actual size range of stable droplets. Figure 4 shows those sizes versus f_s for a particular value of the surface charge. The size

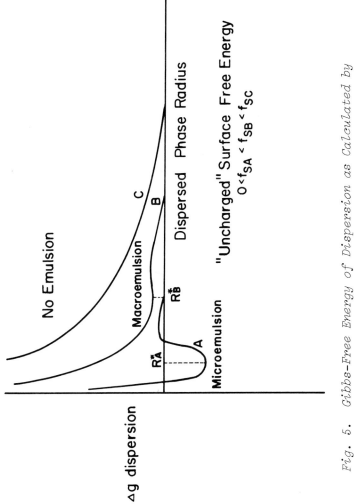

Fig. 5. *Gibbs-Free Energy of Dispersion as Calculated by Ruckenstein and Chi (64) Showing Qualitative Size Behavior for Various "Uncharged" Surface Free Energy*

reaches a lower bound presumably because the surface charge density becomes too great, but in any case, the values are within the usually observed range.

As a result of the parametric studies by Reiss and Ruckenstein and Chi, it can be confidently assumed that micro-emulsions are stable because the entropy of dispersion lowers the free energy sufficiently to overcome the energy of inter-facial formation. However, to establish quantitative models to predict behavior in more complex systems requires a change of parameterization from interfacial free energy and surface charge, which are functions of temperature and composition to molecular quantities which are independent or easily corre-latable with variables of state.

D. Theories of Multiphase Equilibria Involving Micro-structured Phases

Probably the first quantitative model for microstructured phase equilibria was that of Adamson (65) which was further explored by Tosch et al. (66), and by Levine and Robinson (67). It deals with the particular case of multiphase systems where the presence of an ionic amphiphile causes formation of an aqueous microphase within a continuous oil phase, and excess water in the system forms an external aqueous phase which solubilizes little oil and surfactant. A strongly ionized salt such as NaCl and a cosolvent such as isopropyl alcohol are also present, the former being essential to the multiphase system, the latter not.

One fundamental equation for the equilibrium is based on the equality of mean ionic chemical potentials in the dispersed and external aqueous phases, along with the charge neutrality condition. This equation provides the relation to obtain the aqueous salt concentration in the dispersed phase from the total surfactant and external salt concentration when a value for the activity coefficient ratio is assumed or calculated from theory. For completely ionizing amphiphile of formula $C_{\nu_C} S_{\nu_S}$, and a completely ionizing salt of formula $C_{\nu_C} A_{\nu_A}$ the equation becomes

$$q[(\nu_C n_{AM} + \nu_G n_{SM})^{\nu_C} n_{SM}^{\nu_A}]^{1/(\nu_C + \nu_A)} = n_{SE}^{\nu_C/(\nu_C + \nu_A)} \nu_C \quad (60)$$

where the n's are numbers of moles, the q is the ratio of mean ionic activity coefficients (mole fraction scale) in the dispersed to external phases, and the subscripts AM, SM, and SE refer to amphiphile in microstructured phase, salt in microstructured phase and salt in the external phase. Given the ratio of phase volumes, overall surfactant amount (assumed all

in the dispersed phase) and external salt amount, the amount
of salt in the dispersed phase can be calculated. Tosch et al.
(66) show that the equation describes their experimental data
fairly well up to 0.5 molar Na_2SO_4 with alkyl naphthalene
sulfonates and a refinery oil of an average carbon number of
nine if a particular value of q is assumed.

The second fundamental equation of Adamson is an "osmotic
pressure–LaPlace balance" equating the force from the interfacial tension to the force on the dispersed phase from the
osmotic pressure caused by the electrolyte concentration differences in the dispersed and external phases. While Adamson
assumed the simplest possible relation appropriate for an
undissociated amphiphile or uniform charge distribution
throughout the dispersed phase, Levine and Robinson (67) attempted to quantitatively improve upon this approximation.
In either case, there results a relation between the surface
free energy and the concentrations of water, salt and amphiphile in the various phases. While Tosch et al. (66) could
not compare their results directly with experiment, since the
surface free energy is not directly measurable, their equations imply relationships similar to those cited by Robbins
(61) for nonionic surfactants. That is, the ratio of the
amount of dispersed water to surfactant in the microstructured
phase is fixed by the salinity of the external phase. While
the relation is not unique because it depends on the interactions of the surfactant with the aqueous dispersed phase
containing salt, nevertheless for a given system, the dispersed phase salt content itself is uniquely related by
Equation (60) to the external salt concentration and the interfacial tension is a specific function of the salinity of
the dispersed phase. Robbins' work shows precisely such a
relation and indicates to us the basic validity of Adamson's
approach. Unfortunately, it is limited to the multiphase
situation only and is of lesser practical value because of
the unknown interfacial tension parameter.

Much more recently, Robbins (61) has developed a theory
to correlate the phenomena associated with multiphase pseudo
ternary systems (see Reed and Healy (70)). The basis is the
geometry of a dispersed phase which has an amphiphile at its
interface with the continuous phase. If the hydrophilic head
and lipophilic tail volume and "compressibility" (as determined from the structure and conditions of the system) are
such that the head groups can be packed closer together than
the tails can, the aqueous phase will be dispersed. Otherwise
the oil phase will be dispersed. This analysis represents a
quantitative correlation of the "R-theory" of Winsor (57) and
appears to be potentially quite useful in predicting properties, such as interfacial tension, of these solutions from

the surfactant structure and very few simple measurements of
relative volumes of microemulsion and external phases as functions of temperature and salinity. Much of this work has not
yet been published but it should be looked for because, of the
several theories we have described here, it is the only broadly applicable one which does not require values of dispersed
phase interfacial tension to make predictions.

VIII. CONCLUSIONS

The above discussion has covered considerable breadth of
phenomena and theory. We have attempted to emphasize the
general relationships which must be considered in developing
models and theories for systems containing amphiphilic compounds in addition to reviewing the successful theories for
micellization, liquid crystal phase transitions, microemulsion
formation and microstructured phase equilibria. These include
the dependence of the CMC on carbon number and added salt concentration and the application of rigid-body theories for the
excluded volume (hydrophobic) effect in micelles and the entropy of dispersion in microemulsions. Finally, it may be
that the anomalous phase rule behavior in some microstructured
fluid systems is caused by fluctuations of the lamellar interface.

IX. ACKNOWLEDGEMENT

The authors are grateful to Professor D. O. Shah for
helpful discussions and encouragement and to the Industrial
Consortium of the Improved Oil Recovery Project at the
University of Florida for financial support.

X. NOTATION

A_1	surface area of monomer in solution, nm^2
A^1	surface area for cylindrical model of monomer, defined by Equation (33), nm^2
A_{HM}	surface area of hydrocarbon core of micelle, nm^2
A_M	surface area of micelle, nm^2
A_{RM}	surface area per head group in micelle, nm^2
A_{sphere}	surface area for spherical model of monomer, defined by Equation (33), nm^2
a	constant in Equation (30)
a_c	cylinder radius for fibrous liquid crystals, nm
$a_c(m)$	activity of counterions in micelle
$a_c(s)$	activity of counterions in solution
a_c	thickness of lamellar liquid crystals, nm
a_p	activity of amphiphile in solution
$a^\dagger$	constant in Equation (36b)
b	constant in Equation (30)
b'	constant in Equation (36b)

C   number of system components
c   constant in Equation (30)
c_1, c_2 constants in Equation (34), nm
c   constant in Equation (36b)
c_1', c_2' constants in Equation (35), nm^2
f   variance
f_s uncharged surface free energy, J/mol.
G total Gibbs free energy, J/mol.
g   molar Gibbs free energy of transition between fibrous and lamellar liquid crystal phases, J/mol.
g_c "charging" energy contribution to the Gibbs free energy of transition between fibrous and lamellar liquid crystal phases, J/mol.
g_s "surface" energy contribution to the Gibbs free energy of transition between fibrous and lamellar liquid crystal phases, J/mol.
Δg^o standard state Gibbs free energy of micelle formation, defined by Equation (1), J/mol.
Δg_d Gibbs free energy change on dispersion, J/mol.
Δg_{HS}^o "hydrophobic" standard Gibbs free energy of micelle formation calculated from hard sphere model, defined by Equation (31), J/mol.
Δh^o standard state enthalpy change on micellization, defined by Equation (10a), J/mol.
K   equilibrium constant for micelle formation, Equation (24)
K'   constant in Equation (40)
k_1, k_2, k_3 constants in Equation (28)
k_1', k_2', k_3' constants in Equation (36a)
l cylinder length, nm
$\bar{N}$ average micelle number
$\bar{N}^o$ average micelle number in absence of added salt
n   equilibrium number of droplets in dispersed phase of microemulsions
n_{AM} number of moles of amphiphile in microstructured phase
n_{SM} number of moles of salt in microstructured phase
n_{SE} number of moles of salt in external phase
n_c number of carbon atoms in monomers chain
n_i number of moles of species i
n_1 Stigter's number of amphiphile molecules in a spherical micelle
P   pressure, N/m^2
Q   number of counterions "bound" to micelle
q   ratio of mean ionic activity coefficients in the dispersed to external phases
R   gas constant, J/mol. °K
R   droplet radius, nm

S	number of types of surface
S_c	numbers of curved interfaces
Δg^o	standard state entropy change on micellization, defined by Equation (11a), J/mol. °K
T	absolute temperature, °K
x_c	mole fraction of counterions
x_n	mole fraction of micelles
x_w	mole fraction of alkanes in water
x_o	total amphiphile mole fraction, Equation (2)
x_1	monomer mole fraction
y	reduced density, Equation (29)

Greek Symbols

α	apparent fraction of micelle amphiphiles whose change is neutralized by bound counterions
γ	interfacial tension, J/m^2
γ_1	activity coefficient of ionic amphiphile in solution
δ_i	constant in Equation (28), nm^2
$\mu_c(m)$	chemical potential of counterions in micelle, defined by Equation (17), J/mol.
$\mu_c(s)$	chemical potential of counterions in solution defined by Equation (17), J/mol.
$\mu_c^o(m)$	standard state chemical potential of counterions in micelles, J/mol.
$\mu_c^o(s)$	standard state chemical potential of counterions in solution, J/mol.
μ_m	micelle chemical potential, defined by Equation (14), J/mol.
μ_{MQ}	chemical potential of micelle in presence of added salt, J/mol.
μ_m^o	standard state uncharged micelle chemical potential, J/mol.
μ_{MO}^o	standard state chemical potential of the micelle in absence of salt, J/mol.
μ_{MQ}^o	standard state chemical potential of the micelle in the presence of added salt, J/mol.
μ_m^{el}	chemical potential associated with charging the micelle from zero to full value, J/mol.
μ_1	monomer chemical potential, defined by Equation (13), J/mol.
μ_1^o	standard state uncharged monomer chemical potential, J/mol.
μ_1^{el}	chemical potential associated with charging a monomer from zero to full value, J/mol.
$\Delta\mu_{HS}$	"hydrophobic" chemical potential charge upon addition of a solute molecule to a solvent, J/mol.
σ	hard sphere solute diameter, nm
σ_m	hard sphere micelle diameter, nm

σ_s hard sphere solvent diameter, nm
σ_1 hard sphere monomer diameter, nm
ϕ_c denotes "ideal colligative property"
ϕ number of bulk phases
ϕ_v volume fraction of amphiphile
ψ number of surface phases

Superscripts
o   in absence of added salt
+   value at the critical micelle concentration

XI.   LITERATURE CITED
1.   Gibbs, J. W., "The Scientific Papers of J. Willard Gibbs;
       Volume One: Thermodynamics," Dover Publications, New
       York, N.Y., 1961.
2.   Defay, R., Prigogine, I., Bellemans, A. and Everett, D.
       H., "Surface Tension and Adsorption," John Wiley and
       Sons, New York, N.Y., 1966.
3.   Melrose, J. C., "Thermodynamic Aspects of Capillarity,"
       Ind. Eng. Chem. 60, 53 (1968).
4.   Melrose, J. C., "Thermodynamics of Surface Phenomena,"
       Pure Applied Chem. 22, 273 (1970).
5.   Hill, T. L., "Thermodynamics of Small Systems," Vol. 1
       and 2, Benjamin, New York, 1963.
6.   Buff, F. P., "Spherical Interface. II. Molecular
       Theory," J. Chem. Phys. 23, 419 (1955).
7.   Buff, F. P., "Curved Fluid Interfaces I. The General-
       ized Gibbs-Kelvin Equation," J. Chem. Phys. 25, 146
       (1956).
8.   Toxvaerd, S., "Statistical Mechanics of Surfaces" in
       "Statistical Mechanics" (K. Singer, ed.), 2, Chap. 4,
       Chemical Society (London), 1975.
9.   Gubbins, K. E. and Haile, J. M., "Molecular Theories of
       Interfacial Tension," in "Improved Oil Recovery by
       Surfactant and Polymer Flooding" (D. O. Shah and R. S.
       Schechter, eds.), Academic Press, New York, in press,
       1976.
10.   Mukerjee, P., "The Nature of the Association Equilibrium
       and Hydrophobic Bonding in Aqueous Solutions of Associa-
       tion Colloids," Advan. Colloid Interface Sci. 1, 241
       (1967).
11.   Anacker, E. W., "Micelle Formation of Cationic Surfact-
       ants in Aqueous Media," in "Cationic Surfactants"
       (E. Jungermann, ed.), Chap. 7, Marcel Dekker, New York,
       1970.
12.   Lin, I. J. and Somasundaran, P., "Free-Energy Changes on
       Transfer of Surface Active Agents between Various Col-
       loidal and Interfacial States," J. Colloid Interface
       Sci. 37, 731 (1971).

13. Poland, D. C. and Scheraga, H. A., "Hydrophobic Bonding and Micelle Stability," J. Phys. Chem. 69, 2431 (1965).

14. Poland, D. C. and Scheraga, H. A., "Hydrophobic Bonding and Micelle Stability; The Influence of Ionic Head Groups," J. Colloid Interface Sci. 21, 273 (1966).

15. Debye, P., "Light Scattering in Soap Solutions," Ann. N.Y. Acad. Sci. 51, 575 (1949).

16. Ben-Naim, A., "Statistical Mechanical Study of Hydrophobic Interaction. I. Interaction between Two Identical Nonpolar Solute Particles," J. Chem. Phys. 54, 1387 (1971).

17. Hall, D. G. and Pethica, B. A., "Thermodynamics of Micelle Formation," in "Nonionic Surfactants" (Martin J. Schick, ed.), Chap. 16, Marcel Dekker, New York, 1970.

18. Phillips, J. N., "The Energetics of Micelle Formations," Trans. Faraday Soc. 51, 561 (1955).

19. Hall, D. G., "Exact Phenomenological Interpretation of the Micelle Point in Multi-component Systems," Trans. Faraday Soc. 68, 668 (1972).

20. Chung, H. W. and Hielweil, I. J., "A Statistical Treatment of Micellar Solutions," J. Phys. Chem. 74, 488 (1970).

21. Mijnlieff, P. F., "Thermodynamics of Micellar Equilibrium for Ionic Detergents. An Alternative Description and Some Relations Derived from it," J. Colloid Interface Sci. 33, 255 (1970).

22. Stigter, D., "On the Adsorption of Counterions at the Surface of Detergent Micelles," J. Phys. Chem. 68, 3603 (1964).

23. Stigter, D., "Micelle Formation by Ionic Surfactants. I. Two Phase Model, Gouy-Chapman Model, Hydrophobic Interactions," J. Colloid Interface Sci. 47, 473 (1974a).

24. Stigter, D., "Micelle Formation by Ionic Surfactants. II. Specificity of Head Groups, Micelle Structure," J. Phys. Chem. 78, 2480 (1974b).

25. Stigter, D., "Micelle Formation by Ionic Surfactants. III. Model of Stern Layer, Ion Distribution, and Potential Fluctuations," J. Phys. Chem. 79, 1008 (1975a).

26. Stigter, D., "Micelle Formation by Ionic Surfactants. IV. Electrostatic and Hydrophobic Free Energy from Stern-Gouy Ionic Double Layer," J. Phys. Chem. 79, 1015 (1975b).

27. Emerson, M. F. and Holtzer, A., "On the Ionic Strength Dependence of Micelle Number," J. Phys. Chem. 69, 3718 (1965).

28. Emerson, M. F. and Holtzer, A., "On the Ionic Strength Dependence of Micelle Number, II," J. Phys. Chem. 71, 1898 (1967a).

29. Emerson, M. F. and Holtzer, A., "The Hydrophobic Bond in Micellar Systems. Effects of Various Additives on the Stability of Micelles of Sodium Dodecyl Sulfate and of n-Dodecyltrimethylammonium Bromide," J. Phys. Chem. 71, 3320 (1967b).

30. Mukerjee, P., "Hydrophobic and Electrostatic Interactions in Ionic Micelles. Problems in Calculating Monomer Contributions to the Free Energy," J. Phys. Chem. 73, 2054 (1969).

31. Shinoda, K. and Hutchinson, E., "Pseudo-Phase Separation Model for Thermodynamic Calculations on Micellar Solutions," J. Phys. Chem. 66, 577 (1962).

32. Sexsmith, F. H. and White, H. J., "The Absorption of Cationic Surfactants by Cellulosic Materials. I. The Uptake of Cation and Anion by a Variety of Substrates," J. Colloid Sci. 14, 598 (1959a).

33. Sexsmith, F. H. and White, H. J., "The Absorption of Cationic Surfactants by Cellulosic Materials. III. A Theoretical Model for the Absorption Process and a Discussion of Maxima in Absorption Isotherms for Surfactants," J. Colloid Sci. 14, 630 (1959b).

34. White, H. J., "Absorption of Cationic Surfactants by Cellulosic Substrates," in "Cationic Surfactants" (E. Jungermann, ed.), Chap. 9, Marcel Dekker, New York, 1970.

35. Molyneaux, P. and Rhodes, C. T., "Calculation of the Thermodynamic Parameters Controlling Micellization, Micellar Binding and Solubilization," Kolloid-Z. u. Z. Polymere 250, 886 (1972).

36. Kaneshina, S., Tanaka, M., Tomida, T. and Matura, R., "Micelle Formation of Sodium Alkylsulfate under High Pressures," J. Colloid Interface Sci. 48, 450 (1974).

37. Mukerjee, P. and Anavil, A., "Adsorption of Ionic Surfactants to Porous Glass. The Exclusion of Micelles and other Solutes from Adsorbed Layers and the Problem of Adsorption Maxima," in "Adsorption at Interfaces" (K. Mittal, ed.), p. 107, Amer. Chem. Soc., Washington, 1975.

38. Mukerjee, P. and Mysels, K. J., "Critical Micelle Concentrations of Aqueous Surfactant Systems," Nat. Stand. Ref. Data Ser., Nat. Bur. Stand., No. 36, 1971.

39. Tanford, C., "The Hydrophobic Effect: Formation of Micelles and Biological Membranes," John Wiley and Sons, New York, 1973.

40. McAuliffe, C., "Solubility in Water of Paraffin, Cyclo-
       paraffin, Olefin, Acetylene, Cycloolefin, and Aromatic
       Hydrocarbons," J. Phys. Chem. 70, 1267 (1966).
41. Tanford, C., "Micelle Shape and Size," J. Phys. Chem.
       76, 3020 (1972).
42. Tanford, C., "Thermodynamics of Micelle Formation:
       Prediction of Micelle Size and Size Distribution,"
       Proc. Nat. Acad. Sci. USA 71, 1811 (1974a).
43. Tanford, C., "Theory of Micelle Formation in Aqueous
       Solutions," J. Phys. Chem. 78, 2469 (1974b).
44. Tartar, H. V., "A Theory of the Structure of the Micelles
       of Normal Paraffin Chain Salts in Aqueous Solution,"
       J. Phys. Chem. 59, 1195 (1955).
45. Tartar, H. V., "On the Micellar Weights of Normal Paraf-
       fin Chain Salts in Dilute Aqueous Solutions," J. Col-
       loid Sci. 14, 115 (1959).
46. Kitahara, A., "Micelle Formation of Cationic Surfactants
       in Nonaqueous Media," in "Cationic Surfactants" (E.
       Jungermann, ed.), p. 289, Marcel Dekker, New York,
       1970.
47. Corrin, M. L. and Harkins, D. H., "The Effect of Salts
       on the Critical Concentration for the Formation of
       Micelles in Colloidal Electrolytes," J. Am. Chem. Soc.
       69, 683 (1947).
48. Hobbs, M. E., "The Effect of Salts on the Critical Con-
       centration, Size and Stability of Soap Micelles,"
       J. Phys. & Colloid Chem. 55, 675 (1951).
49. Shinoda, K., "The Effect of Chain Length, Salts and
       Alcohols on the Critical Micelle Concentration,"
       Bull. Chem. Soc. Japan 26, 101 (1953).
50. Saupe, A., "On Statistical Theories of Nematic Liquid
       Crystals," Berich Physik. Chem. 78, 848 (1974).
51. Wojtowicz, P. J., "Introduction to the Molecular Theory
       of Smectic-A Liquid Crystals," RCA Rev. 35, 388 (1974).
52. Wadati, M. and Isihara, A., "Theory of Liquid Crystals,"
       Mol. Cryst. Liq. Cryst. 17, 95 (1972).
53. Chandrasakhar, S. and Madhusudana, N. V., "Molecular
       Theory of Nematic Liquid Crystals," Mol. Cryst. Liq.
       Cryst. 17, 37 (1972).
54. Parsegian, V. A., "Theory of Liquid-Crystal Phase
       Transitions in Lipid + Water Systems," Trans.
       Faraday Soc. 62, 848 (1966).
55. Lifson, S. and Katchalsky, A., "The Electrostatic Free
       Energy of Polyelectrolyte Solutions II. Fully
       Stretched Macromolecules," J. Polymer Sci. 13, 43
       (1954).
56. Redlich, O., "Fundamental Thermodynamics Since Cara-
       theodory," Rev. Modern Phys. 40, 556 (1968).

57.   Winsor, P. A., "Binary and Multicomponent Solutions of
      Amphiphilic Compounds. Solubilization and the Forma-
      tion, Structure, and Theoretical Significance of
      Liquid Crystalline Solutions," Chemical Reviews 68, 1
      (1968).

58.   Winsor, P. A., "The Influence of Composition and Temper-
      ature on the Formation of Mesophases in Amphiphilic
      Systems. The R-Theory of Fused Micellar Phases,"
      in"Liquid Crystals and Plastic Crystals" (G. W. Gray
      and P. A. Winsor, eds.), Chap. 5, Vol. 1, Ellis
      Horwood, Chichester, England, 1974.

59.   Ekwall, Per., Mandell, Leo and Fontell, Krister,
      "Solubilization in Micelles and Mesophases and the
      Transition from Normal to Reversed Structures,"
      Molecular Crystals and Liquid Crystals 8, 157 (1969).

60.   Anderson, D. R., Bidner, M. S., Davis, H. T., Manning,
      C. D. and Scriven, L. E., "Interfacial Tension and
      Phase Behavior in Surfactant-Brine-Oil Systems,"
      Society of Petroleum Engineers of AIME, Paper No.
      SPE 5811 presented at the Improved Oil Recovery
      Symposium of SPE, Tulsa, Oklahoma (March, 1976).

61.   Robbins, M. L., "Theory for the Phase Behavior of Micro-
      emulsions," Society of Petroleum Engineers of AIME,
      Paper No. SPE 5839 presented at the Improved Oil
      Recovery Symposium of SPE, Tulsa, Oklahoma (March,
      1976).

62.   Lovett, Ronald A., "Statistical Mechanical Theories of
      Fluid Interfaces," Ph.D. Thesis, University of
      Rochester, 1966.

63.   Reiss, H., "Entropy-Induced Dispersion of Bulk Liquids,"
      J. Coll. Interface Sci. 53, 61 (1975).

64.   Ruckenstein, E. and Chi, J. C., "Stability of Micro-
      emulsions," J. Chemical Soc., Faraday Trans. II 71,
      1690 (1975).

65.   Adamson, A. W., "A Model for Micellar Emulsions," J.
      Colloid and Interface Sci. 29, 261 (1969).

66.   Tosch, W. C., Jones, S. C. and Adamson, A. W., "Distribu-
      tion Equilibria in a Micellar Solution System," J. Col-
      loid and Interface Sci. 31, 297 (1969).

67.   Levine, S. and Robinson, K., "The Role of the Diffuse
      Layer in Water-in-Oil Microemulsions," J. Phys. Chem.
      76, 876 (1972).

68.   Verwey, E.J.W. and Overbeek, J.T.G., "Theory of the
      Stability of Lyophobic Colloids," Elsevier, New York,
      1948.

69.   Ninham, B. W., Parsegian, V. A. and Weiss, G. H., "On
      the Macroscopic Theory of Temperature-Dependent van
      der Waals Forces," J. Stat. Physics 2, 323 (1970).

70.   Reed, R. R. and Healy, R. N., "Some Physicochemical
       Aspects of Microemulsions Flooding: A Review," in
       "Improved Oil Recovery by Surfactant and Polymer
       Flooding" (D. O. Shah and R. S. Schechter, eds.),
       Academic Press, New York, in press, 1976.

Note Added in Proof: Equation (31) should contain a term due
to the collection of the monomers to form a micelle. For rigid
cylinders this term varies as the area of the cylinder. This
is basically consistent with the conclusions we give following
Equation (39) but the values given in Table II would be changed
significantly.

SOME PHYSICOCHEMICAL ASPECTS OF MICROEMULSION
FLOODING: A REVIEW

Ronald L. Reed and Robert N. Healy
Exxon Production Research Co.

I. ABSTRACT

Injection compositions for a variety of microemulsion and
surfactant floods can be represented on equilibrium ternary
diagrams with coordinates surfactant-cosolvent, brine, and oil.
That portion of such a diagram having economic significance,
divides into a single-phase region and a multiphase region.
Within the single-phase region, micellar structure is studied
in relation to effects of salinity and cosolvent on viscosity,
optical birefringence and electrical resistivity. Within the
multiphase region, effects on phase behavior, interfacial tension and solubilization parameter are determined as functions
of salinity, brine composition, temperature, surfactant structure, cosolvent, and oil aromaticity. Correlations are found
between interfacial tension and solubilization parameter that
are useful in preliminary screening of surfactants for oil
recovery potential. When, by any means, extent of the multiphase region is reduced, a circumstance favorable to displacement in the miscible mode; a concomitant effect is that interfacial tensions are also reduced, favoring displacement in the
immiscible mode.

II. INTRODUCTION

Oil remaining within the interstices of porous rock in
the regions contacted by a waterflood can exist in a variety
of configurations, determined by wettability that may vary
from completely water-wet to preferentially oil-wet. At least
in the preferentially water-wet case, this oil is discontinuous.
Since capillarity is responsible for resistance to further displacement, a large reduction of interfacial tension may be the
only practical way to recover additional oil. One method of
achieving this is through injection of surface-active chemicals,
a procedure technically feasible for application to reservoirs
in any wettability state; however, our attention will primarily
focus on the preferentially water-wet case.

Regardless of specific formulations injected, once within
a reservoir the fluid system has three primary constituents:
oil, water and surfactant; so it becomes useful to represent

compositions on a ternary diagram. Also, functions of composition can be so represented, as, for example, micellar
structure, interfacial tension, and dilution paths. Since
the ternary diagram divides into miscible and immiscible
regions, a corresponding dichotomy of flooding regimes can be
expected, and does, in fact, occur.

Accordingly, this chapter commences with discussion of
the occurrence of residual oil and the phenomena responsible
for its displacement by fluids containing surface-active
agents (Section III). The next two sections review a sequence of four papers dealing with the miscible (Section IV)
and immiscible (Section V) aspects of microemulsion flooding
(1-4).

One goal of this work has been to develop screening
procedures that identify the member of a family of surfactants preferred for a given application (i.e., a given temperature, brine composition and crude oil), primarily through
simple "test-tube" experiments that collectively serve as a
guide to reduce the number of long-core floods required.
Emphasis here is on alteration of hydrophile-brine and
lipophile-oil interactions through variation of surfactant
and cosolvent structural parameters, salinity and temperature.
Others (5) have emphasized the oil-lipophile interaction and
show how to select the best hydrocarbon for a given surfactant, brine and temperature.

Although these advances have obvious utility, they pre-
suppose a set of surfactants possessing parameterized functionality. In the event none of these surfactants is sufficiently good, *there exists no theory or empiricism for
construction of new surfactants with improved functionality
to serve a specified purpose.*

III. DISPLACEMENT OF RESIDUAL OIL

Oil, saline water and gas occur naturally within, for
example, sandstone porous media usually consolidated with a
variety of cementing materials deposited over long periods
of time (6); often in configurations where water preferentially wets the rock, gas does not, and oil plays an intermediate role. With this model, it is evident that water
will fill the smallest pores and gas the largest (7). If no
gas is present, oil will fill the largest pores, but will,
nonetheless, occupy a distribution of pore sizes, depending
on wettability of the oil-water-rock system and on rock
lithology. In case of extensive cementation of sand grains,
the porous medium resembles a bundle of interwoven and inter-
secting tubes having bulges and constrictions, rather than the
more easily visualized and popular but often misleading random
or ordered packed beds of spheres or unconsolidated sands of
narrow size distributions.

Waterflooding is a widely used secondary recovery technique wherein water is injected into the porous rock, thereby displacing all of the resident brine (8) and part of the oil in the contacted regions. If the rock-oil-water system is preferentially water-wet, *the oil that remains is in the form of discontinuous globules or ganglia surrounded by rock and isolated by water* (9,10). This oil is trapped by capillarity and is referred to as *residual oil*.

It is generally recognized that during the waterflooding process, water and oil flow simultaneously but through separate porous networks (11), so it is not expected to find both water and oil within a single capillary; except, perhaps, where one phase is adsorbed on rock to the extent of a few molecular layers, or within minute crevices (7) where it was trapped by invading oil during a past displacement event. It follows that *pores where residual oil occurs do not contain water to any significant extent*. A scanning electron micrograph of residual oil ganglia appears in Figure 1. Impressions of the irregular confining rock on the ganglion surface are clear and augment evidence that water is substantially excluded.

Formation of residual oil ganglia during the waterflooding process is dependent on pressure gradient (12), oil-water interfacial tension (12), pore geometry (13), and possibly, on interfacial film properties (14). As water invades the porous rock, an oil-water transition zone is created where increasing water saturation causes rupture of the continuous oil filaments. It follows that points of rupture, and hence *pore constrictions adjacent principal ganglion terminii, must communicate directly with water-saturated flow channels* (see Figure 2). The essential problem of tertiary oil recovery is to replace water flowing through pores adjacent to the trapped ganglia with a different fluid that will interact with the oil, cause part of it to be displaced and subsequently coalesce with other similarly displaced oil or with other trapped ganglia, thus causing them to flow as well. In this way a continuous oil bank is created, caused to flow, and can be recovered.

A. Role of Interfacial Tension

The simplest model of trapped oil consistent with the above discussion is illustrated in Figure 3. Water flowing through adjacent and communicating capillaries establishes a pressure gradient across the trapped drop. The drop moves in the direction closest to the gradient direction, subject to pore wall constraints, until a pore constriction is met that is too small to permit further advance.

For simplicity, assume the oil-water-rock system is completely water-wet, advancing and receding contact angles, θ_2

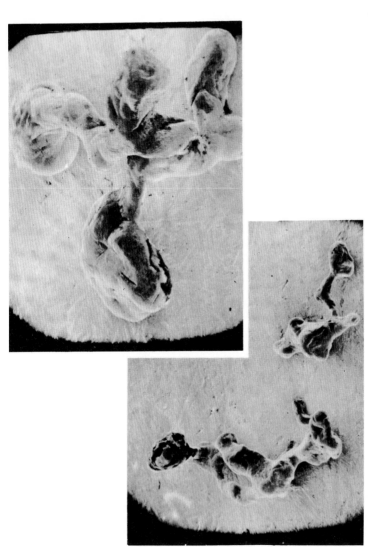

Fig. 1. Electroscan micrograph of a residual "oil" ganglion. A Berea sand-stone core was saturated with water and paraffin at elevated tempera-ture and water flooded to the residual state. The entire core was then quick-frozen in liquid N_2 and the sandstone matrix was sub-sequently dissolved in HF. (Technique developed by J. J. Taber; these ganglia prepared by R. A. Humphrey.)

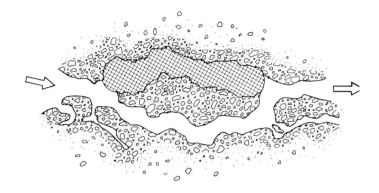

☒ RESIDUAL OIL
☐ WATER
☒ ROCK

*Fig. 2. Approximate configuration of trapped oil ganglion,
showing exclusion of water and accessibility of
water to ganglion terminii*

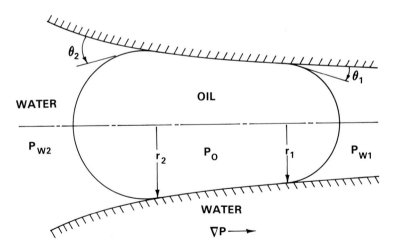

*Fig. 3. Model of an elementary oil ganglion trapped by
capillarity and a pore constriction*

and θ_1, respectively, are zero, interfacial tension is the
same at all oil-water interfaces, and capillaries are sufficiently small that these interfaces are spherical.
Laplace's Equation can then be used to calculate the pressure decrement, or capillary pressure, P_c, across the curved

interface,[*] i.e., with reference to Figure 3:

$$P_{c1} = P_o - P_{w1} \cong \frac{2\gamma}{r_1}$$

and

$$P_{c2} = P_o - P_{w2} \cong \frac{2\gamma}{r_2}$$

It follows that a pressure difference in excess of

$$\Delta P = P_{w2} - P_{w1} \cong 2\gamma \left(\frac{1}{r_1} - \frac{1}{r_2} \right) \tag{1}$$

will be required to exceed the capillary force retaining the drop, and cause it to flow (15,16).

It is instructive to calculate the pressure gradient necessary to move the drop shown in Figure 1, using a typical oil-water interfacial tension of 30 dyne/cm, $r_1 = 9 \times 10^{-4}$ cm, $r_2 = 4 \times 10^{-3}$ cm, and the drop length $= 4 \times 10^{-2}$ cm, all dimensions estimated from the largest drop in the electroscan micrograph. The result is the aqueous phase must develop ~ 573 psi/ft to displace this residual oil ganglion, whereas a practical limit achievable in real field situations is two orders of magnitude less, about 1-2 psi/ft (2,17). Consequently, an interfacial tension reduction to about 0.1 dyne/cm is needed for incipient residual oil production. *Substantial oil production may require less than 0.01 dyne/cm.*

B. Capillary Number

It follows that for a system of length L, a critical value of pressure drop across that length must be exceeded before residual oil can be displaced. More generally, Taber (16) showed *there is a critical value of $\Delta P/L\gamma$ for each porous medium,* and these critical values increase with decreasing permeability (18). Two consequences are that, in a preferentially water-wet system, *all of the residual oil can be recovered by waterflooding at a sufficiently high pressure gradient;* and *no matter how low a non-zero interfacial tension may be, there is always a positive gradient sufficiently small that no oil is displaced.*

$\Delta P/L\gamma$ can be made dimensionless by including K, the effective permeability to the displacing phase, and in other ways (12,19-21). In fact, many pertinent groups can be obtained through dimensional analysis. Among these are

---

[*] Definitions of terms can be found in "Nomenclature".

$$\frac{(\Delta P/L)K}{\gamma} \ , \quad \frac{|v|\mu}{\gamma} \ , \quad \text{and } \cos\theta \ ;$$

where $\Delta P/L$ is the pressure gradient measured over the finite length L at the displacement front, and $|v|$ is the magnitude of the average displacing phase velocity there. Although the first and second groups are equivalent in the case of steady flow, so that the integral form of Darcy's law obtains, the first and third are phenomenologically related to the displacement process; and hence we prefer to define the capillary number as

$$N_c = \frac{(\Delta P/L)K}{\gamma \cos\theta} \ .$$

Caution must be exercised in judging which of two floods has the larger value of N_c. Thus, in Figure 4, flood (a) has the larger value if N_c is calculated for the total core length whereas flood (b) has the larger value if N_c is calculated, as it should be, over the displacement front, other variables being the same.

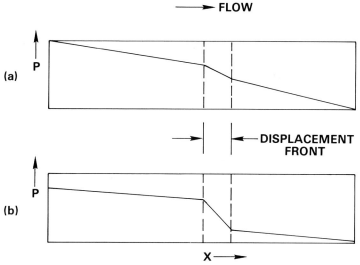

Fig. 4. Simplified pressure vs distance functions for two core floods of differing character

Adsorption must also be considered. If adsorption is significantly different for two floods, capillary number effects may be obscured.

A practical observation is: a necessary but not sufficient scaling criterion is that *laboratory velocities and pressure gradients must be comparable in magnitude to those expected in the reservoir,* presuming invariance of rock and resident fluids.

C. Oil Bank

In the case of anionic surfactants, once residual ganglia are effectively mobilized, a continuous oil bank is observed to form ahead of the flood front and flows at an oil saturation that depends primarily on the ultimate saturations and the fractional flow curve for the resident oil-brine-rock system (22, 23). Although details of the mechanism of oil bank formation are unknown, once formed, the oil bank gathers up residual ganglia at its front, continually extending its length. This coalescence of disjoint ganglia apparently requires establishing a temporal oil saturation considerably in excess of the residual oil saturation, which may account for the spike in fractional oil flow often observed at the leading edge of the oil bank. Fortunately, *criteria for mobilizing residual ganglia are sufficient to maintain and propagate an oil bank*; however, *the converse is not true* (24).

From the onset of oil bank formation, *the function of the surfactant is to maintain continuity of the flowing oil filaments to as low a saturation as possible* before they rupture and are irretrievably trapped.

Since the ultimate residual saturation is established within an interfacially active environment, oil remaining behind may contain surfactant (2,25), and this loss, in addition to that lost through adsorption, contributes to total surfactant retention.

D. Polymer Bank

Since economics severely limits the total quantity of surfactant that can be injected, it is necessary to displace a surfactant containing bank with a much less expensive fluid. Ordinary brine is precluded in view of mobility considerations; i.e., integrity of the surfactant bank requires that the mobility, K/μ, of each bank be less than that of its predecessor (26). This is usually achieved by displacing the surfactant bank with water containing a high molecular weight polymer at low concentration. This provides the necessary mobility reduction through increase in viscosity, μ, and decrease in effective permeability, K. Although a variety of polymers are available, all we have studied suffer from one or more of the following: mechanical (27,28), chemical (29), thermal (30), or bacterial (31) degradation, injection face plugging (32,33), excessive adsorption or entrapment (34), excessive inaccessible pore volume (35,36), or undesirable phase behavior when mixed with surfactants (2,37). Suffice it to say there is need for considerable improvement in polymers for use in oil recovery.

E. Total Flooding System
    Oil saturation, S_o, for the total flooding system is
graphed in Figure 5 and the various banks identified. Often

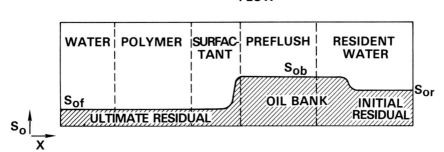

*Fig. 5. Oil saturation is graphed as a function of distance
          from the injection point, and the various banks and
          characteristic regions identified*

a preflush is used to condition the reservoir and provide an
environment more nearly optimal for the surfactant system that
follows. However, it must be remarked that, in view of heterogeneity, changes in conformance attendant mobility reduction
of successive banks, and uncertainties regarding ion exchange
between injected chemicals and interstitial clays; it has not
been established that a preflush is a practical way to substantially and sufficiently reduce total salinity (38-43).
    In this chapter attention will be confined primarily to
phase behavior, micellar structure, interfacial tension and
optimal properties of the surfactant bank, where the surfactant is contained in a microemulsion.

F. Ternary Diagram
    There are a variety of interfacially active liquids that
will displace residual oil from a porous medium. Independent
of what type of surfactant system is injected, once within the
porous medium, the liquid system can be considered composed of
three components: oil, water and surfactant. The "oil" can
be a pure hydrocarbon or as complex as a crude oil; the "water"
can vary from fresh water to an oilfield brine containing a
dozen or so different ions, and it may additionally contain
chemicals injected in a preflush; and the "surfactant" can be
a pure compound, a distribution of homologues (3), a petroleum
sulfonate (45-48), etc., and include cosurfactants such as
ethoxylated alcohols (49,50), sulfated ethoxylated alcohols
(51,52), etc., and cosolvents such as alcohols, ethers, glycols,
etc. The variety is endless, but properties of all of these
can be usefully represented as functions on a pseudo-ternary
diagram (53) such as shown in Figure 6.

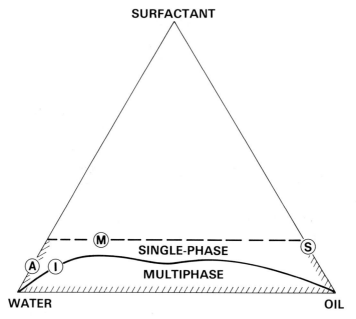

Fig. 6. *Pseudo-ternary representation of the oil-water-*
surfactant system, showing an upper economic
limit and various injection compositions of
interest

The dashed line is an upper bound for injection compositions having economic significance; and, therefore, surfactant
concentrations in excess of about 15% are, for the most part,
of no interest in relation to tertiary oil recovery.

The binodal curve divides the diagram into a single-phase
region above, and a multiphase region below. Although micellar structure varies, no interfaces can be observed when
following any path lying entirely within the single phase
(miscible) region or along the binodal curve. In the multiphase region there are always at least two phases, commonly
three, and on one occasion seven phases were observed in
equilibrium.

Shading along the coordinate axes distinguishes areas
where concentrations of one or two components are so low the
fluid takes on a different character in some respect. On the
bottom there is, in some circumstances, a change in character
of phase behavior and there is the question of achieving a
CMC. On the left or right sides there may be insufficient oil
or water, respectively, to stabilize the system. Phase behavior in these areas has not yet been studied in detail. However, see Reference 54 in regard to the region of low surfactant concentration.

G. Types of Floods
     Examples of injection compositions for all reported sur-
factant flooding systems are indicated by the letters A, M,
S and I in Figure 6.
     Aqueous surfactant flooding (55,56) (A), the oldest of
these processes, has no oil in the material injected except
for that unreacted and not separated from the surfactant
during manufacture. *Conventional microemulsion floods* (57,58)
(M), variously called *micellar* (23), *miscible-type* (59),
Maraflood (60,61) and *high concentration* (62), may have an injection composition anywhere considerably above the binodal
curve and away from the shaded areas. Usually, *soluble oil
floods* (63,64) (S), comprise a substantially anhydrous composition, high in surfactant content. *Uniflooding* (65) includes
floods of type S, but may also involve injection compositions
that include a significant quantity of water (47), in which case
they would be indistinguishable from floods of type M. *Immiscible microemulsion flooding* (4,66) (I) refers to any injection
composition on or in the neighborhood of a binodal curve. In
this chapter, only microemulsions such as those designated I
or M will be considered.
     Idealized paths for flooding compositions at the front
and rear of a surfactant bank can be constructed on a ternary
diagram. Thus, if the phase diagram has the character illustrated in Figure 7, and the composition injected is X, then
the overall composition at the front of the surfactant bank
will follow the dilution path XOB, where OB is the composition
of the oil bank. The displacement will be locally miscible
(2) along XC and immiscible along COB. In view of tie line
behavior the separate phases are oil and a sequence of microemulsions along CW, as shown. At the back of the surfactant
bank, displacement will be locally miscible everywhere along
the dilution path XW. A great variety of other path configurations obtain as phase behavior and injection compositions
change.
     Two criteria for a microemulsion composition of type M
to effectively recover oil are now evident: *The multiphase
region should be minimal so as to prolong locally miscible
displacement, and interfacial tensions in the multiphase
region should be low so as to enhance immiscible displacement* (1,2).

IV. MICROEMULSIONS: THE SINGLE-PHASE REGION
     As remarked earlier in connection with Figure 6, emphasis
will be placed on compositional points above and below the
binodal curve but excluding the shaded regions. In this section we will study some properties of the miscible region
above the binodal curve. Any injection composition within

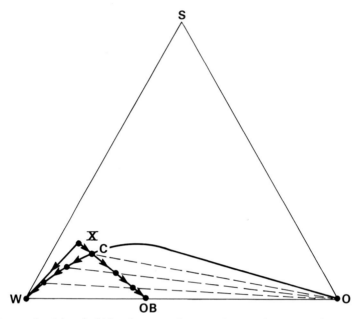

*Fig. 7.   Idealized dilution paths at front (XOB) and rear (XW)
of a surfactant bank are shown for a particular type
of phase behavior*

the miscible region has the advantage of locally miscible displacement until dilution causes the bank to deteriorate into
multiphases. During the locally miscible regime 100% of the
residual oil can be recovered, even at very low flooding rates
(2).

A. Definition
     The term "microemulsion" is a misnomer, as pointed out by
Winsor (67) and others (68,69), but is in common use and will be
retained. In the chemical literature various authors have
pointed out attributes they felt a fluid must have in order to
qualify as a microemulsion. For example, in addition to oil,
water and surfactant, a microemulsion has been required to be
transparent (70-72), thermodynamically stable (73), has almost
always included a cosolvent (an alcohol, for example); and salt
has been specified as essential to existence of an upper-phase
microemulsion in equilibrium with an aqueous phase (74). However, in spite of these qualifications, no concise definition
of a microemulsion was found. The authors introduced the
definition that *a microemulsion is a stable, translucent,
micellar-solution of oil, water that may contain electrolytes,
and one or more amphiphilic compounds*. Accordingly, a microemulsion need not be transparent, is not an emulsion (macro-

emulsion), and is not *required* to contain salts, cosolvents or
cosurfactants.

Winsor (67) points out that certain non-transparent but
translucent and often opalescent micellar solutions are stable.
Further, we have found that many of the most interfacially
active micellar solutions have these characteristics and have
maintained them for two years. The *degree of translucency is*
merely *a measure of average micelle size and configuration and*
can be caused to vary continuously from completely transparent
to nearly opaque simply by varying, for example, salinity.
Although *inclusion of the constraint that a microemulsion must*
be transparent is a matter of choice, it *excludes the pre-*
ponderance of systems that have utility for tertiary oil re-
covery. It is an experimental fact that when compositions
lead to opaque fluids, these fluids are usually unstable,
separate on standing and hence were macroemulsions; so translucency is an essential aspect.

The proposed definition requires a micellar-solution, and
this will survive, but the notion of a micelle has become
broader with time and study. Thus, according to Winsor (67),
there were initially but three kinds: spherical water-external
(Hartley micelles), spherical oil-external, and lamellar, all
shown in dynamic equilibrium in Figure 8. Later Winsor (75)

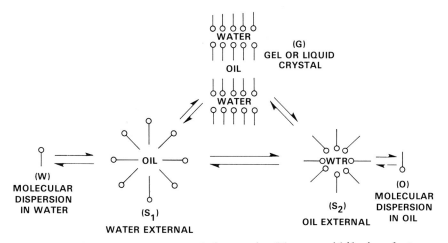

Fig. 8.   *Winsor's concept of intermicellar equilibrium between*
          spherical and lamellar micellar structures.   "It should
          be (re-) emphasized that the conception of intermicel-
          lar equilibrium does not mean that perfect S_1, S_2 and
          G micellar forms are present in equilibrium, but rather
          that a fluctuating micellar form may be regarded as
          resolved into S_1 and S_2 forms in equilibrium with G
          micelles." ... *Winsor (67), p. 56*

revised and expanded his intermicellar equilibrium concept to include spherical micelles in cubical array and cylindrical micelles in hexagonal array as shown in Figure 9. The lamellar micelle is at once the most interesting and also the most difficult to visualize as coextensive with an entire phase, and can evidently exist in many forms consistent with bilayer and multilayer substructures (76).

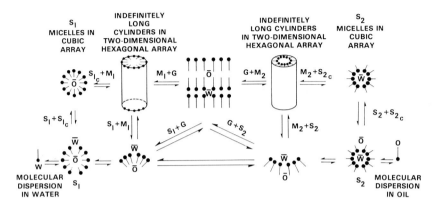

Fig. 9. Winsor's intermicellar equilibrium among spherical, cylindrical and lamellar micelles (75)

B. Stability

     Although to date all investigators have agreed that a microemulsion should be thermodynamically stable, it has not been customary to experimentally validate the fact; rather, it seems to have been taken for granted or confused with the notion of equilibrium alone. There is some difficulty with an experimental approach, because, although stability is well defined in terms of differentials of thermodynamic functions, how this is applied when it is desired to ascertain, in the absence of any such functions, whether or not a specific translucent micellar solution is stable, may not be obvious. The authors have discussed this problem in Reference 1 and the result is largely reproduced in what immediately follows.

     What is implied by "stable" is stable equilibrium. An isolated system has reached equilibrium when no further macroscopic changes occur (77). This statement raises questions: What are the macroscopic variables? How long does one wait? Among the variables that can be checked are temperature, pressure, number and volumes of phases, color, pH, translucency, viscosity, electrical conductivity, and optical birefringence. If several of these were measured as functions of time and found to be convergent, probability of an equilibrium state would be high, but stability would not be established.

From a practical view there are two questions. Is there a chemical reaction occurring? Will there be a change in the number or volumes of phases in the course of time? It may not be possible to establish stability in a rigorous thermodynamic sense; however, it is important to make clear what is meant by stability in this discussion; and this will, of necessity, be arbitrary. In the case at hand, the surfactant–oil–water system is sealed within a rigid container and placed in contact with a heat reservoir so as to maintain a constant temperature. Under these constraints the natural thermodynamic function is the Helmholtz free energy. Necessary and sufficient conditions for a minimum of this function determine criteria for stable equilibrium. A consequence is LeChatelier's principle (78), according to which *"the criterion for stability is that the spontaneous processes induced by a deviation from equilibrium be in a direction to restore the system to equilibrium."*

A convenient way to cause deviation from equilibrium is to increase the temperature. If there are chemical reactions occurring, their rates will increase, chemical potentials will change, and the probability of large local density fluctuations will increase, thereby improving the likelihood for nucleation of a new phase (78). Further, in view of temperature gradients, properties that were independent of time and position at equilibrium, become dependent on both. Nevertheless, upon return to the original temperature, the same equilibrium values of all the properties should be resumed had the state been a stable one.

It seems reasonable then to specify that *if after a temperature cycle the system returns to its original equilibrium state, that state is stable*. As a practical matter, we shall mean an increase in temperature of 10°C for 1 day, and the state will be determined by number, volume, translucency, and optical birefringence of phases measured at least 3 days after returning the system to its original temperature. There is nothing special about the temperatures or times selected. In fact a fluid may *appear* metastable to a 10° perturbation, but stable to a 1° increment. This will be discussed further later.

C. Intermicellar Equilibrium

Winsor's original concept of intermicellar equilibrium is illustrated in Figure 8. Spherical water-external micelles with oil cores are labelled S_1; spherical oil-external micelles with water cores are labelled S_2 and bilayer or multilayer lamellar micelles are designated G. In case a phase is composed of lamellar micelles it may not be possible to identify an external phase. Since these micellar structures are considered in equilibrium they can coexist within a single phase; thus (S_1, G) represents a phase having some attributes of S_1 and some of G. Further discussion appears later.

D. Micellar Structure Maps

The system studied was:

Surfactant: Monoethanol amine salt of C_N o-xylene sulfonic acid (MEACNOXS), vol. %; N = 9, 12, 15.

Cosolvent: Tertiary amyl alcohol (TAA), vol. %.

Oil: 90% Isopar-M plus 10% Heavy Aromatic Naptha (90/10 I/H), vol. %.

Brine: Distilled water plus X% NaCl, where X = (gm NaCl/100 ml solution) x 100.

Molecular weight distribution of the surfactant alkyl chain and other properties of chemicals used are found in Reference 3.

For each ternary system studied, over 100 compositions were prepared at intervals of 2% surfactant and 10% brine or oil, sealed in glass vials with teflon lined caps and equilibrated at constant temperature. Phase boundaries were drawn mid-way between adjacent samples that proved single-phase vs multiphase. This procedure will be called the *grid-point method*.

Viscosity, electrical resistivity and optical birefringence were measured for every single-phase sample. Procedural details are given in Reference 1. Examples of data obtained are given in Figures 10, 11 and 12.

Viscosity data when the brine component is 2% NaCl are shown in Figure 10. Since the oil component has a viscosity of only 2.5 cp, a remarkable feature is that high viscosities appear everywhere except toward the right corner, and there is a completely gelled region toward the left. This suggests predominantly lamellar structures and transitions to other configurations as oil content increases. Samples corresponding to points labeled "gel" are translucent (sometimes transparent), thus qualifying as microemulsions (in this case, microgels).

Resistivity data for the same system are shown in Figure 11. Resistivities are low and fairly constant throughout the region of high viscosity. These data also show a transition occurring at high oil concentration.

Each sample was tested for optical birefringence. If birefringence was exhibited while the sample was stationary, the point was labeled B. If birefringence was not exhibited while the sample was stationary, but was evident during agitation (streaming birefringence), the point was labeled S.

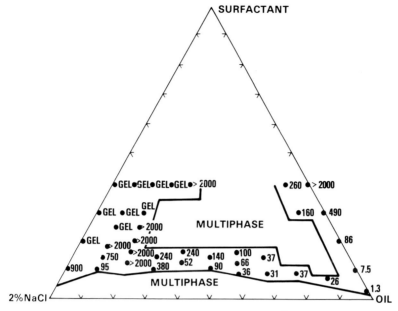

Fig. 10. Viscosity, cp @ 46 sec⁻¹, 2% NaCl

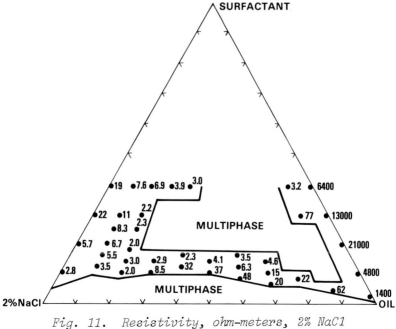

Fig. 11. Resistivity, ohm-meters, 2% NaCl

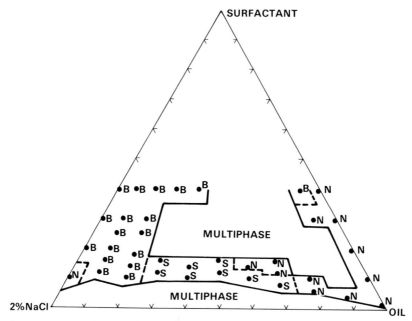

Fig. 12. Optical birefringence; birefringent (B), streaming birefringent (S), neither B nor S (N), 2% NaCl

If birefringence was not exhibited in either case the point was labeled N. B indicates a lamellar structure coextensive with the entire phase and therefore implies a G-phase. S shows that lamellar structures are present, but shear is needed to induce the required degree of anisotropy. It is inferred that S is in a transition region corresponding to (S_1, G) or (G, S_2). N implies no crystalline structure and corresponds to S_1, S_2, or (S_1, S_2). Results are shown in Figure 12. Photomicrographs of birefringent microemulsions appear in Reference 1.

Birefringence and resistivity data were combined to define the micellar structure map shown in Figure 13. Noteworthy features are disjoint pairs of water-external, gel, and multiphase regions. Structural transitions of large variety appear. In particular, *it is possible to move from an oil-external phase to a water-external phase without passing through an intervening gel region.*

1. *Salinity*

Reducing salinity to 1% NaCl drastically alters configuration of the multiphase region, increases viscosity everywhere, increases resistivity somewhat, and expands the region of birefringence. The resulting micellar structure map (Figure 14) is dominated by G, and S_1 has disappeared.

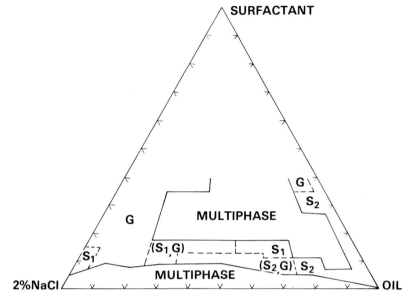

Fig. 13. *Micellar structure map, 2% NaC1; showing disjoint*
G, S₁, and multiphase regions

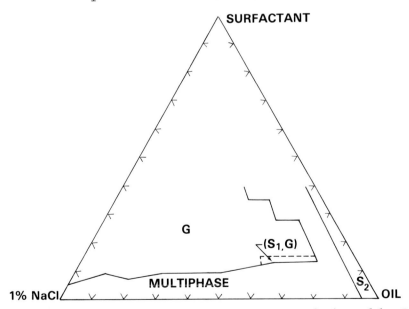

Fig. 14. *Micellar structure map, 1% NaC1 -- dominated by G,*
and S₁ is absent

2. *Cosolvent*

The very high viscosities that can occur in the G and
(G, S) regions of the 1- and 2-percent NaCl ternary diagrams
would prohibit application of these compositions to tertiary
oil recovery. One way to adjust viscosity is to add a cosolvent, such as an alcohol, to the surfactant (79). Effects
of adding 37% TAA to the surfactant for the case of 1% NaCl
are to dramatically improve phase behavior, and reduce viscosity. Resistivity is reduced except for an abrupt jump to
very high values at high oil content. The structure map
(Figure 15) is dominated by S_1 and provides a second instance
of $S_1 \rightleftarrows S_2$ without an intervening G. We conclude therefore
that although it is possible for the transition from S_1 to S_2
to pass through lamellar and cylindrical forms (70), it is by
no means essential. Indeed Winsor's diagram of intermicellar
equilibrium (Figure 8) makes it clear there are two routes
between S_1 and S_2; one of them passes through G; the other
does not.

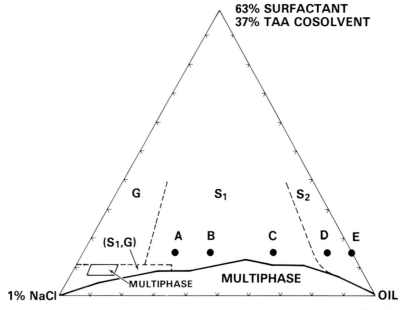

*Fig. 15. Micellar structure map, 1% NaCl, 37% TAA, showing
            simplified phase behavior dominated by S_1; and a
            direct $S_1 \rightleftarrows S_2$ transition without intervening G.
            Also shown are injection compositions A-E used to
            evaluate the effect of external phase on flooding
            results*

3. *Optimal Salinity*

As remarked earlier, locally miscible displacement is favored by a ternary diagram having the largest possible miscible region, i.e., a minimal multiphase region. Since salinity is one of the variables that can strongly affect the ternary diagram it is reasonable to introduce the following definition: *optimal salinity for miscibility*, C_m, *is the salinity that minimizes height*, C_s, *of the multiphase region at 50/50 water-oil ratio* (WOR). In Figure 16, C_s is graphed vs salinity for the 63/37 surfactant/cosolvent system. Evidently, C_m = 1.25% NaCl. A micellar structure map at optimal salinity appears in Figure 17, where it can be seen that only S_1, S_2 and G remain, and the multiphase region is small.

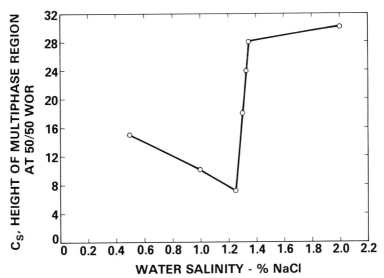

Fig. 16. Determination of optimal salinity for the system 63/37 surfactant/TAA cosolvent

4. *Divalent Ion*

When Ca^{++} was added to the NaCl brine in the ratio 0.91 $NaCl/0.09$ $CaCl_2$, optimal salinity was reduced from 1.25% NaCl to 1.1% total dissolved solids (1.0% NaCl, 0.1% $CaCl_2$), as might be expected; but there was very little effect on the micellar structure map (1).

E. Biopolymer

Since it is necessary to add a polymer to drive water that displaces a microemulsion bank (see IIID), polymer will mix with the microemulsion to an extent that depends, among other things, on the distance traveled. The average polymer

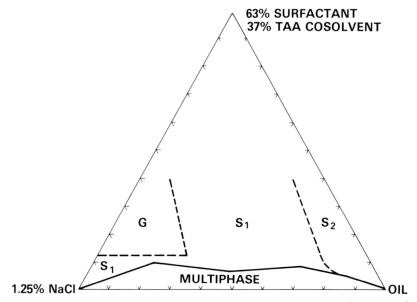

Fig. 17. Micellar structure map at optimal salinity. Only S_1, S_2 and G structures remain

concentration in the polymer bank may vary from a few hundred ppm to one thousand ppm or more, depending on the application and the type of slug grading used (80). If the concentration were 1000 ppm of XC Biopolymer, then some consequences of interaction with the microemulsion are illustrated in Figure 18 where this concentration of polymer was added to the brine. The effect is drastic. Any economic injection composition of type M will separate out another liquid phase or a solid precipitate. In the former case, local regions of low mobility may be generated in situ; in the latter case, the possibility of partial plugging must be considered.

F. Temperature
    The system 63/37 MEAC12OXS/TAA, 90/10 I/H, 1% NaCl was studied as a function of temperature to determine effects on extent of the miscible region. The multiphase boundary was located using the grid-point method (see IV D) at temperatures of 75, 120, 150 and 180°F. Figure 19 shows the miscible region is extensively reduced at 120° and reduced somewhat further at 150 and 180°F. Had the temperature been lowered sufficiently below 75°F (if freezing did not intervene), the multiphase region would again expand thereby defining an optimum temperature analogous to C_m.

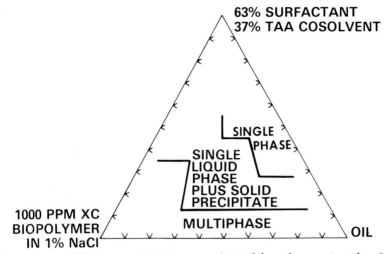

Fig. 18.   *Addition of 1000 ppm of XC biopolymer to the 63/37*
surfactant/cosolvent system at 1% NaCl diminishes
the miscible region

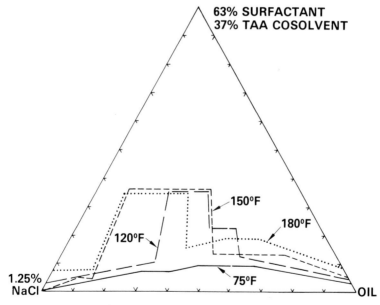

Fig. 19.   *Increasing temperature diminishes the miscible*
region. A sufficient reduction in temperature
would have a similar effect

G. Micellar Structure and Flooding
    Simplicity of phase behavior and the rather sharp $S_1 \rightleftarrows S_2$
transition makes the system illustrated in Figure 15 ideal to
study the effect of external phase on oil recovered from microemulsion flooding of type M. Accordingly, a sequence of floods
in 4 ft Berea cores was conducted using the injection compositions A, B, C, D and E (Figure 15), all containing 15% of
63/37 surfactant/TAA, but with varying oil content. A, B and
C were S_1; D and E were S_2. Table I shows final oil saturation, S_{of}, accounting for injected oil, for each flood. S_{of}
varied somewhat with oil content; however, there appears to
be no obvious advantage attributable to either oil-external
or water-external microemulsions. Complete descriptions of
these floods and all pertinent data can be found in Reference
2.

TABLE I

EFFECT OF EXTERNAL PHASE ON FINAL OIL SATURATION

Slug Composition	External Phase of Injected Slug	S_{of} (percent PV)
A	Water	7.4
B	Water	8.0
C	Water	11.1
D	Oil	6.4
E	Oil	10.6

Significance of the external phase has been a matter of interest (81,82), but to our knowledge, no other studies of oil
recovery in relation to micellar structure have been published.

V. MICROEMULSIONS: THE MULTIPHASE REGION
    As a microemulsion flood of Type M (see III G) progresses,
surfactant adsorption, as well as mixing with brine and oil at
the front and with polymer-water at the rear, cause gradual
deterioration of the bank. Eventually, even the highest surfactant concentration present in the bank will fall below the
multiphase boundary, one or more phases will break out; and
thereafter the displacement assumes an immiscible character.
In this section we will consider properties of the multiphase
region that bear on a microemulsion flood of Type M or an immiscible microemulsion flood of Type I. Overall compositional
points below the multiphase boundary correspond to multiple
phases comprising microemulsions of various kinds, excess oil
and excess water; and therefore opaque macroemulsions of these

are observed upon mixing. Under some circumstances these
macroemulsions completely separate into distinct phases very
rapidly, and under other circumstances they are resistant to
separation, sometimes requiring many years to separate in the
gravitational field alone. It should be evident that understanding the microemulsion displacement process requires com-
prehension of phase behavior. Establishing *equilibrium* phase
behavior seems a logical first step.

A. Phase Behavior and Micellar Structure
     A system will be called *simple* when it behaves as though
composed of three pure components having ternary diagrams
similar to those illustrated in Figure 20. Part (a) of this

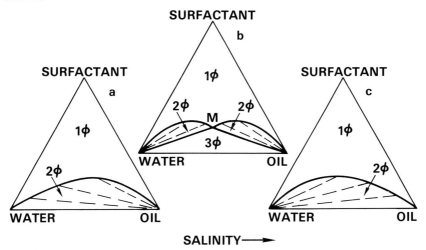

*Fig. 20.   Illustration of simple phase behavior, and Winsor's
            type I, III, and II systems*

figure shows a two-phase region wherein microemulsions along
the binodal curve are in equilibrium with oil containing
molecularly dispersed surfactant (excess oil). This is
Winsor's Type I system (67). Part (c) shows a two-phase region wherein microemulsions along the binodal curve are in
equilibrium with excess water, i.e., Type II. More recently,
Type II systems have been considered by Adamson (83), and by
Tosch, Jones and Adamson (74). Usually a Type I multiphase
region is skewed to the right and a Type II region is skewed
to the left, as illustrated. Part (b) shows Type II in the
upper-left node, Type I in the upper-right node, and Type III
in the lower triangle. Any composition within this triangle
equilibrates into three phases: microemulsion corresponding
to compositional point M, excess water, and excess oil. In
this chapter, Winsor's Type I, II, and III systems will be
called lower-phase, ℓ, upper-phase, u, and middle-phase, m,

respectively. Among the variables that affect the type of diagram observed are salinity, oil composition, surfactant molecular structure, alcohol cosolvent, and temperature.

Not all microemulsions qualitatively conform to simple multiphase behavior; it sometimes happens that where one phase was expected, two or more immiscible microemulsion phases appear. However, these extra phases frequently occur in minor quantities, and so the expected phases predominate. Therefore simple behavior appears to be a good approximation for numerous microemulsion systems having utility for tertiary oil recovery.

1. *Equilibration*

In Section IV D, the multiphase boundaries were determined using the grid-point method. A different method involves preparation of several multiphase samples having constant overall surfactant concentration but varying water-oil ratios. The assumption of negligible surfactant concentration in excess water and oil phases allows calculation of equilibrium microemulsion compositions from equilibrated phase volumes, and construction of a binodal curve (3). *If the overall surfactant concentration* of the samples *is changed, the phase diagram will change,* reflecting the multicomponent nature of surfactant, cosolvent, brine, and oil used. Figure 21 shows binodal curves at two different overall surfactant concentrations for a lower-phase microemulsion system.

2. *Real Systems*

Figure 22 exhibits phase diagrams for a real system that approximates simple behavior. Increasing salinity causes successive appearance of lower-, middle-, and upper-phase diagrams. A notable difference between real and simple behavior is the locus of middle-phase compositions rather than a single point (Figure 20). As the base surfactant concentration is increased, lateral extent of this locus should decrease.

Figure 23 is a photograph of samples having identical overall compositions except that salinity varies from 0.5 to 2.5% NaCl. At this writing, all phase volumes have been constant for over a year. These samples illustrate the effect of salinity on phase behavior when water-oil ratio is maintained constant at 1/1, and surfactant concentration is also held constant. Evidently, increasing salinity causes the microemulsion system to undergo the transitions $\ell \rightarrow m \rightarrow u$. This remarkable, systematic behavior of complex, multicomponent microemulsion systems is essential to all that follows.

3. *Micellar Structure*

Micellar structure of microemulsion phases that undergo the transition $\ell \rightarrow m \rightarrow u$ was studied using the approach of Section IV D, and results are shown in Figure 24. Since none

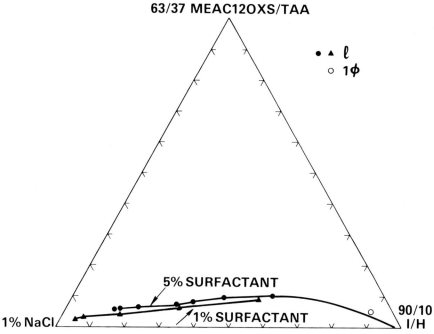

Fig. 21. Effect of surfactant concentration on location of the multiphase boundary. The far right hand portion of the binodal curve for 1% surfactant was not determined as it would require overall compositions of less than 5% brine, in which case minor amounts of salt in the surfactant component can significantly influence phase behavior

of the microemulsions was birefringent or streaming bire-fringent, and since resistivity increased continuously, neither of these properties identifies $\ell \to m$ or $m \to u$. However, viscosity changes abruptly at these transitions and is related to microemulsion phase volume (Figure 24b). Winsor's model would imply S_1 at low salinity, S_2 at high salinity and (S_1, S_2) for the middle phases. A continuous shift from S_1 to S_2 is supported by birefringence and resistivity, but does not appear to account for the abrupt changes in viscosity. These latter phenomena suggest $\ell \to m$ and $m \to u$ are, in a sense, sharp transitions. We accept that S_1 and S_2 predominate in ℓ and u phases, respectively, but are uncertain concerning the middle phase. However, the following middle-phase properties are evident from the data:

  1. If there is an external phase it is neither oil nor water
  2. If there are micelles, they are not, on the average spherical, cylindrical or lamellar.

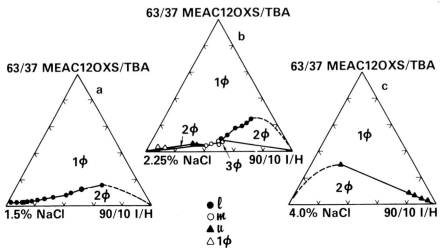

Fig. 22.   *An example of real ternary diagrams approximating*
simple behavior. All points were determined using
a total surfactant component concentration of 3%.
Note the locus of middle-phase points in (b)

B. Stability Revisited

A microemulsion was defined in a way that seems satisfactory for compositions well into the miscible region; but
suppose the composition is very close to, or on the multiphase boundary; and let us then look into the question of
stability (see IV B) again.

For the anionic surfactants studied, if a middle-phase
system is heated, oil will be spontaneously rejected. At the
elevated temperature the microemulsion is undersaturated with
respect to water, but water will not spontaneously solubilize
to any measurable extent, even with mild stirring of the
microemulsion phase. If the system is then cooled to the
original temperature, it will be undersaturated with respect
to oil, but oil will not spontaneously solubilize to any
measurable extent, even over extremely long periods of time,
even with fairly rapid stirring of the microemulsion. Further,
if the oil is slowly re-injected into the bottom of the microemulsion phase, it will thread its way back into the oil phase
in a filament so minute as to be hardly visible, with only a
very small loss of oil. If, however, the oil is rapidly
flushed into the microemulsion it will mostly disappear;
i.e., work is required.

Evidently, resistance to diffusion afforded by oriented
layers of surfactant at the interface (84) and augmented by
the gravitational field, permits only very minute net rates
of oil and water transport. Of course, if the system is

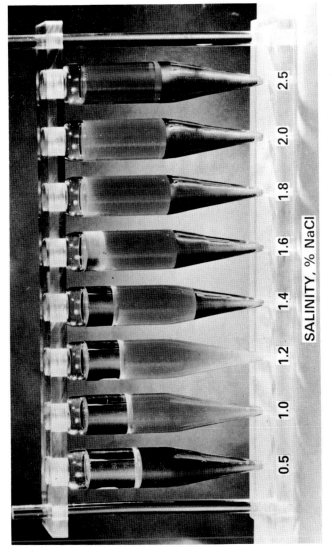

SALINITY, % NaCl

Fig. 23. Effects of increasing salinity for the system 4% 63/37 MEAC12OXS/TAA, 48% 90/10 I/H, 48% X% NaCl, are shown here. The transitions ℓ → m → u are evident; a remarkably systematic result for a compositionally complex system

411

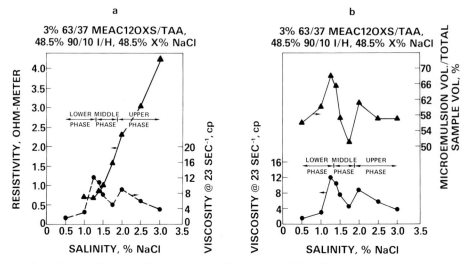

Fig. 24. Abrupt changes in viscosity identify the transitions ℓ → m and m → u whereas resistivity varies continuously

thoroughly mixed and allowed to equilibrate after each perturbation, it will return to the appropriate state. We conclude that although immiscible microemulsions may be stable
in the pure thermodynamic sense, they are not *stable in a
practical sense*; i.e., once perturbed the system does not
spontaneously return to its original state in a period of
time that has practical utility. Nevertheless, these
"immiscible microemulsions" are translucent micellar solutions and serve all the same purposes as the previously defined microemulsions (IV A), except they are saturated with
oil and/or water. Accordingly, we introduce the *operational
definition: a microemulsion is a persistent translucent
combination of oil, water that may contain electrolytes and
one or more amphiphilic compounds.* "Persistence" can then
be defined according to requirements of the specific application.

C. Interfacial Tension and Solubilization Parameter

Figure 23 reveals there are two kinds of interfaces and
hence up to three interfacial tensions can be measured depend-
ing on salinity: γ_{mo} (microemulsion-oil), γ_{mw} (microemulsion-
water) and γ_{ow} (excess oil-excess water)[*]; and further, the
excess oil and water volumes, and hence volumes of oil and

---

[*]γ_{ow} was high, ~ 0.1 dyne/cm, and was not routinely
measured.

water within the microemulsion phase, V_o and V_w, respectively, depend on salinity.

Figure 25 relates interfacial tensions, γ_{mo} and γ_{mw}, and *solubilization parameters*, V_o/V_s (V_s = vol. surfactant in microemulsion phase not including cosolvent), to salinity. As salinity increases, γ_{mo} decreases and γ_{mw} increases. As either tension decreases, the appropriate solubilization parameter increases. Hence in ℓ and u phases *low interfacial tensions correspond to micelles swelled with internal phase.* This is consistent with Robbins' model (85,86).

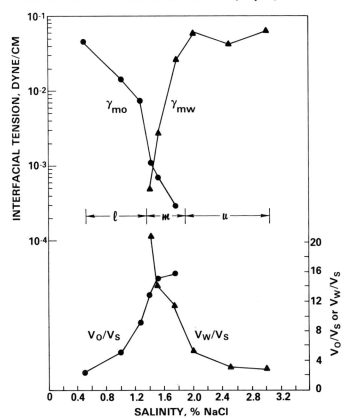

Fig. 25.   *Interfacial tensions and solubilization parameters for the system 3% 63/37 MEAC12OXS/TAA, 48.5% 90/10 I/H, 48.5% X% NaCl. Their respective intersections define C_γ and C_ϕ*

Recognition that *volumes of oil or water solubilized in a microemulsion phase in relation to the amount of surfactant there are measures of interfacial activity* will prove to have practical as well as conceptual value. Although solubility

of one aqueous or organic phase in another is quite different
from the notion of water or oil solubilized within surfactant
micelles, it is interesting to note some parallel, supporting
developments.

It has long been recognized that solubility measurements
provide a means of studying molecular interactions. Hansen's
approach (87), using the solubility parameter concept of
Hildebrand and Scott (88,89), is well known. Other investigators of nonsurfactant systems have reasoned that, for the
liquid-liquid case, cohesional and adhesional molecular forces
that determine the magnitude of interfacial tension, also
determine the extent to which two liquids are soluble. In
1913, Hardy (90) showed that interfacial tension between
aqueous and organic liquid phases may be reliably considered
as a linear function of the log of the "degree of miscibility"
of the liquids. More recently, Donahue and Bartell (91) made
a further study of the relationship between γ and solubility.
None of these studies involved surfactants.

*The salinity, C_γ, where γ_{mw} intersects γ_{mo} is called the
interfacial tension optimal salinity.* Similarly, C_ϕ, *the
phase behavior optimal salinity is defined by the intersection
of* V_o/V_s *with* V_w/V_s. Puerto and Gale (92) have developed
methods for predicting C_ϕ and V_o/V_s at C_ϕ for mixtures of alkyl orthoxylene sulfonates.

Interfacial tension and solubilization parameter graphs
(Figure 25) suggest the correlations shown in Figure 26.
Although these correlations were obtained through variation
of salinity alone, similar results obtain when overall composition is a variable (3). Data scatter implies there are
additional parameters (see Reference 85 in this regard);
nevertheless, correlations of this kind reduce the number of
interfacial tension measurements required to evaluate a surfactant, and show that phase volumes can replace interfacial
tensions as a preliminary measure of interfacial activity.
This becomes particularly useful in the case of some black
crude oils where equilibrated tension measurements are
extremely difficult or impossible, but phase boundaries may
be visible under ultraviolet light (93).

The curves shown result from fitting these data with
the empirical equations:

$$\log (\gamma_{mo}/\gamma'_{mo}) = \frac{a}{m_o (V_o/V_s) + 1} \tag{1}$$

and

$$\log (\gamma_{mw}/\gamma'_{mw}) = \frac{b}{m_w (V_w/V_s) + 1} ; \tag{2}$$

using the parameters,

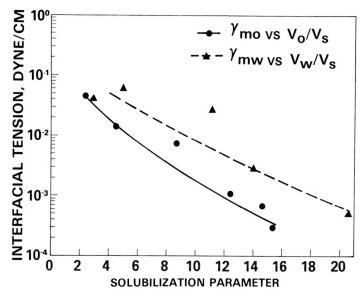

*Fig. 26. Correlation of interfacial tension with solubiliza-
tion parameters for the system 3% 63/37 MEAC120XS/TAA,
48.5% 90/10 I/H, 48.5% X% NaCl*

$$a = 6.285 \qquad\qquad b = 12.167$$

$$\log \gamma'_{mo} = -7.058 \qquad \log \gamma'_{mw} = -12.856$$

$$m_o = 0.04477 \qquad\qquad m_w = 0.01280$$

D. Optimal Salinity

Three optimal salinities have been so far defined, C_m (see
IV D3), C_γ and C_ϕ. For the system 3% 63/37 MEAC120XS/TAA,
48.5% 90/10 I/H, 48.5% X% NaCl, C_m = 1.25% (Figure 16),
C_γ = 1.4%, and C_ϕ = 1.5% (Figure 25); i.e., they are about the
same. All of these were determined using a constant WOR = 1.
A natural question concerns the possible dependence of optimal
salinity on WOR and C_s.

An unusual ternary diagram appears in Figure 27 where increasing WOR results in $\ell \to m \to \ell$. Behavior of C_ϕ in relation
to WOR and C_s was determined for this system and results appear
in Figure 28. Dependence of C_ϕ on C_s is moderate except for
C_s < 3%. Dependence on WOR is also moderate. These results
provide an explanation for the complex phase behavior illustrated in Figure 27. At high water-oil ratios on the left
side of the diagram, salinity is less than C_ϕ; and hence
surfactant will reside in an aqueous phase (ℓ). In the
center, salinity is near optimal so the expected middle-
phase appears there. At low water-oil ratios, salinity

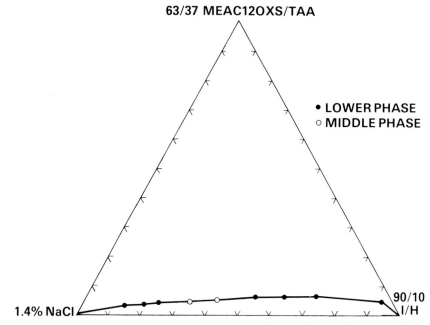

63/37 MEAC12OXS/TAA

• LOWER PHASE
○ MIDDLE PHASE

1.4% NaCl

90/10 I/H

Fig. 27.   The unusual behavior shown here, i.e., l → m → l results from the dependence of optimal salinity on water-oil ratio

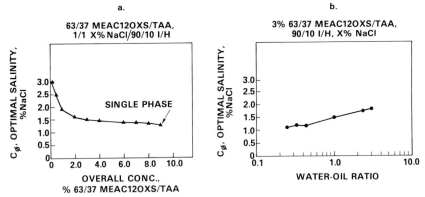

a.

63/37 MEAC12OXS/TAA, 1/1 X% NaCl/90/10 I/H

C_ϕ, OPTIMAL SALINITY, %NaCl

SINGLE PHASE

OVERALL CONC., % 63/37 MEAC12OXS/TAA

b.

3% 63/37 MEAC12OXS/TAA, 90/10 I/H, X% NaCl

C_ϕ, OPTIMAL SALINITY, %NaCl

WATER-OIL RATIO

Fig. 28.   The dependence of C_ϕ on C_s is moderate except for C_s < 3%. C_ϕ is almost linear in log WOR. Qualitative character of these effects depends on the system

is still close to optimal; hence the expected lower-phase microemulsions are found. Although there are insufficient data to analyze a complete flooding situation, it is evident that dependence of optimal salinity on surfactant concentration and water-oil ratio could have important effects.

E. Cohesive Energy Ratio

Winsor viewed interfacial activity in terms of the *cohesive energy ratio*, $R \equiv E_{lo}/E_{hw}$, where E_{lo} and E_{hw} are the lipophile-oil and hydrophile-water interaction energies, respectively. High interfacial activity occurs when E_{lo} and E_{hw} are both large and, in addition, $R \sim 1$. When $R \ll 1$, S_1 is favored; and when $R \gg 1$, S_2 is favored.

The concept of optimal salinity can be interpreted in terms of R and the salinity dependence of E_{hw} and E_{lo}. It seems reasonable to assume E_{hw} is a monotonically decreasing function of salinity, whereas E_{lo} is independent of salinity. Then C_γ corresponds to the salinity where $E_{lo} \sim E_{hw}$; i.e., $R \sim 1$.

F. Interfacial Tension Optimal Salinity

Interfacial tensions between microemulsions and equilibrated excess oil or excess water phases were determined as functions of salinity, temperature, surfactant structure, cosolvent structure, oil composition, and composition of dissolved solids in the aqueous phase.

1. *Surfactant Structure*

Figure 29 shows interfacial tension vs salinity graphs for three alkyl chain lengths, all measured at 112°F. Similar graphs were obtained at 74 and 150°F (3). Increasing N from 9 to 15 decreases C_γ from 4.4 to 0.2% and $\gamma(C_\gamma)$ from 0.01 to 0.001 dyne/cm. Although increasing N decreases $\gamma(C_\gamma)$, the range of salinity over which the surfactant is effective is also decreased.

These results can be interpreted in terms of cohesive energy ratio. Since E_{hw} is a decreasing function of salinity, and $E_{hw} \sim E_{lo}$ at C_γ; both $E_{hw}(C_\gamma)$ and $E_{lo}(C_\gamma)$ increase as C_γ decreases, resulting in higher interfacial activity and hence lower γ. Increasing N at constant salinity causes γ_{mo} and V_w/V_s to decrease, γ_{mw} and V_o/V_s to increase, and phase behavior to move in the direction $\ell \rightarrow m \rightarrow u$.

2. *Temperature*

Figure 30 shows, for N = 12, interfacial tension as a function of salinity at three temperatures. Similar graphs were obtained for N = 9 and 15 (3). Both C_γ and $\gamma(C_\gamma)$ increase with temperature. Increasing temperature causes γ_{mw} and V_o/V_s to decrease, γ_{mo} and V_w/V_s to increase, and phase behavior to change in the direction $u \rightarrow m \rightarrow \ell$.

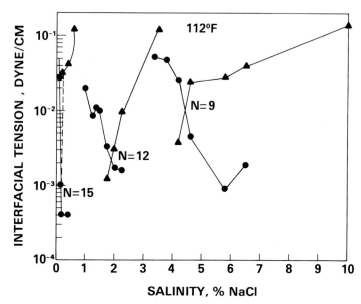

*Fig. 29. Dependence of γ on salinity for the system 3% 63/37
MEACNOXS/TAA, 48.5% 90/10 I/H, 48.5% X% NaCl, where
N = 9, 12, and 15 at 112°F, showing improved inter-
facial tension as N increases*

If the data of Figure 30, augmented with solubilization
parameter data (3), are graphed as functions of temperature
at constant salinity, Figure 31 is obtained. Here it can be
seen that *optimal temperature* can be defined analogously to
optimal salinity. In the case at hand, optimal temperature
is 118 or 130°F depending on γ or solubilization parameter,
respectively.

A convenient summary of relations among C_γ, optimal interfacial tension, temperature and the surfactant structural
parameter, N, appears in Figure 32. An application of these
results is selection of a surfactant structure that gives
$\gamma_{mo} = \gamma_{mw}$ at a given temperature and salinity, and estimation
of the value of interfacial tension there; thus providing a
guide for surfactant structure required to be effective in a
particular reservoir environment.

3. *Cosolvent*

Changing the cosolvent from TAA to tertiary butyl alcohol
(TBA) causes optimal salinity, interfacial tension at optimal
salinity, and the range of salt tolerance to increase (see
Figure 33). Another interesting feature is the symmetry
between the γ_{mo} and γ_{mw} curves. Symmetry is further reflected in Figure 34 where all γ_{mo} and γ_{mw} data are correlated with a single curve, and in phase behavior (Figure 22),

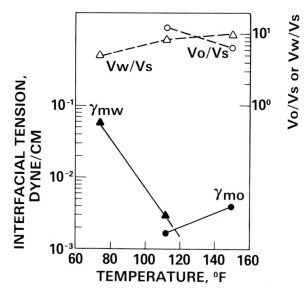

Fig. 30. *Dependence of interfacial tension on salinity for the system 3% 63/37 MEAC120XS/TAA, 48.5% 90/10 I/H, 48.5% X% NaCl showing an increase in optimal salinity and in the value of γ at optimal salinity as temperature increases*

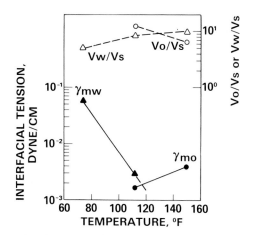

Fig. 31. *Interfacial tension and solubilization parameter determine optimal temperatures for the system 3% 63/37 MEAC120XS/TAA, 48.5% 90/10 I/H, 48.5% 2% NaCl*

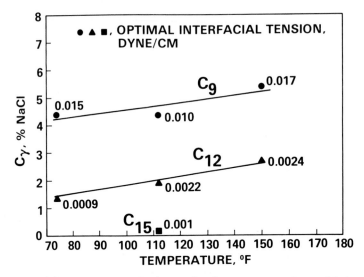

Fig. 32.   This summary of the relations among C_γ, $\gamma(C_\gamma)$, temperature and N is useful to find the value of N most suitable to a given reservoir environment

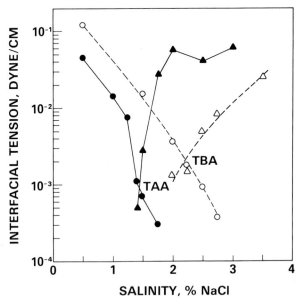

Fig. 33.   Effect of cosolvent on the system 3% 63/37 MEAC120XS/ cosolvent, 48.5% 90/10 I/H, 48.5% X% NaCl. Note the symmetry of the TBA data

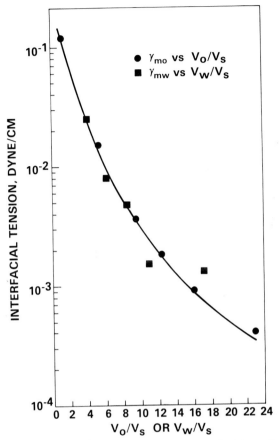

Fig. 34. γ-solubilization parameter correlation for the data
of Fig. 33. Both γ_{mo} and γ_{mw} can be fit with the
same curve. Parameters for Eqs. 1 and 2 (see V C)
are: a = b = 4.5784, log γ'_{mo} = log γ'_{mw} = -5.1381,
and $m_o = m_w = 0.075101$

which is nearly simple. These data, together with physical
reasoning, provide the rationale for fitting solubilization
parameter correlations with curves rather than straight lines.

Increasing alcohol molecular weight at constant salinity
causes γ_{mo} and V_w/V_s to decrease, γ_{mw} and V_o/V_s to increase,
and $\ell \rightarrow m \rightarrow u$.

4. *Other Variables*

Interfacial tension and solubilization parameter were
measured as functions of salinity to ascertain effects of
several other variables (3), all with respect to 3% 63/37
MEAC12OXS/TAA, and WOR = 1.

Increasing aromatic content of the oil from 0% (100% Isopar M) to 10% (90/10 I/H) decreased both optimal salinity and tension. Increasing aromaticity at constant salinity resulted in $\ell \to m \to u$.

Neither optimal tension nor salinity were significantly affected by presence of 750 ppm of XC biopolymer in the aqueous phase. This is in marked contrast to the detrimental effect this polymer has on extent of the miscible region (see IV E).

Addition of Ca^{++} to the aqueous phase in the ratio 10/1 $NaCl/CaCl_2 \cdot 2H_2O$ decreased optimal salinity from 1.4% NaCl to 1.1% total dissolved solids, but the effect on optimal tension was very small. This agrees with a previous finding that Ca^{++} caused a decrease in C_m, but extent of the multiphase region was not significantly affected.

Finally, addition of both 750 ppm XC biopolymer and Ca^{++} to the brine produces interfacial tensions that are very nearly the same as those found for NaCl brine.

G. Summary

For the systems studied, Table II shows trends in γ and phase behavior that result from increasing a given variable of interest, providing the overall composition is constant.

TABLE II

SUMMARY OF INFLUENCE OF SOME VARIABLES ON PHASE BEHAVIOR, INTERFACIAL TENSION AND SOLUBILIZATION PARAMETER

		Resulting Trends				
Increasing Variable	Phase Behavior	γ_{mo}	V_o/V_s	γ_{mw}	V_w/V_s	C_γ
Salinity	$\ell \to m \to u$	−	+	+	−	
Alkyl Chain Carbon No. (N), Molecular wt. of alcohol cosolvent, oil aromaticity $Ca^{++}/NaCl$ Ratio	$\ell \to m \to u$	−	+	+	−	−
Temperature	$u \to m \to \ell$	+	−	−	+	+
XC Biopolymer Conc.		Insignificant Changes				

(−) indicates a decrease
(+) indicates an increase

Tabulated results show that *whenever phase behavior changes
in the direction $\ell \to m \to u$, associated trends are that* γ_{mo}
and V_w/V_s *decrease, while* γ_{mw} *and* V_o/V_s *increase. When the
direction of phase behavior change is opposite* ($u \to m \to \ell$),
opposite trends are observed. It is conjectured these results will hold for all surfactant systems.

H.   Immiscible Microemulsion Floods

It will be recalled that an immiscible microemulsion
flood has an injection composition on or close to the multiphase boundary (see IIIG and Figure 6, Type I). Now that
considerable detail of the multiphase region has been ac-
cumulated, it affords an opportunity to study this type of
flood and determine the extent to which concepts we have
introduced bear a relation to oil recovery. For this purpose it is necessary to further introduce *controlling inter-
facial tension,* γ_c, as the larger of γ_{mo} and γ_{mw}; i.e.,
$\gamma_c \equiv \max (\gamma_{mo}, \gamma_{mw})$. Thus $\gamma_c = \gamma_{mo}$ for lower-phase microemulsions; $\gamma_c = \gamma_{mw}$ for upper-phase microemulsions; and for
middle-phase microemulsions, γ_c may be either γ_{mw} or γ_{mo},
depending on which is greater. γ_c *is minimized when* $\gamma_{mo} = \gamma_{mw}$.

1.   *Continuous Injection*

Lower-, middle-, or upper-phase microemulsions having
N = 9 or 12 were employed in core flooding experiments wherein microemulsion was continuously injected. The floods were
conducted at constant rate in the range 0.5-2.3 ft/day, and
fractional flows of oil, f_o, and water, f_w, during production
of the stabilized oil bank were measured. (See Reference 4
for details.)

For N = 9 or 12, Figure 35 shows fractional flow data
correlated with $N_c(\gamma_{mo})$ or $N_c(\gamma_{mw})$. Since f_o is independent
of γ_{mw}, it suggests nearly all resident water was displaced;
and f_o is determined, in part, by the influence of γ_{mo} on oil
saturation left behind the microemulsion front. A similar
interpretation applies to $f_w(\gamma_{mw})$.

2.   *Bank Injection*

Banks of lower-, middle-, and upper-phase microemulsions
were injected at constant rates in the range 0.5-1.3 ft/day
and final oil saturation determined from

$$S_{of} = (V_{o,inj} + V_{o,initial} - V_{o,prod})/PV_{core} .$$

Figure 36 shows that S_{of} broadly decreases with N_c, but the
correlation depends on which tension is controlling. The
scatter was anticipated in view of changing injection composition, mobility ratio and surfactant retention. Other
considerations are that the correlating group of Figure 36

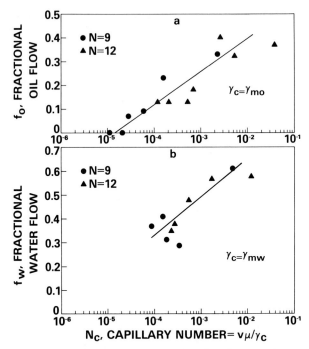

Fig. 35.   *Fractional oil flow for floods where microemulsion is continuously injected correlates with capillary number based on the controlling interfacial tension*

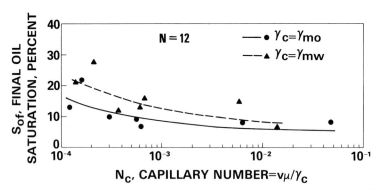

Fig. 36.   *Final oil saturation after floods using immiscible microemulsion banks broadly correlates with capillary number based on the controlling interfacial tension*

does not contain cos θ as suggested in Section III B; and, as remarked earlier (III C), displacement of residual oil is not the problem once an oil bank is formed; rather it is to maintain oil filament continuity to as low a saturation as possible, so that perhaps other dimensionless groups come into play. (See, however, Reference 12.) Although the correlation is expected to depend on properties of the specific system of interest, it is conjectured that S_{of} *decreases with N_c for all systems having favorable mobility, provided surfactant retention does not dominate oil recovery* behavior.

A possible physical interpretation of results from these slug floods is that γ_{mo} *determines the effectiveness of the displacement of oil by microemulsion at the slug front; while displacement of microemulsion by drive water at the slug rear is controlled by* γ_{mw}. *The least effective of these displacements determines the outcome.*

3. *Oil Recovery in Relation to Several Variables*

By this time the reader should be aware that the concepts of optimal salinity, temperature, etc., were introduced in expectation that the best oil recoveries would be obtained at these conditions. By way of verification, some effects on oil recovery of salinity, surfactant and cosolvent molecular structures, temperature, and composition, were determined using immiscible microemulsion slugs. Details are given in Reference 4.

An example of the kind of result obtained is shown in Figure 37 where both interfacial tension and final oil saturation are graphed as functions of salinity. If C^* is defined as the *optimal salinity for oil recovery,* then it can be seen that $C^* = 1.5\%$ and $C_\gamma = 1.4\%$ NaCl. Figure 38 shows C^* is in good agreement with $C_\phi = C_\gamma$ for all variables studied. This means, *the salinity that determines the best oil recovery can be estimated from solubilization parameter data alone*; i.e., all that is required to find the best salinity for a given surfactant system is to make a few volume measurements on equilibrated multiphase systems as a function of salinity (see Figure 23). This procedure can be used for variables other than salinity, and this is discussed in the following section.

I. Screening

Microemulsion flooding may be unique in that so many different variables are available and all functionally related. Consequently, any procedure that can systematically point to variable combinations that give good oil recovery is useful.

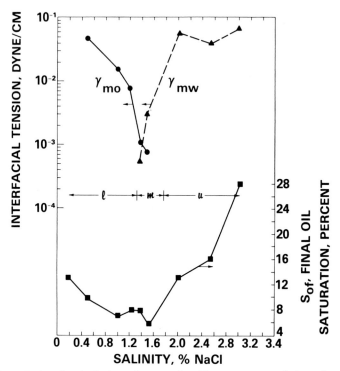

Fig. 37. Interfacial tension and oil recovery data show that interfacial tension optimal salinity very nearly corresponds to the best oil recovery

Let a variable X represent, for example, temperature, salinity, oil composition, or surfactant and cosolvent structural parameters. The screening method assumes $X^* = X_\phi$ and oil recovery correlates with $N_c(\gamma_c)$. Situations may arise wherein variables such as mobility control or surfactant retention influence oil recovery to an extent that these assumptions are invalid; nevertheless, we have found the approach applicable to a number of anionic surfactant systems.

Samples are prepared where X varies monotonically, other composition variables remaining fixed. A water-oil ratio of unity is preferred. Samples are thoroughly mixed and allowed to remain undisturbed at constant temperature until the initial opaque emulsion completely disappears, and distinct translucent phases remain. Graphs of V_o/V_s and V_w/V_s as functions of X are then prepared, X_ϕ determined, and viscosity of each microemulsion phase measured. Values of $N_c = v\mu/\gamma_c$ for the various microemulsion phases are determined and used to provide estimates of oil recovery. If N_c is sufficiently

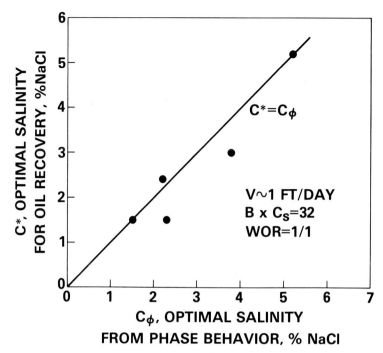

*Fig. 38.   Optimal salinity for oil recovery is about equal to
phase behavior optimal salinity for a variety of
floods using immiscible microemulsion banks*

large in the neighborhood of X_ϕ, laboratory core floods are
run to determine oil recovery as a function of X. A graph of
S_{of} vs X determines X^* and the minimum value of S_{of}. The
value of N_c that is "sufficiently large" depends on the specific system being investigated. For one system studied here,
a value greater than 10^{-3} was necessary for good oil recovery.

A modification of the screening method can sometimes be
applied to develop effective high water content microemulsions,
which have an economic advantage. In this case, X_ϕ is determined as above; but the best oil recovery is established from
core floods using microemulsions equilibrated at high WOR in
the overall mixture. In one case, reported elsewhere (4),
maximal oil recovery and C^* were independent of WOR.

J.   Connection Between Locally Miscible and Immiscible
     Microemulsion Flooding
     The immediate vicinity of the multiphase boundary is the
demarcation between injection compositions for miscible-type,
high concentration, or soluble oil microemulsion floods, on
the one hand, and immiscible microemulsion floods on the

other. It has been pointed out (III G) that minimizing height of the multiphase boundary prolongs locally miscible micro-emulsion displacement, whereas decreasing controlling inter-facial tension enhances immiscible microemulsion displacement. A question arises as to whether these two considerations are related or quite independent. The following developments answer this question and make use of the result.

Along any line passing through $C_s = 1$, i.e., 100% sur-factant, V_w/V_o is a constant, say ξ. Since $C_o + C_w + C_s = 1$, $C_w/C_o = V_w/V_o$, $C_o/C_s = V_o/V_s$ and $C_w/C_s = V_w/V_s$. It follows from Equations 1 and 2 that

$$\log \ (\gamma_{mo}/\gamma'_{mo}) \ = \ \frac{a}{\left(\dfrac{m_o}{1 + \xi}\right)\left(\dfrac{1 - C_s}{C_s}\right) + 1} \ , \tag{3}$$

$$\log \ (\gamma_{mw}/\gamma'_{mw}) \ = \ \frac{b}{\left(\dfrac{m_w \ \xi}{1 + \xi}\right)\left(\dfrac{1 - C_s}{C_s}\right) + 1} \ . \tag{4}$$

Equation (3) applies to a lower- or middle-phase microemulsion. Equation (4) applies to an upper- or middle-phase microemul-sion. If a middle-phase occurs, then C_s is the same in Equa-tions (3) and (4) so that γ_{mo} and γ_{mw} are related through the expression

$$\frac{m_w \ \xi}{m_o} = \frac{-1 + b/\log \ (\gamma_{mw}/\gamma'_{mw})}{-1 + a/\log \ (\gamma_{mo}/\gamma'_{mo})} \ .$$

For a fixed water-oil ratio, ξ, Equations (3) and (4) relate height of the multiphase region $C_s(\xi)$ to γ_{mo} and γ_{mw}. These equations are graphed for $\xi = 1$ in Figure 39, using parameters estimated from Section V C. Also shown in Figure 39 is the $\gamma - C_s$ path followed when a variable is changed monotonically so the microemulsion phase undergoes the tran-sition $\ell \to m \to u$. Every point on the path ABCD corresponds to a different ternary diagram. Along AB, all microemulsions are lower-phase; and interfacial tension, γ_{mo}, decreases as height C_s decreases. At B, the multiphase region having the least height at $\xi = 1$ is achieved, and a middle phase forms having the two tensions $\gamma_{mo}(B)$ and $\gamma_{mw}(C)$. Once past the optimum, $C_{s,min}$, microemulsions along CD are upper-phase; and interfacial tension, γ_{mw}, increases as height C_s increases.

In summary, for surfactants studied here, interfacial tensions are decreasing functions of solubilization parameters.

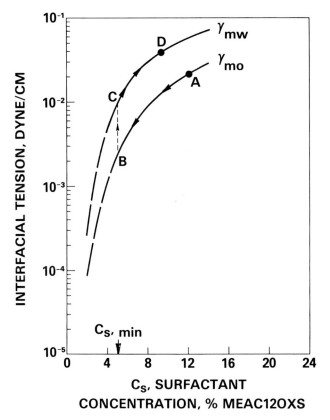

Fig. 39.   *Each point on the graph corresponds to a different*
ternary diagram. The best of these has the least
extensive multiphase region and the lowest inter-
facial tension at WOR of unity

When this is the case, *at any fixed water-oil ratio, inter-*
facial tension decreases as height of the multiphase region
decreases. It follows that *ternary diagrams favorable to*
immiscible microemulsion floods are also favorable to locally
miscible microemulsion floods; and therefore the screening
method developed here is applicable to both approaches to
oil recovery.

VI.   ACKNOWLEDGEMENTS
      The authors are pleased to acknowledge generosity of the
Society of Petroleum Engineers of A.I.M.E. for permission to
include portions of four papers, three published in *Society*
of Petroleum Engineers Journal (1,2,3), and one accepted for
publication (4).

VII.   NOMENCLATURE

a	— constant, dimensionless
b	— constant, dimensionless
c^*	— optimal salinity for oil recovery, % TDS (total dissolved solids)
C_γ	— optimal salinity for interfacial tension, % TDS
C_o	— oil concentration in microemulsion, vol. %
C_ϕ	— optimal salinity for phase behavior, % TDS
C_m	— optimal salinity for miscibility, % TDS
C_s	— surfactant concentration in microemulsion, vol. %
$C_{s,min}$	— surfactant concentration in microemulsion corresponding to minimum height of multiphase region, vol. %
C_w	— water concentration in microemulsion, vol. %
E_{hw}	— hydrophile-water interaction energy
E_{lo}	— lipophile-oil interaction energy
f_o	— fractional oil flow, dimensionless
f_w	— fractional water flow, dimensionless
G	— gel
K	— effective permeability, darcys or md
ℓ	— lower-phase microemulsion
L	— length, cm
m	— middle-phase microemulsion
m_o	— constant, dimensionless
m_w	— constant, dimensionless
N	— carbon number of alkyl side chain
N_c	— capillary number, dimensionless
P	— pressure, dyne/cm^2 or psi
P_c	— capillary pressure, dyne/cm^2
PV_{core}	— core pore volume, ml
r	— pore radius, cm
R	— cohesive energy ratio (E_{lo}/E_{hw}), dimensionless
S_1	— spherical water-external micelles with oil cores
S_2	— spherical oil-external micelles with water cores
S_o	— oil saturation, dimensionless
S_{of}	— final oil saturation, dimensionless
u	— upper-phase microemulsion
v	— velocity, cm/sec or ft/day
V_o	— volume of oil in microemulsion, ml
$V_{o,initial}$	— volume of oil initially contained in core, ml
$V_{o,inj}$	— volume of oil injected, ml
$V_{o,prod}$	— volume of oil produced, ml
V_o/V_s	— solubilization parameter for oil in microemulsion, volume ratio of oil to surfactant in microemulsion phase
V_s	— volume of surfactant in microemulsion, ml

V_w	— volume of water in microemulsion, ml
V_w/V_s	— solubilization parameter for water in micro-emulsion, volume ratio of water to surfactant in microemulsion phase
WOR	— water-oil ratio, dimensionless
X	— a variable
X^*	— optimal X for oil recovery
X_ϕ	— optimal X for phase behavior
γ	— interfacial tension, dyne/cm
γ_c	— controlling interfacial tension, dyne/cm
γ_{mo}	— microemulsion-oil interfacial tension, dyne/cm
γ'_{mo}	— a constant, dyne/cm
γ_{mw}	— microemulsion-water interfacial tension, dyne/cm
γ'_{mw}	— a constant, dyne/cm
ΔP	— pressure drop, dyne/cm^2 or psi
μ	— viscosity, poise or cp
θ	— contact angle, degrees
ξ	— water-oil ratio in microemulsion, dimensionless

VIII.   LITERATURE CITED

1.   Healy, R. N. and Reed, R. L., "Physicochemical Aspects of Microemulsion Flooding," Soc. Pet. Eng. J. Vol. 14, 491-501 (1974); Trans. AIME, Vol. 257.

2.   Healy, R. N., Reed, R. L. and Carpenter, C. W., "A Laboratory Study of Microemulsion Flooding," Soc. Pet. Eng. J. Vol. 15, 87-103 (1975); Trans. AIME (1975).

3.   Healy, R. N., Reed, R. L. and Stenmark, D. G., "Multiphase Microemulsion Systems," Soc. Pet. Eng. J. Vol. 16, 147-160 (1976); Trans. AIME (1976).

4.   Healy, R. N. and Reed, R. L., "Immiscible Microemulsion Flooding," SPE 5817 presented at SPE Improved Oil Recovery Symposium, Tulsa, Okla., Mar. 22-24, 1976.

5.   Cash, R. L., Cayias, J. L., Fournier, R. G., Jacobson, J. K., Schares, T., Schechter, R. S. and Wade, W. H., "Modeling Crude Oils for Low Interfacial Tension," SPE 5813 presented at SPE Improved Oil Recovery Symposium, Tulsa, Okla., Mar. 22-24, 1976.

6.   Muskat, M., Physical Principles of Oil Production, McGraw-Hill Book Co., Inc., New York, 1949.

7.   Wyllie, M.R.J., in Petroleum Production Handbook, 2, 25-1 to 25-14, McGraw-Hill Book Co., Inc., New York, 1962.

8.   Brown, W. O., "Mobility of Connate Water During a Waterflood," J. Pet. Tech. Vol. 9, 190-95 (1957); Trans. AIME, Vol. 210, 190-95.

9.   Jordan, J. K., McCardell, W. M. and Hocott, C. R., "Effect of Rate on Oil Recovery by Waterflooding," Oil and Gas Journal Vol. 55, 98-130 (1957).

10. Moore, T. F. and Slobod, R. L., "The Effect of Viscosity and Capillarity on the Displacement of Oil by Water," Prod. Monthly Vol. 20, 20-30 (1956).

11. Chatenever, A., "Microscopic Behavior of Fluids in Porous Systems," API Research Project Report 47b, April, 1957.

12. Stegemeier, G. L., "Relationship of Trapped Oil Saturation to Petrophysical Properties of Porous Media," SPE 4754 presented at SPE Symposium on Improved Oil Recovery, Tulsa, Okla., April 22-24, 1974.

13. Dullien, F.A.L., Dhawan, G. K., Gurak, N. and Babjak, L., "A Relationship Between Pore Structure and Residual Oil Saturation in Tertiary Surfactant Floods," Soc. Pet. Eng. J. Vol. 12, 289-96 (1972).

14. Kimbler, O. K., Reed, R. L. and Silberberg, I. H., "Physical Characteristics of Natural Films Formed at Crude Oil-Water Interfaces," Soc. Pet. Eng. J. Vol. 6, 153 (1966); Trans. AIME, Vol. 237, II-153.

15. Gardescu, I. I., "Behavior of Gas Bubbles in Capillary Spaces," Trans. AIME, Vol. 136, 351-69 (1930).

16. Taber, J. J., "Dynamic and Static Forces Required to Remove a Discontinuous Oil Phase from Porous Media Containing Both Oil and Water," Soc. Pet. Eng. J. Vol. 9, 3-12 (1969).

17. Parsons, R. W., "Velocities in Developed Five Spot Patterns," J. Pet. Tech. Vol. 26, 550 (1974).

18. Taber, J. J., Kirby, J. C. and Schroeder, F. U., "Studies on the Displacement of Residual Oil: Viscosity and Permeability Effects," Paper 47b presented at Symposium on Transport Phenomena in Porous Media, 71st National AIChE Meeting, Dallas, Texas, Feb. 20-23, 1972.

19. Foster, W. R., "A Low-Tension Waterflooding Process," J. Pet. Tech. Vol. 25, 205-10 (1973); Trans. AIME, Vol. 255.

20. Melrose, J. C. and Brandner, C. F., "Role of Capillary Forces in Determining Microscopic Displacement Efficiency for Oil Recovery by Waterflooding," J. Can. Pet. Tech. Vol. 13, 54-62 (1974).

21. MacDonald, I. F. and Dullien, F.A.L., "Correlating Tertiary Oil Recovery in Water-Wet Systems," Soc. Pet. Eng. J. Vol. 16, 7-9 (1976).

22. Taber, J. J., Kamath, I.S.K. and Reed, R. L., "Mechanism of Alcohol Displacement of Oil From Porous Media," Soc. Pet. Eng. J. Vol. 1, 195-212 (1961); Trans. AIME, Vol. 222.

23. Davis, J. A. and Jones, S. C., "Displacement Mechanisms of Micellar Solutions," J. Pet. Tech. Vol. 20, 1415-28 (1968); Trans. AIME, Vol. 243.

24. Carpenter, C. W., Jr., Private communication, 1971.

25. Wilchester, H. L., Malmberg, E. W., Shepard, J. C., Schultz, E. F., Parmley, J. B. and Dycus, D. W., "Laboratory Studies on Oil Recovery with Aqueous Dispersions of Oil-Soluble Sulfonates," SPE 4742 presented at SPE Symposium on Improved Oil Recovery, Tulsa, Okla., April 22-24, 1974.

26. Gogarty, W. B., Meabon, H. P. and Milton, H. W., Jr., "Mobility Control Design for Miscible-Type Waterfloods Using Micellar Solutions," J. Pet. Tech. Vol. 22, 141-47 (1970).

27. Maerker, J. M., "Shear Degradation of Partially Hydrolyzed Polyacrylamide Solutions," Soc. Pet. Eng. J. Vol. 15, 311-22 (1975); Trans. AIME, Vol. 259.

28. Bilhartz, H. L., Jr. and Charlson, G. S., "Field Polymer Stability Studies," SPE 5551 presented at Fall SPE Meeting, Dallas, Texas, Sept. 28-Oct. 1, 1975.

29. Knight, B. L., "Reservoir Stability of Polymer Solutions," J. Pet. Tech. Vol. 25, 618-26 (1973); Trans. AIME, Vol. 255.

30. Hill, H. J., Brew, J. R., Claridge, E. L., Hite, H. J. and Pope, G. A., "The Behavior of Polymers in Porous Media," SPE 4748 presented at SPE Improved Oil Recovery Symposium, Tulsa, Okla., April 22-24, 1974.

31. Gilliland, H. E. and Conley, F. R., "Pilot Flood Mobilizes Residual Oil," Oil and Gas J. 43-48 (1976).

32. Yost, M. E. and Stokke, O. M., "Filtration of Polymer Solutions," J. Pet. Tech. 161 (1975).

33. Tinker, G. E. and Bowman, R. W., "Determination of In-Situ Mobility and Wellbore Impairment from Polymer Injectivity Data," SPE 4744 presented at SPE Symposium on Improved Oil Recovery, Tulsa, Okla., April 22-24, 1974.

34. Vela, S., Peaceman, D. W. and Sandvik, E. I., "Evaluation of Polymer Flooding in a Layered Reservoir with Crossflow, Retention and Degradation," SPE 5102 presented at Fall SPE Meeting, Houston, Texas, Oct. 6-9, 1974.

35. Trushenski, S. P., Dauben, D. L. and Parrish, D. R., "Micellar Flooding--Fluid Propagation, Interaction and Mobility," Soc. Pet. Eng. J. Vol. 14, 633-42 (1974); Trans. AIME, Vol. 257.

36. Dawson, R. and Lantz, R. B., "Inaccessible Pore Volume in Polymer Flooding," Soc. Pet. Eng. J. Vol. 12, 448-52 (1972).

37. Al-Rikabi, H. and Osaba, J. S., "Data on Microemulsion Displacement of Oil," Oil and Gas J. 87-92 (1973).

38. Pursley, S. A., Healy, R. N. and Sandvik, E. I., "A Field Test of Surfactant Flooding, Loudon, Illinois," J. Pet. Tech. Vol. 25, 793-802 (1973).

39. Kossack, C. A. and Bilhartz, H. L., Jr., "The Sensitivity of Micellar Flooding to Certain Reservoir Properties," SPE 5808 presented at SPE Symposium on Improved Oil Recovery, Tulsa, Okla., Mar. 22-24, 1976.

40. Whiteley, R. C. and Ware, J. W., "Low Tension Waterflood Pilot at the Salem Unit, Marion County, Illinois, Part 1--Field Implementation and Results," SPE 5832 presented at SPE Symposium on Improved Oil Recovery, Tulsa, Okla., Mar. 22-24, 1976.

41. Widmyer, R. H., Satter, A., Frazier, G. D. and Graves, R. H., "Low Tension Waterflood Pilot at the Salem Unit, Marion County, Illinois, Part 2--Performance Evaluation," SPE 5833 presented at SPE Symposium on Improved Oil Recovery, Tulsa, Okla., Mar. 22-24, 1976.

42. Strange, L. K. and Cloud, W. B., "Displacement of Reservoir Brine by Fresh Water--Four Field Case Histories," SPE 5834 presented at SPE Symposium on Improved Oil Recovery, Tulsa, Okla., Mar. 22-24, 1976.

43. Dabbous, M. K. and Elkins, L. E., "Preinjection of Polymers to Increase Reservoir Flooding Efficiency," SPE 5836 presented at SPE Symposium on Improved Oil Recovery, Tulsa, Okla., Mar. 22-24, 1976.

44. Strange, L. K. and Talash, A. W., "Analysis of Salem Low Tension Waterflood Test," SPE 5885 presented at SPE Symposium on Improved Oil Recovery, Tulsa, Okla., Mar. 22-24, 1976.

45. Ahearn, G. P. and Gale, W. W., "Surfactant Waterflooding Process," U.S. Patent 3,302,713, February 7, 1967.

46. Gale, W. W. and Sandvik, E. I., "Tertiary Surfactant Flooding: Petroleum Sulfonate Composition--Efficacy Studies," Soc. Pet. Eng. J. 191-99 (1973).

47. Holm, L. W., "Use of Soluble Oils for Oil Recovery," J. Pet. Tech. Vol. 23, 1475-83 (1971); Trans. AIME, Vol. 251.

48. Wilson, P. M., Murphy, C. L. and Foster, W. R., "The Effects of Sulfonate Molecular Weight and Salt Concentration on the Interfacial Tension of Oil-Brine-Surfactant Systems," SPE 5812 presented at SPE Symposium on Improved Oil Recovery, Tulsa, Okla., Mar. 22-24, 1976.

49. Froning, H. R. and Askew, W. S., "Straight Chain Sul-
     fonates for Use in Solubilized Oil-Water Solutions
     for Miscible Waterflooding," U.S. Patent 3,714,062,
     January 30, 1973.

50. Dauben, D. L. and Froning, H. R., "Development and
     Evaluation of Micellar Solutions to Improve Water
     Injectivity," J. Pet. Tech. Vol. 23, 614-20 (1971).

51. Hill, H. J., "Secondary Oil Recovery Process with
     Incremental Injection of Surfactant Slugs," U.S.
     Patent No. 3,638,728, February 1, 1972.

52. Froning, H. R. and Treiber, L. E., "Development and
     Selection of Chemical Systems for Miscible Water-
     flooding," SPE 5816 presented at SPE Symposium on
     Improved Oil Recovery, Tulsa, Okla., Mar. 22-24,
     1976.

53. Marsden, S. S., Jr. and McBain, J. W., "Aqueous Systems
     of Nonionic Detergents as Studied by X-ray Diffrac-
     tion," J. Phys. Chem. Vol. 52, 110-29 (1948).

54. Anderson, D. R., Bidner, M. S., Davis, H. T., Manning,
     C. D. and Scriven, L. E., "Interfacial Tension and
     Phase Behavior in Surfactant-Brine-Oil Systems,"
     SPE 5811 presented at SPE Symposium on Improved Oil
     Recovery, Tulsa, Okla., Mar. 22-24, 1976.

55. Degroot, M., "Flooding Process for Recovering Oil from
     Subterranean Oil Bearing Strata," U.S. Patent No.
     1,823,439, 1929.

56. Degroot, M., "Flooding Process for Recovering Fixed Oil
     from Subterranean Oil Bearing Strata," U.S. Patent No.
     1,823,440, 1929.

57. Gogarty, W. B. and Tosch, W. C., "Miscible-Type Water-
     flooding: Oil Recovery with Micellar Solutions,"
     J. Pet. Tech. Vol. 20, 1407-14 (1968); Trans. AIME,
     Vol. 243.

58. Cooke, C. E., Jr., "Microemulsion Oil Recovery Process,"
     U.S. Patent No. 3,373,809, March 19, 1968.

59. Gogarty, W. B. and Olson, R. W., "Use of Microemulsions
     in Miscible-Type Oil Recovery Procedure," U.S. Patent
     No. 3,254,714 (1962).

60. Danielson, H. H., Paynter, W. T. and Milton, H. W., Jr.,
     "Tertiary Recovery by the Maraflood Process in the
     Bradford Field," J. Pet. Tech. Vol. 16, 129-38 (1976).

61. Bleakley, W. B., "How the Maraflood Process Performs,"
     Oil and Gas J. Vol. 69, 49-54 (1971).

62. Gogarty, W. B., "Status of Surfactant or Micellar
     Methods," J. Pet. Tech. Vol. 28, 93-102 (1976).

63. Holm, L. W. and Bernard, G. G., "Secondary Recovery
     Water-Flood Process," U.S. Patent No. 3,082,822,
     1959.

64.   Csaszar, A. K., "Solvent–Waterflood Oil Recovery Process," U.S. Patent No. 3,163,214, 1961.

65.   Knight, R. K. and Baer, P. J., "A Field Test of Soluble-Oil Flooding at Higgs Unit," Soc. Pet. Eng. J. Vol. 25, 9–15 (1973).

66.   Reed, R. L., Healy, R. N., Stenmark, D. G. and Gale, W. W., "Recovery of Oil Using Microemulsions," U.S. Patent No. 3,885,628, May 27, 1975.

67.   Winsor, P. A., Solvent Properties of Amphiphilic Compounds, Butterworth's Scientific Publications, London, 1954.

68.   Ekwall, P., Mandell, L. and Fontell, K. J., "Binary and Ternary Aerosol OT Systems," J. Coll. Int. Sci. Vol. 33, 215 (1970).

69.   Shinoda, K. and Kunieda, H., "Conditions to Produce So-Called Microemulsions: Factors to Increase the Mutual Solubility of Oil and Water by Solubilizer," J. Coll. Int. Sci. Vol. 42, 381–387 (1973).

70.   Shah, D. O., Tamjeedi, A., Falco, J. W. and Walker, R. D., Jr., "Interfacial Instability and Spontaneous Formation of Microemulsions," AIChE Journal Vol. 18, 1116–20 (1972).

71.   Schulman, J. H., Stoeckenius, W. and Prince, L. M., "Mechanism of Formation and Structure of Microemulsions by Electron Microscopy," J. Phys. Chem. Vol. 63, 1677–80 (1959).

72.   Cooke, C. E. and Schulman, J. H., in Surface Chemistry (Proc. of the 2nd Scandinavian Symp. on Surface Activity, 1964), Academic Press, New York, 1965.

73.   Osipow, L. I., Surface Chemistry, Reinhold Publishing Corp., New York, 1962.

74.   Tosch, W. C., Jones, S. C. and Adamson, A. W., "Distribution Equilibria in a Micellar Solution System," J. Coll. Int. Sci. Vol. 31, 297–306 (1969).

75.   Winsor, P. A., "Liquid Crystallinity in Relation to Composition and Temperature in Amphiphilic Systems," Third International Liquid Crystal Conference, Berlin, Aug. 24–28, 1970. See also: Gray, G. W. and Winsor, P. A., Liquid Crystals and Plastic Crystals, 1, 199 ff, Ellis Horwood, Ltd., Chichester, England, 1974.

76.   Shah, D. O., Walker, R. D., Hsieh, W. C., Shah, N. J., Dwivedi, S., Nelander, J., Pepinsky, R. and Deamer, D. W., "Some Structural Aspects of Microemulsions and Co-Solubilized Systems," SPE 5815 presented at SPE Symposium on Improved Oil Recovery, Tulsa, Okla., Mar. 22–24, 1976.

77.   Fay, J. A., Molecular Thermodynamics, Addison Wesley, Reading, Mass., 1965.

78. Callen, H. B., Thermodynamics, John Wiley and Sons, Inc., New York, 1960.

79. Jones, S. C. and Dreher, K. D., "Cosurfactants in Micellar Systems Used for Tertiary Oil Recovery," SPE 5566 presented at Fall SPE Meeting, Dallas, Texas, Sept. 28-Oct. 1, 1975.

80. Mungan, N., "Improved Waterflooding Through Mobility Control," Canad. J. Chem. Eng. Vol. 49, No. 1, 32 (1971).

81. Dreher, K. D. and Sydansk, R. D., "On Determining the Continuous Phase of Microemulsions," J. Pet. Tech. Vol. 23, 1437-38 (1971).

82. Dreher, K. D. and Sydansk, R. D., "Observation of Oil Bank Formation During Micellar Flooding," SPE 5838 presented at SPE Symposium on Improved Oil Recovery, Tulsa, Okla., Mar. 22-24, 1976.

83. Adamson, A. W., "A Model for Micellar Emulsions," J. Coll. Int. Sci. Vol. 29, 261-67 (1969).

84. England, D. C. and Berg, J. C., "Transfer of Surface Active Agents Across a Liquid-Liquid Interface," AIChE Journal Vol. 17, 313 (1971).

85. Robbins, M. L., "The Theory of Microemulsions," presented at 76th National AIChE Meeting, Tulsa, Okla., Mar. 7-14, 1974.

86. Robbins, M. L., "Theory For Phase Behavior of Microemulsions," SPE 5839 presented at SPE Symposium on Improved Oil Recovery, Tulsa, Okla., Mar. 22-24, 1976.

87. Hansen, C. M., "The Universality of the Solubility Parameter," I&EC Product Research and Development Vol. 8, 2-11 (1969).

88. Hildebrand, J. and Scott, R., Solubility of Non-Electrolytes, 3rd Ed., Reinhold, New York, 1949.

89. Hildebrand, J. and Scott, R., Regular Solutions, Prentice-Hall, Englewood Cliffs, New Jersey, 1962.

90. Hardy, W. B., Proc. Royal Soc. (London) Vol. 88A, 313 (1913).

91. Donahue, D. J. and Bartell, F. E., "The Boundary Tension at Water-Organic Liquid Interfaces," Journal of Physical Chemistry Vol. 56, 480-84 (1952).

92. Puerto, M. C. and Gale, W. W., "Estimation of Optimal Salinity and Solubilization Parameters For Alkyl Orthoxylene Sulfonate Mixtures," SPE 5814 presented at SPE Symposium on Improved Oil Recovery, Tulsa, Okla., Mar. 22-24, 1976.

93. Fink, T., Private communication, Feb., 1976.

FLOWS OF POLYMERIC SOLUTIONS AND EMULSIONS
THROUGH POROUS MEDIA--CURRENT STATUS

Arthur B. Metzner
University of Delaware

I. ABSTRACT

The abundant literature on this subject is subdivided into separable technical questions in order to focus upon an
orderly progression of the problems in fluid mechanics, thermodynamics and interfacial mechanics deserving further study.

II. SCOPE

Much of the technical literature in this area has been
of a "scouting" nature and has served to define the distinct
problems which may be encountered during studies of complex
multiphase flows through porous media--but not to solve them.
The thesis of this paper is that a sufficient variety of real
problems deserving scientific study have now been defined and
that the primary goal of further research should be to provide precise analyses of these.

III. CONCLUSIONS AND SIGNIFICANCE

The progression of problems warranting understanding include all of the following:
(a) Flowrate-pressure gradient-velocity profile studies
for the flow of homogeneous and isotropic fluids through the
interstices of both "model" and real porous media.
(b) Thermodynamics and fluid mechanics of polymer retention in porous media.
(c) Complications in the above due to:
    (i) Anisotropy of the fluid.
    (ii) Adsorption and pore blockage.
    (iii) Interfacial effects in multiphase systems.
    (iv) Electrical charge effects.
    (v) Heterogeneities in the porous medium.
A summary of the state-of-the-art for the first of these
problems is provided in some detail since it must be understood fully before one may turn productively to a consideration of the other more complex questions. Secondly, an analysis of non-adsorptive polymer retention and concentration in
porous media is provided. Thirdly, non-continuum interfacial

phenomena are listed for reasons of completeness, but this listing is terse with little description and no analysis.

IV. CONTINUUM FLOWS THROUGH POROUS MEDIA

A. Pressure Drop-Flowrate Relationships

Analyses of fluid flows through porous media under laminar flow conditions have almost universally been developed by coupling a specific model of the pore structure of the medium with a specific model of the rheological properties of the fluid being employed in the flow process. Most commonly the pore structure is modeled by means of the "cylindrical equivalent capillary": a cylindrical duct of length and diameter such that it exhibits the same resistance to flow as the actual interstices in the real porous medium. Thoughtful and clear descriptions of this model are presented by Bird, Stewart and Lightfoot (1), in the research papers of Christopher and Middleman (2) and of Gaitonde and Middleman (3) and in the comprehensive summary of the subject by Savins (4). This latter paper shows how a particular rheological model of the viscosity function of the fluid may be replaced with a generalized analysis similar to that employed in laminar and turbulent flows of non-Newtonian fluids through tubes.

Separately, there has been a growing awareness of the fact that the actual pores in a granular bed are, of course, not isolated circular cylinders of constant cross-sectional area: they are interconnected and non-circular, and the changes in cross-sectional area to which a fluid element is exposed as it moves through the bed may occur rapidly and be of large magnitude. An early attempt to apply some of these considerations to the flow of viscoelastic fluids was published by Marshall and Metzner (5); the subject of the flow of Newtonian fluids through pores of complex geometry has recently been extended substantially and its current status is presented in a series of papers by Dullien (6), Payatakes, Tien and Turian (7,8,9), Dullien and Azzam (10,11) and Batra, Fulford and Dullien (12). At the present time, therefore, a designer may choose between analyses which consider either non-linear fluid properties or some of the complexities of real pore geometries, but generally not both.

If we assume the Reynolds and Deborah numbers of flows in petroleum reservoirs to be extremely low one simple approximation for the pressure drop-flowrate relationship in a single pore of local diameter D_p is given by Poiseuille's law and its generalization for non-linear (power law) fluids as

$$- \frac{\partial p}{\partial x} \, \alpha \, \frac{Q_p^n}{(D_p)^{3n+1}} \tag{1}$$

The pressure drop over some finite pore length L is obtained by integration as:

$$\Delta p = - \int_0^L dp = Q_p^n \int_0^L \frac{dx}{D_p^{3n+1}} \tag{2}$$

The integral in the last term of Equation 2 shows that the "mean diameter" of the pore is obtained by integrating or averaging the diameter raised to the (3n+1) power. Whatever the diameter-distance relationship may be for real porous media it is clear that this average will be appreciably different for shear-thinning fluids such as most emulsions and polymeric solutions ($0.1 < n < 0.6$) than it is for linear Newtonian fluids ($n = 1.0$). Sheffield and Metzner (13) have considered this problem in substantial detail; the salient results appear to be the following.

(a) As indicated by Equation 2 the pressure drop-flowrate relation is very sensitive to the precise form of the diameter-distance relationship chosen. This D_p-x behavior is not known for most porous media of interest and is likely to change as the porosity of the medium changes. Design procedures at present must therefore be empirical and different for fluids having differing power law indices n. This is equivalent to stating that the rate at which a polymer slug or emulsion penetrates a formation cannot be predicted at present without having experimental data on samples of the actual porous solids of interest; a correlation has been developed for beds composed of spherical particles and having a void fraction of 0.36, but sufficient data do not yet appear to be available for other porosity levels to permit any confident generalization of these results.

(b) For conditions under which Equations 1 and 2 apply a logarithmic plot of pressure drop vs. flowrate will exhibit a slope of n, equal to the viscometric stress-shear rate slope at equivalent shear rates. This is frequently found not to be the case in practice even when the pores are sufficiently large to render non-continuum effects negligible (14,15,16,17). This may be due to a separation of the flow from the solid boundaries of the pores at Reynolds numbers which decrease as the power law index n decreases. If this explanation proves to be correct, this complication would be expected to disappear at sufficiently low Reynolds numbers. Unfortunately, none of the available laboratory data extend to sufficiently low Reynolds numbers to verify this hypothesis, though many flows under reservoir conditions would. This mismatch between laboratory and field conditions may thus be of rather substantial importance.

(c) Since displacement of the fluid in a porous stràtum
may be adversely affected by separation of the pusher fluid
from the surfaces of the pore one might expect that the displacement efficiency would be inferior whenever the logarithmic pressure drop-flowrate slope in the porous medium exceeds
the viscometric power law slope n at comparable fluid deformation rates. No verification or refutation of this suggestion
appears to be available.

If flows occur through parallel strata of differing permeabilities of pore diameters a rearrangement of Equation 1
gives, for a given pressure gradient applied equally to all
strata:

$$V_p \propto \left(D_p\right)^{\frac{n+1}{n}} \tag{3}$$

The exponent changes numerically from a value of 2 for
Newtonian fluids to a value of 3 when n = 0.5 and to 6 for
n = 0.2. Thus, shear thinning fluids exhibit a far greater
tendency to channel through the most permeable portions of
a formation. This effect may sometimes be exploited by means
of an appropriate slug injection strategy; it may also, in
principle, be offset by viscoelastic effects which increase
with increasing values of the Deborah number $\Theta V_p/D_p$ (5). The
several analyses of this problem by Marshall and Metzner,
Dauben and Menzie (18), Savins (4), Wissler (19) and James
and McLaren (20) are all informative but preliminary: they
require further analysis which considers precisely the simultaneous shearing and stretching of the fluid elements flowing
in a converging-diverging channel. Exploitation of these
effects in practice may be dependent on development of fluids
having sufficiently great relaxation times Θ for the Deborah
number to exceed values of about unity at flowrates of interest.

In summary, even this simplest possible problem of flows
of non-linear continua through porous media is not understood
at all well. Design predictions must therefore be based on
experimental data on flows of fluids having the same power
law index through the actual porous medium of interest (to
insure comparability of pore geometry) and within the same
range of Reynolds numbers. The research problems requiring
study to overcome this scale-up inability include the
following:

(a) A study of the actual pore geometries encountered
in practice and development of simple means of modeling these
is a first and indispensable requirement. This is essentially
an applied mathematics problem in solid geometry.

(b) A study of flows, at low Reynolds numbers, of purely viscous non-linear fluids in pores having the various geometries identified in (a) above is needed. This research represents extensions of the studies of Payatakes and Dullien noted earlier--to non-linear fluids and to pore geometries which may model those encountered in practice. The extension of the numerical work of Payatakes et al. (7,8,9) to non-linear fluids is not trivial; whether their choice of model pore geometries is realistic or somewhat too idealized is also not entirely clear at present.

(c) A study of separational phenomena in flows through orifices or converging-diverging pores is required to confirm or reject the Sheffield-Metzner hypothesis that this factor may be of major importance.

(d) A continued study of viscoelastic fluid flow effects in porous media is needed to determine whether such fluids may be employed to improve the uniformity of flows through parallel strata exhibiting differences in permeability.

B. Thermodynamics and Fluid Mechanics of Polymer Retention in Porous Media

It has been discovered (21-24) that the average concentration of polymer in the interstices of a porous medium may be larger than the concentration of the same dissolved polymer in the fluid stream entering and leaving the porous medium; the enrichment of the solution in the interstices increases with increasing flowrate. The recent definitive study of Dominguez and Willhite (25) shows that such polymer retention in the porous medium need not be due to adsorption on the pore walls but may appear as a continuum phenomenon.

This enrichment process must, of course, conform to known thermodynamic and fluid mechanical principles. The fact that there appears to be no analog of this effect in solutions of non-polymeric species may provide the key to the phenomenon. Let us consider a porous medium made up of two groups of interconnected pores, as follows: Pore Group 1 will be designated as that set of pores through which the major part of the flow process occurs and Group 2 the "inaccessible" or "dead-end" pores which serve as appended but relatively stagnant sinks into which the flowing fluid diffuses or flows slowly. At equilibrium the fluid entrapped in Region 2 must be at the same activity or thermodynamic free energy levels as that in Region 1. Considering the polymeric species dissolved in the fluid we may write:

$$G_2 - G_1 = \Delta G = 0$$
$$= (\Delta G)_{config} + (\Delta G)_{conc} \qquad (4)$$

Assuming dilute solutions we may replace the last term in Equation 4 with its equivalent:

$$(\Delta G)_{conc} = RT \ln c_2/c_1 \tag{5}$$

The configurational free energy change will consist, in general, of changes due to both enthalpy and entropy differences. In the case of flexible polymeric molecules the enthalpy may be largely independent of any stretching or molecular alignment occurring in the fluid as a result of both the shearing and extensional deformations in the flow process, but the entropy decreases sharply with increasing molecular alignment. Thus,

$$(\Delta G)_{config} = (\Delta H)_{config} - T (\Delta S)_{config}$$
$$\approx - T (S_2 - S_1)_{config} \tag{6}$$

Combining Equations 4-6 one obtains:

$$\ln \frac{c_2}{c_1} = \frac{(S_2 - S_1)}{R} \tag{7}$$

The difference $(S_2 - S_1)$ is a positive quantity which increases progressively with increasing molecular alignment (flowrate). Thus, Equation 7 enables computation of the concentration ratio c_2/c_1 as a function of flowrate if the entropy of the flowing stream, S_1, can be computed. The entropy of the stagnant solution in the dead-end pores, S_2, may be assigned a base or datum value of zero.

Marrucci (26) has shown that a simple approximation for the entropy change is given by:

$$S_2 - S_1 = \frac{1}{2c_1 T} \operatorname{tr} \underset{\sim}{\zeta} \tag{8}$$

in which $\underset{\sim}{\zeta}$ denotes the deviatoric stress in Region 1 as computed from the contravariant Maxwell rheological description of polymer solutions and c_1 is the molar concentration of polymer in this region.

Combining Equations 7 and 8:

$$\ln \frac{c_2}{c_1} = \frac{\operatorname{tr} \underset{\sim}{\zeta}}{2RTc_1} \tag{9}$$

Using Equation 9 with the assumption of a steady laminar shearing flow through cylindrical capillaries leads to concentration ratios c_2/c_1 which are smaller than observed experimentally by Maerker or by Dominguez and Willhite. On

the other hand, if one chooses stress levels characteristic
of dilute solutions undergoing extensional deformations as
well as shearing [see, for example, Baid and Metzner (27)]
then Equation 9 yields concentration changes comparable to
those observed. This analysis may provide, therefore, additional confirmation of the inadequacies of the cylindrical
pore model of a porous medium when flows of non-linear fluids
are being considered.

    If these thermodynamic arguments are correct and complete
the magnitude of the excess polymer retained in the porous
medium (though not its concentration) will depend upon the
volume of the dead-end pores and would be zero in the case of
flow studies using idealized single pores. However, in this
case, as well as in the 2-region model just considered, it
is possible that an additional phenomenon may also occur.
Giesekus (28) and, later, Kanel (29) have shown that many
dilute polymeric solutions crystallize when subjected to
elongational deformations. Using the same reasoning embodied
in Equations 4 and 5 we have, if we denote by G_2 the free
energy of a polymer solution at rest and by G_1 that of the
same solution being stretched:

$$\Delta G = G_2 - G_1 = H_2 - H_1 - T(S_2 - S_1) \qquad (10)$$

    For reasons noted earlier $H_2 \approx H_1$ while the solution at
rest exhibits the greater entropy and $(S_2 - S_1)$ is a positive
quantity. Thus, ΔG is negative and the free energy of the
system being deformed, G_1, is greater than that of the stagnant fluid. This free energy change will promote crystalli-
zation of the flowing stream and may be likened to the free
energy change accompanying evaporation of the stagnant solution to its saturation concentration. In fact, in Kanel's
experiments employing a ductless siphon (the Fano effect) to
stretch dilute solutions, it was usually impossible to deform
the fluid without evidence of the formation of a separate
phase appearing in the polyacrylamide solutions being used.

    Flows in the interstices of a porous medium are more
complex than the simple extension studied by Kanel, but there
is no doubt that some molecular alignment will occur during
flow. If this is sufficient to produce crystallites, these
may, in principle, be separated from the surrounding fluid
as it flows through the convergent sections of a pore either
by filtration or as a result of the Uebler effect (Uebler,
30; Metzner, 31; Gohlke, 32). Thus, under the right hypothetical circumstances, a homogeneous and dilute polymer
solution entering a porous medium could form separate
crystalline phases under isothermal conditions, and these
crystallites could be selectively retained by the porous

medium. Since the Uebler effect restricts the motion of the
largest particles selectively, it is, however, somewhat ques-
tionable whether it is likely to be operative in the very
fine pores of interest herein. Likewise, filtration of a
separate phase would cause progressive permeability reductions with increasing time of flow, and this appears to
conflict with the experimental observations of both Maerker
and Dominguez and Willhite though it is consistent with data
reported by Chauveteau and Kohler.

Further and definitive exploration of polymer retention
effects is clearly in order. Whether fractionation of the
polymeric species by molecular weight or molecular linearity
is possible using a chromatographic process exploiting this
effect may be an interesting related consideration.

V.   NON-CONTINUUM EFFECTS IN POROUS MEDIA

At least 4 different phenomena which could effect the
flowrate-pressure gradient characteristics of flows through
porous media are known. In most cases our knowledge is quite
fragmentary and, in a sense, appropriately so: until the
simpler continuum phenomena discussed above are understood
quantitatively it will be impossible to develop quantitative
descriptions of the ancillary complications due to non-
continuum effects unless these are of such a magnitude as
to overshadow the others completely. These will now be
described very tersely.

A.   Anisotropic Fluid Flows

The effects of fluid anisotropy--arising out of rigid
macromolecular structures, anisotropic structures of emulsified phases or anisotropic suspended particles--are felt
primarily at solid surfaces when the ratio of the length of
the structural feature to the pore diameter (ℓ/D_p) is no
longer small. Very readable accounts have been presented
by Stokes (33) and Spearot (34). In principle, anisotropic
fluid structures may also influence continuum fluid behavior
but the theory of non-linear viscoelastic simple fluids is so
general in its formulation that any additional complications
due to continuum anisotropy would appear to be difficult to
find and to ascribe properly--hence do not appear to be worth
considering at this time.

The principal results of Stokes and Spearot may be stated
as follows:

(a) The viscous response of anisotropic fluids in
shearing flows may be dependent upon both the deformation
rate and its gradient. Consequently viscous responses
measured, say, in a Couette viscometer are not simply
translatable into prediction of pressure drop-flowrate
relationships in flow through channels. Additionally the

vorticity of the velocity field may be an explicit parameter
of importance and hence the flow in a channel with converging
or diverging portions may be appreciably different from that
in a duct with parallel walls.

(b) All of the above complications and matters of principle may be ignored, at least in the case of the cholesteric
liquid crystalline structures studied by Spearot, and the
flow process may be instead described by means of a tradi-
tional and very simple empirical construct of physical
chemistry: the flow of a non-linear but isotropic fluid in
a channel whose diameter or width is decreased by the presence
of an immobile adsorbed film. This is not to say that the
continuum effects will always be negligible but rather that
the immobile film is so dominant, at least in some systems,
that its modeling by the simplest means possible should be
the first objective of studies in this area. The proper
means for consideration and interpretation of such surface
effects in flows through tubes are well developed (Mooney,
35). The presence of adsorbed films changes the geometry
of the apparatus being used to study the flow process and
instruments having variable small clearances, such as the
cone-and-plate or biconal viscometers, may be unsuited for
use in these systems.

B.   Adsorption and Pore Blockage

Problems of adsorption on solid surfaces and constriction
of the pore structure by the adsorbed layer appear to involve
relatively few conceptual difficulties but deserve much further study in view of their economic importance: if controlled the adsorption process may have a pronounced effect
upon petroleum recovery; if uncontrolled it may represent a
"sink" for loss of enormous quantities of costly additives,
as noted in other papers in this volume.

C.   Interfacial Effects in Multiphase Systems

The possible role of interfacial viscosity when multiphase flows occur through porous media is considered by other
contributors to this volume and is only mentioned here for
reasons of completeness. Melrose and Brandner (36) have
presented a readable account of the role of capillary forces
in multiphase flows.

Characterization of the physical properties of interfaces
between immiscible phases would appear to require, as a matter
of principle, not only interfacial tensions and viscosities
as relevant physical properties but also elastic moduli:
especially in the case in which macromolecular surface
active species are present one might expect the interface
to exhibit an appreciable rigidity. The elastic properties

of planar interfaces and their influence on wave generation and propagation have been described in detail by Mayer (37). This writer is unaware of published applications of these considerations to multiphase flows through porous media.

D. Electrical Charge Effects, Streaming Potential

The dramatic effects of ionic species on the pressure drop-flowrate behavior of polymer solutions and emulsions may include surface effects as well as simply influencing the conformation of the molecules in solution. Shah (38) has undertaken a study of this problem.

VI. COMPLICATIONS DUE TO THE HETEROGENEITY OF THE POROUS MEDIUM

Most of the references quoted above tacitly assume that flows through porous media may be described adequately using a study of flows through a single pore of appropriate complexity. In fact the pore structure in porous media is intercommunicating and fluid elements which are adjacent at one point in the pore space may become widely separated as the fluid moves downstream--due to the random and interconnecting structures of the pores. This suggests that a stochastic model of the flow process must be developed ultimately. Such a model would appear to require the introduction of additional (adjustable) coefficients into any equations used to describe the flow process; this reviewer believes such complications may be best postponed until the insights obtainable through simpler models have been exhausted.

VII. ACKNOWLEDGEMENT

The writer has profited especially from discussion and correspondence on this subject with two learned industrial colleagues: J. G. Savins and J. M. Maerker.

VIII. NOTATION

c   concentration of polymer in solution
D_p local diameter of a pore
G   Gibbs' free energy; ΔG_{config} refers to the contributions to free energy change arising out of configurational molecular considerations only, and ΔG_{conc} refers to changes arising out of concentration differences only
H   enthalpy
ℓ length of "flow unit" in an anisotropic medium or thickness of an adsorbed film
L   length of porous medium
n   flow behavior index (power law exponent) of non-linear fluid being considered
p   pressure

Q_p volumetric flowrate through a single pore
R gas constant
S entropy
T absolute temperature
V_p mean velocity of flow through a single pore
x distance through the porous medium in the direction of flow
Θ relaxation time of viscoelastic fluid

IX. LITERATURE CITED

1. Bird, R. B., Stewart, W. E. and Lightfoot, E. N., "Transport Phenomena," Wiley, New York, 1960.

2. Christopher, R. H. and Middleman, S., "Power-Law Flow Through a Packed Tube," Ind. Eng. Chem. Fund. 4, 422 (1965).

3. Gaitonde, N. Y. and Middleman, S., "Flow of Viscoelastic Fluids Through Porous Media," Ind. Eng. Chem. Fund. 6, 145 (1967).

4. Savins, J. G., "Non-Newtonian Flow Through Porous Media," Ind. Eng. Chem. 61, #10, 18 (1969).

5. Marshall, R. J. and Metzner, A. B., "Flow of Viscoelastic Fluids Through Porous Media," Ind. Eng. Chem. Fund. 6, 393 (1967).

6. Dullien, F.A.L., "New Network Permeability Model of Porous Media," A.I.Ch.E. J. 21, 299 (1975).

7. Payatakes, A. C., Chi Tien and Turian, R. M., "A New Model for Granular Porous Media. Part I. Model Formulation," A.I.Ch.E. J. 19, 58 (1973).

8. _____. Ibid. "Part II. Numerical Solution of Steady State Incompressible Newtonian Flow Through Periodically Constricted Tubes," A.I.Ch.E. J. 19, 67 (1973).

9. _____. "Further Work on Flow Through Periodically Constricted Tubes--A Reply," A.I.Ch.E. J. 19, 1036 (1973).

10. Dullien, F.A.L. and Azzam, M.I.S., "Flow Rate-Pressure Gradient Measurements in Periodically Nonuniform Capillary Tubes," A.I.Ch.E. J. 19, 222 (1973).

11. Dullien, F.A.L. and Azzam, M.I.S., "Effect of Geometric Parameters on the Friction Factor in Periodically Constricted Tubes," A.I.Ch.E. J. 19, 1035 (1973).

12. Batra, V. K., Fulford, G. D. and Dullien, F.A.L., "Laminar Flow Through Periodically Convergent-Divergent Tubes and Channels," Can. J. Chem. Eng. 48, 622 (1970).

13. Sheffield, R. E. and Metzner, A. B., "Flows of Non-Linear Fluids Through Porous Media," A.I.Ch.E. J. 22, 736 (1976).

14. Burcik, E. J., "A Note on the Flow Behavior of Poly-
     acrylamide Solutions in Porous Media," Producer's
     Monthly 29(6), 14 (1965).

15. Gogarty, W. B., "Rheological Properties of Pseudoplastic
     Fluids in Porous Media," Trans. Soc. Pet. Eng. 240,
     Part 1, 149 (1967).

16. Gogarty, W. B., "Mobility Control with Polymer Solutions,"
     Trans. Soc. Pet. Eng. 240, Part 1, 161 (1967).

17. Pye, D. J., "Improved Secondary Recovery Control of
     Water Mobility," J. Pet. Tech. 231, Part 1, 911
     (1964).

18. Dauben, D. L. and Menzie, D. E., "Flow of Polymer Solu-
     tions Through Porous Media," Trans. Soc. Pet. Eng.
     240, Part 1, 1065 (1967).

19. Wissler, E. H., "Viscoelastic Effects in the Flow of
     Non-Newtonian Fluids Through a Porous Medium," Ind.
     Eng. Chem. Fund. 10, 411 (1971).

20. James, D. F. and McLaren, D. R., "The Laminar Flow of
     Dilute Polymer Solutions Through Porous Media," J.
     Fluid Mech. 70, 733 (1975).

21. Desremaux, L., Chauveteau, G. and Martin, M., "Com-
     portement des Solutions de Polymeres en Milieu
     Poreux," Communication #28, L'Association de
     Recherches sur les Techniques d'Exploitation du
     Petrole, Paris (1971).

22. Maerker, J. M., "Dependence of Polymer Retention on
     Flowrate," J. Pet. Tech. 25, 1307 (1973).

23. Rhudy, J. S., Fullinwider, J. H. and VerSteeg, D. J.,
     U.S. Patent 3,734,183 (1973).

24. Chauveteau, G. and Kohler, N., "Polymer Flooding: the
     Essential Elements for Laboratory Evaluation," SPE
     Paper 4745 presented in Tulsa, Oklahoma (1974).

25. Dominguez, J. G. and Willhite, G. P., "Polymer Reten-
     tion and Flow Characteristics of Polymer Solutions
     in Porous Media," SPE Paper 5835 presented in Tulsa,
     Oklahoma (1976).

26. Marrucci, G., "The Free Energy Constitutive Equation for
     Polymer Solutions from the Dumbbell Model," Trans.
     Soc. Rheology 16, 321 (1972).

27. Baid, K. M. and Metzner, A. B., "Viscoelastic Properties
     of Dilute Polymer Solutions," Trans. Soc. Rheology 20,
     000 (1976).

28. Giesekus, Hanswalter, "Verschiedene Phänomene in
     Strömung viskoelastischer Flüssigkeiten durch
     Düsen," Rheologica Acta 8, 411 (1969).

29. Kanel, F. A., "The Extension of Viscoelastic Materials,"
     Ph.D. Thesis, University of Delaware, Newark, Delaware
     (1972).

30. Uebler, E. A., "Pipe Entrance Flow of Elastic Liquids," Ph.D. Thesis, University of Delaware, Newark, Delaware (1966).

31. Metzner, A. B., "Behavior of Suspended Matter in Rapidly Accelerating Fluids: The Uebler Effect," A.I.Ch.E. J. 13, 316 (1967).

32. Gohlke, D. J., "The Uebler Effect," B.Ch.E. Thesis, University of Delaware, Newark, Delaware (1971).

33. Stokes, V. K., "Couple Stresses in Fluids," Phys. of Fluids 9, 1709 (1966).

34. Spearot, J. A., "The Rheological Behavior of Liquid Crystals in Viscometric Flows," M.Ch.E. Thesis, University of Delaware, Newark, Delaware (1970).

35. Mooney, Melvin, "Explicit Formulas for Slip and Fluidity," J. Rheology 2, 210 (1931).

36. Melrose, J. C. and Brandner, C. F., "Role of Capillary Forces in Determining Microscopic Displacement Efficiency for Oil Recovery by Waterflooding," J. Can. Pet. Tech. 13, #4, 54 (1974).

37. Mayer, Ernest, "The Inference of Interfacial Properties from Capillary Wave Experiments," Ph.D. Thesis, University of Delaware, Newark, Delaware (1969).

38. Shah, D. O., Private communication (1976).

SOLUBLE OILS FOR IMPROVED OIL RECOVERY

L. W. Holm
Union Oil Research Center

I. ABSTRACT

Soluble oils are oleic micellar solutions which, when injected as a small slug and driven by thickened water, are
capable of displacing all the oil they contact in an oil reservoir.

During a soluble oil slug flood, oil and water are displaced from reservoir rock by one or more of the following
mechanisms:
  a) miscible-type displacement of the oil by the soluble
      oil.
  b) miscible-type displacement of resident water by in-
      jected water and/or soluble oil.
  c) reduction of interfacial tension between oil and water
      phases.
  d) formation of microemulsions by the intermingling of
      soluble oil and injected water.
The efficiency of oil displacement by soluble oils is a
function of the amount of soluble oil injected, the phase
relationships between the surfactant system, crude oil and
water, the viscosities of the fluids involved, the salinity
of the water (resident and injected) contacted by the soluble
oil, the rock composition and heterogeneity, and the mobility
ratio maintained between the soluble oil slug and the driving
fluid. The efficiency also depends (to lesser degrees) upon
flow rate, flow path length, reservoir temperature, and the
type of thickening agent added to the drive water.

II. SCOPE

In the investigation of the use of surface active agents
to remove oil from porous rock, it has been observed that oil
left behind by conventional water or gas flooding may be mobilized and produced from the rock under several conditions.
These conditions include the following:
  1. Miscible-type displacement using a fluid (usually an
      oleic material) which is miscible in all proportions
      with the oil to be displaced.

2. Miscible-type displacement using a fluid (water) normally immiscible with the oil but which contains components at sufficient concentrations to cause the fluid to be miscible (Interfacial Tension < 10^{-3} dynes/cm) with that oil.

3. Immiscible displacement under conditions at which the interfacial tension between water and oils is low enough that the normal capillary retention forces between oil and water in the rock do not restrict the movement of the oil by the injected fluid.

The flooding processes under which these conditions prevail have been termed respectively.

1) oil-external micellar (microemulsion) or soluble oil flooding (1-4)

2) water-external micellar (microemulsion) flooding (5-6), and

3) low tension micellar surfactant flooding (7-8) (Figure 1A).

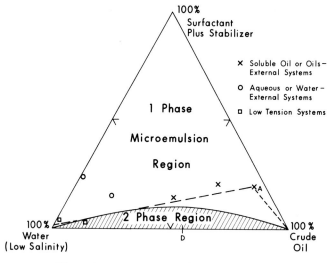

Fig. 1A.   Ternary Phase Diagram

A micellar solution is a dispersion of surfactant in a solvent in which the surfactant ions or molecules are arranged in oriented aggregates, or micelles. Many, but not all, micellar dispersions can spontaneously take up (solubilize) water or oil to form either water-in-oil or oil-in-water microemulsions, respectively. In such cases, the internal phase is solubilized into the centers of the surfactant aggregates.

There are some distinct differences noted in the performance of micellar solutions in laboratory experiments. For instance, the addition of certain waters to a properly compounded soluble oil, or oil-external micellar solution, gives visual evidence of complete miscibility--single phase dispersion of oil and water. Similarly, single phase solutions or dispersions also are obtained by the addition of oil to properly compounded water-external or aqueous micellar solutions. On the other hand, the low concentration, low-tension aqueous surfactant solutions exhibit definite multiphase relationships with oil, even though interfacial tension values as low as 10^{-3} dynes per cm. may be measured between the solution and oil (7). The addition of a few percent oil to these aqueous micellar solutions produces a phase separation. Nevertheless, essentially all of the oil which is immobile to conventional water flooding can be mobilized and removed from laboratory porous media by each of these techniques. The oil, retained by capillary retention forces, is displaced from the pores by a combination of reduction in interfacial tension and high pressure gradients. Healy et al. (6) and Foster (7) have shown that high flow rates (large capillary number $v\mu/\gamma$) enable partially miscible systems to behave like miscible displacements.

The efficient recovery of reservoir oil is dependent upon two other conditions:

  1)   The relative mobilities of the driving and driven
       fluids during the flood from one well to another.
  2)   The maintenance of low interfacial tension during
       the flood.

The mobility of the resident oil and water mobilized is a function of the viscosities of these fluids and the permeability characteristics of the porous rock. The mobility of the displacing micellar fluid is a function primarily of the viscosity of that fluid. (There may be some relative permeability to microemulsions and soluble oils.) The mobility of the micellar fluid must be equal to or less than the mobility of the mobilized oil and water bank to prevent the injected fluid from bypassing the oil-water bank. Because the mobility of the mobilized oil and water bank is low due to relative permeability effects in the rock, the viscosity of the injected micellar solution usually must be much higher than either the resident oil or water in order to achieve the same or lower mobility for these fluids. Total relative mobility = k_{ro}/μ_o + k_{rw}/μ_w = 1/equivalent viscosity.

Gogarty et al. (9) have discussed the basis for the mobility design of micellar fluids.

The interfacial tension between oil and micellar fluid is a function of the concentration of surface active agents and of certain inorganic salts. Loss of surface active agents from the displacement front, or a large increase in salinity of the micellar fluid, can increase the interfacial tension to the extent that little oil is mobilized.

The advantage of a low-tension flood is that large volumes of relatively low concentraton (< 3% active) surfactant fluid may be used to recover the oil, economically. Not all of the resident oil is mobilized at reservoir flow rates, however, because of the immiscible conditions which exist. Also, the viscosity of the low concentration surfactant fluid may not be high enough to achieve favorable mobility conditions during the flood so that much of the oil may be bypassed and left behind the flood front.

In the case of miscible type floods, high concentrations (> 4% active) of surfactants are used so that small volumes must be used to recover oil economically. Aqueous solutions with high concentrations of the low-priced surface active agents (petroleum sulfonates) are very sensitive to inorganic salts. If high concentration brine contacts the aqueous surfactant solutions, the sulfonates may be rejected (precipitated) from the solution and lost for use in mobilizing additional oil. Trushenski et al. (10) have reported on high concentration aqueous surfactant solution in which large amounts of cosurfactant, isopropyl alcohol, are used to help keep the sulfonates in solution. However, the resident brine used in their systems had low (1.3%) salinity and a very low divalent in content. Even more costly cosurfactants such as ethoxylated alcohols and sulfates have been suggested to enhance the efficiency of the sulfonates in the presence of brines.

Soluble oils are designed to overcome the problems associated with the aqueous surfactant solutions (4). The surface active agents are injected as an oleic solution and are completely soluble in the carrier oil. The soluble oils can be miscible with the resident oil regardless of the salinity of the brine associated with that oil and sulfonates are not precipitated from the oils by the brine. Through the formation of single phase microemulsions, soluble oils are essentially miscible with the driving water. Miscibility with water is dependent upon salinity so the water injected to displace soluble oil must be maintained within a certain salinity range.

Soluble oils are mixtures of hydrocarbons, surfactants and other organic liquids (stabilizers or co-solvents) which are miscible with the reservoir oil (Table I). In most cases they are made with the crude oil from the reservoir into which they will be injected. They may contain as much as 60% water as injected, and are referred to as oil-external microemulsions. The basic principle of soluble oil flooding is similar to that of solvent flooding--a miscible slug is maintained in the reservoir during the displacement of oil from one well to another. Through the formation of microemulsions, the soluble oil can be essentially miscibly displaced by water.

TABLE I

COMPONENTS OF SOLUBLE OILS

==

Hydrocarbons	(Gasoline - Gas Oils - Crude Oils)
Surfactant	(Sulfonates - Sulfates - Stearates)
Stabilizer	(Alcohols - Glycol Ethers)
Water	(Containing Inorganic Salt)

III. CONCLUSIONS AND SIGNIFICANCE

Small slugs (3 to 5% PV) of properly constituted soluble oils, driven by polymer solutions, recover 60 to 100 percent of the crude oil remaining in laboratory porous systems after waterflooding. Under reservoir conditions (k_w, S_{or}, crude oil composition, clay, water salinity) which are favorable, the soluble oil-polymer flood should mobilize similar amounts of oil from those portions of a reservoir contacted by the flood.

In soluble oil-polymer solution floods, the soluble oil slug is designed to have a viscosity high enough to match the mobility of the oil-water bank which it mobilizes in the reservoir. Polymer, added to low-salinity drive water, maintains the favorable mobility ratio and thus the integrity of the soluble oil slug through a large portion of the reservoir. Overall, a favorable mobility ratio is maintained to avoid fingering of injected fluids.

Alternate injections of low salt-content water (11), or inclusion of low salt-content water in the soluble oil, provides the proper salinity in the vicinity of the soluble oil as it moves through the reservoir. This water also increases the reservoir pore volume swept by the soluble oil.

Studies of the flow of fluids injected during soluble oil-polymer solution floods through porous media show that:
1. The soluble oil flows at a rate lower than the injection rate.
2. Water injected in or with the soluble oil flows faster than the soluble oil.
3. The polymer flows at a rate lower than the injection rate.

Miscible displacement of oil occurs as long as the surfactant concentration in the soluble oil slug is high enough to maintain low interfacial tensions ($< 10^{-3}$ dynes/cm) with the driving water. Displacement of resident water occurs primarily by the water injected with, dehydrated from, and following the soluble oil. Water flows separately from the soluble oil because the soluble oil picks up excessive metal

cations by mixing with the reservoir brine and through cation
exchange (e.g., sodium/calcium exchange) from reservoir rock.
These cations reduce the ability of soluble oil to solubilize
water.

During miscible displacement by properly compounded soluble oils, a stabilized bank of oil and water is formed; the
ratio of water to oil flowing is related to saturations in the
bank by the equation:

$$\frac{q_{wb}}{q_{ob}} = \frac{S_{wb}}{S_{ob}} \cdot \frac{v_{wb}}{v_{ob}}$$

The presence of excessive polymer from the drive water in
the micellar slug generally enlarges the multiphase region of
the ternary phase diagram, thus leading to loss of slug miscibility more quickly than would be expected from phase diagrams
showing fresh water as the water phase. However, the slower
movement of polyacrylamide through porous media prevents it
from invading the entire soluble oil slug and thereby reducing
the oil recovery efficiency.

Economics dictate that a soluble oil slug size, less than
that required for complete miscible displacement throughout
the reservoir, be used in field application. The distance
that the miscible-type displacement is maintained in the
reservoir by a given soluble oil slug is dependent upon:
1. Dispersion (mixing) of the slug at the front and
    rear.
2. The phase relationship between the soluble oil and
    the driving fluid, and the viscosity characteristics
    of mixtures of these fluids.
3. The degree of loss of surfactant and polymer by
    adsorption and other retention by the reservoir
    rock.

In miscible type displacement, improvement in displacement efficiency occurs with increased floodpath length.
After slug breakdown, low-tension oil displacement occurs;
such displacement is rate sensitive, and more oil is recovered at high flow rates. Also, surfactant retention increases after miscible slug breakdown.

IV. DISCUSSION OF LABORATORY RESULTS
Laboratory studies have been conducted to determine the
effects of the many variables associated with the soluble
oil-polymer flooding process. The results of these studies
are summarized and discussed below under headings which
describe the factors investigated. Laboratory procedures and

materials used in the investigation are included in the
Appendix.

A.   Phase Behavior of Soluble Oils
     In this presentation of the ternary phase behavior of the
crude oil-water-surfactant-stabilizer systems, we have grouped
the surfactant and stabilizer (usually a mutual solvent) as
one component, all the hydrocarbons as another, and water
having a specific salinity as the third phase (Figure 1a).
Furthermore, we consider the one-phase region as any combination of these components which produces a stable, isotropic
mixture; an isotropic mixture is one which forms a film that
appears clear to the eye. The one-phase mixtures are microemulsions of water and oil with a colloidal particle (micelle)
size of less than about .5 microns. This is not true miscibility as would be obtained with a molecular solution because
the microemulsion is more ordered than a true solution.
Within the micelle of a water-oil microemulsion, for example,
many water molecules are grouped to the exclusion of the
hydrocarbon molecules. The interfacial tension between the
oleic and aqueous groups have values approaching zero, well
below 10^{-3} dynes/cm.
     In the soluble oil recovery process (Figure 2), the
soluble oil slug is contacting reservoir oil at its front
end and thickened, low salinity water at its trailing end.

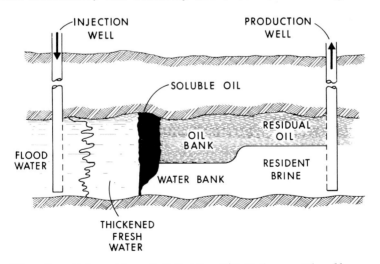

Fig. 2. Schematic of Soluble Oil-Polymer Flooding

The reservoir brine usually is not soluble in the soluble oil
to any appreciable extent but, when favorable mobility exists,
it is pushed ahead by or flows concurrent with the soluble oil.
The soluble oil and brine are in 2-phase equilibrium similar

to that observed by Tosch et al. (12) and by Holm. The interfacial tensions between these phases have been measured at
10^{-1} and 10^{-2} dynes/cm (Table II). The soluble oil slug (composition X_A, Figure 1a) is diluted by reservoir oil in the
front, so the composition of the front end of the slug changes

TABLE II

INTERFACIAL TENSIONS BETWEEN SOLUBLE OILS
AND WATER OR BRINES (dynes/cm)

Phases	IT @ 72-75°F	IT @ 120°F
Soluble Oil C and 0.07% "brine"[a]	$< 10^{-4}$	--
Soluble Oil C and 1.5% brine[a]	2.5×10^{-2}	--
Soluble Oil C and 2.9% brine[a]	2.6×10^{-2}	2.2×10^{-2}
Soluble Oil C and 9.4% brine[a]	1.2×10^{-1}	1.2×10^{-1}
Soluble Oil C and fresh water[b] (3000 ppm NaCl)	single phase	
Soluble Oil C and fresh water[b] 10,000 ppm NaCl)	3.2×10^{-4}	

[a]Brine contained Na, Ca and Mg ions--see Appendix.
 Soluble oil/Water Ratio = 1 to 1
[b]Soluble Oil/Fresh Water Ratio = 1 to 10

as shown by the dashed line between X_A and the 100% oil corner.
Dilution of the front by both reservoir oil and water (path
X_A to D) occurs only when that water has the salinity needed
to maintain single phase conditions. This would seldom occur
at the soluble oil front because of the adverse salinity associated with most reservoir rock. The soluble oil slug is
also diluted at the rear by the low salinity water injected
so the composition of the tail end changes is shown by the
dashed line between X_A and the 100% water corner. It is
obvious that unless the two-phase (immiscible) area is very
small, or the concentration of the surfactant high, continued
dilution of the tail end of the soluble oil slug with water
yields a composition that falls within the two-phase region.
If this occurs, a part of the displaced fluid (oil and/or
water) may be left in the porous medium during the displacement process. The amount of fluid left will depend upon the
amount of surfactant-stabilizer present, the interfacial tension between the phases, the displacement rate, and the

characteristics of the emulsion flow in the porous medium.
This displacement has been described by Healy as locally
miscible and, after slug breakdown, immiscible emulsion displacement.

With a properly formulated soluble oil, very little surfactant-stabilizer is required to achieve single-phase microemulsions (Figure 1B) and, miscible displacement can be
achieved throughout the flood path. Favorable phase relationships (low interfacial tensions) are primarily a function of
the surfactant type, the stabilizer or co-surfactant used, and
the salinity of the water mixing with the soluble oil (Table
II).

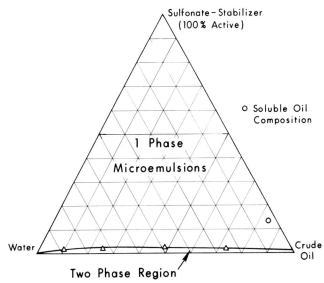

Fig. 1B. Ternary Phase Diagram

B. Dispersion of the Soluble Oil Slug and Polymer Solution
    As shown in Figure 3, when a 3% PV slug of an anhydrous
soluble oil (4% water) was injected into a sand pack which
contained only 9.4% brine, the peak concentration of soluble
oil occurred in the effluent at about 1.08 PV total volume
throughput. When a larger slug of soluble oil was injected,
the peak appeared earlier and was higher in concentration.
When a water-containing soluble oil (42% water) was injected,
the dispersion of the slug and the arrival of the peak concentration was similar to that of the anhydrous slug and
depended upon the amount of anhydrous soluble oil (oil phase)
present in the slug. The data obtained indicated that the
amount of sulfonates in the soluble oil determines the peak
concentration of the slug appearing in the effluent.

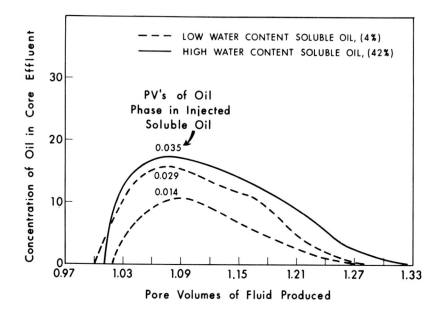

Fig. 3. *Concentration of Oil Produced During Soluble Oil-*
Polymer Floods (Six-Foot Long Sand Packs)

A detailed analysis of the composition of the effluent
from a sand pack during a soluble oil—polymer solution flood
is shown in Figure 4. A low viscosity, fresh water bank
appears prior to and along with the front portion of the
soluble oil slug. In this flood the soluble oil injected
contained 45% water. However, the effluent profiles for the
latter flood were about the same as those obtained when water
was alternately injected with the soluble oil or was injected
immediately after the anhydrous soluble oil. The high viscosity polymer solution appears in the effluent immediately
following the peak concentration of the soluble oil.

In Figure 5 the results obtained from flowing soluble oil
followed by polyacrylamide polymer through a consolidated
Berea sandstone are shown. At the start of the flood, the
sandstone pores contained only 9.4% brine. In these experiments a larger amount of low viscosity, fresh water appeared
ahead of the soluble oil than during the flood of the sand
pack. Also, the anhydrous soluble oil slug produced was more
dispersed, particularly at the trailing end. Note that the
resident water (brine) was not displaced even after about 1.25
PV had been injected. Davis et al. (13) and Healy (6) have
noticed that resident brine is displaced more efficiently as
flood path length increases.

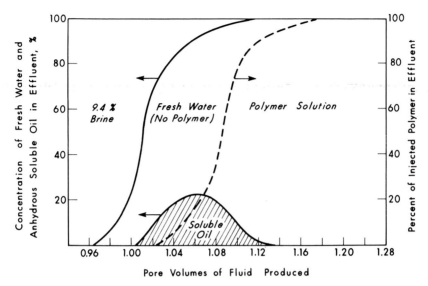

Fig. 4. Production of Fluids During Soluble Oil-Polymer Solution Displacement (Based on Clean Sand Pack Results)

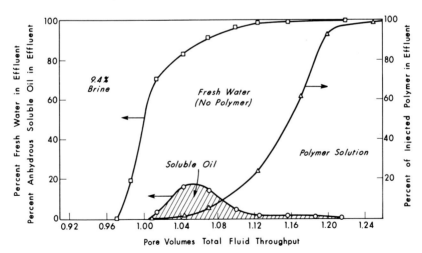

Fig. 5. Production of Fluids During Soluble Oil-Polymer Solution Flood in Berea Sandstone (Injected 3% PV Soluble Oil Slug Contained 45% Water)

The effect of residual crude oil in the rock prior to soluble oil-polymer floods was also investigated, and the effluent profiles of the fresh water, soluble oil, and polymer were similar (Figure 6) to those indicated above. Less

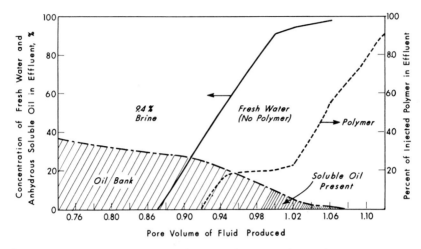

Fig. 6.   Production of Fluids During Soluble Oil-Polymer
Solution Flood in Berea Sandstone (Both Brine and
Residual Oil in Place)

accurate analyses for the slug ingredients were obtained because of the interference of oil that became mixed with the
chemicals.

C.   Viscosity and Density Characteristics of Soluble Oils
      The change in viscosity of the soluble oil slug as it is
diluted with water driving it in a porous medium will determine how efficiently the slug will be displaced. Soluble Oil
A, Figure 7, as it picked up water, formed microemulsions
having viscosities higher than 500 cp. Such high viscosities
in a porous medium would cause unfavorable mobility ratios
and bypassing of the soluble oil by the drive fluid. Soluble
Oil B increased in viscosity only a small amount as water was
added, and the viscosity was relatively constant at water con-
tents up to about 60%. And the microemulsions formed at each
water content shown was clear and single phase (IT < 10^{-3}
dynes/cm).
      The density of a soluble oil also changes as it picks up
water and forms microemulsions during a soluble oil-polymer
solution flood. The density of the soluble oil is close to
that of the displaced oil at the front and becomes closer to
that of the injected water (polymer solution) at its rear as
it takes up water as a microemulsion. Gravity segregation is
minimized under these conditions.

D.   Effect of Slug Size and Surfactant Concentration
      The relationship between crude oil recovery and soluble
oil slug size for soluble oil-polymer floods in 4-foot long,

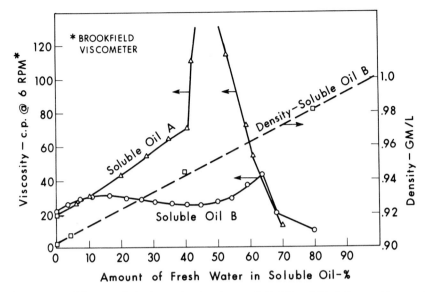

Fig. 7. Effect of Water Content on Soluble Oil Viscosity and Density

watered-out porous systems is presented in Figure 8. The optimum slug size for a homogeneous sand pack and a Berea core appears to be about 3% PV. Lower oil recoveries were obtained from stratified sandstone or sand packs and from heterogeneous Dundee sandstone cores, compared to the homogeneous systems. Nevertheless, the optimum slug size remains at about the 3% PV level.

The amount of oil recovered by soluble oil flooding appears to be a linear function of concentration of the sulfonate in the soluble oil until that concentration gets below about 5%, Figure 9. The results shown in this figure are from floods in which only the concentration of sulfonates and solvent in the soluble oil was changed. A 3% PV soluble oil slug (followed by polymer solution) was used with about the same ratio of sulfonate products and stabilizer in each blend as used for Soluble Oil C. (Some slight changes were made to maintain uniform viscosity characteristics for the soluble oil.)

E. Effect of Including Water with the Soluble Oil

Alternate injection of low salt-content water and soluble oil, or inclusion of low salt-content water with the soluble oil, serves to provide the proper salinity in the vicinity of that soluble oil as it moves through the reservoir rock. The presence of low salt-content water also causes the soluble oil

466   L. W. HOLM

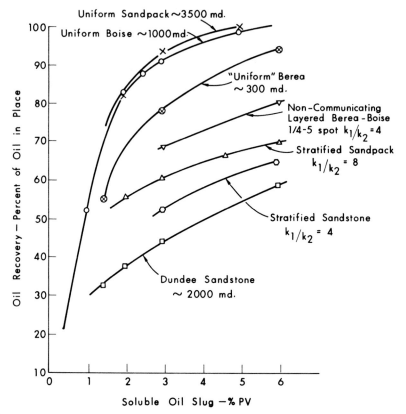

Fig. 8. *Effects of Heterogeneity on Soluble Oil Slug Size
(Soluble Oil Contained 45% Water; Sandstones were
not Fired or Otherwise Pretreated)*

to sweep a larger portion of the reservoir. The four procedures for injecting soluble oils include:

   1.   Injection of a single slug of low water-content
        soluble oil, followed by polymer solution (3).

   2.   Injection of a single slug of a low water-content
        soluble oil, followed by a single slug of water,
        followed by polymer solution.

   3.   Injection of a single slug of a high water-content
        soluble oil, followed by polymer solution (14,15).

   4.   Alternate injection of small slugs of water and
        soluble oil, followed by polymer solution (11).

     The results of laboratory floods on sandstone cores using
these injection procedures are compared in Table III. The oil
recovered by applying each of these procedures was about the
same from the relatively uniform (Berea) sandstone rock. However, the oil recovered during laboratory floods of a more

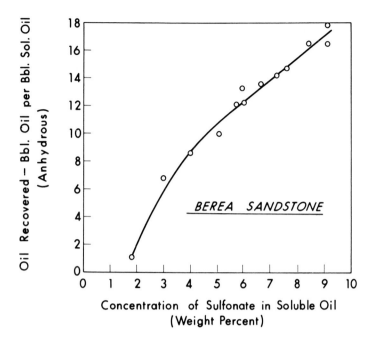

Fig. 9. Effect of Concentration of Sulfonate on Recovery by Soluble Oil Polymer Floods (Equivalent Floods Using 3% PV Soluble Oil Slugs Containing 40% Water)

heterogeneous sandstone (Dundee) was significantly greater when alternate slugs of soluble oil and water were injected. Apparently, improved distribution of the soluble oil was achieved by the alternate injection technique.

F. Effect of Flood Rate

Displacement experiments in which small slugs of soluble oil (less than that required to recover all the oil) were used were sensitive to flow rate. The results showed earlier oil breakthrough and somewhat lower fractional flow of oil (f_o) in the stabilized oil-water bank at the 2 foot per day flood, compared to the bank in the 15 foot per day flood, Figure 10; f_o, of course, is unique for each rock system,

$$f_o = \frac{1}{1 + \dfrac{k_{rw}\mu_o}{k_{ro}\mu_w}} ,$$

and cannot be used for comparison of the performance of the two floods in different rock systems. As shown in Figure 11, lower oil recoveries were obtained at lower flooding rates.

TABLE III

COMPARISON OF OIL RECOVERY OBTAINED BY VARIOUS FLOODING
PROCEDURES UTILIZING SOLUBLE OIL AND WATER[*]
(IN WATERED-OUT 4 FT. BEREA SANDSTONES)

Flooding Procedure	Amount of Anhydrous Soluble Oil Used	Oil Recovered, % OIP
Alternate injection of slugs of low water-content soluble oil and water	1.9[a]	79.5
Injection of a single slug of low water-content soluble oil	1.9[a]	79.3
Injection of a single slug of low water-content soluble oil followed by a single slug of water	1.8[a]	74.9
Injection of a single slug of a high-water content slug of soluble oil	1.9[a]	79.9
(IN WATERED-OUT 4 FT. DUNDEE SANDSTONES)		
Alternate injection of slugs of low water-content soluble oil and water	1.6[b]	49.9
Injection of a single slug of a high-water content soluble oil	1.8[b]	41.4

[*]Polymer solution used to push soluble oil through core

[a]Soluble Oil C used

[b]Soluble Oil B used

Healy et al. (6) reported about the same degree of rate sensitivity for microemulsion floods at rates as low as one-half
foot per day. They found that the rate sensitivity occurred
after the miscible slug broke down and the low-tension flooding mechanism was in effect. It has been shown that the S_{or}
can be correlated with capillary number, N_C (7,16), and Healy
found that the micellar floods are rate insensitive for N_C
greater than 10^{-3}. Oil recovery from a low-tension flood

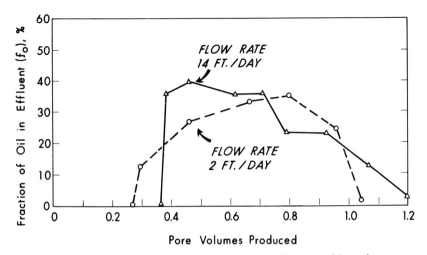

Fig. 10.   *Fractional Oil Flow During Soluble Oil-Polymer Floods (Soluble Oil D Containing 50% Water; Watered-out Berea Sandstone)*

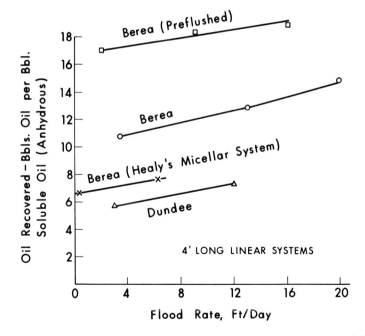

Fig. 11.   *Effect of Flood Rate on Oil Recovered by Soluble Oil Flooding*

approaches that of a miscible flood when interfacial tension
is low and frontal velocity is high.

G.   Effect of Flood Path Length

     Increasing the length of the flooding path in Berea and
Dundee cores resulted in an increase in oil recovered by
soluble oil-polymer solution floods (Figure 12). Healy <u>et al.</u>
(6) found a similar trend for micellar fluids through porous
systems as long as 16 feet. They obtained 100% recovery of
oil from a 16-foot Berea system as compared to 78% recovery
under similar conditions from a 4-foot Berea system. The
improvement is attributed to the longer contact times for
mixing of miscible fluids; more time is afforded for compositional equilibrium to be approached between the soluble
oil and the resident oil, the soluble oil and the resident
water, and between the polymer solution and the soluble oil.
Later arrival of surfactant in the effluent from the longer
systems also indicated a sharper mixing zone at the front of
the soluble oil displacement. Healy found the resident water
to be displaced more efficiently from the longer systems.
When the resident water has a high concentration of inorganic
salts which have unfavorable phase relationships with the
soluble oil, more efficient displacement of this brine permits
more efficient displacement of the oil also.

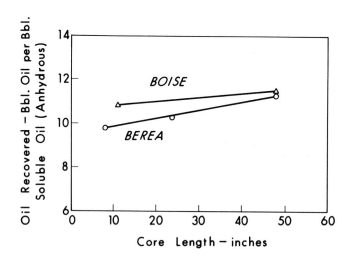

*Fig. 12.   Effect of Core Length on Oil Recovery by Soluble
             Oil Flooding*

     In summary, when a soluble oil slug is used which is in-
sufficient in size for miscible displacement of oil from the
entire reservoir, miscible displacement should dominate the

flood to a greater extent in longer flood path systems. However, this may not be the case in heterogeneous reservoirs
where transverse dispersion and crossflow can occur between
differing permeability zones.

H.   Effect of Reservoir Brine-Clays

In order to determine whether the mixing of soluble oil
with brine within a porous rock would result in the transfer
of metal ions, soluble oil C was passed through a Berea core
which contained only brine. Analyses of the effluent showed
that the soluble oil does pick up bivalent cations as it
passes through this rock. Mixing of soluble oil and brine
was not the sole cause of this inclusion. Further investigation showed that multivalent cations were removed by soluble
oil from other type rock samples (17). Several volumes of
soluble oil were passed through the various rock samples
which contained brine or fresh water and the results are
shown in Figures 13 and 14. Cation exchange occurred when
the rock contained the brine and when it contained fresh
water prior to the soluble oil injection. Such exchange is
not unexpected considering the presence of divalent cations
on most reservoir rock and the high concentration of monovalent cations in these micellar fluids. Only dolomite rock
containing fresh water and sand packs without clay did not
show evidence of cation exchange.

It was further noted in these experiments that the effluent soluble oil from the various rock systems appeared as
2-phase mixtures, oil-phase and water-phase, even though the
soluble oil was injected as a water-containing (45%) clear,
single phase microemulsion. Similar results were obtained
when alternate slugs of anhydrous soluble oil and low salt-
content water were injected into similar rock samples. These
results indicate that water-containing soluble oils may not
move through the reservoir rock as high water content, clear,
single phase microemulsions, but at least partially, as two
phases. The high water-content slugs are dehydrated to low
water-content soluble oils. However, the interfacial tension
measured between these produced phases was very low (10^{-1}
dyne/cm to less than 10^{-3} as the water salinity was reduced--
Table I).

High concentration brines (greater than 2%), particularly
those containing large amounts of multivalent cations such as
calcium, magnesium and iron, make the surfactants in a micellar solution less effective in reducing the interfacial tensions between it and the driving water. They cause deterioration of a micellar slug and early loss of the miscible displacement. Micellar floods in sand systems containing these
brines recover less oil than when fresher waters are present

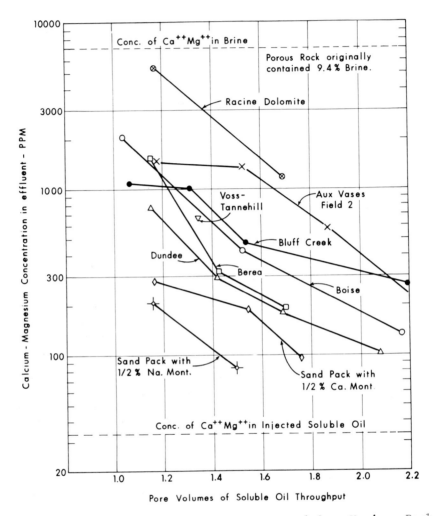

Fig. 13. *Calcium-Magnesium Cations Removed from Various Rock Samples by Soluble Oil C (Containing 50% Water)*

(17,18). Low salt-content water, alternately injected with the soluble oil, tends to displace the reservoir brine ahead of the soluble oil as shown in Figure 5, and as mentioned previously, this displacement is more efficient in long flood paths. Nevertheless, reservoir rock contains clay and other fines, so that sufficient exchangeable metal cations remain to contaminate the soluble oil. Mungan (19) determined that a 250 md. Berea sandstone contained about 6 to 10 milli-equivalents of cation charge per 100 grams of clay. The data shown in Figure 14 indicate that this property varies with the permeability of this rock.

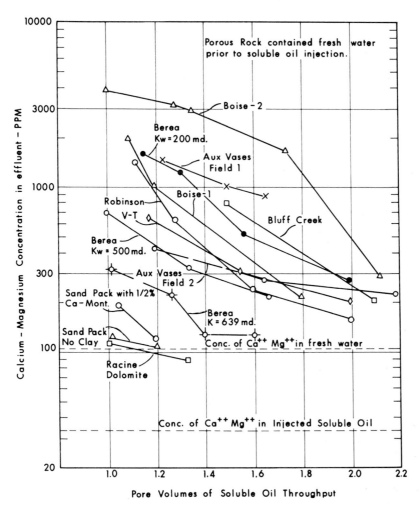

Fig. 14. Calcium-Magnesium Removed from Various Porous Rocks by Soluble Oil C (Containing 50% Water)

Soluble oils have a distinct advantage over miscible aqueous surfactant solutions in the presence of brine and high divalent cations. The sulfonates present in soluble oil do not precipitate when mixed with brine, Table IV. Calcium and magnesium cations are soluble in the soluble oil; even a mixture of oil and water soluble sulfonates remain in solution when these multivalent cations are present. With divalent ions present, the soluble oil becomes less effective in reducing interfacial tension between oil and water. However, we have found that fresh water from the driving polymer solution moves through the soluble oil and leaches divalent

TABLE IV

PARTITIONING OF SULFONATE BETWEEN SOLUBLE OIL AND
FIELD BRINE

	Ca^{++}, Mg^{++}, PPM	Sulfonate, Wt. Percent	Sulfur, Wt. Percent
Exp. #1			
Soluble Oil (Anhydrous)	11	10.9	1.2
Field Brine[*]	4180	0	24 PPM
Oil Phase after mixing Soluble Oil and Brine (equal parts)	1475	11.9	0.95
Water Phase after mixing Soluble Oil and Brine (equal parts)	2524	0.28	0.18
Exp. #2			
Soluble Oil (Anhydrous)	11	10.9	--
Aqueous surfactant solution	27	6.6	--
Field Brine[*]	4180	0	--
Oil Phase after mixing above three (1/3 of each)	1730	14.7	--
Water Phase after mixing above three (1/3 of each)	1376	0.1	--

[*]Total dissolved solids = 94,000 PPM

cations from the slug, tending to reestablish its effectiveness. It is important that the elution waters have a low
inorganic salt content, particularly multivalent ion salts.

Another method to handle the unfavorable divalent cations
is by removing them from the porous medium prior to the micellar flood. The results presented in Figure 11 show the effect
of a sodium chloride solution preflush on the oil recovery
efficiency of soluble oil-polymer floods. (Paul and Froning
(18) presented similar results in similar rock but different
micellar systems.) Large amounts of bivalent ions were

removed from Berea cores by this preflush treatment even after
the resident brine had been thoroughly removed by fresh water
flushing. Obviously, cation exchange was occurring between
the preflush material and the clays and fines in the rock.
Increased oil recovery from a subsequent soluble oil flood
was the result of the preflush treatment of the core.

I.   Effect of Rock Heterogeneity

The results presented in Figure 8 show the lower oil
recovered from stratified sand packs, sandstones, and other
heterogeneous rock, compared to that from uniform sandstone
and sand packs (20). The study included systems in which
crossflow and transverse dispersion of fluids could occur
between sands of different permeabilities.

The curves in Figure 15 are an attempt to correlate the
effect of rock heterogeneity on oil recovered by soluble
oil-polymer solution floods. The stratified sand packs and
Berea-Boise sandstone models were constructed with known
permeability strata so that the permeability ratio k_1/k_2

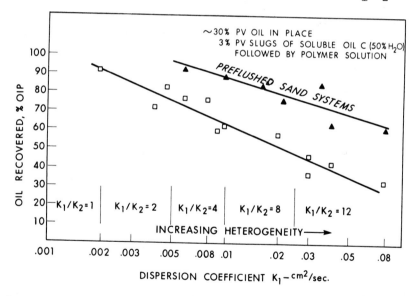

*Fig. 15.   Effect of Heterogeneity on Oil Recovered by Soluble
Oil Flooding*

could be calculated. Tracer tests (using ammonium thiocyanate)
were conducted on each of the stratified systems in order to
provide a correlation between the dispersion (K) of the tracer
and the k_1/k_2 ratio. The dispersion factors of tracers in the
individual sand packs, and Berea, Boise and Dundee sandstone
models were calculated directly from the results of the tracer

tests. The oil recovered from the soluble oil-polymer floods
on these models was then plotted versus K, and k_1/k_2 ratios
estimated on the basis of the stratified model results. This,
for example, means that a consolidated rock having a K = 0.03
responds to a flood similarly to a stratified sand pack having
a k_1/k_2 ratio of about 12.

The heterogeneity of the porous media adversely affects
the soluble oil displacement mechanism in two principal ways:

1. It causes greater dissipation of the soluble oil slug
    so that the miscible slug remains in contact for a
    shorter displacement distance in the porous media.
2. It causes a lower rate of polymer movement through
    the porous media, either by loss of polymer (or
    polymer solution viscosity) through more mixing
    with resident brine or because of excessive reten-
    tion in low permeability rock.

J. Sulfonate Retention in Porous Media

Sulfonate is retained in a porous rock by 1) adsorption,
2) precipitation, and 3) through by-passing. High energy
surfaces (clays, fines, etc.) adsorb sulfonates. Polyvalent
metal cations form insoluble sulfonate salts which tend to
precipitate from aqueous solutions that exist at the trailing
end of the soluble oil slug. Healy et al. (6) has shown that
unfavorable phase behavior can be caused by the presence of
polymer in the slug. They flushed cores in which a micellar-
biopolymer solution flood had been conducted, and produced an
emulsified-multiphase material which apparently had separated
out during polymer-surfactant interaction. Finally, unfavorable mobility ratios may develop to cause part of the sulfonate solution to be bypassed.

In order to determine the retention of sulfonates in
porous rock, we used the techniques explained in the Appendix.
The results of these studies in Berea and Dundee sandstone
have indicated that between 1 and 1.3 lbs. of active sulfonate
were retained per barrel of pore volume under conditions where
a high salinity brine was originally present. This retention
was reduced to 0.4 to 0.7 lbs. of sulfonate per barrel of pore
volume when the Berea rock was flushed with a monovalent, inorganic salt solution prior to the soluble oil-polymer flood
(Figure 16). The amount of sulfonate retained in the last
quarter of the sandstone cores indicates that increased retention of sulfonate occurs after the slug breaks down. The
above magnitude of sulfonate loss was supported by two other
observations:

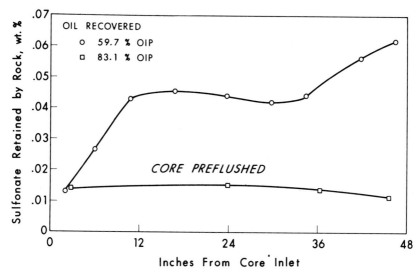

Fig. 16. Sulfonate Retained in Berea Sandstone After Soluble
Oil Floods (Berea originally contained residual
crude oil and 9.4% brine; 3% soluble oil slug fol-
lowed by polyacrylamide polymer solution)

1. The oil recovered by floods in which the sulfonate
    retention was above 1 pound per barrel PV was 50 to
    60% of the oil in place. The oil recovered during
    floods in which a preflush was used and the retention
    was less than 0.8 lb. per barrel PV, was between 70
    and 90% of the oil in place.

2. The amount of sulfonate retained by a Berea core
    which contained only brine originally and which
    was flushed with a monovalent salt solution prior
    to the soluble oil-polymer flood, was 0.64 lb. per
    barrel PV after 1.2 PV of total fluid throughput,
    and 0.43 lb. per barrel PV after 1.4 PV of total
    fluid throughput.

We have not distinguished quantitatively between the
amount of surfactant retention which is due to adsorption
and the amount which is due to precipitation as a calcium or
magnesium sulfonate or left behind because of sulfonate-
polymer interaction. It appears that the role of the divalent
metal cations in the adsorption of sulfonates is important.
Trushenski et al. (10) have indicated that a NaCl preflush
reduces sulfonate adsorption losses from aqueous surfactant
solutions. However, their results were based upon material
balance calculations and not direct rock adsorption measurements. Ion exchange or electrostatic attraction promotes
the adsorption of sulfonates by solids. If it can be

assumed that solids are negatively charged in water, ion exchange of the sodium (or other monovalent) cation would control the adsorption of the sulfonate by bringing the negatively charged sulfonate ion into the molecular layers
surrounding the solid. On the other hand, Van der Waal's
and hydrogen bonding forces can cause the negatively charged
sulfonate molecules to be adsorbed directly on the solids.

V.   POLYMER RETENTION IN POROUS MEDIA

The flow of a polymer solution through a Berea core was
compared to its flow through a similar core when preceded by
a soluble oil slug (Figure 17). The polymer moved through
the core with less retention when following a soluble oil
slug. The loss of soluble oil components lessened the loss

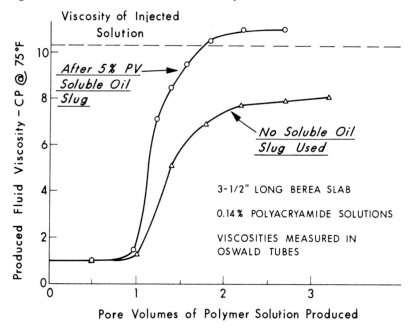

Fig. 17. Viscosity of Polymer Solution Effluent

of polymer from the following polymer solution. The loss of
polymer was determined by comparison of the viscosities of
the core effluent compared with the viscosity of the injected
polymer solution. Analytical determinations of polymer showed
that somewhat more polymer was present than was indicated by
the viscosity; however, the mobility (our prime concern) of
the effluent corresponded more to the viscosity. As shown in
Figure 5, the water originally used to make up the polymer
solution but soon denuded or greatly reduced in polymer concentration, moved faster than the polymer or soluble oil slug.

An interesting effect of polymer type on the viscosity
of the polymer solutions produced behind a soluble oil slug
is shown in Figure 18. The polyacrylamide moved at a slower
rate through the rock than the biopolymer. Although the
biopolymer flowed more readily than the polyacrylamide, the
soluble oil-biopolymer flood resulted in less oil recovery
than the soluble oil-polyacrylamide polymer flood. The explanation of these differences may be as follows:

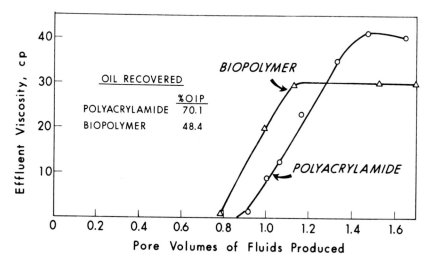

*Fig. 18. Differences in Viscosity of Effluent Polymer Solu-
tion for Different Types of Polymers Following
Soluble Oil Slugs (Residual Oil and 9.4% Brine in
Place in Berea Sandstone)*

Loss by adsorption will be low for both polymers following a micellar flood because of the prior adsorption of surfactant on the rock. Inaccessible pore volume (21) causes
polymer to penetrate the soluble oil slug because the large
polymer molecules can move only through pores above a certain
minimum size, whereas the micellar solution can flow through
smaller pores. As much as 15-20% of the pore volume of Berea
core systems have been found to be inaccessible to both polyacrylamide and biopolymer. On the other hand, retention of
polymer molecules, due to size and shape, causes the polymer
to lag behind the soluble oil. This type of retention occurs
to a greater degree with polyacrylamide than with biopolymer.
This fact has been verified in studies by Healy and by
Trushenski. Healy also showed that when flooding with a
solution containing 1000 PPM biopolymer, a semi-solid precipitate formed indicating surfactant-polymer interaction.
Because of the unfavorable phase behavior caused by the

presence of polymer in a micellar slug, lower oil recoveries are obtained using biopolymer because it mixes with more of the slug than does the polyacrylamide.

VI. THE EFFECT OF TEMPERATURE UPON SOLUBLE OIL FLOODING

Soluble oil-polymer solution floods generally gave higher oil recoveries at higher flood temperatures in the temperature range 75 to 177°F. Table V shows the results of floods on 2 feet long Berea cores containing residual crude oil and water. Only the temperature of the flood was changed; however, there was a difference in k_w between the cores used. These floods were conducted at a rate of about 10 ft/day and extended over a period of hours.

TABLE V

EFFECT OF TEMPERATURE ON RECOVERY BY
SOLUBLE OIL FLOODING

	S_{OR}	Permeability k_w md	Flood Temperature, °F	Oil Recovered, % OIP
Soluble Oil D (containing 53% water)*	31.4	412	75	78.6
Soluble Oil D (containing 53% water)*	30.9	309	177	87.6

*3% PV slug followed by 0.15% polyacrylamide solution. (Slug and polymer viscosity 28-30 cp at 6 RPM and 75°F)

VII. EFFECT OF ROCK PERMEABILITY ON SOLUBLE OIL FLOODS

In general, soluble oil-polymer solution floods recover more oil from high permeability porous systems than from low permeability systems. For example, compare the oil recovered from the 1000 md. Boise cores and 3500 md. sand packs with that recovered from 300 md. Berea cores, Figure 8. This comparison is valid only for systems of similar homogeneity. Floods involving high permeability (2000 md.) Dundee sandstones gave much lower oil recoveries than those involving the Berea cores because the Dundee rock is much more heterogeneous than any of the other systems used (see Appendix).

The results from floods involving Berea cores of different permeability but with similar homogeneity, are shown in Table VI. The trend toward less oil recovery from lower

TABLE VI

OIL RECOVERED BY SOLUBLE OIL-POLYMER FLOODS
FROM VARIOUS PERMEABILITY BEREA SANDSTONES[a]

(0.038 PV slugs of soluble oil containing
50% water followed by fresh water
polyacrylamide solutions)

Flood No.	Permeability to Water-- k_w	Permeability to Water at Residual Oil-- $k_{w_{ro}}$	Oil Recovered From Soluble Oil Flood--% OIP
1	530	53.0	95.5
2	449	35.7	89.5
3	328	28.7	83.0
4	286	25.0	83.5
5	91.6	4.3	58.3
6	86.3	5.4	72.4[b]

[a]Sandstones contained 31 to 34% PV residual oil and 9.4% brine.

[b]Used a lower molecular weight polymer (~ 6 MM vs. 10 MM)

permeability rock could not be attributed to variation in the homogeneity of these cores as measured by tracer tests prior to the soluble oil floods. Furthermore, a favorable mobility ratio was assured in each flood by adjusting the polymer concentration in relation to the k_{rw} of each core. More oil was recovered from the low permeability rock when a lower molecular weight polymer was used to drive the soluble oil (Flood 6 vs. 5). This indicates that a larger cross-section of this tight rock was penetrated by the polymer and, as a consequence, the soluble oil displacement was effective in more of the rock. As shown in Figure 14 more divalent metal cations were present in the low permeability Berea than the high permeability Berea. As these ions and the clays they are associated with have an adverse effect on micellar and polymer solutions, this also accounts for the poorer oil recovery in the low permeability rock.

VIII.   ACKNOWLEDGMENT
The author wishes to recognize D. H. Ferr and K. L. Collins who performed much of the experimental work and to thank L. J. O'Brien for his contributions to this paper.

IX. NOTATIONS

f_o = functional oil flow
k_{ro} = relative permeability to oil, dimensionless
k_{rw} = relative permeability to water, dimensionless
L = length, ft.
N_c = capillary number, dimensionless
PV = pore volume
q = injection rate
S_{ob} = volumetric flow rate of the oil bank
S_{wb} = volumetric flow rate of the water bank
S_{or} = residual oil saturation prior to soluble oil flood
v_{ob} = frontal velocity of oil bank
v_{ow} = frontal velocity of water bank
ΔP = pressure drop, PSI
γ = interfacial tension, dynes/cm.
μ_o = viscosity of oil, cp.
μ_w = viscosity of water, cp.

X. APPENDIX

A. Flooding Procedures

     In preparation for the flooding experiments using the
various sand packs and sandstones listed on the table below,
the systems were first saturated with fresh water or the
brines specified. The sandstones were used as quarried; they
were dried at 250°F but not fired. They were not treated with
any fluids to stabilize or inactivate clays, or to remove
divalent metal ions except as indicated for those floods in
which a preflush was used. Permeability (k_w) was determined
and then the system was flooded with filtered crude oil to
irreducible water saturation. At this point, the system was
waterflooded to residual oil with fresh water or brine and
then a soluble oil flood made.

     A soluble oil flood consisted of injection of a 3% PV
slug of soluble oil (containing 4 to 50% water) followed by
60% PV of polymer solution and then fresh water, unless
otherwise specified. The displacements were performed at
room temperatures, 72–75°F and at a frontal velocity of about
2 to 20 ft. per day.

     Produced fluids were collected and their volumes measured.
Where required, the fluids were analyzed for sulfonate using
ASTM D855 and/or the Hyamine titration method UTM 482; for
water, by the Fisher method, and for polymer, by viscosity
measurements. In some cases, the sandstone sections were
flushed with large volumes of a light hydrocarbon or isopropyl alcohol after the experimental flood to determine the
residual fluids left in the systems. In some cases, small
sections of the porous rock were cut along the length of the

system. These sections were then extracted in order to determine residual fluids. In some cases the entire section was
put into solution using strong acids. Then the weight of sulfonate in these solutions was measured.

B. Fluids and Porous Media
1. *Crude Oils*   38-41°API   4-6 cp. @ 70°F
2. *Compositions of Brines and Tap Water - mg/L*

	Fresh Tap Water	2.9% Brine	9.4% Brine
Sodium	86	9,000	30,794
Calcium	86	1,012	5,510
Magnesium	23	240	1,278
Chloride	81	17,000	52,300
Sulfate	195	10	460
pH	7.1	8.1	7
TDS	700	29,200	94,300

3. *Properties of Porous Media*

Sandstone	k_w, md	ϕ, %	k_{rw}, md.	Dimensions
Berea	85-600	19.7-20.9	5-70	1 " x 1 " x 1' to 4'
Boise	1100-6000	24.0-26.0	300-1000	2" x 2" x 1' to 4'
Dundee	1000-2500	22.5-22.9	200-400	2" x 2" x 4'
Dolomite	50-100	16.0-20.0	4-6	2" x 2" x 3'
Sand packs (American Graded #16)	4000-5000	32-34	–	2" Diam. x 6' Long
Sand packs (with added clay)	2500-4000	30-34	–	2" Diam. x 6' Long

4. *Polymer Used*

   Nalco Q-41F Polyacrylamide
   Dow Pusher 700 Polyacrylamide
   Kelzan MF - Biopolymer

5. *Soluble Oil Compositions - volume percent at 75°F*

Soluble Oil A	Soluble Oil B
69.3 Illinois Crude Oil	71.4 Illinois Crude Oil
13.1 Sulfonate, avg. MW-500	7.4 Sulfonate, avg. MW-500
6.9 Mineral Oil (Sulfonate diluent)	2.7 Sulfonate, avg. MW-340
	4.8 Mineral Oil (Sulfonate diluent)
6.4 Ethylene Glycol Monobutyl Ether (EGMBE)	6.4 Ethylene Glycol Monobutyl Ether
4.3 Water	7.3 Water

Soluble Oil C

76.0 Texas Crude Oil
 7.0 Sulfonate, avg. MW-500
 2.2 Sulfonate, avg. MW-340
 5.7 Mineral Oil (Sulfonate
     diluent)
 1.9 Ethylene Glycol
     Monobutyl Ether
 7.2 Water

Soluble Oil D

71.4 Colorado Crude Oil
13.0 Sulfonate, avg. MW-420
 3.5 Mineral Oil (Sulfonate
     diluent)
 1.6 EGMBE
 3.9 Water

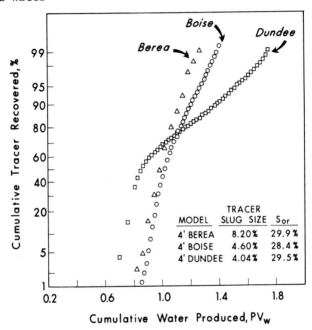

MODEL	TRACER SLUG SIZE	S_{or}
4' BEREA	8.20%	29.9%
4' BOISE	4.60%	28.4%
4' DUNDEE	4.04%	29.5%

Appendix: Tracer Dispersion

XI. LITERATURE CITED

1. Csaszar, A. K., "Solvent-Waterflood Oil Recovery Process,"
     U.S. Patent No. 3,163,214, December 29, 1964.

2. Gogarty, W. B. and Tosch, W. C., "Miscible-Type Water-
     flooding: Oil Recovery with Micellar Solutions,"
     J. Pet. Tech. 1407-1414 (Dec. 1968); Trans., AIME,
     Vol. 243.

3. Holm, L. W., "Anhydrous Soluble Oil Followed by Polymer
     Solution," U.S. Patent No. 3,537,520, November 3, 1970.

4. Holm, L. W., "Use of Soluble Oils for Oil Recovery,"
     J. Pet. Tech., 1475-1483 (Dec. 1971); Trans., AIME,
     Vol. 251.

5. Cooke, C. E., Jr., "Microemulsion Oil Recovery Process,"
     U.S. Patent No. 3,373,809, March 19, 1968.

6.   Healy, R. N., Reed, R. L. and Carpenter, C. W., Jr., "A
     Laboratory Study of Microemulsion Flooding," Soc. Pet.
     Eng. J., 15, No. 1, 87-103 (Feb. 1975).

7.   Foster, W. R., "A Low Tension Waterflooding Process," J.
     Pet. Tech., 205-210 (Feb. 1973); Trans., AIME, Vol. 255.

8.   Hill, H. T., Reisberg, J. and Stegemeier, G. L., "Aqueous
     Surfactant Systems for Oil Recovery," J. Pet. Tech.,
     186-194 (Feb. 1973); Trans., AIME, Vol. 255.

9.   Gogarty, W. B., Meabon, H. P. and Milton, H. W., Jr.,
     "Mobility Control Design for Miscible-Type Waterfloods
     Using Micellar Solutions," J. Pet. Tech., 141-147
     (Feb. 1970).

10. Trushenski, S. P., Dauben, D. L. and Parrish, D. R.,
     "Micellar Flooding--Fluid Propagation, Interaction
     and Mobility," Soc. Pet. Eng. J., 14, No. 6, 633-645
     (Dec. 1974).

11. O'Brien, L. J. and Holm, L. W., "Flooding Process Using
     a Substantially Anhydrous Soluble Oil," U.S. Patent
     No. 3,500,922, March 17, 1970.

12. Tosch, W. C., Jones, S. C. and Adamson, A. W., "Distribu-
     tion Equilibria in a Micellar Solution System," J.
     Colloid and Interface Science, 31, No. 3 (Nov. 1969).

13. Davis, J. A., Jr. and Jones, S. C., "Displacement Mecha-
     nisms of Micellar Solutions," J. Pet. Tech., 1415-1428
     (Dec. 1968); Trans., AIME, Vol. 243.

14. Gogarty, W. B. and Olson, R. W., "Use of Microemulsions
     in Miscible-Type Oil Recovery," U.S. Patent No.
     3,254,714, June 7, 1966.

15. Gogarty, W. B. and Olson, R. W., "Petroleum Recovery
     Materials and Process," U.S. Patent No. 3,301,325,
     January 31, 1967.

16. Taber, J. J., "Dynamic and Static Forces Required to
     Remove a Discontinuous Oil Phase from Porous Media
     Containing both Oil and Water," Soc. Pet. Eng. J.,
     3-12 (March 1962).

17. Holm, L. W., "Reservoir Brines Influence Soluble Oil
     Flooding Process," Oil and Gas J., 158-168 (Nov. 1972).

18. Paul, G. W. and Froning, H. R., "Salinity Effects of
     Micellar Flooding," J. Pet. Tech., 957-958 (Aug. 1973).

19. Mungan, N., "Permeability Reduction Due to Salinity
     Changes," 19th Annual Pet. Soc. CIM, Tech. Meeting,
     May 7, 1968, Preprint #6815.

20. Holm, L. W., "Comments on Finding the Most Profitable
     Slug Size," JPT, 77-79 (Jan. 1973).

21. Dawson, R. and Lantz, R. B., "Inaccessible Pore Volume
     in Polymer Flooding," Soc. Pet. Eng. J., 448-452
     (Oct. 1972); Trans., AIME, Vol. 253.

FLOW OF POLYMERS THROUGH POROUS MEDIA
IN RELATION TO OIL DISPLACEMENT

B. B. Sandiford
Union Oil Research Center

I. ABSTRACT

This paper provides an indepth analysis of the variables
affecting the flow of polymers through porous media in relation to oil displacement.

We have conducted laboratory and field studies of some
polymers that have the potential to improve oil displacement
by reducing water mobilities and by reducing water flow
through high-permeability channels. The variables that we
have studied include rock composition, formation heterogeneity,
water salinity, polymer flow rate, and oil recovery. The
polymers that we are reporting on are partially hydrolyzed
polyacrylamide, polyethylene oxide, hydroxyethyl cellulose,
and a biopolymer. Such polymers reduce the flow rate of water
by increasing its viscosity and, in some cases, by creating
a resistance to flow through a reaction between the polymer
and the formation.

II. SCOPE

Oil displacement resulting from the addition of a water-
soluble polymer to flood water has been investigated for over
20 years. Laboratory studies easily show that, under proper
conditions, polymer flooding, or the injection of a polymer
solution behind a chemical slug, will significantly increase
oil recovery. However, due to a variety of problems, it is
difficult to design effective polymer projects for field
applications. For example, the oil-displacement efficiency
of a polymer solution can be significantly reduced if the
polymer is degraded by shear during mixing or injection into
the reservoir, or if the physical or chemical characteristics
of the reservoir have not been properly taken into account.

There are several types, modifications, and combinations
of polymers available for injection in oil reservoirs. The
selection of the most efficient polymer, or combination of
polymers and chemicals, depends on the reservoir properties
and the oil-recovery process. Polymers can be injected in
large volumes for mobility control, or as relatively small
slugs for improving sweep patterns. Changes in sweep pattern

result from a reduction of flow in high-permeability channels; the reduction in flow can be caused by a reaction between the reservoir rock and the polymer, or between the polymer and other reactive chemicals.

III.   CONCLUSIONS AND SIGNIFICANCE
1.   The use of polymer solutions for flooding reservoirs can increase oil recovery by improving displacement and sweep efficiencies. The selection of a suitable polymer to achieve such improvements is guided by the following general considerations:
    a)   Polyacrylamide and polyethylene oxide polymers improve mobility control by increasing the viscosity of water and by reducing the permeability of reservoir rock to water.
    b)   Polysaccharides and hydroxyethyl cellulose polymers improve mobility control by increasing the viscosity of water.
    c)   Laboratory flow studies demonstrate that controlling mobility by adding polymer to the flood water will increase oil recovery after a waterflood in linear and radial heterogeneous core systems.
    d)   Field studies demonstrate that, under certain conditions, a small slug of polyacrylamide polymer solution will significantly change the sweep pattern and increase oil recovery.
2.   The selection of the most effective polymer for a particular application depends on the reservoir properties and the oil recovery process.
    a)   Polyacrylamide polymers are adversely affected by salty water and, in particular, by divalent cations.
        1)   Polyacrylamide polymers have low viscosities, and they easily lose viscoelastic properties at relatively high shear rates in salty waters.
        2)   In some enhanced recovery projects, the deleterious salts in the reservoir are removed by injection of a chemical solution prior to the injection of the mobility control slug.
    b)   Polysaccharide and hydroxyethyl cellulose polymer solutions are not adversely affected by salty water or by high shear rates.
    c)   It is necessary to simulate reservoir conditions in laboratory studies in order to select the most efficient polymer.
3.   Polymers can be injected alone or in combination with other reactive chemicals to form plugs in water channels and cause changes in flow patterns.

a)   Polymers can be cross-linked before injection into the reservoir, or cross-linked in the formation, by reactions with multivalent cations.
b)   Polymers can be used to slow down the flow of one reactant so that a second reactant can catch up and cause a plug.
c)   Such plugging treatments have been successfully tested in field operations.

IV.   BACKGROUND INFORMATION

The addition of certain polymers to injection water can significantly increase oil recovery by providing mobility control and by reducing channeling. Polymers can be injected at different stages of enhanced recovery· projects in order to improve the efficiency of oil recovery, and they can be injected in combination with other reactive chemicals for the purpose of restricting flow through high-permeability channels.

In order to obtain mobility control in a flood, it is necessary for the displacing phase to have a mobility equal to or lower than the mobility of the oil. This mobility relationship can be expressed as follows:

$$M = \frac{\lambda_w}{\lambda_o} = \frac{K_w/\mu_w}{K_o/\mu_o}$$

The addition of the selected polymers to water increases the viscosity, and in some cases the polymers reduce the permeability of reservoir rock to water. The total reduction in mobility is defined by Pye (1) as a "resistance factor". The resistance factor increases as the mobility of the water is reduced.

$$R = \frac{\lambda_w}{\lambda_p}$$

After the polymer solution has been displaced by water, a residual resistance to water flow remains, and this resistance is defined as:

$$R_{(Residual)} = \frac{\lambda_w \ (initial)}{\lambda_w \ (after \ polymer)}$$

Polymers can also be used to reduce mobility or to cause plugging in water channels by combining the polymers with other reactive chemicals. In these processes, the polymers

can interact in the formation to form gels, or they can be
used to reduce the mobility of one reactant so that a second
reactant can catch up and form a gel plug. The purpose of
such uses of polymers is to improve sweep efficiency. Most
oil reservoirs are heterogeneous; in some cases the high-
permeability channels dominate the flow pattern, but these
channels can be relatively small in cross section. In such
cases, a successful plug in the small channel can greatly
improve the sweep efficiency of a flood.

The physical and chemical properties of four water-soluble
polymers will be discussed in this review of polymer flow in
oil reservoirs. The polymers can be used in different types
of processes for improving oil recovery, and several uses will
be described in this paper. It is possible to present only
a small portion of the available information on these subjects.
Also, since most of the published data, and, in particular,
field data, concerns polyacrylamide type polymers, we will
concentrate on this particular type. Often the other polymers have properties similar to those of the polyacrylamides.
In some cases it is necessary to modify the process so that
a polymer other than polyacrylamide can be used because, under
some conditions, the other polymers have definite advantages.
In all cases a complete evaluation of the different polymers
should be made to select the best for field application.

V.   LABORATORY STUDIES

Some of the physical and chemical properties of four
types of water-soluble polymers (polyacrylamide, polyethylene
oxide, hydroxyethyl cellulose and a polysaccharide) are reviewed in this paper. The general formulas for the polymers
are shown in Figure 1. These polymers are effective in oil
recovery processes because they increase the viscosity of
water, and in some cases they cause an added resistance to
water flow; these two properties are the primary contributors
to mobility control, and they can be evaluated in the laboratory. In order to select the most efficient polymer for an
oil recovery process, it is necessary to simulate reservoir
conditions in the laboratory study.

Figure 1

POLYMER TYPES

1. POLYACRYLAMIDE

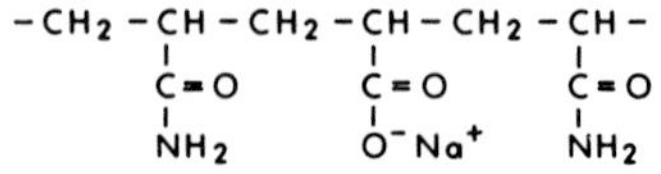

$$-CH_2-CH-CH_2-CH-CH_2-CH-$$
$$\begin{array}{ccc} C=O & C=O & C=O \\ NH_2 & O^-Na^+ & NH_2 \end{array}$$

2. POLYETHYLENE OXIDE

$-CH_2-CH_2-O-CH_2-CH_2-O-CH_2$

3. HYDROXYETHYL CELLULOSE

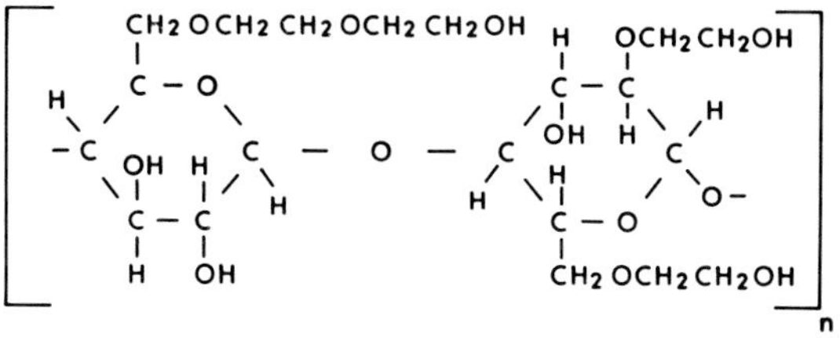

4. POLYSACCHARIDE

HIGH MOLECULAR WEIGHT - FERMENTATION
PRODUCT OF GLUCOSE

The resistance factor needs to be determined during flow studies in cores, but in preliminary studies we sometimes measure screen factors instead of resistance factors. The screen factor can be determined more rapidly, and it can be correlated with resistance factors. These correlations are true only for a given polymer system as shown by Jennings et al. (2). The definition of screen factor is:

$$\text{Screen factor:} \quad \frac{\text{Polymer Flow Time through a Screen Viscometer}}{\text{Water Flow Time through a Screen Viscometer}}$$

In general, the polyacrylamide and polyethylene oxide polymers are quite sensitive to salt and shear, whereas the hydroxyethyl cellulose and polysaccharide polymers are less sensitive to these variables. Selected data that demonstrate the typical effects of salt and polymer concentration on solution viscosities and screen factors for the four polymers are shown in Figures 2 and 3. Within each polymer type it is possible to obtain a very wide variation in physical characteristics, depending on molecular size and chemical substitution. The data that are presented in these figures should only be used as a guideline in polymer selection. The viscosities and screen factors of the polyacrylamide and polyethylene oxide type polymer solutions are greatly reduced by the divalent cations that are present in oil reservoir brine. In general, the greater the cation concentration, the greater the reduction in these properties. The polysaccharide and hydroxyethyl

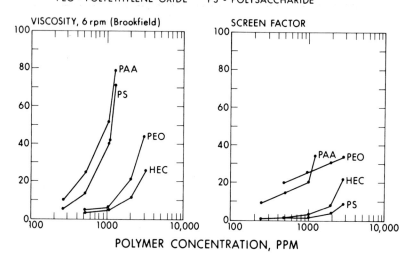

FIGURE 2

EFFECT OF POLYMER CONCENTRATION ON VISCOSITY AND
SCREEN FACTOR

POLYMERS MIXED IN 500 ppm NaCl₂

PAA = POLYACRYLAMIDE   HEC = HYDROXYETHYL CELLULOSE
PEO = POLYETHYLENE OXIDE   PS = POLYSACCHARIDE

FIGURE 3

EFFECT OF POLYMER CONCENTRATION ON VISCOSITY AND SCREEN FACTOR

POLYMERS MIXED IN 500 ppm CaCl₂

PAA = POLYACRYLAMIDE   HEC = HYDROXYETHYL CELLULOSE
PEO = POLYETHYLENE OXIDE   PS = POLYSACCHARIDE

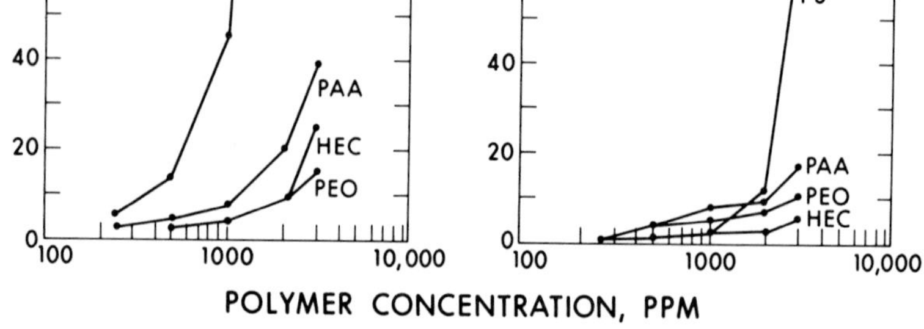

cellulose polymer solutions are not seriously affected by
divalent cations, but neither of these polymers develops
significant screen factors in fresh of salty waters except
in relatively high concentrations (Figure 3). Knight et al.
(3) presented data to show that polysaccharide polymers can
reduce flow rates through a residual resistance mechanism
when saline waters are involved. These data are supported
by the results shown in Figure 3. It is possible to degrade
the polyacrylamide and polyethylene oxide polymers during
mixing and injection operations in the field by using high
shear rates. The results in Figures 4 and 5 show differences
in shear stability between polyacrylamide and polyethylene
oxide polymers under two different salt conditions. These
data demonstrate the desirability of using soft water for
injection of polymer solutions at high rates. Multivalent
cations have a great effect on polymer stability. These
studies are limited to two different types of polymers, but
also very significant differences in polymer stability occur
within each polymer type depending on their molecular configuration. The larger molecules are usually the most
unstable. The studies reported in Figures 4 and 5 were
conducted on Berea cores at several different advance rates.
Jennings (2) showed that flow rate can be related to shear
rate as follows:

$$\text{Shear Rate} = \frac{\bar{V}}{[\frac{1}{2}\,(K/\phi)]^{1/2}}$$

The effect of several variables on polymer stability
have been published by Jennings et al. (2), White et al. (4),
Hill et al. (5), and by Maerker (6). All of these studies
of polyacrylamide polymers show that, under shearing conditions, the screen factor is the primary variable that is
degraded and not the viscosity. Under some conditions,
polymer solution viscosity should be the primary criterion
for mobility control for all four of these types of polymers,
but this decision can be made after the laboratory evaluation
study. Under reservoir flow rates these polymers impart a
residual resistance to water flow, but laboratory studies in
the Berea cores at high flow rates did not show this property
to be of significant importance. The calculated residual
resistance factors were usually in the range of 1.2 in value.
Also the calculated resistance factors decreased with an increase in flow rate at conditions that simulate injection
rates. Other investigators have usually shown the opposite
effect at lower rates.

Under many field conditions, the polysaccharide or hydroxyethyl cellulose polymers will be preferred for mobility control

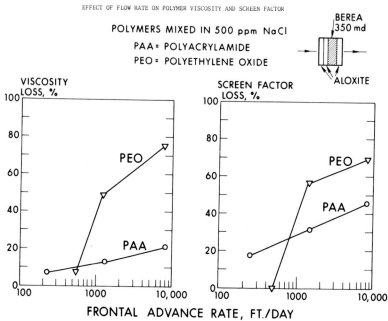

FIGURE 4

EFFECT OF FLOW RATE ON POLYMER VISCOSITY AND SCREEN FACTOR

POLYMERS MIXED IN 500 ppm NaCl

PAA = POLYACRYLAMIDE

PEO = POLYETHYLENE OXIDE

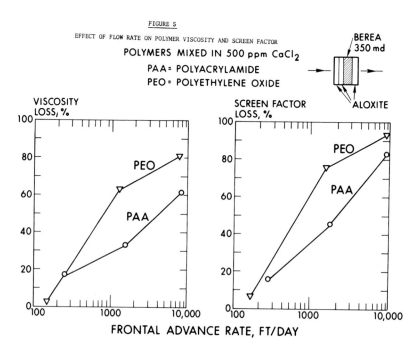

FIGURE 5

EFFECT OF FLOW RATE ON POLYMER VISCOSITY AND SCREEN FACTOR

POLYMERS MIXED IN 500 ppm $CaCl_2$

PAA = POLYACRYLAMIDE

PEO = POLYETHYLENE OXIDE

Since they are not seriously affected by salt and shear, it is not necessary to take special precautions in the selection of the injection water, mixing facilities, or injection rates. In the case of the polysaccharide polymer, it is often necessary to filter the polymer solution in order to remove bacterial debris that results from the fermentation process. These problems and their solution have been discussed in the literature by Lipton (7).

The performance of hydroxyethyl cellulose type polymers in oil recovery processes has not been extensively reported in the literature, but they deserve consideration because of their physical and chemical properties. These polymers are relatively insensitive to shear and salt, and they do not have to be filtered, which gives them an advantage over the polysaccharide type polymers. They have the disadvantage of not increasing the viscosity of water as much as the other three polymers at equivalent concentrations (see Figures 2 and 3). Since hydroxyethyl cellulose type polymers are more stable than other types, the cost of obtaining mobility control for these polymers could be competitive with the others under some reservoir conditions. These variables need to be evaluated for each project.

All of the water-soluble polymers described in this paper are attacked by bacteria and oxygen, and it is necessary to take special precautions during mixing and injection of these solutions (8). Aerobic bacteria readily grow in polysaccharide polymer solutions, and they require a relatively high concentration of preservative as described by Lipton (7). The polyacrylamide and polyethylene oxide polymers are very unstable under oxidizing conditions, and it is necessary to remove dissolved oxygen during the mixing and injection processes. Also, since the polyethylene oxide polymers are unstable in the presence of light, they should be stored in the dark at least for laboratory studies.

Temperature is another variable that can cause stability problems with all of these polymers. Simulated reservoir studies need to be made before a polymer is selected for a high temperature reservoir; under some reservoir conditions all of these polymers start to break down at about 180°F. Thermal-oxidative degradation of polyacrylamides occurs very rapidly at 190°F as demonstrated by Herr and Routson (9) by the use of electron micrographs.

VI.   LABORATORY FLOW STUDIES

The proper application of polymers to field operations is complicated by the many variables encountered in a reservoir, but it is easy to demonstrate large increases in oil recovery by adding polymers to water floods in laboratory studies.

Such a laboratory flow study, conducted with two linear sand packs mounted in parallel, is used in this paper as an example of the oil displacement efficiency that can be gained from mobility control. In this study, the sand was obtained from two unconsolidated field cores, each of different permeability. The sand packs were simultaneously flooded by water from one source and then, near the completion of the waterflood, they were flooded with a polymer solution. The laboratory set-up and injection sequence are shown in Figure 6. The flood results are also shown in Figure 6, and they indicate that the polymer flood almost doubled the oil recovery of 34% by waterflood, bringing it to 62%. The primary

FIGURE 6

WATER FLOOD - POLYMER FLOOD

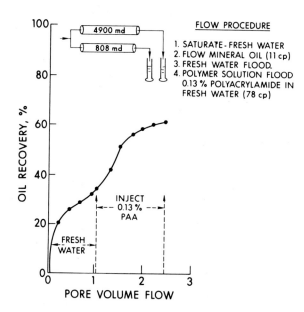

advantage of the polymer solution was that it diverted flow from the high permeability sand pack to the tighter pack (Figure 7), but the high viscosity (72 cp) polymer solution slug also increased recovery from the high-permeability core. At the start of the polymer injection, the polymer solution preferentially entered the high-permeability sand. During this stage of the flood, the polymer solution had a greater effect on flow in the high-permeability sand than in the low. After about 0.7 total pore volume had been injected, however, the relative effect of the polymer on flow rate became greater in the low permeability system. Sufficient diversion of flow

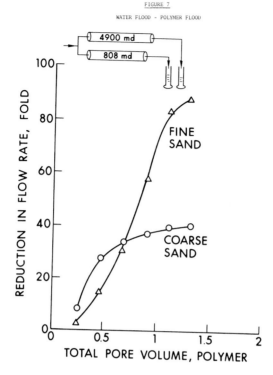

FIGURE 7

WATER FLOOD - POLYMER FLOOD

into the tighter core had occurred before this point so that
there was an increase in oil recovery.

The polymer flood results were evaluated separately for
the high-permeability core (Figure 8), and they show that the
polymer solution caused a significant increase in oil recovery
after the waterflood. In this case the increase in recovery
was due to an improvement in the displacement efficiency. The
polymer solution slug size (1.3 pore volume) was larger than
we would normally propose for field application, but these
results illustrate the mobility control mechanism that can
take place. Similar results have also been observed in core
studies with the other polymers.

The polyacrylamide polymers affect the flow properties
of some reservoir rock more than others. To illustrate, two
cores from different oil fields were mounted in parallel for
flow studies. The laboratory procedure and results are shown
in Figure 9. In this case, the flow properties of the high-
permeability core were greatly affected by a 30% pore volume
slug of polymer solution, whereas the low permeability core
was not. The results show that the polymer slug evened out
the flow of water in this heterogeneous system. During the

FIGURE 8

WATER FLOOD - POLYMER FLOOD

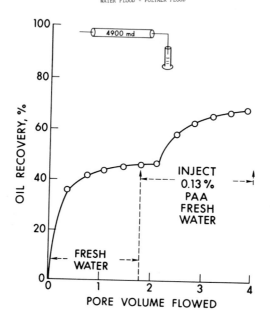

FIGURE 9

POLYMER SLUG DURING WATER FLOOD

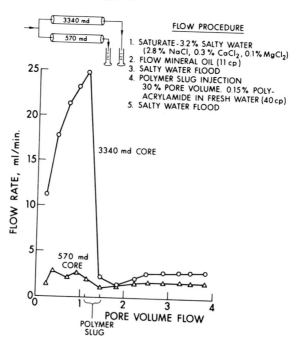

polymer injection, 72% of the solution entered the high-permeability core, but following the polymer the water flow rates in these cores were about the same. The reduction in flow rate in the high-permeability core is typical for this reservoir, and cores from the other reservoir do not exhibit residual resistance factors following polymer flow as is also shown in this study. Cores containing large amounts of clays are often easily affected by polymer solutions. These results illustrate the need for simulating reservoir conditions in the laboratory evaluation of this process. Polymers have been used for years in many industrial processes for flocculating particles. This flocculation reaction is probably important in the reaction between reservoir rock and these polymers.

VII. FIELD STUDIES

Field data on six relatively large polymer floods are shown in Table I. In these applications, the polymer concentrations were below 1000 ppm, and the slug sizes ranged from 4 to 45% of a pore volume. The extra oil produced varied from 0 to 200 bbl/acre-ft.

In the most successful project, conducted in the Vernon Field, Kansas, by Brazos Oil and Gas Company, the polymer cost about one dollar for each extra barrel of oil. The polymer concentration and the slug volumes were larger in the Vernon project than in the other tests. In all six applications, the quantities of polymers used were probably smaller than the quantities required to obtain maximum displacement efficiency. In particular, the polymer floods that were conducted in the Wilmington and Pembina fields did not use sufficient polymer to obtain mobility control. In these floods the polymer slugs equaled 5% or less of a pore volume. Another serious problem that occurred in these floods resulted from reservoir heterogeneity. Polymer solutions severely channeled to nearby producers which caused a loss in mobility control.

It is of value to note that polymer solutions that measured about one-half in screen factor as compared to their original values were produced in some cases. These results show that the polyacrylamide polymers can retain a significant part of their viscoelastic properties during flow through the reservoir.

The requirements for mobility control in a polymer flood are similar to those described for a surfactant flood by Gogarty (16); he presented data to show that, in laboratory studies, the mobility buffer should equal about 50% of the pore volume. But he also pointed out that larger pore volumes of these buffers are probably needed in actual

TABLE I

FIELD STUDIES OF POLYACRYLAMIDE POLYMER FLOODS

Field	State or Province	Status	Polymer Conc., ppm	Slug Size PV	Evaluation- Extra Oil	Literature Reference
Vernon	Kansas	Complete	500	2 zones 33% 45%	200 B/AF	Jones (10)
Taber South	Alberta	Complete but evaluating	350	18%	88 B/AF (computer model)	Shaw and Stright (11)
Brelum	Texas	Complete but estimated final results	Graded 390 to 75	25%	89 B/AF	Rowalt (12)
North Burbank Unit	Oklahoma	Not Complete	Graded 77%-250 15%- 50 8%- 25	23% in Center of Pattern	16 B/AF in Center Pattern	Clampitt and Reid (13)
Pembina	Alberta	Complete	Graded 1000 to 100	4%	About 0	Groenveld (14) Melrose and George
Wilmington	California	Complete	213	5%	About 0	Krebs (15)

reservoirs where geometries are more complex. The mobility
control slugs can be graded in polymer concentration from a
high value, which gives complete mobility control at the
start of the polymer injection, to a low value at the end.
Different systems for grading the polymer slug have been
used by several investigators. The quantity of polymer that
is required for a given process depends on the reservoir
properties, in particular on polymer adsorption. In a polymer flood, economics might favor a smaller than 50% pore
volume slug, but this needs to be determined for each project. In a successful surfactant-polymer flood that was
conducted in Illinois by Marathon (Gogarty), the polymer
slug equaled 100% of a pore volume. We need to consider
large volumes of polymer solution for mobility control in
polymer floods.

When a larger slug of either polyacrylamide or polyethylene oxide is used for mobility control, and when displacement efficiency is of primary importance, it is necessary to
consider the effect of shear rate on polymer stability.
These polymers can become unstable at even moderate shear
rates in a) low permeability reservoirs, b) gun perforated
completions, and c) high salt content injection waters.
They can also be made unstable when they are prepared in
improper mixing and injection equipment. Most of these
problems were discussed in the section on Polymer Characteristics, but field accounts that substantiate polymer instability under shearing conditions have also been published.
B. L. Knight (8) and Schurz (17) separately published on
recommended mixing conditions, and Bilhartz and Charlson (18)
recently presented data on polymer degradation during injection in an oil reservoir. As expected from laboratory data
(Figures 4 and 5), Bilhartz and Charlson found that, under
field conditions, the polyacrylamide polymer also degraded
primarily in screen factor and not viscosity. If screen
factor degradation is a problem, it might be necessary to use
viscosity as the primary mobility control characteristic.

Polyacrylamide and polysaccharide polymers have been used
in several field applications for mobility control in combination with surfactant slugs. In surfactant polymer floods, the
injection sequence is usually 1) preflush, 2) surfactant slug
and 3) polymer slug (19). The selection of the polymer depends on the reservoir condition and the application. Polymer
adsorption is one of the variables that needs to be considered.
Most surfactants are adsorbed by the reservoir to a greater
extent than are polymers, and studies have shown that the
polymer in the mobility control slug behind the surfactant is
adsorbed less than during a polymer flood (20). According to
Jones (21), if polymer is added to the preflush slug, the

surfactant and polymer slugs that follow will be more stable. Sarem and Holm (22) have shown that the injection of an aqueous solution of a water-soluble polymer dissolved in substantially salt-free water as a preflush ahead of a miscible flooding process will increase oil recovery in heterogeneous reservoirs. If the channels in heterogeneous reservoirs are very severe, it might be necessary to form more permanent plugs in order to prevent loss of expensive chemicals. Bernard (23) has demonstrated that cross-linked polymers can be used for these treatments.

Another use for polymers is in aqueous surfactant floods. It is often necessary to add a polymer to the surfactant slug in order to gain mobility control in this phase of the process. French et al. (24) used a polyacrylamide polymer, and Pursley et al. (25) used a polysaccharide polymer for this purpose. The polysaccharide polymer is relatively insensitive to shear and salt contamination as opposed to the problems we have discussed with polyacrylamide polymers. But the complex nature of the surfactant processes and the reservoir conditions that exist in different oil fields makes the selection of the polymer dependent on these variables.

In a large number of polymer injection projects reported in the literature, the polymer slug sizes were very small. In many cases extra oil production was obtained, but it probably did not result from increased displacement efficiency. A small slug of polymer solution can reduce flow through channels and cause a change in the flood pattern.

As an example of this process, a very successful, small-slug polyacrylamide injection treatment was conducted in two injection wells in the Cut Bank Field in Montana. This reservoir was very reactive to the polymer solution injection as evidenced by rapid increases in injection pressures. The pressures in the two treated wells increased by several hundred pounds during the injection of only about 100 pounds of polymer mixed at a concentration of about 0.03% in salty reservoir water. It was necessary to stop the injection of polymer and to return the wells to regular water after the injection of each small slug of polymer solution. The total polymer injected equaled 500 pounds in well No. 1 and 800 pounds in well No. 2. The wells were treated in 1969, and increases in oil production in nearby wells were noted in about six months for well No. 3 and after one year in well No. 4 with a total of extra oil for the two wells equal to 38,000 barrels. The production from wells 3 and 4 are shown in Figures 10 and 11.

The polymer-reservoir reactivity that we observed in this test was greater than normal, but it can occur, as we

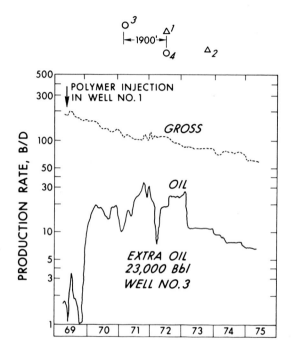

FIGURE 10

PRODUCTION HISTORY - WELL NO. 3

showed in the section on Laboratory Studies. The success of
this project demonstrates the stability of polyacrylamide
under relatively high shear-rate conditions. The average
permeability of the Cut Bank reservoir is 35 md, the injection rate was 200 B/D, and the sand thickness about 13 feet.
The effective sand thickness was probably less than the 13
feet due to a high permeability zone. Channeling was observed
prior to the injection of the polymer solutions by the use of
chemical tracers. Under the conditions of the treatment
described, laboratory studies normally indicate that serious
polymer degradation should have occurred, and also in Berea
cores, residual resistance to high rate water flow following
the polymer was not a significant factor. The field results
at Cut Bank show successful diversion of injection water that
lasted several years. The results demonstrate the need for
using field cores and for measuring resistance factors in
core flow studies.
     Polymers also can be used in the treatment of producing
wells. Field data from several investigators, White et al.
(4), Sandiford et al. (26), and Sparlin (27) show that the

504   B. B. SANDIFORD

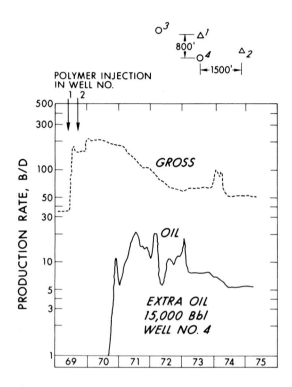

FIGURE 11

PRODUCTION HISTORY - WELL NO. 4

injection of polyacrylamide polymers in producing wells can
reduce water and increase oil production. The success of
these treatments again depends on the reactivity of the polymer and the reservoir rocks. These polymers normally reduce
water flow without seriously restricting oil production. A
laboratory study that demonstrates these permeability effects
is shown in Figure 12. In this study a core from the Dominguez
field in California was mounted in plastic and flow studies
that simulated a reservoir production history were conducted.

The core was saturated with oil and water at restored
state conditions before the start of a regular waterflood.
The injection of a polyacrylamide polymer solution significantly reduced the permeability to water but not the permeability to oil. The fact that polymers can reduce water flow
rates without seriously affecting oil flow rates is important
to oil field operations. The reduction in water production
often means that a well can be pumped off. A lower fluid
level in the treated well will reduce the back pressure on
the formation, and this reduction in pressure differential

FIGURE·12

EFFECT OF POLYACRYLAMIDE POLYMER
ON SANDSTONE CORE PERMEABILITY

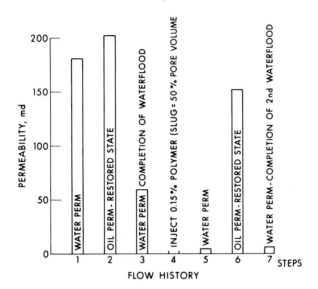

CORE FROM DOMINGUEZ
FIELD, CALIFORNIA

POROSITY = 25 %
WATER = 2.3 % NaCl, 0.3 % CaCl$_2$, 0.2 % MgCl$_2$
OIL = MINERAL OIL - 11 cp

can cause an increase in oil production. The water that is diverted from the channel in the treated well can enter another interval, push oil or water towards nearby producing wells or flow, to the aquifer. Thus, the results of these treatments can mean a reduction in water production, an increase in oil production, or both.

In some waterfloods the channels are too severe to be corrected by the injection of polymer alone. Many types of plugging treatments have been tested ever since the start of waterflooding. In all cases, the success of a given type of treatment depends on the reservoir conditions, and failures are often due to inadequate knowledge of the problem. Several plugging processes depend on reactions with polymers.

One of the first blocking agents was developed by Routson et al. (28), and it is commercially sold under its trademark Channelblock.* Channelblock depends on cross-linking polyacrylamide polymers with multivalent cations,

---

* Trademark of the Dow Chemical Company.

and then injecting the plugging mixtures into the channel. A similar process has been tested in the laboratory by Needham et al. (29) and by Fitch and Canfield (30) in field treatments. In this process the polymers are cross-linked in the reservoir, and they depend on a reaction between a first-stage cationic polymer and the reservoir rock and a second-stage cross-linking mixture that contains an anionic polyacrylamide polymer.

Eaton (31) used polymers to reduce the mobility of a first stage reactant slug so that a second chemical slug that is required in the reaction can penetrate and mix with the first slug. We have combined a polymer with ammonium sulfate in the first slug and sodium silicate in the second slug for plugging channels in injection and production wells. Also, we usually follow these reactive materials with one or more reactive chemicals in order to increase the size or severity of the plug, depending on the reservoir problem. An example of a treatment in a steamed producer is shown in Figure 13. In this case the producing well was in its sixth steam cycle, and in the previous cycle all of the steam entered the bottom interval. The injection of the polymer-plug treatment effectively changed the steam-injection pro-file and allowed a new interval to be heated. In another field, a producing well in a a normal waterflood project was treated with this same sequence of reactive chemicals. In this case a fracture channel from the bottom of the oil reservoir to a water bearing zone was partially plugged by the treatment. The results of this plug caused a reduc-tion in water flow and allowed a high fluid level in the well to be pumped down. The lower fluid level reduced the back pressure on the oil zone, and the oil production increased from zero to over 50 B/D. The success of these treatments depends on the reservoir prob-lem and the potential for oil production.

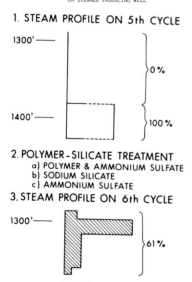

FIGURE 13

POLYMER-SILICATE DIVERSION TREATMENT
ON STEAMED PRODUCING WELL

1. STEAM PROFILE ON 5th CYCLE

1300' —   ⎱ 0%

1400' —   ⎱ 100%

2. POLYMER-SILICATE TREATMENT
   a) POLYMER & AMMONIUM SULFATE
   b) SODIUM SILICATE
   c) AMMONIUM SULFATE

3. STEAM PROFILE ON 6th CYCLE

1300' —   ⎱ 61%

1400' —   ⎱ 39%

VIII.   FUTURE
    Enhanced recovery projects depend on uniformly sweeping
the reservoir with chemical or solvent slugs. In order to
prevent channeling, improve sweep efficiencies, and gain
mobility control, a combination of polymer injection processes will be necessary. In each case, the type of polymer
that will most efficiently perform will depend on the reservoir conditions and the particular enhanced-recovery process.

IX.   LITERATURE CITED
1.   Pye, D. J., "Improved Secondary Recovery by Control of
        Water Mobility," J. Pet. Tech. 911-916 (August 1964).
2.   Jennings, R. R., Rogers, J. H. and West, T. J., "Factors
        Influencing Mobility Control by Polymer Solutions,"
        J. Pet. Tech. 391-401 (March 1971).
3.   Knight, B. L., Jones, S. C. and Parsons, R. W., "Discus-
        sion" (see Trushenski in Reference), Soc. Pet. Eng. J.
        643-644 (December 1974).
4.   White, J. L., Goddard, J. E. and Phillips, H. M., "Use
        of Polymers to Control Water Production in Oil Wells,"
        J. Pet. Tech. 143-150 (February 1973).
5.   Hill, H. J., Brew, J. R., Claridge, E. L., Hite, J. R.
        and Pope, G. A., "The Behavior of Polymers in Porous
        Media," SPE Paper No. 4748 presented at Soc. Pet. Eng.
        meeting in Tulsa, Oklahoma, April 22-24, 1974.
6.   Maerker, J. M., "Shear Degradation of Partially Hydrolyzed
        Polyacrylamide Solutions," Soc. Pet. Eng. J. 311-322
        (August 1975).
7.   Lipton, D., "Improved Injectability of Biopolymer Solu-
        tions," SPE Paper No. 5099 presented at Soc. Pet. Eng.
        meeting in Houston, Texas, October 6-9, 1974.
8.   Knight, B. L., "Reservoir Stability of Polymer Solutions,"
        Paper No. 45d, AIChE Meeting, Dallas, Texas, February
        20-23, 1972.
9.   Herr, J. W. and Routson, W. G., "Polymer Structure and
        its Relationship to the Dilute Solution Properties of
        High Molecular Weight Polyacrylamide," SPE Paper No.
        5098 presented at Soc. Pet. Eng. meeting in Houston,
        Texas, October 6-9, 1974.
10. Jones, M. A., "Waterflood Mobility Control: A Case
        History," J. Pet. Tech. 1151-1156 (September 1966).
11. Shaw, R. A. and Stright, O. H., Jr., "Performance of the
        Taber South Polymer Flood," Pet. Soc. of CIM, June 11-
        13, 1975.
12. Rowalt, R. J., "A Case History of Polymer Waterflooding-
        Brelum Field Unit," SPE Paper No. 4671 presented at
        Soc. Pet. Eng. meeting in Las Vegas, Nevada, September
        30-October 3, 1973.

13. Clampitt, R. L. and Reid, T. B., "An Economic Polymer Flood in the North Burbank Unit, Osage County, Oklahoma," SPE Paper No. 5552 presented at Soc. Pet. Eng. meeting in Dallas, Texas, September 28-October 1, 1975.

14. Groenveld, H., Melrose, J. C. and George, R. A., "Pembina Polymer Pilot Flood," SPE Paper 5829 presented at Soc. Pet. Eng. meeting in Tulsa, Oklahoma, March 22-24, 1976.

15. Krebs, H. J., "Wilmington Field California Polymer Flood--A Case History," SPE Paper No. 5828 presented at Soc. Pet. Eng. meeting in Tulsa, Oklahoma, March 22-24, 1976.

16. Gogarty, W. B., "Status of Surfactant or Micellar Methods," J. Pet. Tech. 93-102 (January 1976).

17. Schurz, G. F., "Mobility Control, A New Engineering Tool," SPE Paper No. 986 presented at Soc. Pet. Eng. meeting in Houston, Texas, October 11-14, 1964.

18. Bilhartz, H. L., Jr. and Charlson, G. S., "Field Polymer Stability Studies," SPE Paper No. 5551 presented at Soc. Pet. Eng. meeting in Dallas, Texas, September 28-October 1, 1975.

19. Knight, R. K. and Baer, P. J., "A Field Test of Soluble-Oil Flooding at Higgs Unit," J. Pet. Tech. 9-15 (January 1973).

20. Trushenski, S. P., Dauben, D. L. and Parrish, D. R., "Micellar Flooding-Fluid Propagation, Interaction, and Mobility," Soc. Pet. Eng. J. 633-645 (December 1974).

21. Jones, S. C., "Secondary Recovery Process Utilizing a Pre-Slug Prior to a Displacing Fluid," U.S. Patent 3,482,631, December 9, 1969.

22. Sarem, A. M. and Holm, L. W., "Process for Recovering Oil from Heterogeneous Reservoirs," U.S. Patent 3,704,990, December 5, 1972.

23. Bernard, G. G., "Process for Recovering Oil from Heterogeneous Reservoirs," U.S. Patent 3,882,938, May 13, 1975.

24. French, M. S., Keys, G. W., Stegemeier, G. L., Weber, R. C., Abrams, A. and Hill, H. J., "Field Test of an Aqueous Surfactant System for Oil Recovery, Benton Field, Illinois," J. Pet. Tech. 195-204 (February 1974).

25. Pursley, S. A., Healy, R. N. and Sandvik, E. I., "A Field Test of Surfactant Flooding, London, Illinois," J. Pet. Tech. 793-802 (July 1973).

26. Sandiford, B. B. and Graham, G. A., "Injection of Poly-
     mer Solutions in Producing Wells," Paper No. 45C,
     AIChE Meeting, Dallas, Texas, February 20-23, 1972.
27. Sparlin, D., "An Evaluation of Polyacrylamides for
     Reducing Water Production," SPE Paper No. 5610
     presented at Soc. Pet. Eng. meeting in Dallas, Texas,
     September 28-October 1, 1975.
28. Routson, W. G., Neale, M. and Penton, J. R., "A New
     Blocking Agent for Waterflood Channeling," SPE Paper
     No. 3992 presented at Soc. Pet. Eng. meeting in San
     Antonio, Texas, October 8-11, 1972.
29. Needham, R. B., Threlkeld, C. B. and Gall, J. W., "Con-
     trol of Water Mobility Using Polymers and Multivalent
     Cations," SPE Paper No. 4747 presented at Soc. Pet.
     Eng. meeting in Tulsa, Oklahoma, April 22-24, 1974.
30. Fitch, J. P. and Canfield, C. M., "Field Performance
     Evaluation of Crosslinked Polymers to Increase Oil
     Recovery in the Wilmington Field, California," SPE
     Paper No. 5366 presented at Soc. Pet. Eng. meeting
     in Ventura, California, April 2-4, 1975.
31. Eaton, B. A., "Selective Plugging of Permeable Water
     Channels in Subterranean Formations," U.S. Patent
     3,396,790, August 13, 1968.

X.   NOMENCLATURE

K =   permeability

M =   mobility ratio

R =   resistance factor

$\bar{V}$ =   rate of advance

λ =   mobility

ϕ =   porosity

Subscripts

o   =   oil

w   =   water

MECHANISMS OF POLYMER RETENTION IN POROUS MEDIA

G. Paul Willhite
Jose G. Dominguez
University of Kansas

I. ABSTRACT

Polymers used as mobility control agents in oil recovery processes are retained or "lost" as they flow through porous materials. Several mechanisms have been identified which cause polymer retention. These include adsorption on the solid surfaces, mechanical entrapment in crevices and narrow pores and hydrodynamic retention, observed after changes in the flow rate. Mechanisms of retention are described and the relative importance of each mechanism is assessed from the available experimental data. Models of polymer retention and the effect of polymer retention on flow through porous media are explored.

II. INTRODUCTION

Certain high molecular weight polymers exhibit large increases in solution viscosity at concentrations as low as a few hundred parts per million. This property has led to extensive use of polymers to control the movement of fluids in porous media. Polymer solutions improve the displacement efficiency of waterfloods (1). They have important applications in controlling the movement of the displacing fluids in micellar or low tension displacement processes (2,3).

The effectiveness of a polymer solution in controlling mobility is a function of the polymer concentration. Polymer molecules are retained as a polymer solution flows through porous media, reducing the concentration of the polymer and its solution viscosity. This paper contains an analysis of the mechanisms which polymer molecules are retained in porous materials. We begin by reviewing the properties of these large molecules in aqueous solution. We examine the methods used to determine polymer retention, studies which identified specific retention mechanisms, and models which have been proposed to represent the polymer retention mechanisms. The paper concludes with a discussion of the effects of polymer retention on flow through porous media.

III.   PROPERTIES OF POLYMER SOLUTIONS

A.   Polymer Molecule in Solution
      Figures 1 and 2 depict the chemical structure of the two
polymers, a partially hydrolyzed polyacrylamide and a biopolymer, which are used in oil displacement processes. Partially
hydrolyzed polyacrylamides are made by hydrolyzing polyacril-
amide with sodium or potassium hydroxide. The degree of
hydrolysis ranges from 15 to 35%. Xanthan gum is a biopolymer
produced by the microbial action of the oragnism xanthomonas
campestris (4) on a carbohydrate. Both polymers are large
macromolecules with molecular weights ranging from 1×10^6 to
10×10^6.

Structure of Polyacrylamide

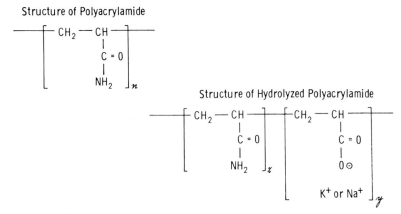

Structure of Hydrolyzed Polyacrylamide

Fig. 1.   *Molecular Structure of High Molecular Weight Poly-*
          acrylamides

      As expressed by Flory (6), the usual chemical formulas
do not convey the most significant structural characteristic
of a polymer molecule. It has the capacity to assume a variety of three dimensional configurations. Each segment
dangles around a bond and the probability of any one of these
configurations is related to the free energy of the molecules
at the given position.
      The simplest representation of a polymer molecule from
a structural viewpoint is to consider the molecule as a
string of pearls which is constantly changing its configuration. Vollmert (7) proposed that the polymer molecules are
more appropriately described as random statistical coils
which have become saturated with solvent. According to
Vollmert, the solvent within the polymer coil is considered
bound to the polymer chain because the polymer coil cannot
be separated from the solvent during a sedimentation run in
an ultracentrifuge. Vollmert's (7) representation of a dilute
macromolecular solution is presented in Figure 3.

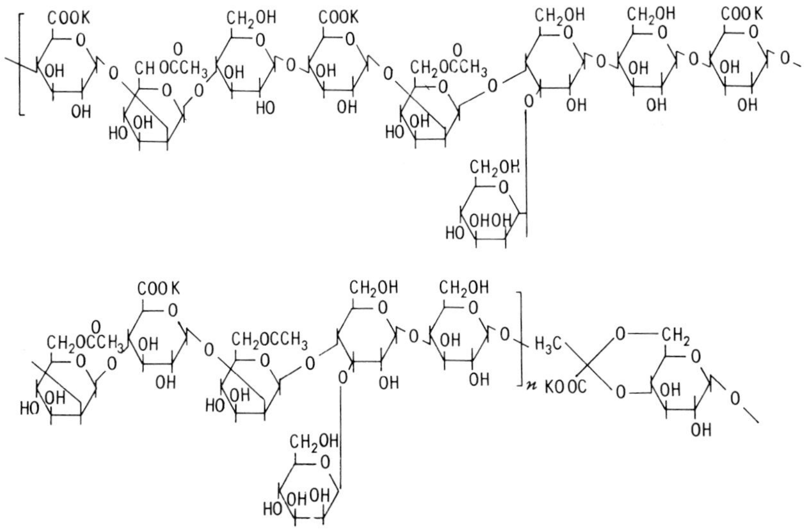

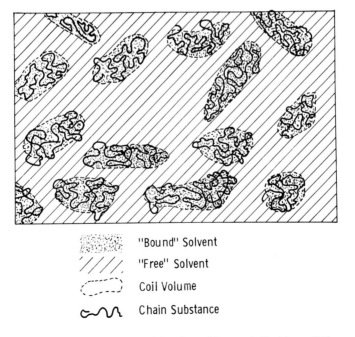

Fig. 2.   *Structure of Xanthan Biopolymer (11)*

"Bound" Solvent
"Free" Solvent
Coil Volume
Chain Substance

Fig. 3.   *Polymer Coils in Dilute Solution (7)*

B.   Apparent Size of Polymer Molecules in Dilute Solutions

The size of polymer molecules in dilute solutions is influenced by polymer-solvent interactions. If the solvent is
a good solvent, the polymer segments prefer to be surrounded
by solvent molecules rather than by other polymer segments.
If the solvent is poor, the polymer molecule tries to minimize the area of contact with the solvent molecules. Therefore, polymer solvent interactions control the extension of
the polymer molecule.

Solvent-polymer interactions are particularly noticeable
when the polymer is a polyelectrolyte and the solvent is water.
Polyelectrolyte molecules in dilute salt-free water solutions
extend largely due to the interactions between the charges on
the backbone of the polymer (7a). If an electrolyte is added
to water, the repulsion between backbone charges is screened
by a double layer of simple electrolytes. As the simple
electrolyte concentration is increased, the extension of the
polymer molecules decreases. Divalent ions (Ca^{++}, Mg^{++}) bond
to a negatively charged macroion in preference to monovalent
ions. Their effect is more pronounced since even small
amounts of polyvalent ions locate themselves in strategic
positions where they are able to screen the charge of the
polyion more effectively.

Every factor which affects the extension of the polymer
molecule in solution affects the viscosity of the solution.
Effects of electrolyte concentration (NaCl) and degree of
hydrolysis are shown in Figure 4 for polyacrylamide solutions (8).

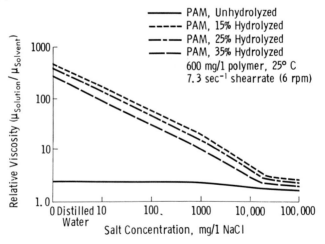

Fig. 4.   Relative Viscosity of Hydrolyzed Polyacrylamides
in Sodium Chloride Waters (8)

Two parameters, the root-mean-square end to end distance $\sqrt{\overline{r^2}}$ and the root-mean-square distance of the elements of the chain from its center of gravity, also called radius of gyration, $\sqrt{\overline{s^2}}$, are used to characterize the effective size of polymer molecules in solution. For a linear polymer, the two parameters are related by the equation $\sqrt{\overline{s^2}} = \sqrt{\overline{r^2}}/6$ (6). The root mean square end to end distance for a polymer may be estimated from the intrinsic viscosity [μ] and the molecular weight, M. Lynch and MacWilliams (9) applied Equation 1 to a polyelectrolyte in a good solvent. This relationship was developed by Flory (6) for noncharged polymers. The intrinsic viscosity is defined by Equation 2

$$\sqrt{\overline{r^2}} = 8(M[\mu])^{1/3} \tag{1}$$

$$[\mu] = \lim_{c \to 0} \frac{\mu - \mu_s}{c\mu_s} \tag{2}$$

where μ_s is the viscosity of the solution and c = concentration of polymer.

Lynch and MacWilliams calculated a mean square end to end distance of approximately 0.28 microns for a hydrolyzed polyacrylamide molecule (M = 3 x 10^6) in a 3.0% NaCl solution.

High molecular weight polymers are molecular mixtures rather than chemical individuals. The molecular weight distribution depends upon the kind of polymerization. Figure 5 shows the molecular weight distribution of a polyacrylamide polymer similar to those used in polymer floods (10). A dispersion in molecular weight as shown in Figure 5 implies a similar dispersion in molecular size. If the average size of a polymer is of the order of magnitude of the pore size of a porous medium, this dispersion in molecular size should be expected to influence the behavior of polymer solutions in such a porous medium.

C. Non-Newtonian Behavior

Polymer solutions frequently behave as non-Newtonian fluids. Shear thinning or pseudoplastic behavior shown in Figure 6 (11) is typical. Since partially hydrolyzed polyacrylamide is a polyelectrolyte, its size in solution and consequently its solution viscosity are sensitive to electrolyte concentration. Figure 7 (11) shows the dependence of solution viscosity on sodium chloride concentration and shear rate. Xanthan biopolymers have few ionizable groups. Thus,

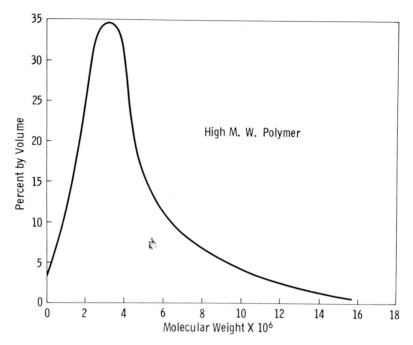

*Fig. 5.   Frequency Distribution of High MW Polyacrylamide at
1 ppm Concentration*

the viscosity of aqueous biopolymer solutions is relatively
insensitive to salinity changes as shown in Figure 8 (5).

Nouri (12) studied the rheological behavior of 88 solutions of twelve different polymers frequently used in the
petroleum industry and found that the power-law model fitted
the data. The power law model is defined by Equation 3.

$$\tau = H\gamma^n \qquad (3)$$

where τ is shear stress
       H is a constant
       γ is the shear rate
       n is a constant with a value less than 1

The constants H and n depend on the molecular weight of
the polymer, and polymer concentration. When the polymer is
a polyelectrolyte, both parameters are also dependent on
simple electrolyte concentration.

Polyacrylamides and polyethylene polymer solutions have
been found to exhibit shear thickening or viscoelastic behavior in porous media at high shear rates (13,14).

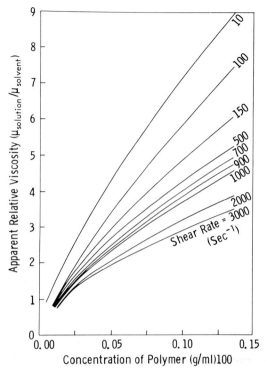

Fig. 6. Relationship between Relative Viscosity, Polymer Concentration and Shear Rate. Pusher 700 in 2% NaCl at 20°C (10)

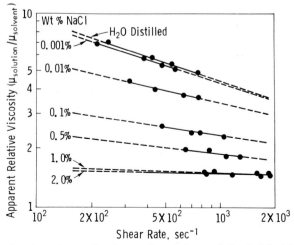

Fig. 7. Variation in Relative Viscosity with Salinity and Shear Rate for 250 ppm Pusher 700 Solutions at 25°C (10)

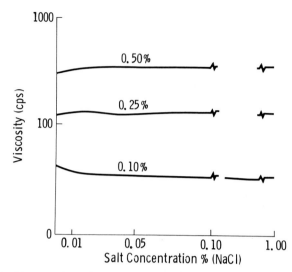

Fig. 8.   Effect of Salt Concentration on Viscosity of Xanthan Biopolymer Solutions at a Shear Rate of 7.35 sec^{-1} (5)

IV.   POLYMER RETENTION IN POROUS MEDIA

The quantity of polymer retained when a polymer solution flows through a porous material can be determined from displacement experiments. Effluent concentration profiles representing typical displacement experiments are shown in Figure 9. In these experiments, a polymer solution of known concentration is displaced through porous media at a constant flow rate. Effluent polymer concentrations are determined as a function of pore volumes injected. The effluent concentration profile in Figure 9a corresponds to a step change in polymer concentration from $C = 0$ to $C = C_0$ at the inlet of the porous media at time $t = 0$. Figure 9b shows the effluent concentration profile when a slug of polymer solution is displaced through the porous material by solvent. Retained polymer is determined for either displacement process from an overall material balance.

Polymer retention data obtained from material balance experiments are summarized in Table I. Most of the available data were measured using solutions of partially hydrolyzed polyacrylamides. Comparison of the data should be limited to general concepts due to the wide differences in the properties of the porous media, polymer and fluids. Retention varies from 35 lb/acre ft. to about 1000 lb/acre ft. The most consistent trend in the data is the increase in retention with decreasing permeability which is depicted in Figure 10 from the data presented by Vela, Peaceman and

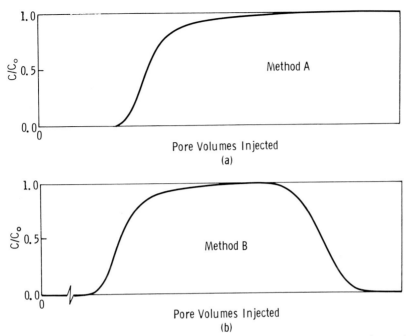

Fig. 9.   Effluent Concentration Profiles for Computation of
           Polymer Retention from Overall Material Balances

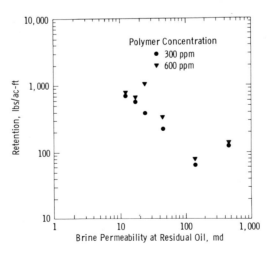

Fig. 10.   Variation of Polymer Retention with Permeability
            in Sandstone Containing Residual Oil Saturations
            (15)

TABLE I

SELECTED DATA
RETENTION OF HIGH MOLECULAR WEIGHT POLYMERS
DURING FLOW THROUGH POROUS MEDIA

Polymer	Concentration mg/l	Salinity or Total Dissolved Solids ppm	Porous Media	Permeability md	Retention lb/AF Bulk Volume	Reference
Hpam (MW ~3 x 10^6)	500	Brine	Miocene Sand	53 @ S_{or}	684.0	Jennings, Rogers, and West (14)
Hpam (MW ~3 - 10 x 10^6)	500	Distilled Water	Muffled Berea Sandstone		201.0*	Mungan (20)
	500		Berea Sandstone		316.0*	"
	500		Ottawa Sand		747.0**	"
Kelzan M		Brine	Nevada Sand	6,000	38.0**	Patton, Coats, and Colegrove (25)
Pusher 700 (MW ~5 x 10^6)	500	Natural Brine with Surfactant	Reservoir Cores	359	34.9	Hirasaki and Pope (31)
	500	"	"	117	44.0	"
	400	"	"	97	46.9	"
	500	"	"	80	75.4	"
Hpam (MW ~5 x 10^6)	300	133,000	"	45 @ S_{or}	224.0	Vela, Peaceman, and Sandvik (15)
	300	133,000	"	30 @ S_{or}	561.0	"
	300	13,340	"	17 @ S_{or}	580.0	"
	300	20,000	"	137 @ S_{or}	64.0	"
Biopolymer		Brine	Berea Sandstone		78.0*	Trushenski, Dauben, and Parrish (24)

Hpam   Hydrolyzed Polyacrylamide
* Assumed 20% Porosity
** Assumed 35% Porosity

Sandvik (15). Polymer retention appears to be slightly higher for higher polymer concentrations but the increase is not proportional to the concentration increase. Data for biopolymers are limited.

Evaluation of polymer retention by material balance is straightforward in concept. In Method A, the injected polymer not found in the column effluent or resident in the pore space when the effluent concentration reached the injected concentration ($C/C_o = 1$) is assumed to be retained. Method B shown in Figure 9b differs from Method A in that injection is switched to solvent at a specified time to flush all mobile polymer from the pore space. The quantity of injected polymer not recovered in the effluent stream is considered retained.

Method A appears to be the desirable approach because fewer data points are required. This advantage is offset by another property of polymer solutions in porous media. Dawson and Lantz (16) have shown that part of the pore space can be inaccessible to both polyacrylamide and biopolymer solutions. In their study, a Berea sandstone core was contacted with a polymer solution containing 2% NaCl until no further retention occurred. Then, the composition of the injected fluid was changed by reducing the polymer and salt concentration.

Concentration profiles showing the arrival of the polymer and
salt concentration changes in the effluent stream are shown
in Figure 11. The polymer solution reached the end of the
core considerably earlier than the NaCl concentration change.
They interpreted the early arrival of the polymer solution to
indicate that a portion of the pore volume (about 24%) was
not accessible to the flow of polymer molecules. The salt
concentration change arrived at 1 PV injection indicating
that the water injected with the polymer contacted all of the
pore space. Inaccessible pore volume has also been observed
by other investigators (15,17).

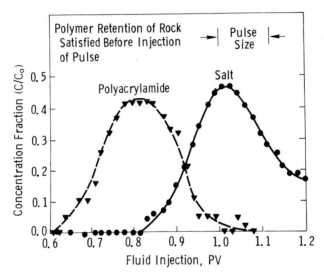

Fig. 11. *Early Arrival of Polymer Front Due to Inaccessible*
          Pore Volume (16)

The presence of inaccessible pore volume in porous media
introduces uncertainty in the values of polymer retention
determined from Method A. Some investigators consider inaccessible pore volume to be confined to pores which have pore
diameters less than the average size of the polymer molecules.
Analysis of the concentration data obtained in Method A requires determination of the inaccessible pore volume (15)
because retention does not occur in the inaccessible pore
volume. Computation of polymer retention from displacement
data obtained using Method B does not require the knowledge
of the inaccessible pore volume as all mobile polymer is
flushed from the pore space. Polymer retention from Method
B will be lower than present during polymer flow if the
retention mechanisms are reversible.

V.   MECHANISMS OF POLYMER RETENTION

A.   Adsorption
1.   *Static Equilibrium Adsorption*
     Polymers adsorb on the materials which are found in res-
ervoir rocks (16,18-22). Adsorption data for polyvinyl acetate, polystyrene, polymethyl methacrylate, polyethylene oxide
and partially hydrolyzed polyacrylamide on several porous
materials are summarized in Table II. Adsorption data have
not been published on biopolymers but these polymers are known
to adsorb (16).
     A typical polymer adsorption isotherm is shown in Figure
12 (16). The quantity of adsorbed polymer increases rapidly
with concentration to about 100 - 200 ppm. Adsorption levels
off at higher concentration and eventually becomes relatively
insensitive to polymer concentration. Adsorption of polyelectrolytes such as partially hydrolyzed polyacrylamides is
particularly sensitive to the nature of the dissolved salt
and the solid surface. Szabo's (22) static adsorption data
(Curves 1 and 2) and desorption data (Curve 3) presented in
Figure 13 show a strong dependence of the adsorption of partially hydrolyzed polyacrylamide on salt concentration between
0% and 2%. Szabo attributed the increased adsorption to the
reduction in the hydrodynamic size of the polymer molecule
with increased electrolyte concentration.
     Smith (18) found that calcium ions were more effective
in promoting adsorption of partially hydrolyzed polyacrylamide
on negatively charged silica surfaces than sodium ions up to
a concentration of about 10% NaCl. This effect is probably
due to bridging of the calcium ion between the negatively
charged polyelectrolyte and the silica surface. A greater adsorption of polyacrylamide occurred on calcium carbonate powder
than on silica. Smith attributes part of the increased adsorption to calcium carboxylate surface interaction. However,
part of the increase may be due to the decrease in molecular
size caused by calcium ions in solution.
     Michaels and Morelos (23) examined the adsorption of par-
tially hydrolyzed polyacrylamide on kaolinite and proposed the
adsorption mechanism based on hydrogen bonding between the
amide groups and the negatively charged surface. The quantity
of polyacrylamide adsorbed was quite sensitive to pH. Adsorption was not observed on pure hydrogen kaolinite above a pH
of 6.7.
     Mungan et al. (19) observed no adsorption of partially
hydrolyzed polyacrylamide on silica which had contacted crude
oil. Mungan (20) observed no adsorption on Ottawa sand which
had been pretreated with dimethyldichlorosilane. Trushenski
et al. (24) have reported reduced retention of biopolymers

TABLE II

SELECTED DATA

ADSORPTION OF HIGH MOLECULAR WEIGHT POLYMERS ON POROUS MATERIALS

Polymer	Molecular Weight x 10^{-6}	Concentration mg/1	Solvent	Porous Media	Amount Adsorbed μg/g	Reference
Polyvinyl Acetate	1,190	1,000	2-butanone (30°C)	325 Mesh Pyrex 7740 Glass	2,750	Rowland and Eirich (28)
"	"	"	Benzene (30°C)	"	2,300	"
Polystyrene	1,400	"	"	"	1,800	"
Polymethyl-methacrylate	1,430	"	"	"	2,350	"
Partially Hydrolyzed Polyacrylamide	2-3	1,000	Distilled Water	Disaggregated Berea Sandstone BET = 1.70 m^2/g	880	Mungan (20)
"	"	500	"	"	700	"
"	"	250	"	"	370	"
"	"	1,000	2% NaCl	"	450	"
"	"	500	"	"	300	"
"	"	250	"	"	100	"

TABLE II (Continued)

Polymer	Molecular Weight x 10^{-6}	Concentration mg/l	Solvent	Porous Media	Amount Adsorbed μg/g	Reference
Polyacrylamide	1.150	100	Water (25°C)	Hydrogen Kaolinite	16,000	Schamp and Huylebroeck (27)
"	1.150	100	"	Hydrogen Montmorillonite	278,000	"
"	0.128	100	"	Hydrogen Illite	41,000	"
Partially Hydrolyzed Polyacrylamide	3	30–75	10% NaCl	Silica Powder (1 m^2/gm)	50	Smith (18)
"	3	30–75	3% NaCl	"	20	"
"	3	30–150	10% NaCl	Calcium Carbonate Powder (0.46 m^2/gm)	310	"
"	3	10–125	+0.05% Ca^{++}	"	450	"

TABLE II (Continued)

Polymer	Molecular Weight x 10⁻⁶	Concentration mg/l	Solvent	Porous Media	Amount Adsorbed µg/g	Reference
Partially Hydrolyzed Polyacrylamide (50% Hydrolyzed)	4–5	300–1800	2% NaCl	Illite	2,900	Desremaux, Chauveteau, and Martin (21)
Polyethylene Oxide	4	0–50	"	Sandstone	190	"
"	"	0–200	"	Limestone	520	"
"	"	100–1600	"	Illite	5,000–25,000	"

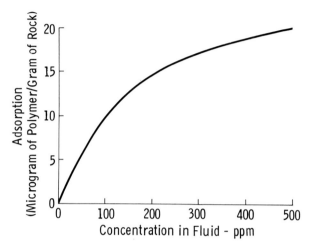

Fig. 12. *Typical Adsorption Isotherm for High Molecular Weight Polymers (16)*

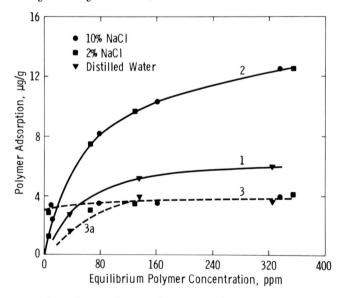

Fig. 13. *Static Adsorption and Desorption Isotherms for Partially Hydrolyzed Polyacrylamide on Silica Sand (22)*

and polyacrylamides in Berea cores following contact with micellar fluids containing surfactants.

Adsorption of polymer molecules appears to be irreversible on some surfaces (26,28). Desremaux et al. (21) observed that polymers could not be desorbed from limestone with 2% NaCl.

Schamp and Huylebroeck (27) found adsorption of polyacrylamide on montmorillonite kaolinite and illite to be irreversible. However, Szabo's data (Figure 13) for one type of partially hydrolyzed polyacrylamide on silica sand indicates substantial reversibility. Curve 3 in Figure 13 is the desorption isotherm after 3 hours of desorption with strong brine.

Dawson and Lantz (16) report instantaneous adsorption of biopolymers on crushed reservoir rock while partially hydrolyzed polyacrylamides had a time constant of about 12 hours. Schamp and Huylebroeck (27) found that adsorption of polyacrylamides on peptized sodium montmorillonite was completed in 10-15 minutes while adsorption on kaolinite and illite was still occurring at slow rates after 2 hours of shaking. Desremaux et al. (21) noted attainment of equilibrium adsorption of polyethylene oxide on silica in a few hours.

2. *Adsorption During Flow Through Porous Media*

The role of adsorption on the retention of polymer solutions during flow through porous media has been investigated by comparing adsorption in static experiments with retention from material balances on flow experiments. Mungan (20) investigated the adsorption of partially hydrolyzed polyacrylamide (M = 3-10 x 10^6) dissolved in distilled water on crushed samples of muffled Berea sandstone, Berea sandstone, and Ottawa sand. Mungan's results are compared with values obtained from flow experiments in Table III. In each case, the quantity of polymer retained in the flow test was considerably less than anticipated from batch adsorption experiments. One can argue that disintegration of a consolidated Berea core exposes binding materials such as clays to polymer solution

TABLE III

POLYMER RETENTION
IN STATIC AND FLOW EXPERIMENTS
Mungan (20)

500 mg/l Partially Hydrolyzed Polyacrylamide
(MW 3 - 10 x 10^6) in Distilled Water

Porous Material	Static Adsorption μg/gm	Retention in Flow Experiment μg/gm
Muffled Berea Sandstone	500	35
Natural Berea Sandstone	610	55
Ottawa Sand	300	160

in a manner which could never occur in the flow experiment. However, this argument fails to explain the difference in adsorption and polymer retention observed for the Ottawa sand.

Mungan attributed this difference to the possibilities that 1) polymer solutions flow through larger pores which have the smallest surface area per unit volume, 2) pores smaller than some critical size as well as sand grain contacts are not accessible to the polymer. Apparently, Mungan considered adsorption on these materials to be irreversible.

Szabo (22) determined the retention of a C^{14} tagged partially hydrolyzed polyacrylamide on silica sand in static adsorption experiments (Curves 1 and 2, Figure 13) and by material balance following displacement of the fluid through porous material. Amounts of polymer retained are presented in Table IV for polymer injection into a 1,200 md unconsolidated sand. More polymer was retained after the brine flush than expected from the desorption data represented by Curve 3 in Figure 13. Table IV also shows the importance of the method used to obtain polymer retention.

TABLE IV

RETENTION OF PARTIALLY HYDROLYZED POLYACRYLAMIDE
IN A 1200 md UNCONSOLIDATED SAND - SZABO (22)

Polymer Concentration (in 2% NaCl) ppm	Experiment	Polymer Retention				
		During Flow - Method A Assuming No Inaccessible Pore Volumes		Static Adsorption	After Brine Flush - Method B		Static Desorption
		μg/gm	lb/AF	μg/gm	μg/gm	lb/AF	μg/gm
300	Single Phase Flow at 100% Water Saturation	3.50	14.19	12.0	7.34	29.72	3.3
600	Single Phase Flow at 100% Water Saturation	6.00	29.29		12.93	52.37	
300	Single Phase Flow in Presence of Residual Oil Saturation	16.65	67.43		11.34	45.93	

3. *Models of Polymer Adsorption in Porous Media*

The available adsorption data suggest polymer adsorption probably occurs in the form of a monolayer in which the density of the adsorbed layer is limited by the tendency of the adsorbed molecular coils to be mutually repelling (28). Data obtained (22) using partially hydrolyzed polyacrylamides support the hypothesis of monolayer adsorption. The molecules are envisioned to adsorb with p monomer segments/molecule adsorbed on the surface with the remaining segments residing in the solvent. The adsorbed polymer molecules occupy less space than expected from the computed size in a dilute solution and consequently, appear to be compressed or have interpenetrating coils (29,30).

Two models of polymer adsorption have been developed. Since polymer molecules adsorb as if the adsorbed molecules are in a monolayer, the adsorption isotherm can usually be represented by Equation 4, the Langmuir isotherm.

$$Ad = \frac{abC}{1+bC} \tag{4}$$

The Langmuir isotherm is an equilibrium relationship and represents polymer adsorption when equilibrium is reached instantaneously. There are two constants (a,b) which can be determined experimentally from batch adsorption experiments for the porous media and solution.

Although there are some data indicating a rate effect, these effects should not be significant outside of the laboratory because of the long time scales. A single paper has been presented which shows polymer desorption. Other investigators apparently consider adsorption to be irreversible as no attempt is made to account for desorption in mathematical models.

There has been widespread application of capillary tube models to represent polymer adsorption in porous media. Rowland and Eirich (28) approximated adsorption in sintered Pyrex discs with the capillary model shown in Figure 14. The radius of the capillary was obtained by using the root-mean-fourth average of the pore size distribution defined by Equation 5.

$$||r|| = \left(\frac{1}{N} \sum_{i=1}^{N} r_i^4 \right)^{1/4} \tag{5}$$

Polymer adsorption was considered to form a monolayer of thickness Δr which reduced the hydrodynamic size of the capillary by $2\Delta r$. The thickness of the adsorbed layer was computed from the increased flow resistance which was observed across the Pyrex disc.

Hirasaki and Pope (31) applied the same model to predict flow resistance in porous media, assuming monolayer adsorption. Their model is based on correlating the area occupied by the adsorbed polymer molecule on the surface with the segment density σ as shown in Equation 6.

$$Ad = \frac{M\sigma}{Am\tilde{N}} \tag{6}$$

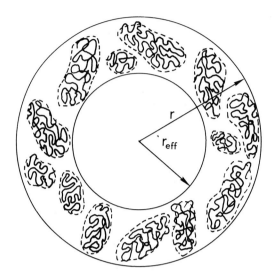

Fig. 14.   Capillary Radius Model of
Polymer Retention

where M = molecular weight of the polymer
      Am = surface area covered by a close packed
           spherical particle of radius r_m
      Ñ = Avogadro's number

When it is assumed that the radius of the molecule before
compression on the surface is equal to the radius of the
molecular coil in the solvent at infinite dilution,

$$Ad = \frac{M\sigma}{3.5 \left[\frac{30\ M[\mu]}{\pi\tilde{N}}\right]^{2/3} \tilde{N}} \qquad (7)$$

Equation 7 has one parameter, the segment density σ which
must be determined from experimental data. While σ appears
to be descriptive of the adsorption mechanism, there does
not appear to be substantial justification for choosing between the Langmuir and capillary models to represent adsorption.

     In both illustrations of the capillary models, all re-
tention was attributed to adsorption. Retention by other
mechanisms such as mechanical entrapment, discussed later
in this paper, is reported as adsorbed polymer. Simulation
of polymer retention would reflect the extent to which the
polymer and porous media used to obtain the coefficients in
the capillary bundle model were similar to the system being
modeled. Thakur (32) applied this approach with some success

on alundum discs. Results on Berea cores were not satisfactory.

Thomas (33) studied the flow of aqueous solutions of polyacrylamides and biopolymers through fused glass arrays of capillary tubes with diameters of 2, 5 and 25 microns. An additional flow resistance was observed after the array was flushed
with brine which was attributed to polymer adsorption on the
walls of the capillary tubes. The effective thickness of the
polymer layer was 0.2 to 0.3 micron using the capillary radius
model. The effective thickness of this layer was independent
of capillary size when the tube diameter was 3 to 4 times the
size of the polymer molecules in solution.

These data appeared to disagree with results reported by
Jennings, Rogers and West (14) in which no additional flow
resistance was observed when similar polyacrylamide solutions
flowed through a 5.5 micron smooth glass capillary. Discussion with Thomas (35) indicated that the data do not conflict.
The fused glass arrays used by Thomas were constructed from
composite glass rods with an etch resistant glass on the outside and an etchable glass in the core. The capillary tubes
were made by etching the soft glass from each tube in the
array. Photomicrographs of individual tubes revealed the
presence of small irregularly shaped particles of soft glass
on the walls of the tube. BET data also indicated considerably more surface area than expected from the same number of
smooth tubes. The data obtained by Thomas substantiate the
mechanism of polymer retention by adsorption and show that
adsorbed polymer molecules cause additional flow reduction in
tubes which are not tortuous.

Perhaps the most serious difficulties in evaluating an
adsorption model lie in the experimental determination of adsorption on crushed core samples and in the estimation of the
surface area in a porous media which is available for adsorption. The former difficulty seems unavoidable in that there
is a probability that crushing a core sample will expose
cementing materials to the polymer solution which would not
be exposed in the consolidated core.

The total surface area in a porous media may be estimated
by the BET method subject to the inherent limitations of the
technique. The measurement is useful only as an upper bound
on the quantity of polymer which might be adsorbed because
of the presence of inaccessible pore volume. In practice, we
would expect less polymer to be retained by adsorption because
it is likely that part of the inaccessible pore volume consists
of small pores with a relatively large surface area. However,
little additional improvement in simulating adsorption can be
expected until the inaccessible pore volume is described in
a quantitative manner.

B.   Mechanical Entrapment
1.   *Effect of Polymer Size and Pore Size*
    Polymer molecules are relatively large in solution. The
possibility of polymer retention due to filtering or mechanical entrapment was investigated by Gogarty (36) and Smith
(18). Gogarty determined the effective size of a high molecular weight hydrolyzed polyacrylamide (the molecular weight
of this polymer was not specified by the author) by filtering
a 400 ppm polymer solution through different size nuclepore
filters. Figure 15 shows that the effective size ranges from
about 0.5 to about 2.0 microns for the simple electrolyte free
solution and from 0.4 to about 1.5 microns for the solution
in 600 ppm NaCl solution.

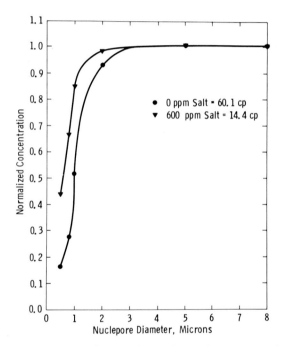

*Fig. 15.   Polymer Filtration with Different Size
            Nuclepore Filters (36)*

    Smith (18) obtained results similar to Gogarty's. The
effective size of partially hydrolyzed polyacrylamide
(M = 3.0 x 10^6) in 0.5 percent NaCl was in the range of 0.3
to 1.0 micron. It is interesting to notice that computations
using Equation 38 from Flory (6) and the viscosity of this
solution give an average size of 0.34 microns. Szabo (34)
also measured similar values to the ones obtained by Gogarty
and Smith.

Filtration experiments show that flow reduction occurs when polymer solutions flow through filters with holes ranging from 0.5 to 2. microns. A pore size distribution of a Berea sandstone determined by mercury porosimetry (33) is presented in Figure 16. The shaded area represents the 14% of the pore volume which would be inaccessible to polymer molecules with radii of 1 micron. These are pores in this sandstone which are small enough to filter or mechanically entrap polymer molecules.

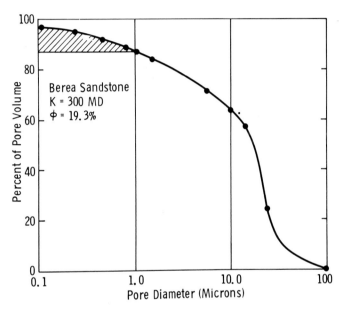

Fig. 16.   Pore Size Distribution in a Berea Sandstone Determined by Mercury Porosimeter (33)

2. *Mechanical Entrapment During Flow*

Most polymer retention data have been obtained by the material balance method on porous materials known to adsorb large quantities of polymer. Three studies have been completed in which it was possible to determine the relative effects of mechanical entrapment and adsorption on polymer retention (17,22,37).

Szabo (22) analyzed samples of sand at distances of 1 cm and 6 cm from the injection face of 12 cm sand packs to obtain the distribution of retained polyacrylamide. These data are presented in Figure 17. Curves 1-4 represent retention in the presence of a residual oil saturation. The quantity of retained polymer is much larger at the 1 cm position and clearly indicate that the mechanical entrapment is the

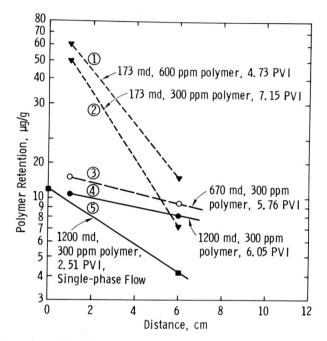

Fig. 17.   Distribution of Retained Polymer in Unconsolidated
Silica Sand Packs (22)

dominant retention mechanism there. Values at 7 cm are probably more representative of the overall retention. Retention
due to adsorption is 1/3 to 1/2 of the total polymer retention represented by curves 1-4. The factor varies because
the 173 md sand was reported to have twice the surface area
as the 1200 md sand.

Dominguez and Willhite (17) determined the retention of
partially hydrolyzed polyacrylamide ($M = 5 \times 10^6$) in an 86 md
Teflon core in which adsorption was shown to be small. Related experiments were conducted by Palmer (37) in a 3.5 darcy
unconsolidated packed bed made using the same Teflon materials.
The porosity in the 3.5 darcy pack was 0.44 compared to 0.21
for the 86 md core. The quantities of polymer retained are
presented in Table V. These data demonstrate that mechanical
entrapment of polymer molecules occurs over a wide range of
permeability and porosity even when there is little adsorption. Sarem (38) also reported retention of polymer in a
packed bed of Teflon particles but considered the polymer to
be retained by adsorption rather than mechanical entrapment.
3. *Mechanisms of Mechanical Entrapment*

Entrapment of high molecular weight polymers occurs in
porous media because the polymer molecules are large relative

TABLE V

RETENTION OF PARTIALLY HYDROLYZED POLYACRYLAMIDE (MW~5 x 10^6)
IN POROUS MEDIA BY MECHANICAL ENTRAPMENT

Porous Media	Polymer Concentration ppm	Polymer Retention After Brine Flush		Reference
		µg/gm	lb/AF	
Compacted Teflon	99*	10. 87	44. 73	17
Core (~86 md)	187*	10. 85	48. 19	17
	489	21. 20	94. 26	17
Packed Bed	100	4. 50	17. 40	37
of Teflon Particles	145	7. 50	29. 30	37
(3, 500 md)	200	15. 60	61. 00	37
	200	10. 60	41. 60	37
	500	16. 90	66. 30	37

* Concentration of Injected Polymer Was Not Reached at the Column Effluent
When Brine Injection Began.

to the size of the pores. Several mechanisms probably contribute to the net effect which is termed mechanical entrapment. Gogarty (36) visualized mechanical entrapment as the
end result of plugging of small pores by polymer molecules
which were too large to enter them. The possibility that
adsorbed polymer molecules could also cause permeability
reduction without completely blocking the pores was discussed by Sandiford, Knight, Sarem and Amott (39).
4. *Analogy to Deep Filtration Mechanism*
        Mechanical entrapment of polymer molecules resembles the
process of deep filtration in which fine particles are mechanically entrapped as they flow through a packed bed.
Some differences should be expected in retention mechanisms
because polymer molecules are flexible and can deform under
shear stress. Herzig, Leclerc and LeGoff (40) present an
excellent analysis of the mechanisms of the deep filtration

process by relating retention to potential retention sites
in the porous media. Our analysis closely follows the work
of Herzig et al. (40) with appropriate modifications for
polymers.

Potential sites where polymer molecules can be retained
are shown in Figure 18 (40). Surface sites are occupied by
adsorbed polymer molecules. Possible interactions between
adsorbed molecules and molecules in solution will be discussed later. The remaining retention sites illustrate
locations where polymer molecules could become entrapped
because of mechanical forces, the size of the constriction
or the size of the polymer coil.

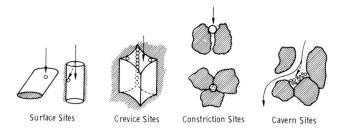

Surface Sites   Crevice Sites   Constriction Sites   Cavern Sites

Fig. 18. Sites for Polymer Retention in Porous Materials (40)

Crevice sites capture polymer coils which become wedged
in the contact areas near two sand grains. Constriction
sites are those which have pores too small for a polymer coil
to penetrate. Cavern sites are visualized as locations where
the flow velocity decreases, possibly due to a change in direction of flow caused by the tortuosity of the flow path or
reduction in permeability in the direction of flow.

An idealized crevice site is depicted in Figure 19 (40)
as the region formed by the contact points of two spheres of
equal diameter. In Figure 19, the polymer coil is represented

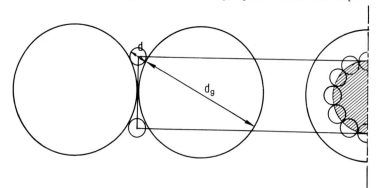

*Fig. 19. Excluded Surface Area for Adsorption Near a Crevice
Site (40)*

as an equivalent sphere of diameter d. Herzig et al. show
that retention in crevice sites is important only if $\dfrac{d}{d_g} \geq 0.05$.
The equivalent grain diameter would be 20 microns for a
polymer coil which is 1 micron in diameter.

The size of the polymer molecule in solution may also
prevent adsorption from occurring in the vicinity of the in-
tergranular contact shown as the shaded area in Figure 19.
The importance of this effect can be estimated for packed beds
of uniform spheres. In this case, it can be shown that the
ratio of the excluded surface area (A_e) per contact point to
the total surface area (A_p) is given by Equation 8.

$$\frac{A_e}{A_p} = \frac{d}{2(d + d_g)} \tag{8}$$

Polymer molecules 1 micron in diameter are excluded from about
1% of the surface area per contact point for spheres which
are 50 microns in diameter. The number of contact points per
sphere is limited to 6 in close packing so that, at most, 6%
of the surface area could be exluded from polymer adsorption
due to the hydraulic size of the polymer molecule. Thus,
this mechanism would appear to be a significant factor in
reducing adsorption for the particle sizes in the range of
10 microns.

Retention at constriction sites is probably large, particularly in porous media with pore diameters of the same
size as the polymer coil. Comparison of the coil diameter
with the pore diameter does not indicate the full extent of
retention. Pores are not capillary tubes of uniform diameter.
They have irregular shapes. A better representation although
still idealized is shown in Figure 20 (40). Shown in Figure
20 are the configurations of polymer coils as they flow
through the pore space. Limiting ratios of polymer coil diameter to grain diameter are 0.154, 0.1 and 0.082 for the
cases in which one, three and four polymer coils pass through
the pore simultaneously. However, retention of one polymer
coil of slightly larger diameter in a large pore could effectively initiate the retention of smaller particles.

The presence of pores which are small enough to retain
polymer molecules is evidence that mechanical entrapment
can occur by this mechanism. Further discussion of retention
by this mechanism must be subjective. In the absence of adsorption, the retention of polymer molecules after the first
molecule retained at a site is related to the hydrodynamic
forces exerted by the flowing fluid. Retention around a
constriction site could occur as shown in Figure 21.

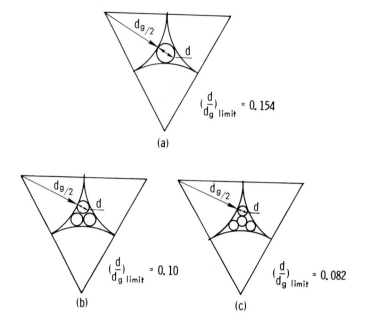

Fig. 20.   *Size of Polymer Molecules Related to Constriction Site (40)*

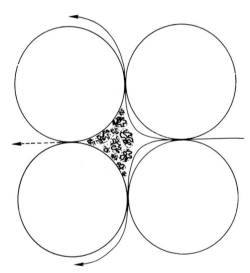

Fig. 21.   *Polymer Retention by Filtering at a Constriction Site*

Solvent flows through the constriction area with a continual increase in the number of polymer molecules retained in front of the site. The permeability of the solvent at the retention site decreases with time which diverts more fluid away from the constriction site, thereby reducing or eliminating further retention. This, in effect, would slowly convert a constriction site into a cavern site.

The preceding discussion assumed that retention was initiated because the polymer molecules were larger than the pores. Retention also occurs because of adsorption. An adsorbed molecule may be somewhat smaller than an entrapped molecule but the effect should be essentially the same.

5. *Inaccessible Pore Volume*

The concept of inaccessible pore volume is closely related to the mechanisms of mechanical entrapment. The simplest relationship is to consider all pores which are less than the hydraulic diameter of the polymer molecule to be inaccessible. This approach can be examined further by developing a relationship between the area open at a constriction site and the equivalent diameter of the pore (d_{pore}) which has the same cross sectional area as the pore area of the constriction site. For the pore area shown in Figure 20a, $d_{pore}/d_g = 0.227$. A rigid spherical polymer molecule would be retained or prevented from entering a pore whenever

$$\frac{\dfrac{d}{d_g}}{\dfrac{d_{pore}}{d_g}} = \frac{d}{d_{pore}} > 0.678 \tag{9}$$

Thus, the number of inaccessible pores should be larger than inferred from a comparison of pore diameter and equivalent polymer diameter. A flexible polymer molecule could deform and enter a pore where the ratio is not limiting.

Other models of inaccessible pore volume are possible. There is reason to believe that the volume occupied by the retained polymer molecules is inaccessible to other polymer molecules. Figure 14 depicts one model of polymer retention by adsorption of a monolayer on the walls of a capillary tube. In this model, the volume occupied by the retained polymer molecules is inaccessible to other polymer molecules. The pore would be accessible to polymer molecules flowing through the center of the capillary. However, a certain fraction of the pore volume would be inaccessible to the flowing polymer stream. The amount of inaccessible pore volume is related to the hydrodynamic volume of the polymer molecule, its effective size when retained and the size of the pore. Inaccessible pore volumes based on this model were calculated as a function of pore size for adsorbed polymer diameters of 0.2 micron and 0.5 micron and are presented in Table VI.

TABLE VI

INACCESSIBLE PORE VOLUME
DUE TO ADSORPTION OF A MONOLAYER OF POLYMER MOLECULES ON THE WALL OF A CAPILLARY TUBE

Diameter of Equivalent Capillary Tube	Inaccessible Pore Volume Diameter of Adsorbed Polymer Molecule	
microns	0.2 microns (2,000 Å)	0.5 microns (5,000 Å)
1.0	0.64	1.00
1.5	0.46	0.89
2.0	0.36	0.75
3.0	0.25	0.55
5.0	0.15	0.36
7.0	0.11	0.26
10.0	0.08	0.19
20.0	0.04	0.10

A discussion of inaccessible pore volume would be incomplete if we did not point out another relationship between inaccessible pore volume and polymer movement through a capillary tube. Di Marzio and Guttman (41) proposed that polymer molecules moving through a thin capillary would have an average velocity greater than the velocity of the solvent. It is well known that the velocity of the solvent reaches a maximum in the center of the tube and is zero at the walls. The polymer molecule is so large that its center cannot get any closer to the tube wall than its radius. They proposed that the polymer molecule is excluded from the portion of the velocity distribution where the velocity is lowest and, consequently,

will move at a velocity which exceeds the mean velocity of the
solvent.   This effect would be dependent on the ratio of polymer diameter to capillary tube diameter.
6.   *Dependence of Polymer Retention on Flow Rate (Hydro-*
     dynamic Retention)
     Several investigators have presented evidence that poly-
mer retention in porous media is related to flow rate.
Desremaux, Chauveteau and Martin (21) displaced solutions of
Pusher 500, a partially hydrolyzed polyacrylamide (M = 3 x 10^6),
through a 140 md silica sand pack at velocities varying from
10 cm to 20 meters per day.   The quantity of polymer retained
increased with decreasing mobility of the polymer solution.
The retained polymer was visualized as consisting of two
fractions, a quasi-irreversible fixed fraction and a mobile
fraction.   Data obtained by Chauveteau and Kohler (42) which
are presented in Figure 22 support the reversibility of polymer retention as the quantity of polymer retained when the
velocity was increased to 10 m/d is approximately the same
as the amount expelled when the velocity was reduced to the
original velocity of 3 m/d.

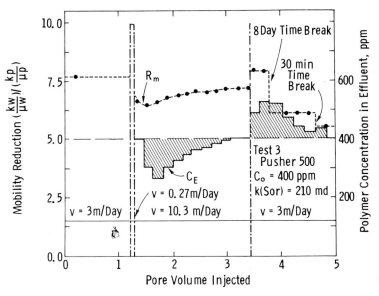

*Fig. 22.   Hydrodynamic Retention of Partially Hydrolyzed
            Polyacrylamide (42)*

     Dominguez and Willhite (17) studied polymer retention
in an 86 md Teflon core, 8.75 cm in diameter and 29.81 cm
long in which adsorption was low.   Analysis of the effluent
concentration following a flow rate change from 3.22 ft/day
to 6.3 ft/day showed that the retention of Pusher 700, a

partially hydrolyzed polyacrylamide with a molecular weight
of about 5 million, increased. Polymer was expelled from the
core when the flow rate was reduced, substantiating the observations of Chauveteau and Kohler which were made in the presence of a residual oil saturation.

Maerker (43) observed a flow rate dependence of a 500 ppm
polysaccharide solution (XC biopolymer, Xanco Div. of Kelco)
in 2% salt solution which was displaced through a 121 md Berea
core. Maerker's data are presented in Figure 23. Points A,
B and C in Figure 23 represent changes in pressure drop during
the experiment. At position A, flow was stopped for 16 hours.
The quantity of polymer expelled from the core when flow was
resumed appears to have been recaptured during the subsequent
flow period. Each increase in flow caused further polymer
retention as indicated by the shaded area under the $C/C_o = 1.0$
line in Figure 23.

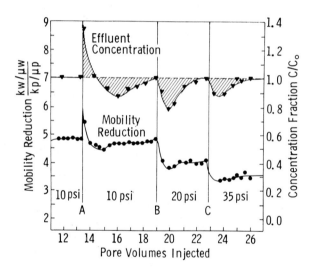

*Fig. 23. Relationship of Hydrodynamic Retention and Mobility
Reduction for a 500 ppm XC Biopolymer Solution with
2% NaCl (43)*

Although the retention which occurs when the flow rate
changes appears to be mechanical, it is also related to the
hydrodynamic forces which exist in the porous media. In
order to identify this dependence, we will refer to the
process in which polymer is retained and expelled in a quasi-
reversible manner as hydrodynamic retention.

The mechanisms which cause hydrodynamic retention are
not completely understood. Maerker (43) proposed the existence of sites in the porous media where polymer molecules
become trapped under the influence of a positive pressure

gradient. These sites are analogous to cavern sites of Herzig and the constriction site shown in Figure 21. Cessation of flow eliminates hydrodynamic drag permitting the trapped molecules to uncoil or diffuse into the pore space. The in situ concentration of the mobile polymer solution increases and can exceed the concentration of the injected fluid as shown in Figures 22 and 23. Each increase in flow rate was accompanied by further retention until the trapping sites were saturated. Thus, the polymer molecules in trapping sites are compacted with a "trapped" concentration which exceeds the injected composition.

The polymers which exhibit a flow dependent retention are quite different in structure and rheological behavior. The porous media are also different. Based on these data, it would appear that polymer retention caused by changes in flow rate may be observed in any solution containing high molecular weight polymers as suggested by Maerker (43). Flow rate dependence would occur even when a reduction in the original permeability is not observed after the polymer solution is displaced from a porous media by the solvent.

7. *Molecular Interaction*

The retention mechanisms described in the previous sections appear adequate for porous materials which have small pores and adsorb polymer. However, the fraction of the pore volume associated with small pores is too small in some porous materials to attribute all inaccessible pore volume and polymer retention to straining or filtration through constriction sites. For instance, Palmer (37) found polymer retention ($5\mu g/g$ to $17\mu g/g$) in a packed bed of 80-100 mesh Teflon particles. Retention by adsorption was not significant and the pore size distribution does not appear to contain enough pores to support the constriction model.

In this case, as well as in the large pores which exist in porous materials, it is possible to visualize a retention mechanism which we will refer to as molecular interaction. This concept is based on interactions between polymer molecules at different polymer concentrations.

In a dilute solution, a polymer coil (molecular weight about 5×10^6) occupies a volume which is equivalent to a sphere with a diameter of 0.4 micron. As the concentration of polymer coils increases, the polymer coils come closer to each other and must interact. If a porosity of 25% is assumed for the packing, the concentration at which the equivalent spherical polymer molecules begin to interact is 182 ppm. Although polymer molecules are not spheres, this estimated concentration should not be too far off. The implications for larger polymer concentrations are obvious.

There are other data supporting molecular interaction as the polymer concentration goes from a dilute or inhomogeneous solution to a concentrated or homogeneous solution. Schamp and Huylebroeck (44) observed that the adsorption isotherm or polyacrylamide on sodium montmorillonite exhibited a depression at certain concentrations for polyacrylamides of molecular weight 110,000, 280,000, 500,000 and 820,000. They interpreted these data to represent the transition from a dilute solution where polymer molecules are separated from each other by pure solvent to the concentrated solution in which polymer coils become entangled. The existence of this transition has been demonstrated by NMR (45) and light scattering (44). A correlation of critical concentrations defining the transition region as a function of molecular weight is presented in Figure 24 (44). Extrapolation of this correlation to a molecular weight of 5×10^6 yields an estimate of the critical concentration of about 140 ppm which is in reasonable agreement with the estimate made in the preceding paragraph.

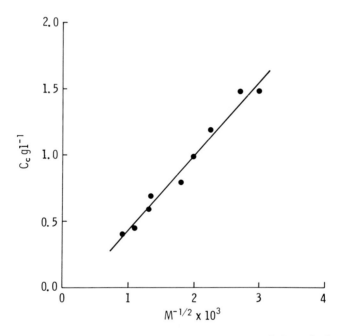

Fig. 24.   Correlation of Molecular Weight with Critical Concentration for Intermolecular Interaction for Polyacrylamides (44)

The mechanism of polymer retention in porous media by molecular interaction can be illustrated by considering the flow of polymer molecules in dilute and concentrated solutions. If a dilute solution of polymer flows through a porous media and one polymer molecule becomes retained in a crevice site, other molecules have enough space in the free solvent to bypass the retained molecule. However, retention of one polymer molecule from a concentrated solution immobilizes several polymer coils if the molecular interaction model is substantially correct. Palmer's data (37) support this mechanism as the polymer retention in his Teflon pack increased from 4.5 µg/gm to 15.6 µg/gm when the concentration of the polymer solution varied from 100 ppm to 200 ppm of a 2% NaCl solution of partially hydrolyzed polyacrylamide (M = 5 x 10^6). Increases in concentration above 200 ppm did not affect the retention appreciably. Palmer's data were taken at the same interstitial velocity (3.2 ft/day). The amount of polymer retained in this manner should vary with the velocity and shear rate. However, the data were not measured to investigate this effect.

Two models of polymer structure in concentrated solutions have been presented which would describe the molecular interaction mechanism. The first model, attributed to Flory (6), considers polymer molecules to be entangled when the polymer concentration exceeds a critical level. Above this concentration, polymer coils overlap in solution as illustrated by the spaghetti-like structure in Figure 25a. Vollmert (7) presents data which show that vinyl polymers exist in concentrated solutions as individual coils like those depicted in Figure 25b. Contact between individual coils is limited to narrow boundary zones where the polymer coils may become lightly cross-linked by covalent bonds. Vollmert states that there is no evidence that polymer coils interpenetrate to any extent and are probably prevented from doing so by steric hindrance.

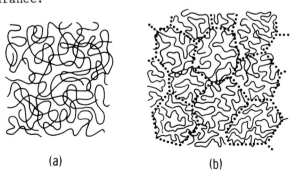

(a)   (b)

Fig. 25. Models of Polymer Structure in Concentrated Solutions (7)

Other investigators (46) have considered the possibility of polymer retention caused by filtration of microgels which were presumed to be originally present in the dry polymer. Burcik and Thakur (48) studied the properties of partially hydrolyzed polyacrylamide and proposed a cross-linking mechanism in which anhydride linkages were formed during the manufacturing process between COOH groups on adjacent polymer molecules by elimination of water. They also proposed that microgels present in the dry polymer disintegrated due to rehydration of hydrolyzable cross-links. The rehydration reaction was found to be a function of pH and hydration time. The proposed microgel has been discussed extensively (9,49). It appears that properly hydrated solutions of partially hydrolyzed polyacrylamide do not contain appreciable quantities of cross-linked polymer. Consequently, the presence of microgels is not a significant factor in the retention of partially hydrolyzed polyacrylamides in porous materials.

VI.   EFFECT OF POLYMER RETENTION ON FLOW THROUGH POROUS MEDIA

Retained polymer molecules occupy some portion of the pore volume in a porous media. The process of polymer filtration in front of small pores should change the permeability of the porous media. There are other effects caused by retention of polymer. The concentration of the retained polymer can be estimated by assuming that it is confined to the inaccessible pore volume (IAPV). Estimates from the data obtained in our research (17,37) are presented in Table VII. The estimated concentration of the retained polymer exceeds

TABLE VII

CONCENTRATION OF POLYMER
RETAINED BY MECHANICAL ENTRAPMENT

Polymer	Porous Media	Permeability md	Porosity (percent)	Injected Polymer Concentration ppm	IAPV	Concentration[*] Retained Polymer mg/l
Partially Hydrolyzed Polyacrylamide (MW 5 x 10^6)	Compressed Teflon Core	86	21	500	0.19	871
In 2% NaCl	Packed Teflon Core	3500	44	200	0.03	1397

[*]Confined to Inaccessible Pore Volume

the concentration of the solution from which the polymer was retained. Since the solution viscosity depends upon both

concentration and shear rate as shown in Figure 6, the retained
polymer coils become more resistant to deformation. Thus, effects of polymer retention on flow reflect a combination of
permeability reduction and the rheological properties of the
retained polymer.

A. Permeability Reduction

Pye (50) reported experimental data showing that solutions
of partially hydrolyzed polyacrylamides exhibited effective
viscosities when flowing through porous media which were 5-15
times larger than measured solution viscosities but offered no
explanation for the additional flow resistance. Burcik (51)
compared the flow rates under a constant pressure drop of 2%
NaCl solutions through a Berea sandstone disc before and after
the disc had been contacted with several pore volumes of a 2%
NaCl solution containing 500 ppm of partially hydrolyzed polyacrylamide. He found that the flow rate of the 2% NaCl solution was 4 times lower than the pre-polymer rate even after
100 pore volumes were injected. The decrease in water mobility was attributed to the strong interaction between water
and polymer molecules retained in the pore space by adsorption or mechanical entrapment.

Burcik (51) demonstrated that the flow resistance was
related to the nature of the solvent. Flow rates were measured through a sintered glass disc for the following sequence of fluids: 2% NaCl, polymer in 2% NaCl, 2% NaCl, pure
water, ethanol and pure water. The resistance to flow of the
pure water was about 40 times larger than the 2% NaCl. This
increase is due to the expansion of the polymer with decreasing salinity as discussed earlier. The ethanol flowed through
the disc without additional flow resistance. This is probably due to the precipitation of the retained polymer molecules and subsequent loss of solution volume due to the
limited solubility of the polymer in ethanol. Subsequent
displacement of ethanol by pure water resulted in a flow
resistance comparable to that observed before ethanol flow.
These experimental data prove that polymer molecules are retained in the pore space after the polymer solution has been
displaced from the porous media.

Gogarty (36) studied the flow of partially hydrolyzed
polyacrylamide solutions through reservoir and Berea cores.
Typical permeability data obtained after flushing polymer
contacted Berea cores with many pore volumes of brine are
presented in Table VIII. He reported permeability reductions between 2 and 3 for reservoir cores. Smith (18) determined the influence of initial permeability on the reduction of permeability to brine in Berea cores which had been
contacted with a 500 ppm solution of partially hydrolyzed

TABLE VIII

FLUSHED PERMEABILITY AFTER CONTACT
OF BEREA CORE WITH POLYACRYLAMIDE SOLUTION

Original Permeability 203 md

Run	Polymer Solution Injected (pv)	Solvent Solution Injected (pv)	Permeability md
1	15.7	67.3	89.1
		92.9	88.8
2	14.1	57.5	92.4
		83.3	92.1
3	11.8	54.5	88.3
		86.5	88.3

polyacrylamide in 3% NaCl. Smith's data are presented in Figure 26. Permeability reduction varies with both initial permeability and concentration of the polymer solution.

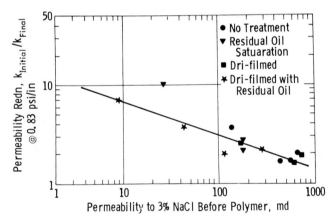

Fig. 26. *Influence of Initial Permeability on Reduction of Permeability of Berea, Sandstone Core by Partially Hydrolyzed Polyacrylamide in 3% NaCl (18)*

Mungan et al. (19) reported permeability reductions for both polyethylene oxide and polyacrylamide in sintered glass discs. Permeability reduction for polyethylene oxide solutions were considerably smaller than those determined for polyacrylamide solutions. Dauben and Menzie (52) found no reduction in permeability in a packed bed of 177 micron glass beads after polyethylene oxide solutions were flushed from the bed with distilled water. Jennings et al. (14) indicate that XAN biopolymers do not interact with porous media to produce a reduction in permeability.

The plugging or reduction in permeability of a porous media following the flow of high molecular weight polymer solutions leads to control of fluid movement with lower polymer concentrations in certain cases. However, some high molecular weight polymers which are known to adsorb on porous media do not appear to cause significant reduction in the permeability and lower fluid mobility when the polymer solution is flushed from the porous media with solvent.

B. Dependence of Flow Resistance on Polymer Retention

In each example where polymer retention changed with flow rate, a corresponding change occurred in the resistance to flow. In Figure 23, the flow resistance (mobility reduction) decreased with increased velocity indicating thinning of the biopolymer solution as the shear rate increased. Gogarty (36) observed shear thinning of partially hydrolyzed polyacrylamide solutions as the velocity increased. The data (R_m) of Chauveteau and Kohler in Figure 22 show a similar response. Several investigators have found an increase in flow resistance of partially hydrolyzed polyacrylamide solutions as the velocity increased (14,21,31). Since partially hydrolyzed polyacrylamide solutions are viscoelastic, increased flow resistance has been considered to occur as a result of viscoelasticity.

Recent experimental data suggest that polymer retention also contributes to increased flow resistance in these situations. Szabo (22,34) observed a linear increase in the flow resistance with time when the velocity exceeded a "critical" flow velocity. In one experiment, a 300 ppm solution of partially hydrolyzed polyacrylamide was injected through a 173 md sand pack in the presence of a residual oil saturation at a velocity of 6 ft/day. The effluent concentration stabilized at about 275 ppm after 0.8 pore volumes of injected polymer and remained at that level until 4 pore volumes of polymer solution were injected. The flow resistance increased linearly with pore volumes injected during this time. Dominguez and Willhite (17) observed the same linear increase in an 86 md Teflon core when the velocity was changed from 3.2 ft/day to 6.4 ft/day. These data show that the linear increase in flow resistance is related to increased polymer retention and corresponds to a time interval where polymer was retained at a constant rate.

The flow resistance (μ/k) for a pseudoplastic fluid which follows the power law model can be expressed by Equation 10, the modified Blake-Kozeny equation (53).

$$\frac{\mu}{k} = \frac{K}{12} \left(\frac{9n + 3}{n}\right)^n (150k\phi)^{\frac{(1-n)}{2}} \left(\frac{q}{A}\right)^{n-1} \tag{10}$$

The power law constant n varies from about 0.5 to 0.97 for polymer solutions used to control water mobility.

Let us examine the ratio of μ/k for two different flow rates.

$$\frac{\frac{\mu_2}{k_2}}{\frac{\mu_1}{k_1}} = \left(\frac{k_1}{k_2}\right)^{\frac{1+n}{2}} \left(\frac{q_1}{q_2}\right)^{1-n} \left(\frac{\phi_2}{\phi_1}\right)^{\frac{1-n}{2}} \tag{11}$$

If we neglect the differences in values of ϕ, we have

$$\frac{\frac{\mu_2}{k_2}}{\frac{\mu_1}{k_1}} = \left(\frac{k_1}{k_2}\right)^{\frac{1+n}{2}} \left(\frac{q_1}{q_2}\right)^{1-n} \tag{12}$$

We can now investigate the size of the permeability change which would be needed to offset the effect of a change in flow rate (i.e., $\mu_2/k_2 = \mu_1/k_1$). Under these assumptions,

$$\left(\frac{k_2}{k_1}\right)^{\frac{1+n}{2}} = \left(\frac{q_1}{q_2}\right)^{1-n} \tag{13}$$

For n = 0.87 which corresponds to a 500 ppm solution of partially hydrolyzed polyacrylamide (M = 5 x 10^6) in 2% NaCl and $q_2 = 2q_1$

$$\frac{k_2}{k_1} = 0.908$$

This example, although simplified somewhat, shows that the effect of permeability change on the ratio μ/k is larger than the effect of changes in flow rate. Furthermore, equal velocity changes will offset equal permeability changes for $n \leq 1/3$.

The presence of appreciable reduction in μ/k with velocity coupled with the fact that retention increases as flow rate increases suggests the existence of two entrapment regions caused by hydrodynamic effects. In the low rate region, polymer retention reduces the permeability but not enough to offset the decrease in viscosity caused by additional shear thinning from permeability reduction. The second region begins when increased retention reduces the permeability significantly so that the effects of additional shear thinning are offset by the permeability reduction. Both polyacrylamide and polyethylene oxide solutions exhibit this behavior which has been attributed to viscoelastic effects (13,14). However, there is evidence to demonstrate that this phenomenon is not solely a viscoelastic effect for solutions containing partially hydrolyzed polyacrylamides (17,21).

VII. SUMMARY

In previous sections, we have presented the experimental evidence of polymer retention in porous media as well as some interpretation of retention mechanisms based on the data. High molecular weight polymers are retained in porous media by mechanical entrapment and adsorption. A portion of polymer retained during flow is not permanently held in the porous media and can be mobilized by a reduction in flow rate. Some of the pore volume is inaccessible to the flow of polymer solutions yet is in hydraulic contact with the solvent solution which is injected with the polymer. Each of these factors has an effect on the retention of polymer molecules in porous media which should be considered in any attempt to develop a model of polymer retention in porous media. There is considerable interaction between the various mechanisms. Models were proposed to represent polymer retention and interactions between retention mechanisms. Much of the analysis is qualitative because of the complexity of retention mechanisms and the absence of data to evaluate proposed models.

VIII. NOTATION

a   =   constant in Langmuir adsorption model
A   =   cross sectional area perpendicular to flow
Ad =   polymer adsorbed/unit mass of porous material
Ae =   excluded surface area of a sphere per contact point
Ap =   surface area for sphere of diameter d_g

Am = surface area covered/adsorbed polymer coil
b = constant in Langmuir adsorption model
c = concentration
C_o = injected concentration
d = equivalent diameter of spherical polymer molecule
d_g = diameter of sphere
d_{pore} = equivalent diameter of pore opening
H = parameter in power law model
k = permeability of porous media
K = power law parameter in Equation 10
k_p = permeability of porous media to polymer solution
k_w = permeability of porous media to aqueous solvent prior to contact by polymer
M = polymer molecular weight
n = power law parameter
N = number of pores of different radii in the pore size distribution
Ñ = Avogadro's number
q = volumetric flow rate
$\sqrt{\overline{s^2}}$ = radius of gyration of polymer molecule
r_i = radius of pore in a pore size distribution
$\sqrt{\overline{r^2}}$ = root-mean-square end to end distance of polymer molecule
$||r||$ = root mean average radius defined by Equation 5
γ = shear rate
μ = viscosity of flowing fluid
[μ] = intrinsic viscosity defined by Equation 2
μ_p = apparent viscosity of polymer solution
μ_s = viscosity of polymer free solvent
μ_w = viscosity of aqueous solvent
τ = shear stress
σ = ratio of segment density of molecular coil on the surface to that in dilute solution
φ = porosity

IX. LITERATURE CITED

1. Jewett, R. L. and Schurz, G. F., J. Pet. Technol. 22, 675 (1970).
2. Davis, J. A. and Jones, S. C., J. Pet. Technol. 20, 1415 (1968).
3. Gogarty, W. B., Meabon, H. P. and Milton, H. W., Jr., J. Pet. Technol. 22, 141 (1970).
4. Lipton, D., Paper SPE 5099 presented at the 49th Annual Fall Meeting of the Society of Petroleum Engineers of AIME, Houston (1974).
5. _____, Xanthan Gum/Keltrol/Kelzan/ Technical Bulletin X474, Kelco Co., San Diego (1974).

6.   Flory, P. J., "Principles of Polymer Chemistry," Cornell
     University Press, Ithaca, New York, 1953.
7.   Vollmert, B., "Polymer Chemistry," p. 493, p. 551,
     Springer-Verlag, New York, 1973.
7a. Oosawa, F., "Polyelectrolytes," Marcel Dekker, Inc.,
     New York, 1971.
8.   Martin, F. D. and Sherwood, N. S., Paper SPE 5339 pre-
     sented at the Rocky Mountain Regional Meeting of SPE-
     AIME, Denver (1975).
9.   Lynch, E. J. and MacWilliams, D. C., J. Pet. Technol.
     21, 1247 (1969).
10. Herr, J. W. and Routson, W. G., Paper SPE 5098 presented
     at the 49th Annual Fall Meeting of SPE-AIME, Houston
     (1974).
11. Mungan, N., Rev. Inst. Fr. Pet. 24, 232 (1969).
12. Nouri, H., Ph.D. Thesis, University of Oklahoma (1971).
13. Marshall, R. J. and Metzner, A. B., Ind. Eng. Chem.
     Fundam. 6, 393 (1967).
14. Jennings, R. R., Rogers, J. H. and West, T. J., J. Pet.
     Technol. 23, 391 (1971).
15. Vela, S., Peaceman, D. W. and Sandvik, E. I., Paper SPE
     5102 presented at the 49th Annual Fall Meeting of SPE-
     AIME, Houston (1974).
16. Dawson, R. and Lantz, R. B., Soc. Pet. Eng. J. 12, 448
     (1972).
17. Dominguez, J. G. and Willhite, G. P., Paper SPE 5835
     presented at the Improved Oil Recovery Symposium,
     SPE-AIME, Tulsa (1976).
18. Smith, F. W., J. Pet. Technol. 22, 148 (1970).
19. Mungan, N., Smith, F. W. and Thompson, J. L., J. Pet.
     Technol. 18, 1143 (1966).
20. Mungan, N., J. Can. Pet. Technol. 8, 45 (1969).
21. Desremaux, L., Chauveteau, G. and Mlle. Martin, Com-
     munication No. 28, Colloque de L'association de
     Recherches, Sur les Techniques D'exploitation du
     Petrole, Paris (1971).
22. Szabo, M. T., Soc. Pet. Eng. J. 15, 323 (1975).
23. Michael, A. S. and Morelos, O., Ind. Eng. Chem. 47, 1801
     (1955).
24. Trushenski, S. P., Dauben, D. C. and Parrish, D. R.,
     Soc. Pet. Eng. J. 14, 633 (1974).
25. Patton, J. T., Coats, K. H. and Colegrove, G. T., Soc.
     Pet. Eng. J. 72 (1971).
26. Silberberg, A., J. Phys. Chem. 66, 1884 (1962).
27. Schamp, N. and Huylebroeck, J., J. Polym. Sci. 42, 553
     (1973).
28. Rowland, F. W. and Eirich, F. R., J. Polym. Sci. 4, 2401
     (1966).

29. Greene, B. W., J. Colloid Interface Sci. 37, 144 (1971).
30. Rowland, F., Bulas, R., Rothstein, E. and Eirich, R. F.,
     Ind. Eng. Chem. 66, 1884 (1962).
31. Hirasaki, G. J. and Pope, G. A., Soc. Pet. Eng. J. 14,
     337 (1974).
32. Thakur, G. C., Paper SPE 4956 presented at Permian Basin
     Oil Recovery Conference of SPE–AIME, Midland, Texas
     (1974).
33. Thomas, C. P., Paper SPE 5556 prepared for the 50th Fall
     Meeting of SPE–AIME, Dallas ·(1975).
34. Szabo, M. T., Paper SPE 4028 presented at the 47th Annual
     Fall Meeting of SPE–AIME, San Antonio (1972).
35. Thomas, C. P., Personal communication.
36. Gogarty, W. B., Soc. Pet. Eng. J. 7, 161 (1967).
37. Palmer, J. S., Unpublished data, Department of Chemical
     and Petroleum Engineering, University of Kansas (1974).
38. Sarem, A. M., Paper SPE 3002 presented at the 45th Annual
     Fall Meeting of the SPE–AIME, Houston (1970).
39. Sandiford, B. B., Knight, R. K., Sarem, A. M. and Amott,
     E., Soc. Pet. Eng. J. 7, 170 (1967).
40. Herzig, J. P., Leclerc, D. M. and Le Goff, P., "Flow
     Through Porous Media," p. 129, American Chemical
     Society (1970).
41. DiMarzio, E. A. and Guttman, C. M., Macromolecules 3,
     131 (1970).
42. Chauveteau, G. and Kohler, N., Paper SPE 4745 presented
     at the Improved Oil Recovery Symposium of SPE–AIME,
     Tulsa (1974).
43. Maerker, J. M., J. Pet. Technol. 25, 1307 (1973).
44. Schamp, N. and Huylebroeck, J., Nature (London) Phys.
     Sci. 242, 143 (1973).
45. Cotton, J. P., Farnoux, B. and Jannink, G., Preprints
     IUPAC, Symposium on Macromolecules, 3, 83 (Helsinki,
     1972).
46. Burcik, E. J. and Walrond, K. W., Prod. Mon. 32, 12
     (1968).
47. Burcik, E. J., J. Pet. Technol. 21, 373 (1969).
48. Burcik, E. J. and Thakur, G. C., J. Pet. Technol. 26,
     545 (1974).
49. Knight, B. L., J. Pet. Technol. 26, 547 (1974).
50. Pye, D. J., J. Pet. Technol. 16, 911 (1964).
51. Burcik, E. J., Prod. Mon. 29, 14 (1965).
52. Dauben, D. L. and Menzie, D. E., J. Pet. Technol. 19,
     1065 (1967).
53. Christopher, R. H. and Middleman, S., Ind. Eng. Chem.
     Fundam. 4, 422 (1967).

MICELLAR FLOODING: SULFONATE-POLYMER INTERACTION

Scott P. Trushenski
Amoco Production Company

I. ABSTRACT

Polymer presence in micellar fluids often causes phase separation. This reduces tertiary oil recovery in porous media and increases sulfonate requirements. Phase studies and core tests show sulfonate-polymer incompatibility can be controlled with careful adjustment of sulfonate, cosurfactant, water, and salt concentrations in the micellar and mobility buffer banks.

II. SCOPE

Micellar flooding is an improved oil recovery method employing multiple banks. Typically, a brine preflush conditions the reservoir, micellar fluid displaces the formation oil and water, a polymer-thickened mobility buffer bank displaces the micellar fluid, and finally, drive water follows the mobility buffer bank. Complex interactions occur between the injected fluid banks; between the crude oil, formation water, and injected fluids; and between the injected fluids and the rock surface. The micellar bank, which is of primary importance, must have: low interfacial tension with oil and water, low sulfonate loss, adequate viscosity, and compatibility with all contacted fluids.

Sulfonate loss is the major factor which determines the optimum volume of the expensive micellar fluid needed to displace the tertiary oil. Sulfonate adsorption by the rock surface usually is assumed to be the only source of loss. Extensive oil displacement tests in consolidated cores show that many micellar fluids are not compatible with the polymer from the mobility buffer bank. When polymer invades or dilutes the micellar fluid, multiple phases often develop (sulfonate-polymer incompatibility). If the interfacial tension between these sulfonate-containing phases is sufficient, phase trapping occurs in the porous media. This sulfonate loss causes reduced oil displacement effectiveness and increased chemical requirements.

An objective of this study is to relate phase trapping in dynamic core tests (using 2" diameter Berea sandstone cores from 4' to 24' long) to static equilibrium phase observations. If the dynamic and static tests can be related, then rapid inexpensive screening tests can be used to select micellar formulations which are compatible with polymer at the conditions encountered in a dynamic core test.

Ternary phase diagrams varying sulfonate, cosurfactant and brine (all containing polymer) concentrations are used to identify injected micellar compositions and dilution paths which eliminate or minimize the development of two phases. The effects of sulfonate, cosurfactant, polymer, water, temperature, and salt concentrations on sulfonate-polymer incompatibility have been examined.

The identification of a previously unreported source of sulfonate loss (sulfonate-polymer incompatibility) and methods to reduce its effects should aid in the design of micellar floods which are more efficient in the displacement of tertiary oil.

III. CONCLUSIONS AND SIGNIFICANCE

When polymer from the mobility buffer bank mixes and invades the micellar bank ahead, two phases often form. When the interfacial tension between these two sulfonate-containing phases is sufficient, one is trapped in the porous media. This sulfonate-polymer incompatibility (SPI) increases sulfonate requirements and reduces oil displacement efficiency. Not only does SPI occur in the micellar-mobility buffer mixing zone, but can occur throughout the entire micellar slug due to invasion of the micellar slug by polymer molecules which move more rapidly than the carrier water. This is due to polymer inaccessible pore volume. For the Mahogany AA sulfonate:isopropyl alcohol system investigated, varying polymer concentration in the range of practical interest (500–1500 ppm) did not affect SPI significantly nor did the polymer type (polyacrylamide or biopolymer). Static phase studies correlate well with the incidence of trapping in dynamic core tests. This reduces the time and expense of screening micellar formulations for those which are compatible with polymer.

Several methods can be used to reduce sulfonate trapping due to SPI. Reducing the salinity of the mobility buffer bank behind the micellar slug reduces sulfonate loss significantly. Increasing the concentration of sulfonate solubilizers (e.g., cosurfactants or cosolvents) in the micellar and mobility buffer banks can eliminate SPI. The addition of a small amount of crude oil (< 5%) to the micellar fluid often prevents polymer invasion and SPI.

IV.   INTRODUCTION

Micellar flooding is a multi-bank tertiary oil recovery process which has many complex fluid-fluid and fluid-rock interactions.   In a micellar flood a brine preflush may be injected first to condition the formation.   Second, a slug (3-20% PV) of micellar fluid (larger slugs are injected in high water content, low tension floods) is injected to displace the oil (and formation water).   Next, a mobility buffer (polymer) bank drives the micellar fluid through the reservoir. Finally, a chase-water bank displaces the mobility buffer bank. Sulfonate (surfactant) loss from the micellar fluid is a major factor which determines the volume of expensive micellar fluid needed to recover all the tertiary oil.   Usually sulfonate loss is associated with adsorption to the rock surface, precipitation by the formation brine, and partitioning into bypassed oil or water.

A frequent source of sulfonate loss, often overlooked, occurs as a result of micellar-mobility buffer bank interaction.   Micellar-mobility buffer interaction can take two forms: sulfonate loss resulting from dilution of the micellar fluid by the mobility buffer bank or loss due to the invasion of the micellar fluid by the polymer molecules in the mobility buffer bank.   Here sulfonate-polymer interaction will be discussed.   However, this work has broader applicability since similar behavior is often observed when micellar fluids are diluted with brine (in the absence of polymer).

V.   BACKGROUND

Early studies showed that during a polymer flood, polymer adsorption caused a denuded zone to form at the leading edge of the polymer bank (1).   However, in a micellar flood, adsorption sites which normally are available to polymer molecules are now occupied by surfactant.   In most cases polymer loss is reduced to insignificant levels, and a polymer denuded zone does not develop (2).   Instead, polymer penetrates the micellar slug.

High molecular weight polymer molecules (M.W. $> 10^6$) are excluded from the small pores of a rock and only propagate through the larger pores.   In contrast, both the small and large pores transport the water.   The net effect is that the pore volume available to the polymer molecules is less than available to the water molecules.   This is termed "polymer inaccessible pore volume" (IPV) (1).   Polymer molecules move faster than the carrier water they are injected with in a manner similar to the molecular separation which occurs in gel permeation chromatography.

In Figure 1 the premature breakthrough of polymer due to IPV is shown.   Here one pore volume of 500 ppm Kelzan MF

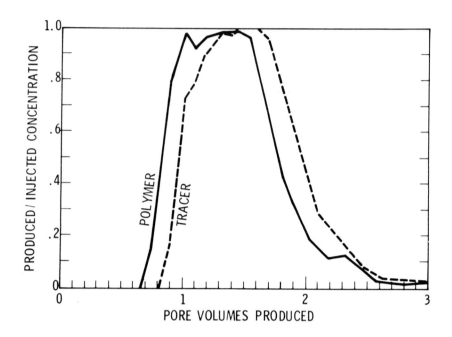

FIGURE 1.   KELZAN MF BIOPOLYMER PROPAGATION ILLUSTRATING INACCESSIBLE PORE
VOLUME IN A BEREA CORE.

biopolymer and traced water were injected into a 2" diameter
by 24" long Berea sandstone core. No oil or micellar fluid
were present. In this case the polymer led the water tracer
by 18% PV at both the front and rear of the mobility buffer
bank. Because of IPV, polymer molecules from the mobility
buffer bank can invade the micellar fluid without having a
dilution effect. Therefore, interactions which occur between
the micellar fluid and mobility buffer bank occur within the
bulk of the micellar bank as well as in the micellar-mobility
buffer mixing zone. Polymer molecules rapidly invade the
small micellar bank. For example, if IPV is 35% PV (which is
typical for the laboratory micellar floods reported here),
then after the injection of only 18% PV of the mobility buffer
bank, the polymer advances to the leading edge of a 10% PV
slug of micellar fluid.

VI.   DISCUSSION

A.   Core Tests: Experimental Procedure
       Sulfonate-polymer interaction (SPI) can be identified
in long-core, oil-displacement tests in two ways: effluent

composition profiles and pressure behavior. Typically, tests
are conducted in 2" diameter Berea sandstone cores up to 24'
long with absolute brine permeabilities of approximately 500 md.
Frontal advance rates are 2 ft/day. The core is wrapped in
epoxy fiberglass and pairs of pressure taps are located across
short intervals. From pressure differential data, the rate of
advance, growth, and intensity of mobility anomalies can be
monitored. The arrival of mobility anomalies at the core exit
can be calculated from the pressure data, and the corresponding effluent fractions can be analyzed to identify the interactions which occur to produce unusual or unfavorable propagation behavior.

In general, the rock properties, fluid compositions, and
volumes injected in all the tests reported here are similar.
Major differences which are important for this discussion
will be identified. For brevity, a complete description and
discussion of each test will not be presented. In all tests
the sulfonate used was Mahogany AA petroleum sulfonate. It
has an average equivalent weight of 444 (E.W. range 435-465)
and is 60% active sulfonate. The polymers used in the studies
reported here are restricted to a polysaccharide and polyacrylamides which have been well characterized in the
literature.

Table I lists the rock properties, fluid compositions,
and volumes of a representative 16' Berea core test when a
large bank of micellar fluid is injected. The volume of the

TABLE I

TYPICAL CORE PROPERTIES AND TEST CONDITIONS (FOR FIGURE 2)
==

Core Length	= 16 feet
Core Diameter	= 2 inches
Porosity	= 0.21
Pore Volume	= 2146 ml
Temperature	= 110°F
Crude Oil	= Second Wall Creek
Crude Oil Viscosity	= 4.0 cp @ 110°F
Connate Water Saturation	= 44.6%
Residual Oil Saturation	= 34.4%
Absolute Permeability	= 571 md
Brine Permeability at Residual Oil	= 37 md

micellar bank (132% PV) is much larger than would be used in a field application; however, this large volume is convenient for interpretation since interactions which occur between the major fluid banks can be readily identified. In Table I, note in particular that the salinity of the mobility buffer bank (0.05 N NaCl) is less than that of the micellar bank (0.23 N NaCl). This will be discussed in more detail later.

A useful basis for comparing pressure drop behavior over short distances along the length of a core is to use comparative mobility (2). Comparative mobility relates the mobility of the fluid flowing past a pair of pressure taps at any time to that of water flowing at residual oil. Therefore, at a constant flow rate, it is the ratio of pressure drop at residual oil to the pressure drop at any time ($\Delta P_{sor}/\Delta P_t$).

B. Sulfonate-Polymer Interaction: Core Tests

The effluent and comparative mobility profiles for the long core test described in Table II are shown in Figure 2. Oil fractions, dimensionless component concentrations in the water phase, and dimensionless oil-water interfacial tension (IFT) are plotted versus the pore volume of effluent produced after the initiation of micellar injection. A comparative mobility plot at the last pair of pressure taps (2" and 14" from the core exit) is also included so that the pressure behavior can be compared to the composition of the core effluent. A comprehensive discussion of this test has been presented previously (2). Here, the micellar-mobility buffer region is of primary interest.

TABLE II

FLUID VOLUMES, COMPOSITIONS, AND SEQUENCE
FOR LONG-CORE TEST OF FIGURE 2

1.8 PV	0.23N NaCl Brine
0.8 PV	Second Wall Creek Crude Oil
0.4 PV	Preflush (Waterflood) 0.23N NaCl
1.3 PV	5:3, Mahogany AA:IPA in 92% 0.23N Brine (Amoco 231)
1.1 PV	1000 ppm Kelzan MF Biopolymer 1% ETOH, 0.05N NaCl
0.5 PV	0.05N NaCl

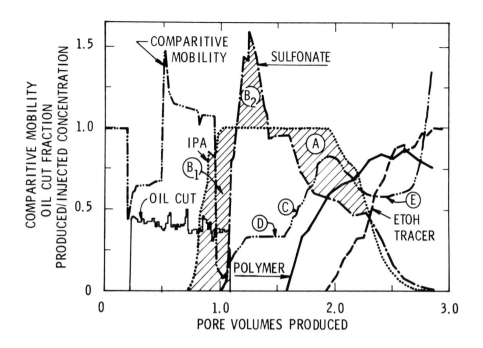

FIGURE 2.   EFFLUENT COMPOSITION AND COMPARATIVE MOBILITY
PROFILES FOR LONG CORE TEST SHOWING EFFECTS OF
SULFONATE-POLYMER INTERACTION.

1. *Sulfonate-Polymer Interaction: Effluent Profile*

As in Figure 1, Figure 2 shows that polymer molecules
move more rapidly than the ethanol tracer (IPV = 35% PV at
$c/c_0 = 0.5$).   Polymer has invaded the micellar fluid without
a dilution effect.   Dilution of the IPA tracer (cosurfactant)
in the micellar bank did not occur until the ethanol (ETOH)
concentration began increasing.   Note, however, when polymer
is first produced the sulfonate concentration decreases.   If
this was a dilution effect the IPA concentration would also
decrease.   With this micellar formulation (5:3:92% by weight,
Mahogany AA sulfonate:IPA:0.23N NaCl brine), the presence of
polymer in the micellar fluid causes an unexpected increase
in sulfonate loss.   This will be termed sulfonate-polymer
interaction (SPI).   Integration between the sulfonate and cosurfactant profiles (IPA) in the zone of SPI (A in Figure 2)
gives a sulfonate loss of 2.12 lb active sulfonate/bbl PV.
This is 3.5 times the adsorption loss at the leading edge of
the micellar slug (area zone B_1 X produced water fraction-area
zone B_2).   Interestingly, the dimensionless sulfonate and IPA
profiles reconverge at 2.25 PV where no further sulfonate loss
occurs.

2. *SPI: Comparative Mobility Plots*

The effects of SPI also are observed in the comparative mobility plot in Figure 2. In the region of micellar-polymer mixing (C) the fluid mobility is greater than that of the stabilized micellar bank (D) and polymer bank (E). High fluid mobility is characteristic of SPI. In the absence of SPI the mobility in Zone B decreases or is unchanged. The comparative mobility increase in the zone of SPI has not been explained; however, it is a useful tool since SPI can be identified without a complete effluent concentration profile. This is especially helpful with experimental sulfonates which are difficult to chemically analyze.

C. Significance of SPI

SPI is detrimental to a micellar flood since sulfonate loss is increased. This increases the volume of micellar fluid needed for optimum oil displacement efficiency. Adsorption loss remains virtually constant with micellar bank size, while SPI loss decreases with decreasing bank size. Studies not reported here show that when low-salinity polymer water is used, a constant fraction of the injected sulfonate is lost due to SPI. For the micellar formulation of Table II, a maximum of one-third of the sulfonate is lost due to SPI. This increases the required micellar design volume 50%.

With this particular micellar fluid-mobility buffer sequence, optimum oil recovery is achieved when a 9% PV micellar slug is injected. Three percent PV is lost due to SPI (1/3 of injected) and 6% PV is lost due to adsorption. Although SPI loss may seem excessive when a large bank of micellar fluid is injected, it may be insignificant in small slug injection tests. If SPI occurs, this does not necessarily imply that oil will not be displaced--only that the efficiency of the displacement can be reduced.

D. Cause of SPI

Long-core tests and bench-top phase studies show that SPI results from the phase separation of the micellar fluid in the presence of polymer. Chemical analyses of these phases show that the turbid, caramel-colored, lower phase has a higher sulfonate concentration than the transparent, amber, upper phase. When polymer is added to the micellar fluid of Table II, the sulfonate concentration in the top phase is two-thirds of the mixture concentration. The turbid lower phase is trapped in the rock. This was verified from the effluent of Figure 2, since the fluid produced in the region of SPI was transparent, amber and its composition was lower than the average mixture composition being injected.

Sulfonate phase trapping is similar to the trapping of oil when driving a system to connate water by simultaneously injecting oil and water. Initially, oil is trapped in the porous matrix until the water saturation is reduced to the connate level. At this point both oil and water are produced in the same ratio as injected. Likewise, when two phases develop in the micellar bank, one phase can be trapped in the rock, if the interfacial tension between the two phases is high enough. When the mobile saturation of this trapped phase is reached (this saturation will be a function of the IFT between the two surfactant phases, flow rate, etc.), both phases begin flowing and are produced in the effluent.

The trapping and remobilization of the second micellar phase is demonstrated in Figure 3. In this long-core test the water content of the micellar fluid was 94% (the Mahogany AA sulfonate:IPA ratio remained 5:3), and the salinity was 0.275N NaCl. A mobility buffer bank was not injected. Instead, after the injection of 2.0 PV of the original micellar fluid, 750 ppm Kelzan MF polymer (plus ETOH tracer) was added to the injected micellar fluid. Now, any decrease in sulfonate concentration in the mixing region between the two micellar slugs is due to polymer presence, not dilution.

Figure 3 shows the sulfonate concentration decreased when polymer was first produced. As the sulfonate concentration decreased in the zone of SPI there was a marked change in effluent appearance. The fluid changed from turbid, caramel-colored to transparent amber. In this case the sulfonate concentration reached a minimum, then began increasing. As it increased, the second turbid, caramel-colored sulfonate phase was produced. The fractional volume of this phase increased as sulfonate concentration increased. This indicates that SPI loss is caused by phase trapping.

The trapped sulfonate phase can be displaced by chase water behind the mobility buffer bank. In the long-core test shown in Figure 4, conditions were similar to those listed in Table II, except the polymer concentration was reduced to 700 ppm. Again, the polymer preceded the ethanol tracer, and SPI occurred. The produced sulfonate concentration in the zone of SPI stabilized at two-thirds of injected. This corresponds to the test tube concentration previously observed in the upper phase for this micellar formulation. Figure 4 shows that the sulfonate concentration peaked again when the polymer concentration decreased. Although the trapped material can be displaced, it is not effective in displacing oil.

E. Phase Studies: Experimental Procedure

Since sulfonate loss due to the presence of polymer is caused by the development of two phases, one of which is

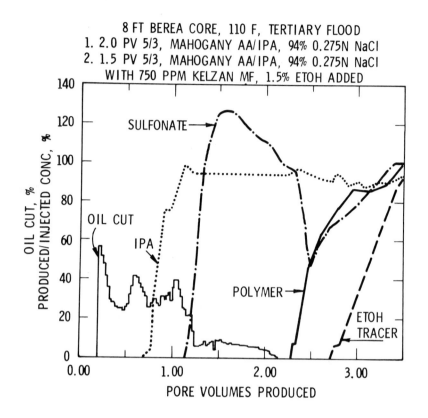

8 FT BEREA CORE, 110 F, TERTIARY FLOOD
1. 2.0 PV 5/3, MAHOGANY AA/IPA, 94% 0.275N NaCl
2. 1.5 PV 5/3, MAHOGANY AA/IPA, 94% 0.275N NaCl
   WITH 750 PPM KELZAN MF, 1.5% ETOH ADDED

FIGURE 3.   TRAPPED SULFONATE IS DISPLACED AT SATURATION LIMIT.

trapped in the porous medium, dynamic long-core tests are not
required to determine when SPI will occur. Instead, static
bench-top phase studies can be conducted.

Polymer IPV complicates phase stability studies since
the polymer penetrates the micellar slug without a diluting
effect. Therefore, phase studies which simply dilute the
micellar fluid with polymer water are not valid. Polymer
rapidly invades the micellar bank so that the polymer concentration is uniform throughout. Therefore, in the dilution
region between the micellar and mobility buffer banks the
variables are water content, salinity, sulfonate and cosurfactant concentration, but not polymer concentration.
Micellar fluids must be developed which are stable with polymer as injected, as well as, when the micellar fluid is diluted
by the brine in the mobility buffer bank. The required phase
studies can be represented by a ternary diagram (Figure 5).
This diagram differs considerably from those typically used

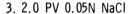

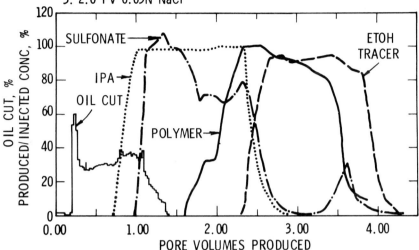

FIGURE 4.   TRAPPED SULFONATE IS DISPLACED WHEN POLYMER CONCENTRATION DECREASES.

to represent micellar floods: water is present at all three
apices; no oil is present; and sulfonate and cosurfactant are
located at separate apices rather than at a fixed ratio. The
phase diagram in Figure 5 can be considered as a small corner
of a much larger diagram (since water content varies from 88
to 100%). In addition, this ternary diagram can be used to
represent the displacement of a micellar fluid by the mobility buffer bank much as other ternary diagrams are used to
represent the displacement of oil and water by the micellar
fluid.

The variables in Figure 5 are: water content (88-100%),
Mahogany AA sulfonate concentration (0-12% bulk), IPA cosurfactant concentration (0-12%), and salinity (0.05-0.336N
NaCl). All fluids contain 1500 ppm Kelzan MF biopolymer.
Each phase diagram represents more than 300 fluid mixtures
throughout regions of practical interest. The mixtures were
prepared by mixing fluids A, B, and C (the fluids at the
apices) in evacuated stoppered "vacutainers". From hours,
to days were required before no change in the phase envelope
was observed at the test temperature (110°F, 190°F).

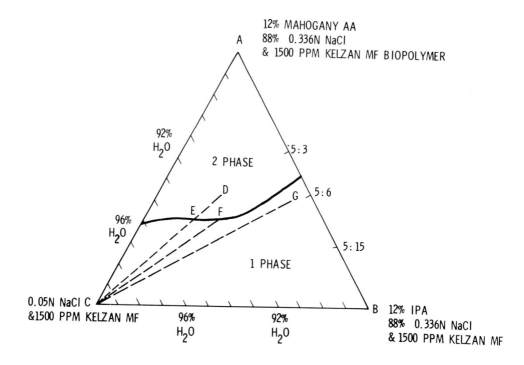

FIGURE 5.   PHASE BEHAVIOR OF MAHOGANY AA, IPA WHEN DILUTED
WITH LOW SALINITY POLYMER WATER (110°F)

F.   Sulfonate-Polymer Interaction: Phase Studies

Composition D in Figure 5 represents the injected micellar
fluid used in the long-core test of Figure 4 (5:3:92% by
weight, Mahogany AA:IPA:0.23N NaCl).   In the presence of 1500
ppm biopolymer the micellar fluid is two-phase.   Trapping of
one of these phases was verified in the test shown in Figure 4.
Dilution of this micellar solution by the low salinity mobility buffer bank can be represented by the line D-C in Figure 5.
Note that at 94% water (E in Figure 5), the fluid is one-phase.
In the long-core test of Figure 4 no further sulfonate was lost
in the micellar-mobility buffer zone at 94% water, since the
sulfonate and cosurfactant (tracer) profiles reconverge.

Compositions F and G in Figure 5 represent stable fluids,
since they are in the one-phase region and remain one-phase
throughout the dilution range (F-C and G-C).   A long-core test
was conducted with the fluid represented by G in Figure 5
(5:6:89, Mahogany AA sulfonate:IPA:0.23N NaCl).   The effluent
profiles for this test are shown in Figure 6.   No SPI occurs
since the surfactant and cosurfactant profiles coincide at the
trailing edge of the micellar bank.   (Note that no polymer IPV

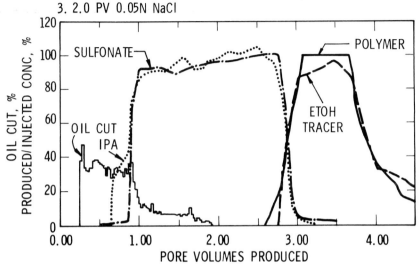

8 FT BEREA CORE, 110 F, TERTIARY FLOOD
1. 2.0 PV 5/6, MAHOGANY AA/IPA IN 89% 0.23N NaCl
2. 1.3 PV 1250 PPM KELZAN IN 0.05N NaCl, 1% ETOH
3. 2.0 PV 0.05N NaCl

FIGURE 6.   NO SULFONATE IS TRAPPED WHEN MICELLAR-POLYMER ARE ONE PHASE
AT ALL DILUTIONS.

is observed. This is unexplained and has only been observed
in one other long-core test (Figure 14).

   In general, Figure 5 shows that phase stability with low
salinity polymer water behind the micellar fluid is promoted
when surfactant:cosurfactant ratio decreases, and salinity
decreases with increasing water content.

1. *Effect of Polymer Water Salinity on SPI*

   Increasing polymer water salinity reduces oil displacement efficiency. Figure 7 shows oil recovery is reduced when
the salinity of the polymer water following the micellar slug
is equal to the micellar fluid (3). (In all cases a 10% PV
slug of the 5:3:92, Mahogany AA:IPA:0.23N NaCl brine was used.)
Interestingly, oil recovery is improved when the polymer water
following the micellar slug has a lower salinity (0.05N NaCl).

   Phase studies indicate that decreased oil recovery probably is caused by increased sulfonate-polymer interaction.
The equal salinity micellar-polymer system (0.23N NaCl) is
represented by the ternary phase diagram of Figure 8. Comparison with Figure 5 shows that the two-phase region is much
larger. Now the 5:3 ratio of Mahogany AA sulfonate:IPA (D)
remains in the two-phase zone at all dilutions. Figure 8 also

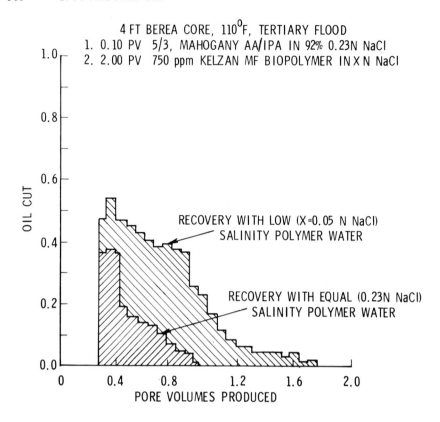

FIGURE 7.   REDUCING POLYMER WATER SALINITY IMPROVES TERTIARY OIL RECOVERY.

shows that the phase envelope is not affected significantly
by the biopolymer concentration (in the range of practical
interest). When equal-salinity polymer water is used, most
of the sulfonate can be lost a short distance from the core
inlet. When low salinity brine is used in the mobility buffer
bank (Figure 5), less of the sulfonate is lost due to SPI.
These observations can explain why oil recovery may be reduced
when equal salinity waters are used.

To verify that sulfonate losses increase when the micellar and mobility buffer banks have equal salinity, a long-core
test was conducted which was nearly identical to that in
Figure 2, except that the polymer water salinity was increased
from 0.05N to 0.23N NaCl. In this case the effluent profiles
show that SPI occurred over the entire dilution region between
the high salinity polymer water and the micellar bank (Figure
9).

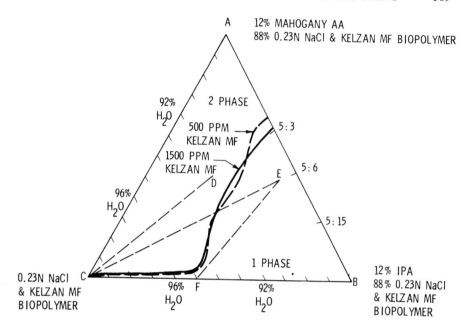

A   12% MAHOGANY AA
88% 0.23N NaCl & KELZAN MF BIOPOLYMER

92% H₂O

2 PHASE

500 PPM KELZAN MF

1500 PPM KELZAN MF

96% H₂O

5:3

5:6

5:15

1 PHASE

0.23N NaCl & KELZAN MF BIOPOLYMER   C

96% H₂O   F   92% H₂O   B

12% IPA
88% 0.23N NaCl & KELZAN MF BIOPOLYMER

FIGURE 8.   PHASE BEHAVIOR OF MAHOGANY AA, IPA WHEN DILUTED
WITH EQUAL SALINITY POLYMER WATER (110°F)

2. *Effect of Dilution on Stability*

Line D-C in Figure 5 shows that a two-phase micellar
fluid becomes one-phase when diluted by low-salinity polymer
water, while line E-C in Figure 8 shows that a fluid which is
initially one-phase (E) can separate into two phases when it
is diluted with equal salinity polymer water. The cosurfactant concentration can play an important role in maintaining
sulfonate solubility when diluted by equal salinity polymer
water, as shown by the E-F in Figure 8.

A long-core test was conducted to verify that a micellar-
polymer system which is initially one-phase can become two-
phase when diluted with equal salinity polymer water. A compatible (one-phase) micellar fluid (5:6:89, AA:IPA:0.23N NaCl)
was injected into a 4' Berea core and followed by 0.23N NaCl
polymer water. Polymer was added to the micellar fluid for
mobility control. The effluent profiles of Figure 10 show
that dilution of the trailing edge of the micellar bank caused
the sulfonate concentration to decrease more rapidly than the
cosurfactant concentration. This is due to the trapping of
the second sulfonate phase which develops upon dilution as
shown by line E-C in Figure 8.

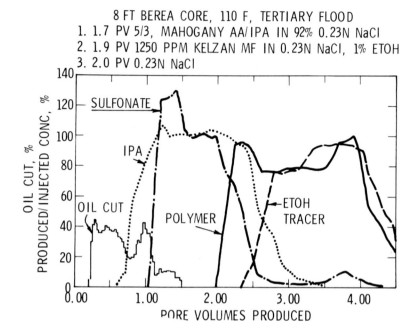

8 FT BEREA CORE, 110 F, TERTIARY FLOOD
1. 1.7 PV 5/3, MAHOGANY AA/IPA IN 92% 0.23N NaCl
2. 1.9 PV 1250 PPM KELZAN MF IN 0.23N NaCl, 1% ETOH
3. 2.0 PV 0.23N NaCl

FIGURE 9.   SULFONATE IS TRAPPED AT ALL DILUTION LEVELS WITH EQUAL SALINITY
POLYMER WATER.

The Mahogany AA sulfonate-IPA system (110°F, 0.23N NaCl) has no practical micellar composition which remains one-phase at all dilutions with equal salinity polymer water (Figure 8). However, Figure 5 shows that reducing the salinity of the mobility buffer bank can eliminate or reduce the effects of SPI.

3. *Effect of Polymer Type and Temperature on SPI*

Polyacrylamides and polysaccharides are the two classes of polymer commonly considered as mobility control agents in tertiary oil recovery applications. Figure 11 compares phase behavior for the equal salinity case at 1500 ppm Dow Pusher 500 polyacrylamide and Kelzan MF polysaccharide. Polymer type does not affect phase behavior significantly.

Figure 12 shows effluent profiles for a long-core test using Dow Pusher 700 polyacrylamide in a low-salinity mobility buffer bank. Comparison of the effluent profiles to Figure 4 (which is an equivalent test except for polymer type) shows similar surfactant loss for both the biopolymer and poly-acrylamide in the region of micellar invasion by the polymer. Interestingly, the polyacrylamide does not penetrate the micellar slug as rapidly as the polysaccharide so net sulfonate loss was less with the polyacrylamide.

4 FT BEREA CORE, 110 F, TERTIARY FLOOD
1. 1.5 PV 5/6, MAHOGANY AA/IPA IN 89% 0.23N NaCl
   1500 PPM KELZAN ADDED
2. 2.4 PV 1500 PPM KELZAN MF IN 0.23N NaCl, 1% ETOH
3. 2.0 PV 0.23N NaCl

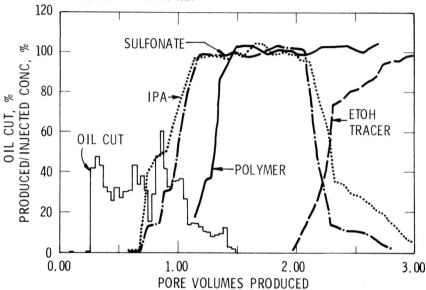

FIGURE 10. SULFONATE IS STABLE WITH POLYMER AT INJECTED MICELLAR CONCENTRATION BUT SEPARATES INTO TWO PHASES WHEN DILUTED.

As temperature increases, the size of the one-phase envelope increases. This is shown for 110°F and 190°F in the ternary diagram for Figure 13. Micellar fluids which are not compatible with polymer at one reservoir temperature may show excellent behavior at higher temperatures. Compatibility studies must be made for each reservoir application and micellar fluid type (4).

G. Eliminating SPI

SPI is observed in three situations: (1) the micellar fluid separates into two phases at its injected concentration, but becomes stable when diluted (D-C, Figure 5); (2) the micellar fluid is stable with polymer as injected, but separates into two phases when diluted by the mobility buffer bank (E-C, Figure 8); and (3) the micellar fluid is unstable with polymer at its injected concentration and at all dilutions with polymer water (D-C, Figure 8).

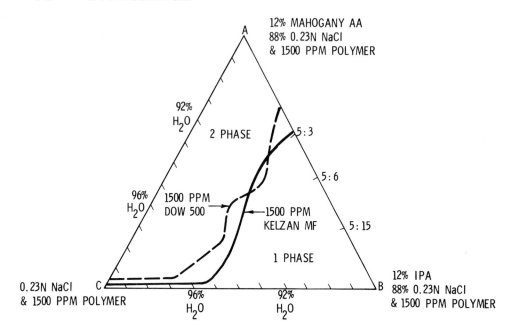

FIGURE 11.   COMPARISON OF PHASE BEHAVIOR FOR DOW 500 POLY-
ACRYLAMIDE AND KELZAN MF BIOPOLYMER (110°F).

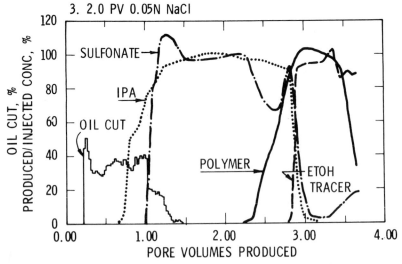

8 FT BEREA CORE, 110 F, TERTIARY FLOOD
1. 1.9 PV 5/3, MAHOGANY AA/IPA IN 92% 0.23N NaCl
2. 1.0 PV 1000 PPM DOW PUSHER 700 IN 0.05N NaCl 1% ETOH
3. 2.0 PV 0.05N NaCl

FIGURE 12.   SULFONATE TRAPPING IN PRESENCE OF POLYACRYLAMINES IS SIMILAR TO
TRAPPING WITH BIOPOLYMER (SEE FIGURE 6)

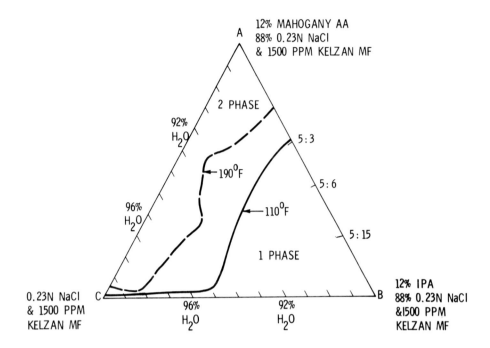

FIGURE 13.   EFFECT OF TEMPERATURE ON PHASE BEHAVIOR OF
MAHOGANY AA, IPA, KELZAN SYSTEM (100°F & 190°F)

    When the micellar becomes two-phase upon dilution, sta-
bility can be achieved with minor modification of the mobility
buffer bank. Figure 5 (G-C) showed that reducing the salinity
of the mobility buffer bank can eliminate unfavorable phase
separation. The addition of a surfactant solubilizer (cosurfactant) to the mobility buffer bank can also eliminate
phase separation when the micellar fluid is diluted by the
mobility buffer bank (E-F, Figure 8).

    Another approach to reduce SPI is to prevent the invasion
of the micellar fluid by polymer. When the water-soluble polymer is insoluble in the micellar fluid, then invasion by the
polymer should not occur. Polymer will not invade an oil-
external micellar fluid. However, the use of an oil-external
system may be economically unattractive in many applications.
Preliminary laboratory tests have shown that the addition of
less than 5% crude oil to the micellar bank can eliminate
polymer penetration. A long-core test was conducted using
conditions similar to those in Table II, except 3% Second
Wall Creek crude oil was added to the micellar fluid. The
e_fluent profile for this test shows that polymer did not

invade the micellar bank so SPI could not occur (again, IPV was not observed) (Figure 14).

Other system modifications can be utilized to eliminate SPI. Some approaches may be to reduce salinity, to use a different cosurfactant or sulfonate, or to blend sulfonates and cosurfactants.

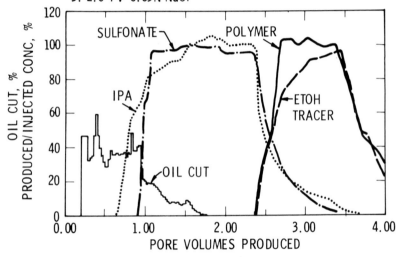

8 FT BEREA CORE, 110 F, TERTIARY FLOOD
1. 1.4 PV 5/3/3, MAHOGANY AA/IPA/2ND WALL CREEK CRUDE   89% 0.23N NaCl
2. 1.4 PV 1000 PPM KELZAN MF 0.05N NaCl, 1% ETOH
3. 2.0 PV 0.05N NaCl

FIGURE 14.   ADDED HYDROCARBON TO MICELLAR FLUID PREVENTS POLYMER INVASION AND ELIMINATES SULFONATE TRAPPING.

VII.   SUMMARY AND CONCLUSIONS
1.   Some micellar formulations separate into two phases in the presence of waterflooding polymers. In a porous matrix one of the sulfonate-containing phases can be trapped.
2.   Sulfonate-polymer interaction (SPI) between the micellar bank and mobility buffer bank can increase sulfonate requirements in a micellar flood.
3.   Inaccessible pore volume (IPV) to the polymer can increase SPI.
4.   Polymer concentration (over the range of practical interest) and polymer type do not significantly affect SPI.
5.   SPI is reduced when the salinity of the mobility buffer bank is lower than the salinity of the micellar fluid.

6. SPI and polymer invasion of the micellar bank is eliminated if the polymer is insoluble in the micellar fluid.
7. Additional sulfonate solubilizers (e.g., cosurfactants or cosolvents) in the micellar or mobility buffer banks can eliminate SPI.
8. Increasing temperature promotes sulfonate-polymer compatibility.
9. In a micellar flood, each of the injected fluid banks must be carefully formulated to eliminate unfavorable interactions which occur when these banks propagate and mix.

VIII. LITERATURE CITED

1. Dawson, R. and Lantz, R. B., "Inaccessible Pore Volume in Polymer Flooding," Soc. Pet. Eng. J. 448-452 (1972).
2. Trushenski, S. P., Dauben, D. L. and Parrish, D. R., "Micellar Flooding--Fluid Propagation, Interaction and Mobility," Soc. Pet. Eng. J. 633-645 (1974).
3. Paul, G. W. and Froning, H. R., "Salinity Effects of Micellar Flooding," J. Pet. Tech. 957-958 (1973).
4. Froning, H. R. and Treiber, L. E., "Development and Selection of Chemical Systems for Miscible Waterflooding," SPE 5816, presented at SPE Symposium on Improved Oil Recovery, Tulsa, Oklahoma, March 22-24, 1976.

Index